ADVANCES IN CHEMICAL PHYSICS

VOLUME LIV

Advances in
CHEMICAL PHYSICS

EDITED BY

I. PRIGOGINE

University of Brussels
Brussels, Belgium
and
University of Texas
Austin, Texas

AND

STUART A. RICE

Department of Chemistry
and
The James Franck Institute
The University of Chicago
Chicago, Illinois

VOLUME LIV

AN INTERSCIENCE® PUBLICATION
JOHN WILEY & SONS
NEW YORK · CHICHESTER · BRISBANE · TORONTO · SINGAPORE

An Interscience® Publication

Library of Congress Catalog Number: 58-9935

ISBN 0-471-89570-9

Printed in the United States of America

10 9 8 7 6 5 4 3 2 1

CONTRIBUTORS TO VOLUME LIV

LIONEL GOODMAN, Department of Chemistry, Rutgers University, New Brunswick, New Jersey

POUL JØRGENSEN, Chemistry Department, Aarhus University, Aarhus, Denmark

PH. KOTTIS, Centre de Physique Moléculaire Optique et Hertzienne, Université de Bordeaux I and C.N.R.S., Talence, France

JEPPE OLSEN, Chemistry Department, Texas A & M University, College Station, Texas

M. R. PHILPOTT, I.B.M. Research Laboratory, San Jose, California

A. K. RAJAGOPAL, Department of Physics and Astronomy, Louisiana State University, Baton Rouge, Louisiana

M. V. RAMANA, Department of Physics and Astronomy, Louisiana State University, Baton Rouge, Louisiana

RICHARD P. RAVA, Department of Chemistry, Rutgers University, New Brunswick, New Jersey

J. M. TURLET, Centre de Physique Moléculaire Optique et Hertzienne, Université de Bordeaux I and C.N.R.S., Talence, France

DANNY L. YEAGER, Chemistry Department, Texas A & M University, College Station, Texas

INTRODUCTION

Few of us can any longer keep up with the flood of scientific literature, even in specialized subfields. Any attempt to do more and be broadly educated with respect to a large domain of science has the appearance of tilting at windmills. Yet the synthesis of ideas drawn from different subjects into new, powerful, general concepts is as valuable as ever, and the desire to remain educated persists in all scientists. This series, *Advances in Chemical Physics*, is devoted to helping the reader obtain general information about a wide variety of topics in chemical physics, which field we interpret very broadly. Our intent is to have experts present comprehensive analyses of subjects of interest and to encourage the expression of individual points of view. We hope that this approach to the presentation of an overview of a subject will both stimulate new research and serve as a personalized learning text for beginners in a field.

ILYA PRIGOGINE

STUART A. RICE

CONTENTS

OPTIMIZATION AND CHARACTERIZATION OF A MULTICONFIGURATIONAL SELF-CONSISTENT FIELD (MCSCF) STATE

JEPPE OLSEN AND DANNY L. YEAGER

Chemistry Department
Texas A & M University
College Station, Texas

AND

POUL JØRGENSEN

Chemistry Department
Aarhus University
Aarhus, Denmark

CONTENTS

I. INTRODUCTION

In multiconfigurational self-consistent field (MCSCF) calculations a simultaneous optimization is performed for the orbitals and configuration expansion coefficients. Until recently, MCSCF optimization procedures considered the orbital and state optimization problems separately.[1, 2] The state optimization problem was treated by performing a configuration interaction (CI) calculation within the MCSCF configuration space, whereas the orbital optimization was carried out using multiconfigurational extensions of the Hartree–Fock (HF) iterative procedure.[1,2] The Hartree–Fock iterative procedure[3,4] has been used very successfully to determine an optimal set of orbitals for a single configuration state. The success of the HF iterative scheme largely relies, however, on the physical justification (an independent set of particles moving in an average potential) and interpretation (Koopmans's theorem[5]) that the orbitals can be given in the HF approximation. The physical justification and interpretation serve in a sense as the "driving force" in getting the HF iterative scheme to converge. When several configurations are used to describe a molecular system, orbitals cannot be given such a direct physical justification and interpretation and MCSCF iterative procedures that represent extensions of the HF iterative procedure often lose their ability to converge reliably.

The MCSCF orbital optimization problem is a standard optimization problem of numerical analysis and has to be attacked with the efficient optimization procedures of numerical analysis to converge reliably.[6–9] It is the purpose of the present review to discuss how the MCSCF optimization problem may be parameterized to a form for which the standard optimization procedures of numerical analysis can be applied and to show how convergence problems essentially disappear when these optimization procedures are correctly used.

In the following discussion (Section II) we will describe how the MCSCF optimization problem can be formulated in terms of unitary operators that are exponentials of nonredundant operators for orbital[10–12] and for state optimization.[11,12] The optimized MCSCF state may then be described by the effect of these unitary operators acting upon an unoptimized state. By applying the variational principle, different optimization procedures may be easily derived.

A widely used optimization procedure in numerical analysis is the Newton–Raphson technique.[6–9] Newton–Raphson[10–37] and other complete second-order[2,19,28–33] and approximate second-order[2,11,13,15,35,37] techniques are now used fairly extensively for SCF and MCSCF optimization. In Section III we demonstrate how the Newton–Raphson technique may be implemented into MCSCF. The Newton–Raphson iterative function is derived from a quadratic expansion of the energy function. When the Newton–Raphson approach is applied in numerical analysis, a linear transformation among the variables is carried out in between each step of the iterative procedure. In Section III, we discuss the Newton–Raphson technique and describe how it is most straightforwardly applied in MCSCF if a nonlinear transformation of the variables is carried out between each step of the iterative procedure.[10–17]

Before proceeding further with an analysis of various optimization procedures, we briefly discuss characteristics of an MCSCF state.[13,15–17] In a CI calculation, the nth root in energy of the CI secular problem is an upper bound to the nth state. The upper bound property is commonly used as a criterion for determining whether a CI state is a valid representation of the desired state. The upper bound property can be established because the configuration expansion coefficients may be used as *linear* variation parameters. In addition to configuration expansion coefficients, an MCSCF calculation also optimizes nonlinear orbital parameters. An alternative to the upper bound criterion therefore has to be established to decide whether an MCSCF state is a proper representation of the state of consideration. In this review we discuss the criteria that an MCSCF state may be required to fulfill in order to be a valid representation of the state of consideration. We consider requirements that may be imposed to ensure that the response of the MCSCF state to an external one-electron perturbation is physically feasible, for example; that the response calculation is stable and gives the appropriate number of negative excitation energies. We also discuss requirements that may be justified based on the MCSCF states' ability to simulate properties of a full CI solution; for instance, the Hessian matrix for an MCSCF calculation on the nth state may be required to have $n-1$ negative eigenvalues as does the full CI Hessian matrix.

The MCSCF state should, of course, desirably fulfill all the discussed criteria to be a proper representation of the state of consideration. In many cases, however, it is not possible to get the MCSCF state to satisfy all criteria when a finite basis set is used and only a limited number of configurations are included in the MCSCF calculation. The criteria that may need to be imposed on the MCSCF state as the minimal requirement may then depend on the physical property that the MCSCF state is to calculate. For example, if the MCSCF state is used to evaluate second-order properties it becomes a requirement that the linear MCSCF response calculation give the desired number of negative excitation energies, whereas if the MCSCF state is used only to evaluate dipole moments the requirement that the response calculation give the correct number of negative excitation energies may be less important. A more mathematical and detailed discussion of these points is presented in Section IV.

When the Newton–Raphson approach is applied far from convergence (global convergence), step size control algorithms have been applied in numerical analysis to get the Newton–Raphson approach to converge reliably.[9] We discuss in Section V how step size and sign control algorithms may be implemented in an MCSCF calculation[13,15,17] and report a series of nonlinear step size and sign controlled Newton–Raphson MCSCF calculations to illustrate the convergence problems that may appear when optimizing an

MCSCF state using a Newton–Raphson technique. The Newton–Raphson approach is found to converge rapidly and reliably in nearly all cases provided the step sizes and signs are controlled when far from convergence. When impediments to convergence are observed, they can usually be associated to an incorrect structure or small eigenvalues of the Hessian matrix.[9,17]

While the most important characteristic for global convergence is the reliability of the approach in bringing the calculation into the local region, the most important property for obtaining local convergence is the (computational) efficiency at which the approach is able to bring the calculation from a point relatively near the stationary point to the stationary point. In the local region the convergence characteristics of an iterative procedure may be expressed in terms of a total order of convergence in an error vector that describes the distance from the point of consideration to the desired stationary point.[8,9] When n steps of such a Newton–Raphson iterative procedure are applied, the Newton–Raphson approach is shown to give a total order of convergence of 2^n in the error vector of the initial iteration. If the Hessian matrix is kept fixed during a sequence of n iterations (the initial fixed Hessian iteration is a Newton–Raphson iteration), the total order of convergence of such a procedure is $n + 1$. Since each iteration in a fixed Hessian sequence requires only construction of the first derivative matrix (the gradient) of the total energy whereas the Newton–Raphson approach requires construction of a Hessian matrix, fixed Hessian procedures may efficiently be used to speed up the local convergence of an MCSCF calculation.[16,22] This is further discussed in Section VI.

When a fixed Hessian series of iterations is performed, a new energy gradient is calculated at each step of the iterative procedure. Since the difference between the new energy gradient and the energy gradient of the previous iteration contains information about the Hessian matrix at the new point, it is desirable also to incorporate this information into the iterative scheme. Such an incorporation is obtained by updating the Hessian matrix during a sequence of iterations. In the numerical analysis literature Hessian update techniques are considered to be the most efficient method for optimizing a function of many variables.[9] In Section VII we demonstrate how Hessian update methods may be implemented in MCSCF and how the Hessian update methods constitute an efficient iterative approach for obtaining local convergence of an MCSCF calculation.[21] Our Hessian update results indicate that the Hessian update methods may also have very promising global convergence characteristics when implemented with a step size control algorithm.

The convergence problem may alternatively be approached using a cubic expansion of the energy.[16,18,20] In Section VIII we derive and study three cubic optimization procedures. These cubic approaches have desirable local convergence properties, whereas the iterative cubic approach is particularly useful when far from convergence. In fact, with the iterative cubic technique, step size and sign constraint procedures are often not required.

Since certain of the optimization procedures to be considered herein are designed specifically for the global problem and others for efficiencies in the local region, in Section IX we discuss the effective implementation and combination of MCSCF procedures. In Section X we conclude and summarize.

II. MCSCF OPTIMIZATION USING EXPONENTIAL UNITARY OPERATORS

The initial step of an MCSCF calculation uses an initial guess of a set of orbitals $\{a_r^+\}$. For example, such a guess may be a set of single configuration Hartree–Fock orbitals with regular virtual orbitals or improved virtual orbitals[38-39] or grand canonical orbitals.[40] In Section V we discuss how to get an initial set of orthogonal orbitals of sufficient quality to ensure that an MCSCF iterative procedure will converge. The initial "guess" of orbitals may then be used to calculate an initial multiconfiguration reference state $|0\rangle$, say, by performing a CI calculation within the configurations of the MCSCF calculation. The orthogonal complement set of states $\{|k\rangle\}$ to $|0\rangle$ may also be determined, for example, through a CI calculation (alternatively, see Section IX.B), and MCSCF interative procedures can then be established that carry out simultaneous unitary transformations within the configuration space $\{|0\rangle, |k\rangle\}$ and orbital space $\{a_r^+\}$. In this next section we describe how to parameterize a calculation to carry out simultaneous transformations in the orbital and configurational spaces and how to implement MCSCF interative procedures based on simultaneous orbital and state transformations.

A. Unitary Transformations of the Multiconfigurational Hartree–Fock State

The multiconfigurational Hartree–Fock (MCSCF) reference state $|0\rangle$ may be regarded as a member of the set of states $\{|j\rangle\}$

$$|0\rangle = \sum_g |\Phi_g\rangle C_{g0} \tag{1}$$

$$|k\rangle = \sum_g |\Phi_g\rangle C_{gk} \tag{2}$$

where the coefficient matrix **C** forms a unitary matrix. The configuration state functions $|\Phi_g\rangle$ are composed of simple linear combinations of determinants. The determinants $|\Phi_f^D\rangle$ are given as

$$|\Phi_f^D\rangle = \prod_{r \in f} a_r^+ |\text{vac}\rangle \tag{3}$$

where $\Pi_{r \in f} a_r^+$ refers to an ordered product of creation operators of electrons in spin orbitals. We assume in the following that all orbitals and states are real. A detailed discussion of how a simultaneous unitary transformation of the orbitals and the state expansion coefficients of the MCSCF reference state may be carried out is given in Refs. 10–12. Below we summarize the results of this derivation.

A unitary transformation among the states $\{|j\rangle\}$ may be described as[11, 12]

$$\exp(i\hat{S})|j\rangle = \sum |k\rangle(\exp(-\mathbf{S}))_{kj} = \sum |k\rangle T_{kj} \tag{4}$$

where

$$\hat{S} = i \sum_{k \neq 0} S_{k0}(|k\rangle\langle 0| - |0\rangle\langle k|) \tag{5}$$

and $\mathbf{T} = \exp(-\mathbf{S})$ is a unitary matrix and $\mathbf{S}$ a real antisymmetric matrix with elements S_{k0} and S_{0k} and zero elsewhere.

A unitary transformation of the orbitals may similarly be described[10-12] as

$$\tilde{a}_r^+ = \exp(i\hat{\kappa}) a_r^+ \exp(-i\hat{\kappa}) \tag{6}$$

where

$$\hat{\kappa} = i \sum_{r > s} \kappa_{rs}(a_r^+ a_s - a_s^+ a_r) \tag{7}$$

From Eqs. (6) and (7) we get

$$\tilde{a}_r^+ = \sum_s a_s^+ (\exp - \boldsymbol{\kappa})_{sr} = \sum_s a_s^+ X_{sr} \tag{8}$$

where $\mathbf{X} = \exp(-\boldsymbol{\kappa})$ is a unitary matrix and $\boldsymbol{\kappa}$ an antisymmetric matrix with the elements κ_{rs} and κ_{sr} of Eq. (7) and zero elsewhere.

A unitary transformation of the reference state $|0\rangle$ that considers simultaneously a unitary transformation in the orbital and in the configurational space may therefore be described[11-13] as

$$|\tilde{0}\rangle = \exp(i\hat{\kappa})\exp(i\hat{S}|0\rangle \tag{9}$$

Thus the states

$$|\tilde{k}\rangle = \exp(i\hat{\kappa})\exp(i\hat{S})|k\rangle \tag{10}$$

together with $|\tilde{0}\rangle$ will form an orthonormal set. That is, through specifying a set of parameters $\boldsymbol{\kappa}$ and $\mathbf{S}$ we may generate an arbitrary state $|\tilde{0}\rangle$ and ensure that this state together with the states $|\tilde{k}\rangle$ form an orthonormal set of states $\{|\tilde{j}\rangle\}$. In what follows, for convenience we sometimes use $\underline{\lambda}$ to denote the set of rotational parameters arranged as a column vector

$$\begin{pmatrix} \underline{\kappa} \\ \mathbf{S} \end{pmatrix}$$

and occasionally refer to $\underline{\lambda}$ as the step length vector. The specific choice of step length vector $\underline{\lambda} = \mathbf{0}$ represents, of course, the untransformed state $|0\rangle$.

The set of vectors $(a_r^+ a_s - a_s^+ a_r)|0\rangle$, $(|k\rangle\langle 0| - |0\rangle\langle k|)|0\rangle$ may be linearly dependent, and some variables may be redundant when an optimization is carried out. The elimination of linear dependencies and redundant operators for a specific choice of configurations in the reference state is discussed in detail in Appendix A.[10, 17]

B. Variations in the Total Energy

The total energy corresponding to the unitary transformation of the reference state given in Eq. (9) may be written[10–12] as

$$E(\underline{\lambda}) = E(\underline{\kappa}, \mathbf{S}) = \langle\tilde{0}|H|\tilde{0}\rangle = \langle 0|\exp(-i\hat{S})\exp(-i\hat{\kappa})H\exp(i\hat{\kappa})\exp(i\hat{S})|0\rangle \tag{11}$$

The total energy is thus defined in terms of a set of rotational parameters, $\underline{\lambda}$. A Taylor series expansion may be carried out in these rotational parameters at $\underline{\lambda} = {}^0\underline{\lambda}$:

$$\begin{aligned} E(\underline{\lambda}) = E({}^0\underline{\lambda}) &+ \left.\frac{\partial E(\underline{\lambda})}{\partial \lambda_i}\right|_{\underline{\lambda}={}^0\underline{\lambda}} [\lambda_i - {}^0\lambda_i] \\ &+ \frac{1}{2}\left.\frac{\partial^2 E(\underline{\lambda})}{\partial \lambda_i \partial \lambda_j}\right|_{\underline{\lambda}={}^0\underline{\lambda}} [\lambda_i - {}^0\lambda_i][\lambda_j - {}^0\lambda_j] \\ &+ \frac{1}{6}\left.\frac{\partial^3 E(\underline{\lambda})}{\partial \lambda_i \partial \lambda_j \partial \lambda_k}\right|_{\underline{\lambda}={}^0\underline{\lambda}} [\lambda_i - {}^0\lambda_i][\lambda_j - {}^0\lambda_j][\lambda_k - {}^0\lambda_k] \\ &+ \frac{1}{24}\left.\frac{\partial E^4(\underline{\lambda})}{\partial \lambda_i \partial \lambda_j \partial \lambda_k \partial \lambda_l}\right|_{\underline{\lambda}={}^0\underline{\lambda}} [\lambda_i - {}^0\lambda_i][\lambda_j - {}^0\lambda_j] \\ &\times [\lambda_k - {}^0\lambda_k][\lambda_l - {}^0\lambda_l] \\ &+ \cdots \end{aligned} \tag{12}$$

Note that here and in all subsequent equations, unless noted otherwise, for convenience we use the Einstein summation convention.

Introducing the supermatrix notation for the partial derivatives

$$\left.\frac{\partial E(\underline{\lambda})}{\partial \lambda_i}\right|_{\underline{\lambda}={}^0\underline{\lambda}} = F_i({}^0\underline{\lambda}) \tag{13}$$

$$\left.\frac{\partial^2 E(\underline{\lambda})}{\partial \lambda_i \partial \lambda_j}\right|_{\underline{\lambda}={}^0\underline{\lambda}} = G_{ij}({}^0\underline{\lambda}) \tag{14}$$

$$\left.\frac{\partial^3 E(\underline{\lambda})}{\partial \lambda_i \partial \lambda_j \partial \lambda_k}\right|_{\underline{\lambda}={}^0\underline{\lambda}} = K_{ijk}({}^0\underline{\lambda}) \tag{15}$$

$$\left.\frac{\partial^4 E(\underline{\lambda})}{\partial \lambda_i \partial \lambda_j \partial \lambda_k \partial \lambda_l}\right|_{\underline{\lambda}={}^0\underline{\lambda}} = M_{ijkl}({}^0\underline{\lambda}) \tag{16}$$

allows us to express the Taylor expansion in a compact form:

$$\begin{aligned} E(\underline{\lambda}) = {} & E({}^0\underline{\lambda}) + F_i({}^0\underline{\lambda})[\lambda_i - {}^0\lambda_i] + \tfrac{1}{2}G_{ij}({}^0\underline{\lambda})[\lambda_i - {}^0\lambda_i][\lambda_j - {}^0\lambda_j] \\ & + \tfrac{1}{6}K_{ijk}({}^0\underline{\lambda})[\lambda_i - {}^0\lambda_i][\lambda_j - {}^0\lambda_j][\lambda_k - {}^0\lambda_k] \\ & + \tfrac{1}{24}M_{ijkl}({}^0\underline{\lambda})[\lambda_i - {}^0\lambda_i][\lambda_j - {}^0\lambda_j][\lambda_k - {}^0\lambda_k][\lambda_l - {}^0\lambda_l] + \cdots \end{aligned} \tag{17}$$

The partial derivatives in Eqs. (13–16) are cumbersome to evaluate at a general point ${}^0\underline{\lambda}$. In Appendix B we have derived an explicit expression for the first partial derivative $F_i({}^0\underline{\lambda})$. The second partial derivative is approximately one order of magnitude more difficult to evaluate than the first partial derivative; the third partial derivative is approximately an order of magnitude more complicated to evaluate than the second partial derivative; and so on. General expressions for the second and higher partial derivatives evaluated at an arbitrary point ${}^0\underline{\lambda}$ will not be given.

The vector **F** and matrix **G** represent the slope and curvature of the energy hypersurface, respectively. **F** is often referred to as the energy gradient vector, while **G** is often denoted the Hessian matrix. The MCSCF reference state represents a stationary point on the energy hypersurface. At such a point, the first order variation in the total energy is zero, that is, $\delta E(\underline{\lambda}) = 0$. The condition that $\delta E(\underline{\lambda}) = 0$ may be used to determine the step length to take on the energy hypersurface in order to reach a stationary point. From Eq. (17), the first-order variation of the total energy may be used to obtain $\underline{\lambda}$:

$$F_i({}^0\underline{\lambda})+G_{ij}({}^0\underline{\lambda})[\lambda_j-{}^0\lambda_j]+\tfrac{1}{2}K_{Ijk}({}^0\underline{\lambda})[\lambda_j-{}^0\lambda_j][\lambda_k-{}^0\lambda_k]+\cdots=0 \tag{18}$$

At the stationary point $\underline{\lambda}=\underline{\alpha}$,

$$\mathbf{F}(\underline{\alpha})=\mathbf{0} \tag{18a}$$

If we could determine a set of $\underline{\lambda}$ parameters from Eq. (18) without truncating this equation at any order of $\underline{\lambda}$, we would reach a stationary point on the energy hypersurface in one iteration. The iterative nature of MCSCF approaches originates from using truncated forms of Eq. (18) to determine the stationary points on the energy hypersurface. These truncated forms require successive applications to determine a stationary point on the energy hypersurface.

If Eq. (18) is truncated after the second term, which corresponds to truncating Eq. (12) after terms quadratic in $[\underline{\lambda}-{}^0\underline{\lambda}]$, second-order approaches are derived. These are discussed in detail in Section III. If Eq. (17) is truncated after the third terms [cubic in $[\underline{\lambda}-{}^0\underline{\lambda}]$ in Eq. (12)] cubic approaches may be developed. These are studied in Section VIII.

III. THE NEWTON–RAPHSON ITERATIVE FUNCTION

As a first example, we consider the Newton–Raphson iterative approach. The Newton–Raphson iterative function is derived by truncating the energy function [Eq. (12)] through the second power in the rotational parameters.[10-12] The Newton–Raphson approach straightforwardly illustrates how the local convergence of an iterative procedure may be rationalized in terms of a total order of convergence in an error vector.[16, 22] When the Newton–Raphson approach is implemented in numerical analysis, a linear transformation of the variables is performed between each step of the iterative procedure.[9] We demonstrate how it often may be more convenient when using the Newton–Raphson MCSCF approach to perform a nonlinear transformation of the variables between each step of the iterative procedure when orbital optimizations are carried out. The local convergence of the Newton–Raphson approach is then discussed, and we demonstrate that a sequence of n linear or nonlinear Newton–Raphson MCSCF iterations will have a total order of convergence of 2^n in the error vector of the initial iteration.

A. The Linear Newton–Raphson Iterative Function

The Newton–Raphson iterative function is derived from Eq. (18) by neglecting the nonlinear terms:

$$F_i(^0\underline{\lambda}) + G_{ij}(^0\underline{\lambda})\left[\lambda_j - {}^0\lambda_j\right] = 0 \tag{19}$$

The Newton–Raphson iterative function therefore becomes

$$^k\lambda_i = {}^{k-1}\lambda_i - G_{ij}^{-1}(^{k-1}\underline{\lambda}) F_j(^{k-1}\underline{\lambda}) \tag{20}$$

where $^k\underline{\lambda}$ denotes the set of $\underline{\lambda}$ parameters determined from the kth iteration of Eq. (20). The **F** and **G** matrices are evaluated at the $(k-1)$st iteration point, $^{k-1}\underline{\lambda}$. A sequence of iterations which uses Eq. (20) as an iterative function will result in a linear transformation of the variables $\underline{\lambda}$ in each step of the iterative sequence.

In the following we discuss in a little more detail how a sequence of iterations may be carried out which uses Eq. (20) as an iterative function. The first step of a Newton–Raphson sequence of iterations from an initial guess of orbitals and states ($^0\underline{\lambda} = \mathbf{0}$) is to apply Eq. (20) to determine a set of parameters

$$^1\underline{\lambda} - {}^0\underline{\lambda} = {}^1\underline{\lambda} = \begin{pmatrix} ^1\underline{\kappa} \\ ^1\mathbf{S} \end{pmatrix}.$$

This parameter set describes how rotations of the orbitals and states are carried out relative to the parameter point $^0\underline{\lambda} = \mathbf{0}$. The partial derivatives in Eq. (20) are evaluated at $^0\underline{\lambda} = \mathbf{0}$ in the initial iteration only. The second step of the Newton–Raphson sequence is then to determine a parameter set

$$^2\underline{\lambda} - {}^1\underline{\lambda} = \begin{pmatrix} ^2\underline{\kappa} - {}^1\underline{\kappa} \\ ^2\mathbf{S} - {}^1\mathbf{S} \end{pmatrix}$$

that describes how rotations are carried out relative to the point $^1\underline{\lambda} - {}^0\underline{\lambda} = {}^1\underline{\lambda}$. In the second step of the iterative procedure, the partial derivatives in Eq. (20) need to be evaluated at $^1\underline{\lambda}$ (no orbital or state transformations are carried out between each step of the iterative procedure). The third step of the Newton–Raphson iterative procedure determines $^3\underline{\lambda} - {}^2\underline{\lambda}$, which requires partial derivatives to be evaluated at $^2\underline{\lambda}$. And so on. Thus when n steps of Eq. (20) are applied, the transformed orbitals after the nth iteration are determined as

$$ {}^{n}\tilde{a}_r^+ = \exp(i\hat{\kappa}(n))\, a_s^+ \exp(-i\hat{\kappa}(n)) \tag{21} $$

where

$$ \hat{\kappa}(n) = i \sum_{r>s} {}^{n}\kappa_{rs} (a_r^+ a_s - a_s^+ a_r) \tag{22} $$

Defining the matrix

$$ {}^{n}\mathbf{X} = \exp({}^{n}\boldsymbol{\kappa}) \tag{23} $$

Eq. (21) may be rewritten using Eq. (8):

$$ {}^{n}\tilde{a}_r^+ = \sum_s a_s^+ ({}^{n}\mathbf{X})_{sr} \tag{24} $$

The transformed MCSCF reference state after n steps of such an iterative procedure becomes

$$ |{}^{n}0\rangle = \exp(i\hat{\kappa}(n)) \exp(i\hat{S}(n)) |0\rangle \tag{25} $$

and the coefficients of the states $\{|{}^{n}0\rangle, |{}^{n}k\rangle\}$ thus are determined as

$$ {}^{n}\mathbf{C} = {}^{0}\mathbf{C} \exp(-{}^{n}\mathbf{S}) \tag{26} $$

where ${}^{0}\mathbf{C}$ is the matrix of the initial guess (note $|{}^{0}0\rangle = |0\rangle$). Equations (24)–(26) demonstrate that a straightforward application of the Newton–Raphson iterative function results in an iterative procedure in which a linear transformation of the vector amplitudes

$$ \underline{\lambda} = \begin{pmatrix} \boldsymbol{\kappa} \\ \mathbf{S} \end{pmatrix} $$

is carried out between each step of the iterative procedure.

We now examine the local convergence properties of a sequence of n Newton–Raphson iterations. Consider initially how the $\mathbf{F}(\underline{\lambda})$ and $\mathbf{G}(\underline{\lambda})$ matrix elements may be expanded around the stationary point $\underline{\alpha}$

$$ F_i(\underline{\lambda}) = G_{ij}(\underline{\alpha})[\lambda_j - \alpha_j] + \tfrac{1}{2} K_{ijk}(\underline{\alpha})[\lambda_j - \alpha_j][\lambda_k - \alpha_k] + \cdots \tag{27} $$

$$ G_{ij}(\underline{\lambda}) = G_{ij}(\underline{\alpha}) + K_{ijk}(\underline{\alpha})[\lambda_k - \alpha_k] + \cdots \tag{28} $$

since $F_i(\underline{\alpha}) = 0$. The kth step of a Newton–Raphson sequence of iterations in Eq. (20) may now be rewritten using Eqs. (27) and (28)

$$\begin{aligned} {}^{k}\lambda_i &= {}^{k-1}\lambda_i - \left(G_{ij}(\underline{\alpha}) + K_{ijm}(\underline{\alpha})\left[{}^{k-1}\lambda_m - \alpha_m\right] + \cdots\right)^{-1} \\ &\quad \times \left(G_{ij}(\underline{\alpha})\left[{}^{k-1}\lambda_j - \alpha_j\right]\right. \\ &\quad \left. + \tfrac{1}{2}K_{ijm}(\underline{\alpha})\left[{}^{k-1}\lambda_j - \alpha_j\right]\left[{}^{k-1}\lambda_m - \alpha_m\right] + \cdots\right) \\ &= \alpha_i + \tfrac{1}{2}G_{in}^{-1}(\underline{\alpha})K_{njm}(\underline{\alpha})\left[{}^{k-1}\lambda_j - \alpha_j\right]\left[{}^{k-1}\lambda_m - \alpha_m\right] + O({}^{k-1}\underline{\lambda} - \underline{\alpha})^3 \end{aligned} \tag{29}$$

where we have expanded the inverse matrix. The kth Newton–Raphson iteration thus contains errors of order 2 in the error vector

$$ {}^{k-1}\mathbf{e} = {}^{k-1}\underline{\lambda} - \underline{\alpha} \tag{30}$$

of the $(k-1)$st iteration. The error term of the kth iteration may be written as

$$ {}^{k}e_i = \tfrac{1}{2}G_{in}^{-1}(\underline{\alpha})K_{njl}(\underline{\alpha})\,{}^{k-1}e_j\,{}^{k-1}e_l \tag{31}$$

When a sequence of two Newton–Raphson iterations are carried out after an initial guess of orbitals, we obtain a total order of convergence of 4 in ${}^{0}\mathbf{e}$ (the error vector of the initial iteration) with an error term

$$ {}^{2}e_i = \tfrac{1}{2}L_{ijk}\,{}^{1}e_j\,{}^{1}e_k = \tfrac{1}{8}L_{ijk}L_{jmn}\,{}^{0}e_m\,{}^{0}e_n\,L_{kpq}\,{}^{0}e_p\,{}^{0}e_q \tag{32}$$

where

$$ L_{ijk} = G_{in}^{-1}(\underline{\alpha})K_{njk}(\underline{\alpha}) \tag{33}$$

After n Newton–Raphson iterations are performed, the total order of convergence is 2^n in ${}^{0}\mathbf{e}$ and the error term becomes

$$ {}^{n}e_i = \tfrac{1}{2}L_{ijk}\,{}^{n-1}e_j\,{}^{n-1}e_k = 2^{-(2^n-1)}L^{(n)}_{ik_1k_2\cdots k_{(2^n)}}\,{}^{0}e_{k_1}\,{}^{0}e_{k_2}\cdots{}^{0}e_{k_{(2^n)}} \tag{34}$$

where L is defined recursively

$$ L^{(n)}_{ik_1k_2\cdots k_{(2^n)}} = L_{il_1l_2}L^{(n-1)}_{l_1k_1k_2\cdots k_{(2^{n-1})}}L^{(n-1)}_{l_2k_{(2^{n-1}+1)}k_{(2^{n-1}+2)}\cdots k_{(2^n)}} \tag{35}$$

and

$$L^{(1)}_{ik_1k_2} = L_{ik_1k_2} \tag{36}$$

We have now described how a sequence of linear Newton–Raphson iterations may be performed and that a sequence consisting of n iterations has a total order of convergence of 2^n in the error vector of the initial iteration.

B. The Nonlinear Newton–Raphson Iterative Function

The Newton–Raphson MCSCF approach of the previous section is just a standard application of the Newton–Raphson technique discussed in elementary textbooks of numerical analysis. The application of the Newton–Raphson approach requires evaluation of the first and second derivatives [Eqs. (13) and (14)] of the energy function at a general point. The second partial derivative is rather cumbersome to evaluate at a general point in MCSCF. Partial derivatives of the energy function are in general much easier to evaluate at the single point $\underline{\lambda} = \mathbf{0}$ than at a general point. This fact may be used to modify the Newton–Raphson approach to a form that is practically applicable in MCSCF.

According to Eq. (9) a transformed state $|\tilde{0}\rangle$ is defined in terms of a set of orbital operators $\{a_r^+\}$ and states $\{|j\rangle\}$ and a set of rotational parameters $\underline{\lambda}$. As previously discussed, the parameter set $\underline{\lambda} = \mathbf{0}$ represents the initial guess of the reference state $|0\rangle$ and orbitals. The initial Newton–Raphson iteration requires only evaluation of partial derivatives at $\underline{\lambda} = \mathbf{0}$. The initial Newton–Raphson iteration determines a parameter set ${}^1\underline{\lambda}$ which through Eqs. (8) and (9) defines the rotations that have to be carried out to determine transformed sets of orbital operators $\{{}^1\tilde{a}_r^+\}$ and states $\{|{}^1\tilde{j}\rangle\}$. These transformed orbitals and states may then be considered as a new initial guess. The Newton–Raphson iterative procedure may now be applied with this new "initial" guess of orbitals and states. The parameter set ${}^2\underline{\lambda}$ determined in the second Newton–Raphson iteration then describes how the rotations of the orbital operators $\{{}^1\tilde{a}_r^+\}$ and states $\{|{}^1\tilde{j}\rangle\}$ have to be carried out [use Eqs. (8) and (9)] to determine a new transformed set of orbitals $\{{}^2\tilde{a}_r^+\}$ and states $\{|{}^2\tilde{j}\rangle\}$, and so on. We note that $\{{}^n\tilde{a}_r^+\}$ and $\{|{}^n\tilde{j}\rangle\}$ are used in iteration $n+1$ (at iteration point n) to determine ${}^{n+1}\underline{\lambda}$ since $n=0$ is the initial guess of orbitals and states (i.e., $\{{}^0\tilde{a}_r^+\} = \{a_r^+\}$ and $\{|{}^0\tilde{j}\rangle\} = \{|j\rangle\}$).

After n applications of such a sequence of Newton–Raphson iterations, n unitary transformations have been carried out on the orbitals and the states. The orbital operators after n such unitary transformations may be described as

$$\begin{aligned}{}^n\tilde{a}_r^+ &= \exp\left[i\,{}^n\hat{\kappa}(n)\right]\cdots\exp\left[i\,{}^2\hat{\kappa}(2)\right]\exp\left[i\,{}^1\hat{\kappa}(1)\right]a_r^+\\ &\quad\times\exp\left[-i\,{}^1\hat{\kappa}(1)\right]\exp\left[-i\,{}^2\hat{\kappa}(2)\right]\cdots\exp\left[-{}^n\hat{\kappa}(n)\right]\end{aligned} \tag{37}$$

where ${}^{p}\hat{\tilde{\kappa}}(m)$ is defined as

$$ {}^{p}\hat{\tilde{\kappa}}(m) = i\sum {}^{m}\kappa_{rs}\left[{}^{p}\tilde{a}_r^{+}{}^{p}\tilde{a}_s - {}^{p}\tilde{a}_s^{+}{}^{p}\tilde{a}_r\right] \tag{38} $$

Henceforth, the tilde˜ and parentheses ( ) will generally be omitted when reference is made to the initial basis. Note that in Eq. (38) m and p are not necessarily equal; that is, p refers to the number of orbital operator transformations and m refers to the parameter set determined in the mth step of the iterative procedure.

The ${}^{n}\tilde{a}_r^{+}$ in Eq. (37) may also be expressed in terms of the initial set of creation and annihilation operators. Using the relation

$$ \exp\left[i\left({}^{2}\hat{\kappa}(2)\right)\right] = \exp[i\hat{\kappa}(1)]\exp[i\hat{\kappa}(2)]\exp[-i\hat{\kappa}(1)], \tag{39} $$

Eq. (37) may be written as[14]

$$ {}^{n}\tilde{a}_r^{+} = \exp[i\hat{\kappa}(1)]\cdots\exp[i\hat{\kappa}(n)]\,a_r^{+}\exp[-i\hat{\kappa}(n)]\cdots\exp[-i\hat{\kappa}(1)], \tag{40} $$

where $\hat{\kappa}(m)$ is expressed in terms of the initial set of creation and annihilation operators:

$$ \hat{\kappa}(m) = i\sum {}^{m}\kappa_{rs}\left[a_r^{+}a_s - a_s^{+}a_r\right] \tag{41} $$

The evaluation of the transformed set of creation operators may be performed as

$$ {}^{n}\tilde{a}_r^{+} = \sum_s a_s^{+}\left[\exp(-{}^{1}\boldsymbol{\kappa})\exp(-{}^{2}\boldsymbol{\kappa})\cdots\exp(-{}^{n}\boldsymbol{\kappa})\right]_{sr} \tag{42} $$

where the exponential matrices are defined through Eq. (8).

When n successive unitary transformations have been carried out on the MCSCF reference state, we obtain

$$ |{}^{n}0\rangle = \exp\left[i\,{}^{n}\hat{\tilde{\kappa}}(n)\right]\exp\left[i\,{}^{n}\hat{\tilde{S}}(n)\right]\cdots\exp\left[i\,{}^{2}\hat{\tilde{\kappa}}(2)\right] \times\exp\left[i\,{}^{2}\hat{\tilde{S}}(2)\right]\exp\left[i\,{}^{1}\hat{\tilde{\kappa}}(1)\right]\exp\left[i\,{}^{1}\hat{\tilde{S}}(1)\right]|0\rangle \tag{43} $$

where again the ${}^{n}\sim$ denotes that the operators are expressed in the basis obtained after n successive unitary transformations of the orbitals and the states. Equation (38) and successive application of the relation

$$ \exp\left[i\left({}^{2}\hat{\tilde{S}}(2)\right)\right] = \exp[i\hat{\kappa}(1)]\exp[i\hat{S}(1)]\exp[i\hat{S}(2)] \times\exp[-i\hat{S}(1)]\exp[-i\hat{\kappa}(1)] \tag{44} $$

allow us to express $|{}^n0\rangle$ in Eq. (43) in terms of the initial set of orbitals and states:[14]

$$|{}^n0\rangle = \exp[i\hat{\kappa}(1)]\exp[i\hat{\kappa}(2)]\cdots\exp[i\hat{\kappa}(n)] \times \exp[i\hat{S}(1)]\exp[i\hat{S}(2)]\cdots\exp[i\hat{S}(n)]|0\rangle \tag{45}$$

The coefficients for the states $\{|{}^n0\rangle, |{}^nk\rangle\}$ therefore may be determined from the relation

$${}^n\mathbf{C} = {}^0\mathbf{C}\exp(-{}^1\mathbf{S})\exp(-{}^2\mathbf{S})\cdots\exp(-{}^n\mathbf{S}) \tag{46}$$

The orbitals and states used in the $(n+1)$st iteration are thus determined by multiplying n exponential matrices each containing a set of rotational parameters.

Equations (42) and (46) clearly display that a nonlinear transformation of the rotational parameters is carried out between each step of the iterative procedure. The basis of orbitals in which we express our energy function (and consequently our iterative function) is changed at each step of the iterative process in such a nonlinear iterative procedure.

We shall now discuss how the partial derivatives of the total energy may be relatively straightforwardly evaluated at the point $\underline{\lambda} = \mathbf{0}$. This fact is used for practical MCSCF implementation of the Newton–Raphson algorithm but also in many other practical implementations of MSCSF iterative schemes. At this initial point $\underline{\lambda} = \mathbf{0}$, the partial derivatives may be evaluated by expanding the exponential operators

$$\begin{aligned}
E(\underline{\lambda}) &= \langle\tilde{0}|H|\tilde{0}\rangle = \langle 0|\exp(-i\hat{S})\exp(-i\hat{\kappa})H\exp(i\hat{\kappa})\exp(i\hat{S})|0\rangle \\
&= E(\mathbf{0}) - i\langle 0|[\hat{S}+\hat{\kappa}, H]|0\rangle - \frac{1}{2}\langle 0|[\hat{S},[\hat{S},H]]|0\rangle - \frac{1}{2}\langle 0|[\hat{\kappa},[\hat{\kappa},H]]|0\rangle \\
&\quad - \langle 0|[\hat{S},[\hat{\kappa},H]]|0\rangle + \frac{i}{6}\langle 0|[\hat{S},[\hat{S},[\hat{S},H]]]|0\rangle \\
&\quad + \frac{i}{6}\langle 0|[\hat{\kappa},[\hat{\kappa},[\hat{\kappa},H]]]|0\rangle + \frac{i}{2}\langle 0|[\hat{S},[\hat{S},[\hat{\kappa},H]]]|0\rangle \\
&\quad + \frac{i}{2}\langle 0|[\hat{S},[\hat{\kappa},[\hat{\kappa},H]]]|0\rangle + \frac{1}{24}\langle 0|[\hat{S},[\hat{S},[\hat{S},[\hat{S},H]]]]|0\rangle \\
&\quad + \frac{1}{24}\langle 0|[\hat{\kappa},[\hat{\kappa},[\hat{\kappa},[\hat{\kappa},H]]]]|0\rangle + \frac{1}{6}\langle 0|[\hat{S},[\hat{S},[\hat{S},[\hat{\kappa},H]]]]|0\rangle \\
&\quad + \frac{1}{4}\langle 0|[\hat{S},[\hat{S},[\hat{\kappa},[\hat{\kappa},H]]]]|0\rangle + \frac{1}{6}\langle 0|[\hat{S},[\hat{\kappa},[\hat{\kappa},[\hat{\kappa},H]]]]|0\rangle \\
&\quad + \cdots
\end{aligned} \tag{47}$$

All terms through fourth order in $\hat{\kappa}$ and $\hat{S}$ are written out explicitly. The n

tupple symmetric commutator for the operators $D_1, D_2, \ldots, D_n$ may be defined[16, 20] as

$$[D_1, D_2, \ldots, D_n, H] = \frac{1}{n!} P(1,2,\ldots,n)[D_1,[D_2, \cdots [D_n, H] \cdots]] \tag{48}$$

where $P(1,2,\ldots,n)$ is a permutation operator which contains the $n!$ permutations of the indices $1,2,\ldots,n$. Equation (47) may then be written as

$$\begin{aligned} E(\hat{\kappa}, \hat{S}) = {} & \langle 0|H|0\rangle - i\langle 0|[\hat{S}+\hat{\kappa}, h]|0\rangle - \frac{1}{2}\langle 0|[\hat{S}, \hat{S}, H]|0\rangle \\ & - \frac{1}{2}\langle 0|[\hat{\kappa}, \hat{\kappa}, H]|0\rangle - \langle 0|[\hat{S},[\hat{\kappa}, H]]|0\rangle + \frac{i}{6}\langle 0|[\hat{S}, \hat{S}, \hat{S}, H]|0\rangle \\ & + \frac{i}{6}\langle 0|[\hat{\kappa}, \hat{\kappa}, \hat{\kappa}, H]|0\rangle + \frac{i}{2}\langle 0|[\hat{S}, \hat{S},[\hat{\kappa}, H]]|0\rangle \\ & + \frac{i}{2}\langle 0|[\hat{S},[\hat{\kappa}, \hat{\kappa}, H]]|0\rangle + \frac{1}{24}\langle 0|[\hat{S}, \hat{S}, \hat{S}, \hat{S}, H]|0\rangle \\ & + \frac{1}{24}\langle 0|[\hat{\kappa}, \hat{\kappa}, \hat{\kappa}, \hat{\kappa}, H]|0\rangle + \frac{1}{6}\langle 0|[\hat{S}, \hat{S}, \hat{S},[\hat{\kappa}, H]]|0\rangle \\ & + \frac{1}{4}\langle 0|[\hat{S}, \hat{S},[\hat{\kappa}, \hat{\kappa}, H]]|0\rangle + \frac{1}{6}\langle 0|[\hat{S},[\hat{\kappa}, \hat{\kappa}, \hat{\kappa}, H]]|0\rangle \\ & + \cdots \end{aligned} \tag{49}$$

We have used the fact that each of the operators $\hat{\kappa}$ and $\hat{S}$ commutes with itself.

By using a supermatrix notation where the variational parameters $\{\kappa_{rs}\}$ and $\{S_{n0}\}$ and the excitation operators

$$\mathbf{Q}^+ = \{a_r^+ a_s\}, r > s; \mathbf{R}^+ = \{|n\rangle\langle 0|\}\} \tag{50}$$

both form column vectors

$$\underline{\lambda} = \begin{pmatrix} \boldsymbol{\kappa} \\ \mathbf{S} \end{pmatrix} \tag{51}$$

$$T^+ = \begin{pmatrix} \mathbf{Q}^+ \\ \mathbf{R}^+ \end{pmatrix} \tag{52}$$

we can write the total energy in Eq. (49) as

$$E(\underline{\lambda}) = E(\mathbf{0}) + F_i(\mathbf{0})\lambda_i + \tfrac{1}{2}G_{ij}(\mathbf{0})\lambda_i\lambda_j + \tfrac{1}{6}K_{ijk}(\mathbf{0})\lambda_i\lambda_j\lambda_k + \tfrac{1}{24}M_{ijkl}(\mathbf{0})\lambda_i\lambda_j\lambda_k\lambda_l + \cdots \tag{53}$$

where the matrices $F_i(\mathbf{0})$, $G_{ij}(\mathbf{0})$, $K_{ijk}(\mathbf{0})$, and $M_{ijkl}(\mathbf{0})$ are identified as

$$F_i(\mathbf{0}) = \langle 0|\left[T_i^+ - T_i, H\right]|0\rangle \tag{54}$$

$$G_{ij}(\mathbf{0}) = \langle 0|\left[T_i^+ - T_i, T_j^+ - T_j, H\right]|0\rangle \tag{55}$$

$$K_{ijk}(\mathbf{0}) = \langle 0|\left[T_i^+ - T_i, T_j^+ - T_j, T_k^+ - T_k, H\right]|0\rangle \tag{56}$$

$$M_{ijkl}(\mathbf{0}) = \langle 0|\left[T_i^+ - T_i, T_j^+ - T_j, T_k^+ - T_k, T_l^+ - T_l, H\right]|0\rangle \tag{57}$$

Again note that the Einstein summation convention is used. All terms in the n-tuple commutator that couple the configuration and orbital space are defined such that the Hamiltonian first operates on the orbital space excitation operators and then on the configuration space excitation operators; for example; all the six terms

$$\begin{aligned}[\mathbf{R}^+,\mathbf{R},\mathbf{Q},\mathbf{Q}^+, H] &= [\mathbf{Q},\mathbf{Q}^+,\mathbf{R}^+,\mathbf{R}, H] = [\mathbf{R},\mathbf{Q},\mathbf{Q}^+,\mathbf{R}^+, H] \\ &= [\mathbf{R}^+,\mathbf{Q},\mathbf{Q}^+,\mathbf{R}, H] = [\mathbf{R},\mathbf{R}^+,\mathbf{Q},\mathbf{Q}^+, H] \\ &= [\mathbf{Q},\mathbf{Q}^+,\mathbf{R},\mathbf{R}^+, H]\end{aligned}$$

are defined to equal

$$[\mathbf{R}^+,\mathbf{R},[\mathbf{Q},\mathbf{Q}^+, H]] \tag{58}$$

Equations (54)–(57) thus give explicit expressions for partial derivatives at $\underline{\lambda} = \mathbf{0}$. When partial derivatives are evaluated at the point ${}^0\underline{\lambda} = \mathbf{0}$ we omit reference to this point in the partial derivative matrices. Explicit expressions for evaluating the matrices $\mathbf{F}$ and $\mathbf{G}$ are given in Section III and Appendix C.

Stationary points on the energy hypersurface satisfy the condition $\delta E = 0$, that is,

$$F_i + G_{ij}\lambda_j + \tfrac{1}{2}K_{ijk}\lambda_j\lambda_k + \tfrac{1}{6}M_{ijkl}\lambda_j\lambda_k\lambda_l + \cdots = 0 \tag{59}$$

The nonlinear Newton–Raphson iterative function therefore becomes

$$\underline{\lambda} = -\mathbf{G}^{-1}\mathbf{F} \tag{60}$$

We now discuss the convergence properties when a sequence of k nonlinear Newton–Raphson iterations is carried out. The nth nonlinear Newton–Raphson iteration formally may be written as

$$ {}^{n}\tilde{\lambda}_i = -\overset{n-1}{\tilde{G}}{}_{ij}^{-1}\ \overset{n-1}{\tilde{F}}_j \tag{61}$$

where here $\overset{n-1}{}$ denotes that the matrices are evaluated for successively transformed orbitals and states involving ${}^{1}\underline{\tilde{\lambda}}, {}^{2}\underline{\tilde{\lambda}}, \ldots, {}^{n-1}\underline{\tilde{\lambda}}$ obtained from Eq. (61). The orbital and state transformations are carried out as described in Eqs. (42) and (46). The error term of the nth iteration may be expressed as

$$ {}^{n}\overset{n-1}{\tilde{e}}_i = \tfrac{1}{2}\overset{n-1}{\tilde{L}}_{ijl}\ {}^{n-1}\overset{n-1}{\tilde{e}}_j\ {}^{n-1}\overset{n-1}{\tilde{e}}_l \tag{62}$$

where we have used Eqs. (31) and (33). The error vector of the $n-1$ iteration may similarly be written as

$$ {}^{n-1}\overset{n-2}{\tilde{e}}_i = \tfrac{1}{2}\ \overset{n-2}{\tilde{L}}_{ijl}\ {}^{n-2}\overset{n-2}{\tilde{e}}_j\ {}^{n-2}\overset{n-2}{\tilde{e}}_l \tag{63}$$

The $\overset{n-1}{}$ basis is related to the $\overset{n-2}{}$ through multiplications involving the exponential unitary matrices $\exp(-{}^{n-1}\boldsymbol{\kappa})$ and $\exp(-{}^{n-1}\mathbf{S})$. The matrices of the $\overset{n-1}{}$ basis may therefore be expressed in the $\overset{n-2}{}$ basis as

$$ \overset{n-1}{\tilde{L}}_{ijl} = \overset{n-2}{\tilde{L}}_{ijl}\left(1 + O({}^{n-1}\underline{\tilde{\lambda}})\right) \tag{64}$$

$$ {}^{n}\overset{n-1}{\tilde{e}}_i = {}^{n}\overset{n-2}{\tilde{e}}_i\left(1 + O({}^{n-1}\underline{\tilde{\lambda}})\right) \tag{65}$$

Since ${}^{n-1}\underline{\lambda}$ contains error terms of the order $O(({}^{n-2}\overset{n-2}{\mathbf{\tilde{e}}})^2)$, Eqs. (64) and (65) may be written as

$$ \overset{n-1}{\tilde{L}}_{ijl} = \overset{n-2}{\tilde{L}}_{ijl}\left(1 + O\left(\left({}^{n-2}\overset{n-2}{\mathbf{\tilde{e}}}\right)^2\right)\right) \tag{66}$$

$$ {}^{n}\overset{n-1}{\tilde{e}}_i = {}^{n}\overset{n-2}{\tilde{e}}_i\left(1 + O\left(\left({}^{n-2}\overset{n-2}{\mathbf{\tilde{e}}}\right)^2\right)\right) \tag{67}$$

and, when we use Eqs. (63), (66), and (67), Eq. (62) becomes

$$ {}^{n}\overset{n-1}{\tilde{e}}_i = \tfrac{1}{8}\ \overset{n-2}{\tilde{L}}_{ijl}\ \overset{n-2}{\tilde{L}}_{jmn}\ {}^{n-2}\overset{n-2}{\tilde{e}}_m\ {}^{n-2}\overset{n-2}{\tilde{e}}_n\ \overset{n-2}{\tilde{L}}_{lqr}\ {}^{n-2}\overset{n-2}{\tilde{e}}_q\ {}^{n-2}\overset{n-2}{\tilde{e}}_r \tag{68}$$

Successive application of Eqs. (63), (66), and (67) then shows that the error term of the nth iteration of a nonlinear Newton–Raphson iteration ${}^{n}\overset{n-1}{\tilde{e}}$ becomes identical to the error term ${}^{n}\mathbf{e}$ of Eq. (34) obtained after n linear Newton–Raphson iterations. We have thus proved that n steps of a Newton–Raphson iteration procedure that performs either a linear or a nonlinear transformation of the variables between each step of the iterative procedure have the same order of convergence and contain the same error terms. Since the nonlinear and linear Newton–Raphson approaches have the same convergence properties and since the nonlinear Newton–Raphson approach only requires evaluation of partial derivatives at $\underline{\lambda} = \mathbf{0}$, practical applications of the MCSCF Newton–Raphson iterative approach have exclusively used a nonlinear approach.[1,2,22] Henceforth, for convenience, we will not use tilde with the nonlinear Newton–Raphson procedure.

Thus, the nomenclature we use is generally as follows. At iteration point $n-1$ the energy is evaluated, $\mathbf{F}$ and $\mathbf{G}$ are constructed, and ${}^{n}\underline{\lambda}$ is determined. Hence at iteration point $n-1$ the nth iteration is performed to obtain the rotational parameters ${}^{n}\underline{\lambda}$.

To understand the local convergence characteristics of an iterative calculation, it is necessary to get an estimate of the magnitude of the error vector at each step of the iterative procedure. The error vector describes the distance between the actual parameter values and the parameter values of the stationary point. Since the parameter values of the stationary point first are determined at convergence, an accurate measure of the error vector in principle cannot be obtained before convergence. However, the error vector at iteration point $n-1$ is

$$ {}^{n-1}\mathbf{e} = {}^{n-1}\underline{\lambda} - \underline{\alpha} = {}^{n-1}\underline{\lambda} - {}^{n}\underline{\lambda} + {}^{n}\underline{\lambda} - \underline{\alpha} = {}^{n-1}\underline{\lambda} - {}^{n}\underline{\lambda} + {}^{n}\mathbf{e} \tag{69} $$

If the length of the error vector at iteration point $n-1$ is much smaller than at iteration point n,

$$ \|{}^{n-1}\mathbf{e}\| \gg \|{}^{n}\mathbf{e}\| \tag{70} $$

then a rough estimate of the size of the error vector at iteration point $n-1$ becomes the length of the set of rotational parameters that are determined in the nth iteration

$$ \|{}^{n-1}\mathbf{e}\| \approx \|{}^{n-1}\underline{\lambda} - {}^{n}\underline{\lambda}\| \tag{71} $$

Equation (70) is relatively well satisfied when a second- or higher-order iterative scheme is applied in the local region, and Eq. (71) then becomes a reasonable measure of the length of the error vector. However, when parameter values are far from convergence, we are not justified in neglecting ${}^{n}\mathbf{e}$ compared to ${}^{n-1}\underline{\lambda} - {}^{n}\underline{\lambda}$ in Eq. (69) and the set of the rotational parameters of the nth iteration cannot be used with confidence for characterizing the global convergence of a sequence of iterations. This is especially true when parameter values are far from convergence and a step sign and size control algorithm is applied (see Section V). When a nonlinear iterative procedure is carried out, the basis of orbitals is changed at each step of the iterative procedure. The set of ${}^{n-1}\underline{\lambda}$ parameter values consequently are zero and $\|{}^{n}\lambda\|$ then directly may be used as an estimate of the magnitude of the error vector at iteration point $n-1$. Thus the initial error $\|{}^{0}\mathbf{e}\|$ is approximately given by $\|{}^{1}\underline{\lambda}\|$, and so on.

C. Implementation of the Nonlinear Newton–Raphson Approach

1. *One-Step Approach*

The nonlinear Newton–Raphson equation is defined in Eq. (60). A computationally more tractable form is obtained by using the right-hand side of Eqs. (51) and (52), respectively, rather than $\underline{\lambda}$ and $\underline{T}^{+}$. Equation (60) then becomes

$$\begin{pmatrix} \underline{\kappa} \\ \mathbf{S} \end{pmatrix} = -(\mathbf{A}-\mathbf{B})^{-1}\begin{pmatrix} \mathbf{W} \\ \mathbf{V} \end{pmatrix} = -\mathbf{G}^{-1}\mathbf{F} \tag{72}$$

where

$$\mathbf{W} = \langle 0|[\mathbf{Q}^{+}, H]|0\rangle \tag{73}$$

$$\mathbf{V} = \langle 0|[\mathbf{R}^{+}, H]|0\rangle \tag{74}$$

and

$$(\mathbf{A}-\mathbf{B}) = \tfrac{1}{2}\mathbf{G}$$

$$\begin{pmatrix} \mathbf{W} \\ \mathbf{V} \end{pmatrix} = \tfrac{1}{2}\mathbf{F}$$

$$\mathbf{A} = \begin{pmatrix} \mathbf{A}^{\mathrm{OO}} & \mathbf{A}^{\mathrm{OC}} \\ \mathbf{A}^{\mathrm{CO}} & \mathbf{A}^{\mathrm{CC}} \end{pmatrix} = -\begin{pmatrix} \langle 0|[\mathbf{Q},\mathbf{Q}^{+}, H]|0\rangle & \langle 0|[\mathbf{R},[\mathbf{Q}^{+}, H]]|0\rangle \\ \langle 0|[\mathbf{R}^{+},[\mathbf{Q}, H]]|0\rangle & \langle 0|[\mathbf{R},\mathbf{R}^{+}, H]|0\rangle \end{pmatrix} \tag{75}$$

$$\mathbf{B} = \begin{pmatrix} \mathbf{B}^{\mathrm{OO}} & \mathbf{B}^{\mathrm{OC}} \\ \mathbf{B}^{\mathrm{CO}} & \mathbf{B}^{\mathrm{CC}} \end{pmatrix} = -\begin{pmatrix} \langle 0|[\mathbf{Q}^{+},\mathbf{Q}^{+}, H]|0\rangle & \langle 0|[\mathbf{R}^{+},[\mathbf{Q}^{+}, H]]|0\rangle \\ \langle 0|[\mathbf{R}^{+},[\mathbf{Q}^{+}, H]]|0\rangle & \langle 0|[\mathbf{R}^{+},\mathbf{R}^{+}, H]|0\rangle \end{pmatrix} \tag{76}$$

Once $\underline{\kappa}$ and **S** are obtained, a new orbital basis [see Eq. (8)] and state expansion coefficients [see Eq. (4)] are defined. The new orbitals and state expansion coefficients are used to construct $\mathbf{A}-\mathbf{B}$, **W**, and **V**. This requires a two-electron transformation (see Section IX.A). Subsequently a new $\underline{\kappa}$ and a new **S** are again obtained from Eq. (72), and so on (as above). This sequence is continued until all elements in **W** and **V** (or $\underline{\kappa}$ and **S**) are smaller than a certain tolerance. The expressions $\mathbf{W}=\mathbf{0}$ and $\mathbf{V}=\mathbf{0}$ are sometimes referred to together as the generalized Brillouin theorem (GBT). When Eq. (72) is used as an iterative function, the calculation is often referred to as the one-step nonlinear Newton–Raphson approach.[11,12]

2. *Two-Step Approach*

Equation (72) may be used as an iterative function in a slightly different way if a configuration interaction (CI) calculation is carried out in the MCSCF configuration space before Eq. (72) is applied. When the configuration interaction calculation is carried out

$$\langle m|H|l\rangle = E_l\delta_{ml} \tag{77}$$

and

$$V_n = \langle 0|[|n\rangle\langle 0|, H]|0\rangle = 0 \tag{78}$$

$$A^{CC}_{nm} = -\langle 0|[R_n, R^+_m, H]|0\rangle = \delta_{nm}(E_m - E_0) \tag{79}$$

$$B^{CC}_{nm} = 0$$

Equation (72) may therefore be partitioned,[11,41] using Eq. (78), as

$$\underline{\kappa} = -\left(\mathbf{A}^{OO} - \mathbf{B}^{OO} - (\mathbf{A}^{OC} - \mathbf{B}^{OC})(\mathbf{A}^{CC})^{-1}(\mathbf{A}^{CO} - \mathbf{B}^{CO})\right)^{-1}\mathbf{W} \tag{80}$$

A similar iterative sequence to the one-step procedure may be used to obtain all elements of **W** (or $\underline{\kappa}$) smaller than a certain tolerance. However, an additional CI is required in each iteration subsequent to the transformation and prior to construction of $\mathbf{A}-\mathbf{B}$ and **W**. When Eq. (80) is used as an iterative function, the calculation is often referred to as the two-step nonlinear Newton–Raphson approach.[10,11,13]

If energy difference $E_m - E_0$ occurring in diagonal elements of $\mathbf{A}^{CC}$ [see Eq. (79)] is relative small, the last term in the inverse matrix may be very large and the two-step procedure is then less tractable for use as an iterative function than the one-step procedure. In a later section we will explicitly compare the convergence characteristics of the one- and two-step nonlinear Newton–Raphson calculations. We will also present an example of where it is dangerous to use the two-step procedure. The nonlinear Newton–Raphson calculations we report will generally, if not otherwise specified, be one-step calculations.

The terms $\mathbf{A}^{\mathrm{CO}} - \mathbf{B}^{\mathrm{CO}}$ which couple the configuration and orbital variation have been neglected in some calculations. When the coupling terms have been neglected, convergence problems have often been encountered.[11,13,15,36] In some cases an iterative scheme which neglects the coupling cannot converge even if the initial guess of orbitals and states are infinitesimally close to the desired stationary point.[13,15]

When converging to the $(n-1)$st excited state of a given symmetry, the Hessian matrix for that state should usually have $n-1$ negative eigenvalues. Since the Hessian matrix appears directly in the one-step Newton–Raphson iterative function, this fact may be used to direct the calculation to the desired stationary point (see Section V). When the two-step Newton–Raphson procedure is applied only eigenvalues of the reduced Hessian,

$$\mathbf{A}^{\mathrm{OO}} - \mathbf{B}^{\mathrm{OO}}(\mathbf{A}^{\mathrm{OC}} - \mathbf{B}^{\mathrm{OC}})(\mathbf{A}^{\mathrm{CC}})^{-1}(\mathbf{A}^{\mathrm{CO}} - \mathbf{B}^{\mathrm{CO}})$$

and of the configuration block of the Hessian, $\mathbf{A}^{\mathrm{CC}}$, are directly available. In Appendix D we show that if the Hessian matrix is required to have n negative eigenvalues and the configurational block of the Hessian matrix has m negative eigenvalues, then the reduced Hessian matrix needs to have $n-m$ negative eigenvalues. The step size and sign control algorithm of Section V may therefore also be used directly on Eq. (80) to ensure convergence to the desired solution. However, it may be very difficult to ensure convergence to the desired state if an energy difference $E_m - E_0$ of Eq. (79) is very small.

D. Matrix Elements in a Nonlinear Newton–Raphson Approach

We now consider the specific case where $\hat{H}$ is the electronic Hamiltonian of an isolated atom or molecule, that is,

$$\hat{H} = \sum_{ij} h_{ij} a_i^+ a_j + \frac{1}{2} \sum_{ijkl} (ik|jl)\, a_i^+ a_j^+ a_l a_k$$

where

$$h_{ij} = \langle \phi_i | \hat{h} | \phi_j \rangle \qquad (\hat{h} \text{ is the sum of one-body operators})$$

and

$$(ik|jl) = \langle \phi_i(1)\phi_j(2) | \frac{1}{r_{12}} | \phi_k(1)\phi_l(2) \rangle$$

where the ϕ's are spin orbitals.

Let us now consider explicitly the evaluation of some of the matrix elements appearing in the one-step Newton–Raphson approach.[11,23,26–27]

$$A_{nm}^{CC} = -\langle 0|[R_n, R_m^+, H]|0\rangle = \langle n|H|m\rangle - \delta_{nm}\langle 0|H|0\rangle, \qquad B_{nm}^{CC} = 0 \tag{81}$$

$$V_n = \langle 0|[|n\rangle\langle 0|, H]|0\rangle = -\langle 0|H|n\rangle \tag{82}$$

Thus, $\mathbf{A}^{CC}$ and $\mathbf{V}$ contain all matrix elements contained in the MCSCF CI calculation. When the iterative procedure has converged, all elements of $\mathbf{V}$ are zero and the interaction between the MCSCF reference state $|0\rangle$ and the residual states are eliminated. The diagonal and off-diagonal matrix elements of the Hamiltonian in the residual space $\{|k\rangle\}$ may, however, all be nonvanishing.

Except for $\mathbf{A}^{CC}$ and $\mathbf{V}$, the form of the matrix elements in the one- and two-step Newton–Raphson approaches are the same. The matrix elements of $\mathbf{W}$, $\mathbf{A}^{OO} - \mathbf{B}^{OO}$ may be derived from Eqs. (83) and (84) by index substitution

$$\begin{aligned}\langle 0|[a_{t\sigma}^+ a_{u\sigma}, H]|0\rangle &= h_{up}\langle 0|a_{t\sigma}^+ a_{p\sigma}|0\rangle - h_{pt}\langle 0|a_{p\sigma}^+ a_{u\sigma}|0\rangle \\ &\quad -(pr|qt)\rho_{qpru}^{(2)} + (ur|qs)\rho_{tqsr}^{(2)}\end{aligned} \tag{83}$$

$$\begin{aligned}\langle 0|[a_{l\sigma}^+ a_{k\sigma}, [H, a_{t\sigma'}^+ a_{u\sigma'}]]|0\rangle &= h_{kt}\langle 0|a_{l\sigma}^+ a_{u\sigma}|0\rangle + h_{ul}\langle 0|a_{t\sigma}^+ a_{k\sigma}|0\rangle \\ &\quad - \delta_{kt} h_{up}\langle 0|a_{l\sigma}^+ a_{p\sigma}|0\rangle - \delta_{lu} h_{pt}\langle 0|a_{p\sigma}^+ a_{k\sigma}|0\rangle \\ &\quad - \delta_{lu}(pr|qt)\rho_{pqkr}^{(2)} \\ &\quad - \delta_{kt}(ur|qs)\rho_{lqsr}^{(2)} + (pl|qt)\rho_{pquk}^{(2)} \\ &\quad -(ur|ks)\rho_{tlsr}^{(2)} + (kr|pt)\rho_{lpur}^{(2)} + (kt|pr)\rho_{plur}^{(2)} \\ &\quad + (ul|qs)\rho_{tqsk}^{(2)} + (us|ql)\rho_{qtsk}^{(2)}\end{aligned} \tag{84}$$

where

$$\rho_{ijkl}^{(2)} = \langle 0|a_{i\sigma}^+ a_{j\sigma'}^+ a_{k\sigma'} a_{l\sigma}|0\rangle \tag{85}$$

In these equations we have explicitly introduced spin (σ and σ' run over the electron spin indices α and β). Note again that the Einstein summation convention is used. The excitation operators in Eqs. (83) and (84) have been coupled to singlet spin symmetry since they appear in the operator $\hat{\kappa}$ which must preserve the symmetry of the reference state when forming $\exp(i\hat{\kappa})|0\rangle$. The matrix elements of $\mathbf{W}$, and $\mathbf{A}^{OO} - \mathbf{B}^{OO}$ can be expressed in terms of one- and two-electron integrals and the one- and two-electron density matrices.

The elements of $\mathbf{A}^{OC} - \mathbf{B}^{OC}$ may be reduced as follows:

$$\langle 0|[|0\rangle\langle n|, [a^+_{t\sigma} a_{u\sigma}, H]]|0\rangle = \langle n|[a^+_{t\sigma} a_{u\sigma}, H]|0\rangle \tag{86}$$

Then an explicit formula for Eq. (86) may be obtained from Eq. (83) by replacing one- and two-electron density matrices with the corresponding transition density matrix elements.

All indices in the two-electron density matrix must refer to occupied or partly occupied orbitals to give a nonvanishing two-electron density matrix element. Since in Eq. (83) three indices are common for the two-electron integral multiplying the two-electron density matrix element, only one index in the two-electron integrals in Eq. (83) can refer to a completely unoccupied orbital. Also, in Eq. (84) there are at least two indices common for a certain two-electron integral and the multiplying density matrix element. Hence, for the two-electron integrals in Eq. (84) only at most two indices can refer to completely unoccupied orbitals.

We note that we have not explicitly written any additional symmetry indices on the formulas. This is, of course, straightforward to do; however, a rather cumbersome nomenclature results. In all calculations we have reported, spatial and spin symmetry has, of course, been explicitly incorporated.[11-22]

Further analysis of the gradient and Hessian matrix elements is given in Appendix C.

IV. CHARACTERIZATION OF AN MCSCF STATE

Before we proceed further we will analyze in more detail the requirements that may be imposed on an MCSCF state to assure that the MCSCF state is a proper representation of the exact nth state. So far the MCSCF state is only required to be variationally correct, that is, to satisfy the GBT (i.e., be at a stationary point on the energy hypersurface). This is, of course, an insufficient condition to ensure that the MCSCF state also is a proper representation of the nth exact state, and additional conditions should be fulfilled.

An MCSCF state optimization is performed both with respect to linear (configuration expansion coefficients) and nonlinear (orbital expansion coefficients) variational parameters. In CI calculations the optimization is only performed with respect to the linear configuration expansion parameter and the Hyleraas–Undheim–McDonald theorem[42] (the nth root in energy of a certain symmetry of the CI secular problem is an upper bound to the nth exact state of a certain symmetry) is often used as a necessary and sufficient condition for identifying the states. The Hyleraas–Undheim–McDonald theorem is only valid when variations are restricted to linear variation parameters. Hence the Hyleraas–Undheim–McDonald theorem can only be applied within the MCSCF configuration space. If the Hyleraas–Undheim–McDonald theorem is applied within the MCSCF configuration space we would require the CI within the MCSCF configuration space to have $n-1$ roots lower in energy than the root for the desired nth state. Since the MCSCF optimization also is performed with respect to the nonlinear orbital expansion parameters, it may be more appropriate to use criteria for characterizing an MCSCF state that are applicable to both the linear and nonlinear part of the variational space.

The MCSCF state is an approximation to the nth root in energy of a certain symmetry of a full CI calculation. The Hessian matrix corresponding to the nth root of a certain symmetry of the full CI solution has $n-1$ negative Hessian eigenvalues. It is reasonable for the MCSCF state to have $n-1$ negative Hessian eigenvalues, thereby simulating the full CI solution.

When characterizing the MCSCF state it is also appropriate to examine the linear response of the MCSCF state to an external one-electron perturbation. The linear response [or multiconfigurational time-dependent Hartree–Fock (MCTDHF)] calculation determines a set of excitation energies.[43-53] When describing the nth state in energy of a certain symmetry it is relevant to require that the response calculation for that symmetry gives $n-1$ negative excitation energies. In addition, the multiconfigurational stability conditions should be fulfilled.[16,43-44,52-54]

In summary, the MCSCF state which represents the nth state of a certain symmetry should ideally have the following characteristics:

1. The full (unpartitioned) MCSCF Hessian should have $n-1$ negative eigenvalues.
2. The MCTDHF calculations using the converged MCSCF orbitals and state expansion coefficients should be stable and have $n-1$ negative excitation energies to states of the same symmetry as the MCSCF state.
3. The CI with the MCSCF configuration state functions should have $n-1$ roots lower in energy.

For the lowest state of a given symmetry all these conditions must be satisfied at convergence. For an excited state of a given symmetry all of the above criteria should also be fulfilled at convergence. However, with a limited MCSCF configuration state space and with finite basis sets some (or, in rare cases, all) of these will not be met. This is particularly true when there are two (or more) states of the same symmetry that are very close in energy. Since it has not been traditional to examine these criteria either when converging or at convergence, we suspect that many of the previously reported MCSCF calculations are erroneous.

A further consideration is, of course, to examine the magnitude of state expansion coefficients. This is particularly relevant when some or all of the foregoing three criteria are not fulfilled. However, assignments based primarily on these magnitudes should be made only with extreme caution.

Another important reason for discussing at this point the characteristics of an MCSCF state is that these characteristics can be used to ensure that an MCSCF calculation is proceeding to the region of the correct stationary point. For example, when far from convergence (the global convergence problem) the Newton–Raphson procedure may give a few large step length amplitudes. These large amplitudes may take a calculation to an undesired place on energy hypersurface. The Hessian may have too many ($>n-1$) or too few ($<n-1$) negative eigenvalues. These may also cause a calculation to proceed to an undesired stationary point. Hence some sort of step size and sign control algorithm (or higher order procedure) needs to be used when far from convergence. Step size and sign control is discussed in the next section. In all cases the algorithm should be *firmly* based on principles which will lead us to the correct stationary point. For example, we have found that the most reliable and efficient constraint procedures should be designed to force the full Hessian to have the proper number, $n-1$, of negative eigenvalues in every iteration.

A. *n*th State of the CI Using the MCSCF Configuration State Functions (CSFs)

It is important that the nth state of a certain symmetry has $n-1$ negative eigenvalues of the converged full MCSCF Hessian. This condition does not, however, imply that the state needs to be the nth root of a certain symmetry of the CI using the MCSCF configuration state functions. The CI matrix [actually $2\langle i|H|j\rangle - 2\delta_{ij}\langle 0|H|0\rangle$; see Eq. (81)] is only one subblock ($\mathbf{G}^{CC}$) of the full Hessian. Even though the full Hessian has $n-1$ negative eigenvalues at convergence, the state we are converging to may not be the nth state of the CI. Such a solution where the state is not the nth root of the CI (root flipping) will be variationally correct (and may, in fact, even be a very good approximation) but, of course, may not be an upper bound to the nth state.[42]

Root flipping occurs when the nth state in the MCSCF CI becomes the ($n-1$)st [or ($n-2$)nd or ($n-3$)rd...] state as convergence progresses. This occurs because the orbitals and state expansion coefficients are being optimized for the original nth state and not for the other states of the same symmetry. When root flipping occurs the Hyleraas–Undheim–McDonald theorem[42] may no longer be used to characterize the MCSCF state and the MCSCF state in general cannot assuredly be taken as an upper bound to the nth exact eigenstate. We have demonstrated through the correct application of the ideas on redundant variables and linear dependency that for many cases root flipping is an artifact of the choice of MCSCF configurations.[17] This is because there is often a freedom of choice between state and orbital variables for MCSCF optimization. Hence, a converged calculation with a smaller MCSCF CI space may have root flipping, whereas a calculation with a larger MCSCF CI space may have no root flipping. However, both converged MCSCF states will have the same total MCSCF energy. Hence, even though root flipping has occurred, the energy may still be an upper bound since certainly the energy of the calculation with no root flipping is an upper bound. We conclude that, in general, criteria 1 and 2 (above) are probably more significant than criterion 3 and that the current rather strong emphasis[29–30, 32, 35–36] on having MCSCF calculations with no root flipping is probably somewhat misplaced.

B. Eigenvalues of the Hessian

Eigenvalues of the converged Hessian matrix of Eq. (14) are positive for the ground state. Similarly, a condition on the first, second,... excited state is that the corresponding Hessian matrix has one, two,... negative eigenvalues. In the limit of the full CI, all orbital optimization variables are redundant and hence the Hessian will just be $\mathbf{G}^{CC}$. The matrix elements of $\frac{1}{2}\mathbf{G}^{CC}$ are

$$\langle n|H|m\rangle - \delta_{mn}\langle 0|H|0\rangle \tag{81}$$

where n and m are CI states. In this case the eigenvalues of $\mathbf{G}^{CC}$ and the full Hessian are the same. Obviously, in the limit of the full CI with an adequate basis set the Hessian will also have the correct number of negative eigenvalues.

With a smaller choice of configurations, the eigenvalues of $\mathbf{G}^{CC}$ and $\mathbf{G}$ are no longer the same. In fact, at convergence $\mathbf{G}^{CC}$ may not have the appropriate number $(n-1)$ negative eigenvalues for the nth state while $\mathbf{G}$ may have $n-1$ negative eigenvalues. Thus, when root flipping occurs, our MCSCF reference state energy may not be an upper bound to the energy of this state. (However, see the discussion following and in Section V and Appendix A concerning reparametrization of an MCSCF problem). With a reasonable choice of configurations and provided the full Hessian has the proper number of negative eigenvalues, the converged MCSCF state may be a good representation of an excited state even when root flipping has occurred. The fact that an energy is an upper bound to the exact energy is, of course, no guarantee that the corresponding state is a good representation of an exact solution to the Schrödinger equation. Techniques which attempt to force $\mathbf{G}^{CC}$ to have $n-1$ negative eigenvalues are variationally constrained and may lead to undesired results.

In the two-step procedure, the eigenvalues of the MCSCF CI problem and the eigenvalues of the reduced Hessian matrix are separately determined. The eigenvalue problem for the full Hessian

$$\begin{pmatrix} \mathbf{G}^{OO} & \mathbf{G}^{OC} \\ \mathbf{G}^{CO} & \mathbf{G}^{CC} \end{pmatrix} \begin{pmatrix} \mathbf{U}_O \\ \mathbf{U}_C \end{pmatrix} = \varepsilon \begin{pmatrix} \mathbf{U}_O \\ \mathbf{U}_C \end{pmatrix} \tag{87}$$

transforms, using partitioning theory[41] to an equation of reduced dimension

$$\left[\mathbf{G}^{OO} + \mathbf{G}^{OC}(\varepsilon\mathbf{1} - \mathbf{G}^{CC})^{-1}\mathbf{G}^{CO}\right]\mathbf{U}_O = \varepsilon\mathbf{U}_O \tag{88}$$

The resolvent can then be expanded

$$\begin{aligned} &\left[\mathbf{G}^{OO} - \mathbf{G}^{OC}(\mathbf{G}^{CC} - \varepsilon\mathbf{1})^{-1}\mathbf{G}^{CO}\right]\mathbf{U}_O \\ &= \left(\mathbf{G}^{OO} - \mathbf{G}^{OC}(\mathbf{G}^{CC})^{-1}\mathbf{G}^{CO} - \varepsilon\mathbf{G}^{OC}(\mathbf{G}^{CC})^{-1}\mathbf{1}(\mathbf{G}^{CC})^{-1}\mathbf{G}^{OC} - \cdots\right)\mathbf{U}_O \end{aligned} \tag{89}$$

Hence, the eigenvalues and eigenvectors of the reduced Hessian are not the eigenvalues and eigenvectors of the full Hessian. They will be similar to the eigenvalues and vectors of the full Hessian only if the third, fourth, etc. terms in Eq. (89) are small and may be ignored.

The eigenvalues of the reduced Hessian are related to the eigenvalues of the Hessian matrix [Eq. (87)] through the nonorthogonal transformation

$$\begin{aligned} &\begin{pmatrix} \mathbf{1} & -\mathbf{G}^{OC}\mathbf{G}^{CC^{-1}} \\ \mathbf{0} & \mathbf{1} \end{pmatrix} \begin{pmatrix} \mathbf{G}^{OO} & \mathbf{G}^{OC} \\ \mathbf{G}^{CO} & \mathbf{G}^{CC} \end{pmatrix} \begin{pmatrix} \mathbf{1} & \mathbf{0} \\ -\mathbf{G}^{CC^{-1}}\mathbf{G}^{CO} & \mathbf{1} \end{pmatrix} \\ &= \begin{pmatrix} \mathbf{G}^{OO} - \mathbf{G}^{OC}\mathbf{G}^{CC^{-1}}\mathbf{G}^{CO} & \mathbf{0} \\ \mathbf{0} & \mathbf{G}^{CC} \end{pmatrix} \end{aligned} \tag{90}$$

This transformation may be used to relate the number of negative eigenvalues of the Hessian matrix to the number of negative eigenvalues of the reduced Hessian matrix. If we define an arbitrary vector $\mathbf{Y}$ as

$$\mathbf{Y} = \begin{pmatrix} \mathbf{1} & \mathbf{0} \\ -\mathbf{G}^{CC^{-1}}\mathbf{G}^{CO} & \mathbf{1} \end{pmatrix} \mathbf{X} \tag{91}$$

for any vector $\mathbf{X}$ satisfying $|\mathbf{X}| > 0$, it is easily seen that if the Hessian matrix is positive definite

$$\mathbf{Y}^T \begin{pmatrix} \mathbf{G}^{OO} & \mathbf{G}^{OC} \\ \mathbf{G}^{CO} & \mathbf{G}^{CC} \end{pmatrix} \mathbf{Y} > 0 \tag{92}$$

then the eigenvalues of

$$\begin{pmatrix} \mathbf{G}^{\text{OO}} - \mathbf{G}^{\text{OC}}\mathbf{G}^{\text{CC}^{-1}}\mathbf{G}^{\text{CO}} & \mathbf{0} \\ \mathbf{0} & \mathbf{G}^{\text{CC}} \end{pmatrix} \tag{93}$$

have to be positive. Similarly, if the eigenvalues of Eq. (93) are positive, the eigenvalues of the Hessian matrix are positive. If the Hessian matrix has one negative eigenvalue, Eq. (93) contains at least one negative eigenvalue, but from the preceding analysis it is difficult to determine the number of negative eigenvalues in Eq. (93).

A more thorough analysis based on a multidimensional partitioning[41] of the full Hessian (see Appendix D) shows that the number of negative eigenvalues of Eq. (93) is the same as the number of negative eigenvalues of Eq. (87). Hence, if there is root flipping, negative eigenvalues will appear in the upper-left-hand block (reduced Hessian matrix) of Eq. (93).

A final point is that since the CI coefficient problem is linear and the orbital optimization problem is nonlinear it is expected that there may be several stationary points with the same number of negative eigenvalues of the full Hessian (e.g., the BO calculations reported in Ref. 17). In such cases, a more detailed examination of the CI coefficients and criteria 1–3, above, is usually required.

C. Excitation Energies in the Multiconfigurational Time-Dependent Hartree–Fock Approximation

We will now describe how excitation energies (positive and negative) may be determined in a MCTDHF [also known as the multiconfigurational random phase approximation (MCRPA)] calculation.[43–50]

In the MCTDHF approximation the linear response of an MCSCF state to a frequency-dependent one-electron perturbation is examined. The MCTDHF approximation has been derived previously and excitation energies are determined as eigenvalues of the generalized eigenvalue problem

$$\begin{pmatrix} \mathbf{A} & \mathbf{B} \\ \mathbf{B} & \mathbf{A} \end{pmatrix}\begin{pmatrix} \mathbf{Z} \\ \mathbf{Y} \end{pmatrix} = \omega \begin{pmatrix} \mathbf{S} & \boldsymbol{\Delta} \\ -\boldsymbol{\Delta} & -\mathbf{S} \end{pmatrix}\begin{pmatrix} \mathbf{Z} \\ \mathbf{Y} \end{pmatrix} \tag{94}$$

where the **A** and **B** matrices are defined in Eqs. (75) and (76) and

$$\mathbf{S} = \langle 0 | [\mathbf{T}, \mathbf{T}^+] | 0 \rangle \tag{95}$$

$$\boldsymbol{\Delta} = \langle 0 | [\mathbf{T}, \mathbf{T}] | 0 \rangle \tag{96}$$

Here $\begin{pmatrix} \mathbf{Z} \\ \mathbf{Y} \end{pmatrix}$ are the eigenvectors and ω is the excitation energy, $\omega_i = E_i - E_0$.

The solution of Eq. (94) may be determined through performing a series of transformations involving matrices of only half the dimension of Eq. (94).[16, 43-44, 52] To achieve this reduction, we write Eq. (94)

$$\mathbf{AZ}+\mathbf{BY}=\omega SZ+\omega\mathbf{\Delta Y} \tag{97}$$

$$\mathbf{BZ}+\mathbf{AY}=-\omega\mathbf{\Delta Z}-\omega\mathbf{SY} \tag{98}$$

Successively adding and subtracting the above two equations gives

$$(\mathbf{A}+\mathbf{B})(\mathbf{Z}+\mathbf{Y})=\omega(\mathbf{S}-\mathbf{\Delta})(\mathbf{Z}-\mathbf{Y}) \tag{99}$$

$$(\mathbf{A}-\mathbf{B})(\mathbf{Z}-\mathbf{Y})=\omega(\mathbf{S}+\mathbf{\Delta})(\mathbf{Z}+\mathbf{Y}) \tag{100}$$

Equation (99) may then be rearranged

$$\mathbf{Z}+\mathbf{Y}=\omega(\mathbf{A}+\mathbf{B})^{-1}(\mathbf{S}-\mathbf{\Delta})(\mathbf{Z}-\mathbf{Y}) \tag{101}$$

and inserted into Eq. (100) to give the nonhermitian eigenvalue problem of half the dimensions of Eq. (94):

$$(\mathbf{S}-\mathbf{\Delta})^{-1}(\mathbf{A}+\mathbf{B})(\mathbf{S}+\mathbf{\Delta})^{-1}(\mathbf{A}-\mathbf{B})(\mathbf{Z}-\mathbf{Y})=\omega^2(\mathbf{Z}-\mathbf{Y}) \tag{102}$$

If $\mathbf{S}-\mathbf{\Delta}$ or $\mathbf{S}+\mathbf{\Delta}$ are singular or near singular, it may be advantageous to rearrange Eq. (102) to be

$$\frac{1}{\omega^2}(\mathbf{Z}-\mathbf{Y})=(\mathbf{A}-\mathbf{B})^{-1}(\mathbf{S}+\mathbf{\Delta})(\mathbf{A}+\mathbf{B})^{-1}(\mathbf{S}-\mathbf{\Delta})(\mathbf{Z}-\mathbf{Y}) \tag{103}$$

which gives $1/\omega^2$ as eigenvalues. If $\mathbf{A}-\mathbf{B}$ is positive definite, Eq. (102) may be arranged to a hermitian eigenvalue problem

$$\begin{aligned}(\mathbf{A}-\mathbf{B})^{1/2}(\mathbf{S}-\mathbf{\Delta})^{-1}(\mathbf{A}+\mathbf{B})(\mathbf{S}+\mathbf{\Delta})^{-1}(\mathbf{A}-\mathbf{B})^{1/2}\\ (\mathbf{A}-\mathbf{B})^{1/2}(\mathbf{Z}-\mathbf{Y})=\omega^2(\mathbf{A}-\mathbf{B})^{1/2}(\mathbf{Z}-\mathbf{Y})\end{aligned} \tag{104}$$

since $\mathbf{\Delta}$ is antisymmetric.

Excitation energies $\{\omega_i\}$ and the corresponding eigenvectors $\begin{Bmatrix} {}^i\mathbf{Z} \\ {}^i\mathbf{Y} \end{Bmatrix}$ are obtained from Eqs. (94), (102), (103), or (104). Another set of solutions is $\{-\omega_i\}$

and $\begin{Bmatrix} {}^i\mathbf{Y} \\ {}^i\mathbf{Z} \end{Bmatrix}$. A comparison with a spectral representation of the polarization propagator shows that the excitation energy is ω if the eigenvector $\begin{Bmatrix} \mathbf{Z} \\ \mathbf{Y} \end{Bmatrix}$ is normalized to 1:

$$(\mathbf{Z}^{\mathrm{T}} \quad \mathbf{Y}^{\mathrm{T}})\begin{pmatrix} \mathbf{S} & \boldsymbol{\Delta} \\ -\boldsymbol{\Delta} & -\mathbf{S} \end{pmatrix}\begin{pmatrix} \mathbf{Z} \\ \mathbf{Y} \end{pmatrix} = 1 \tag{105}$$

while the excitation energy is $-\omega$ if the eigenvector $\begin{pmatrix} \mathbf{Z} \\ \mathbf{Y} \end{pmatrix}$ is normalized to -1:

$$(\mathbf{Z}^{\mathrm{T}} \quad \mathbf{Y}^{\mathrm{T}})\begin{pmatrix} \mathbf{S} & \boldsymbol{\Delta} \\ -\boldsymbol{\Delta} & -\mathbf{S} \end{pmatrix}\begin{pmatrix} \mathbf{Z} \\ \mathbf{Y} \end{pmatrix} = -1 \tag{106}$$

That is, if we want the MCSCF state $|0\rangle$ to represent the third lowest state of a given symmetry, we require the MCTDHF calculation to give *two* negative excitation energies, that is, two negative ω_i's with corresponding positive norm eigenvectors.

D. Stability Condition for a Multiconfigurational Hartree–Fock State

When solving the nonhermitian eigenvalue problem in Eq. (102) negative (ω^2 negative) or complex roots may be encountered. If negative or complex roots occur as eigenvalues of Eq. (102) the MCTDHF approximation is said to have an instability.[16] We analyze in the following the conditions under which such instabilities may occur. If $\mathbf{A}-\mathbf{B}$ is positive definite, instabilities are not encountered if the matrix $(\mathbf{A}-\mathbf{B})^{1/2}(\mathbf{S}-\boldsymbol{\Delta})^{-1}(\mathbf{A}+\mathbf{B})(\mathbf{S}+\boldsymbol{\Delta})^{-1}(\mathbf{A}-\mathbf{B})^{1/2}$ is positive definite [see Eq. (104)], that is,

$$\mathbf{X}(\mathbf{A}-\mathbf{B})^{1/2}(\mathbf{S}-\boldsymbol{\Delta})^{-1}(\mathbf{A}+\mathbf{B})(\mathbf{S}+\boldsymbol{\Delta})^{-1}(\mathbf{A}-\mathbf{B})^{1/2}\mathbf{X}^{\mathrm{T}} \geqq 0 \tag{107}$$

Defining a vector

$$\mathbf{U} = \mathbf{X}(\mathbf{A}-\mathbf{B})^{1/2}(\mathbf{S}-\boldsymbol{\Delta})^{-1} \tag{108}$$

we may write Eq. (107) as

$$\mathbf{U}(\mathbf{A}+\mathbf{B})\mathbf{U}^{\mathrm{T}} \geqq 0 \tag{109}$$

which tells us that $\mathbf{A}+\mathbf{B}$ has to be positive definite to ensure that Eq. (107) is fulfilled. Thus if $\mathbf{A}-\mathbf{B}$ is positive definite and $\mathbf{A}+\mathbf{B}$ is not, an MCTDHF instability is encountered. When determining the solution to Eq. (94) we might alternatively have determined $\mathbf{Z}-\mathbf{Y}$ from Eq. (100) and then inserted $\mathbf{Z}-\mathbf{Y}$ into Eq. (99). A derivation similar to the one described shows than that, if $\mathbf{A}+\mathbf{B}$ is positive definite, then an MCTDHF instability is encountered if $\mathbf{A}-\mathbf{B}$ is not positive definite. If both $\mathbf{A}\pm\mathbf{B}$ are positive definite, instabilities are not encountered. If both $\mathbf{A}\pm\mathbf{B}$ are not positive definite, an explicit solution to Eq. (102) has to be determined before it is clear whether an instability occurs. If both $\mathbf{A}\pm\mathbf{B}$ are not positive definite and instabilities are not encountered, negative excitation energies are obtained in the MCTDHF calculation. We recall that the curvature of the energy hypersurface at a point representing an MCSCF state refers to twice the $\mathbf{A}-\mathbf{B}$ matrix which occurs in the MCTDHF stability condition. Further, if we previously had examined the variations in the energy resulting from purely imaginary orbital variations, the second derivative of the total energy would involve the $\mathbf{A}+\mathbf{B}$ matrix. Hence, the condition that $\mathbf{A}\pm\mathbf{B}$ is positive definite is satisfied if the MCSCF state represents a local minimum both with respect to real and imaginary orbital variations. This point has previously been mentioned by Dalgaard.[44] When $\mathbf{A}\pm\mathbf{B}$ are not positive definite, the MCSCF state represents a saddle point on the energy hypersurface.

The aforementioned stability condition is not always trivially fulfilled for a state that satisfies the GBT. We describe below one example where the stability condition is not fulfilled for a state which satisfies the GBT. Include in the configuration space one main configuration and all singly excited configurations relative to the main configuration. The orbital optimization excitation operators connecting orbitals that both are occupied or both are unoccupied in the main configuration are redundant operators for that particular choice of configurations. The Hartree–Fock (HF) state for the main configuration represents a stationary point in the MCSCF calculation just described. The HF state, however, is surely not the state that represents the ground state in this MCSCF calculation, nor does this state represent an excited state. We have considered cases which cover the above-described choice of configurations and found that the HF ground state was unstable in the MCSCF calculation. We found that $\mathbf{A}-\mathbf{B}$ was not positive definite while $\mathbf{A}+\mathbf{B}$ was positive definite and did thus not have the problem of giving the HF ground state a physical interpretation in the MCSCF calculation. In the aforementioned calculations physical intuition might have told us that the stationary point which represented the HF ground state did not show the right characteristics to represent either an MCSCF ground or an excited state. However, in more complicated cases, it may be very useful to use the stability condition as a constraint the MCSCF state should satisfy to represent the exact state, that is, in addition to having the proper number of negative excitation energies instabilities are not encountered.

V. CALCULATIONS WITH THE NEWTON–RAPHSON APPROACH

We have previously discussed the Newton–Raphson approach for MCSCF optimization (see Section III). This approach demonstrates second order convergence. The order concept is only defined when close to convergence[9, 20, 22]; when farther from convergence, the Newton–Raphson approach may give step length amplitudes which are very large.[13,15,17] The full Hessian **G** may also have an incorrect structure, that is, too few or too many negative eigenvalues,[13,15,17] leading to a Newton–Raphson step in an undesired direction. Large step length amplitudes or step length amplitudes in the wrong direction may introduce large fluctuations in the step length amplitudes of the subsequent iterations. A divergent sequence of iterations may be observed if the large step length amplitudes or step length amplitudes of the wrong sign are used uncritically.

When the Newton–Raphson approach does not straightforwardly converge, in numerical analysis it has been advocated to use restricted step methods for obtaining global convergence.[9] In a restricted step method for the lowest state of a certain symmetry a minimum for the second-order approximation to the Taylor series is determined subject to the constraint that the norm of $\underline{\lambda}$ has to be less than h, where h is a constant defining the trust region of the second order Taylor series expansion

$$\min_{\underline{\lambda}} q(\underline{\lambda}) = E_0 + F_i\lambda_i + \tfrac{1}{2}G_{ij}\lambda_i\lambda_j, \qquad \|\lambda\| \leqq h \tag{110}$$

The value of h is chosen as large as possible subject to the condition that a certain measure of agreement exists between $q(\underline{\lambda})$ and $E(\underline{\lambda})$ in Eq. (47). Fletcher[9] describes an algorithm for quantifying the agreement between $q(\underline{\lambda})$ and $E(\underline{\lambda})$. In iteration k a set of ${}^k\underline{\lambda}$ parameters is determined and

$$\Delta q^{(k)} = q({}^k\underline{\lambda}) - q(\mathbf{0}) \tag{111}$$

$$\Delta E^{(k)} = E({}^k\underline{\lambda}) - E(\mathbf{0}) \tag{112}$$

are evaluated.

The ratio

$$r^{(k)} = \frac{\Delta E^{(k)}}{\Delta q^{(k)}} \tag{113}$$

then measures the accuracy to which $q({}^k\underline{\lambda})$ approximates $E({}^k\underline{\lambda})$ in the sense that the closer $r^{(k)}$ is to unity, the better is the agreement. The algorithm of

Fletcher changes h adaptively, attempting to maintain a certain degree of agreement between $q({}^k\underline{\lambda})$ and $E({}^k\underline{\lambda})$ as measured by $r^{(k)}$ while keeping h as large as possible. The details of the algorithm of Fletcher may be found in Ref. 9. Fletcher proves that a restricted step algorithm is globally converging for the lowest state of a certain symmetry.

An essential part of the Fletcher restricted step method is that large step sizes never are allowed. A region of the energy hypersurface is first examined and never left before assurance is obtained that the region contains no stationary points. The step length amplitudes are then in a sense determined such as to bring the calculation to the part of the restricted region that shows the most promise for determining a stationary point. The stationary point for $q(\underline{\lambda})$ when no restrictions are imposed on the step length amplitude is of course the set of Newton–Raphson parameters.

In our implementation of restricted step methods we have chosen as an alternative to solve Eq. (110) to modify the Newton–Raphson step length amplitudes, ensuring that the modified Newton–Raphson step never becomes large and always is in the right direction.[13,15,17] A comparison of the global convergence properties of the Newton–Raphson step size and sign control algorithm[15] and Fletcher's restricted step method will be reported very soon.[55]

For now we will show how we have implemented the step size and sign control algorithm in MCSCF Newton–Raphson calculations. We then report a series of nonlinear step size and sign controlled Newton–Raphson calculations to illustrate the local and global convergence characteristics of an MCSCF Newton–Raphson calculation. Calculations indicate that the MCSCF Newton–Raphson approach reliably can be used to get an MCSCF calculation to converge. However, the calculations also indicate that alternative iterative procedures may be derived which efficiently may be used to obtain both global[20] and local[20-22] convergence of an MCSCF calculation. These approaches will be discussed in subsequent sections.

A. Step Size and Sign Control in the Newton–Raphson Approach

The MCSCF Newton–Raphson approach is used to determine state wavefunctions for both the lowest and for excited states of a given symmetry. The algorithm we apply is constructed such that deviations from the Newton–Raphson step occur when (1) undesired negative eigenvalues show up in the Hessian matrix or (2) the Newton–Raphson approach gives very large step length amplitudes. In developing this algorithm as well as previously developed algorithms[15] we have been particularly concerned with designing techniques which properly account for the characteristics of an MCSCF state as discussed in Section IV. Excited states represent saddle points on the energy hypersurface. We have discussed how the MCSCF state that represents the nth state of a certain symmetry may be required to have $n-1$ negative Hessian eigenvalues. The constraint algorithm therefore needs to include a feature that ensures convergence to a state that has the proper number of negative Hessian eigenvalues. A further criterion is that, of course, no step length amplitude should be so large that a calculation is moved far away from the stationary point of interest. We note that in usual calculations[11-23] most step length amplitudes are well behaved and only a few amplitudes need to be reduced and/or changed in sign. Hence for an MCSCF problem we do not recommend scaling[31, 32] the length of $\underline{\lambda}$. Furthermore, procedures which may constrain a calculation even in the local region[32] are definitely *not* advocated. Before we describe how step size and sign control modifications are introduced, we analyze a Newton–Raphson step in more detail.

The nonlinear Newton–Raphson iterative function in Eq. (60) may be transformed into a basis where the Hessian matrix is diagonal:[13, 15]

$$\mathbf{U}\,\mathbf{G}\,\mathbf{U}^{+}=\boldsymbol{\varepsilon}, \qquad \varepsilon_{ij}=\delta_{ij}\varepsilon_{j} \tag{114}$$

The Newton–Raphson equation then becomes

$$\bar{\underline{\lambda}}=-\boldsymbol{\varepsilon}^{-1}\bar{\mathbf{F}} \tag{115}$$

where

$$\bar{\mathbf{F}}=\mathbf{U}\mathbf{F};\ \bar{\underline{\lambda}}=\mathbf{U}\underline{\lambda} \tag{116}$$

In the bar basis each mode can in a sense be described independently. The second order energy change in a Newton–Raphson iteration may further be expressed as

$$\Delta E(2)=F_i\lambda_i+\tfrac{1}{2}G_{ij}\lambda_i\lambda_j \tag{117}$$

$$=-F_iG_{ij}^{-1}F_j+\tfrac{1}{2}G_{ij}G_{im}^{-1}F_mG_{jk}^{-1}F_k$$

$$=-\tfrac{1}{2}G_{ij}^{-1}F_iF_j \tag{118}$$

In the bar basis the second order energy change therefore becomes

$$\Delta E(2)=-\tfrac{1}{2}\varepsilon_i^{-1}\bar{F}_i^2=-\tfrac{1}{2}\bar{\lambda}_i^2\varepsilon_i \tag{119}$$

and hence only contains terms which consist of sums of products of uncoupled contributions. The variables in the bar basis may in that sense be considered independent.

In an MCSCF optimization problem we consider optimization of both orbital and state variables. In the original basis these two types of variables are distinct whereas in the bar basis they become mixed. To obtain a measure of a variable's orbital/state character we introduce a function τ, defined as the norm of the part of the eigenvector U_i [see Eq. (114)] which is in the configuration space, that is,

$$\tau_i = \sqrt{\sum_j U_{ij}^2} \tag{120}$$

where j is summed over all state variables. If $\tau_i = 1$ we thus have a pure configurational ith mode and if $\tau_i = 0$ we purely orbital ith mode.

We are now ready to describe how the step size control algorithm may be implemented.

1. *Incorrect Negative Hessian Eigenvalues*

Undesired negative Hessian eigenvalues may originate because the original point on the energy hypersurface is close to a stationary point with a "wrong" number of negative Hessian eigenvalues or the undesired negative eigenvalues may be of more accidental nature (e.g., due to the fact that orbital optimization is a highly nonlinear problem there may be several stationary points with the same number of negative eigenvalues; this may also be caused by basis set limitations). If the Hessian matrix has a number of negative eigenvalues that deviate from the desired number, it is usually necessary to change from the Newton–Raphson step to assure rapid convergence to the proper state. The most simple situation is encountered when optimizing the lowest state of a given symmetry, i.e., a minimization problem. If we ensure that the energy [or even the second-order energy $E(2)$] decreases during each iteration, convergence is very close to being ensured since a monotonic decreasing sequence of numbers has to converge for a lower bound function. If we denote the positive (negative) Hessian eigenvalues by $\varepsilon^p(\varepsilon^n)$, the second-order energy change may be divided into a positive ΔE_p and a negative ΔE_n contribution:

$$\Delta E(2) = \Delta E_p + \Delta E_n = -\tfrac{1}{2}\varepsilon_i^n \bar{\lambda}_i^2 - \tfrac{1}{2}\varepsilon_i^p \bar{\lambda}_i^2 \tag{121}$$

For minimization, deviation from the Newton–Raphson step should be implemented such that the term ΔE_n becomes negative. This might be done in the most simple way by changing signs of the step length amplitudes and eigenvalues which correspond to negative eigenvalues, that is

$$\bar{\lambda}_i \to -\bar{\lambda}_i \quad \text{and} \quad \varepsilon_i^n \to |\varepsilon_i^n|, \qquad \varepsilon_i^n < 0 \tag{122}$$

The energy shift which corresponds to such a change becomes

$$\Delta E_2 = -\tfrac{1}{2}\varepsilon_i^p \bar{\lambda}_i^2 - \tfrac{1}{2}|\varepsilon_i^n|\bar{\lambda}_i^2 \tag{123}$$

which is a sum of two negative terms.

When optimizing an excited state of a given symmetry, it is more difficult to determine an efficient strategy for changing from the Newton–Raphson step if undesired negative Hessian eigenvalues occur. The problem consists of distinguishing between the desired and undesired negative Hessian eigenvalues. This differentiation is usually relatively simple when considering optimization problems in which no root flipping occurs since the desired negative eigenvalues then are associated with the configuration space (that is, the τ values for the corresponding mode are large (> 0.7)). The undesired negative eigenvalues then normally will be primarily of orbital nature ($\tau < 0.1$). In iterations where root flipping occurs, τ cannot usually be used to distinguish between desired and undesired negative eigenvalues since the desired negative eigenvalues may have predominantly orbital character. It is then necessary to use some more ad hoc rules to differentiate between the desired and undesired eigenvalues. One such rule is that incorrect negative eigenvalues often are small compared to the desired negative eigenvalues. If the negative eigenvalues are of the same order of magnitude, one may have to rely on a trial-and-error procedure to determine which negative eigenvalues are desired and which are not. However, this situation appears very seldom in practical calculations. After identification of all undesired negative eigenvalues, the corresponding Newton–Raphson step length amplitudes and the eigenvalues are negated to assure convergence to the desired stationary point.

2. *Step Size Control*

In the initial couple of Newton–Raphson iterations large step size elements $\bar{\lambda}_i$ are often encountered for a few modes.[13,15,17] These large step size elements have to be avoided to get the Newton–Raphson sequence of iterations to reliably converge. In the following we describe how such elements may be constrainted. We will control the step sizes in two parts. Initially we apply the general constraint

$$\text{if} \quad |\bar{\lambda}_i| > K_1 \quad \text{set} \quad |\bar{\lambda}_i| = K_1 \tag{124}$$

where the size of K_1 will be discussed later. Step sizes then will be constrained according to the orbital/configuration nature, that is,

$$\text{if} \quad |\bar{\lambda}_i| > \frac{K_2}{\sqrt{\tau_i}} \quad \text{set} \quad |\bar{\lambda}_i| = \frac{K_2}{\sqrt{\tau_i}} \tag{125}$$

where K_2 is smaller than K_1. The size of K_2 will be discussed in more detail later. The second requirement [Eq. (125)] is motivated by the fact that the orbitals may change form completely during the iterative procedure (e.g., from being diffuse to being fairly tight) while for the most typical reasonable initial guesses the dominant configuration amplitudes seldom change by more than some 10–20% from the initial to the final iteration. This means that changes in the configuration amplitudes should usually be restricted more than changes of the orbital expansion coefficients.

The actual size of K_1 and K_2 also should depend on the number of step size amplitudes that have to be constrained. If many amplitudes have to be constrained $K_1(K_2)$ should be smaller than if just one step element has to be constrained in order to ensure that the total change in the orbital/configuration coefficients is no larger than a maximal value. This requirement is implemented into K_1 and K_2 as

$$K_1 = \frac{K_1^0}{\sqrt{n_1}}$$

$$K_2 = \frac{K_2^0}{\sqrt{n_2}} \tag{126}$$

where K_1^0 (and K_2^0) are constants and $n_1(n_2)$ are the number of step size elements which are larger than $K_1^0(K_2^0)$. The precise values of K_1^0 and K_2^0 are not essential and reasonable regions have been determined[17] to be

$$K_1^0:\quad 0.4\text{–}0.6$$

$$K_2^0:\quad 0.1\text{–}0.15 \tag{127}$$

where the upper bounds are used in relatively easy optimization problems while the lower bounds are used in more difficult cases. The actual values we have used in the nonlinear Newton–Raphson calculations we describe are $K_1^0 = 0.426$ and $K_2^0 = 0.113$. This value of K_1^0 is chosen to allow that a complete change in orbital nature may take place in approximately three iterations. The K_2^0 value ensures that the dominant configuration amplitudes change a maximum of some 10–15% in three iterations.

Finally, we separately constrain very small modes ($|\varepsilon_i| < 0.002$) so that $|\bar{\lambda}_i|$ is at most 0.20.

The step size and sign control algorithm we have described is a slight modification of the restricted step length methods that are used in numerical analysis.[9] The restricted step methods are proven to assure convergence to a local minimum if the sequence of iterations remains in a closed region and if the Hessian matrix is bounded.[9] The step size and sign control algorithm we have described above is therefore expected to reliably converge to the lowest state of a given symmetry. Convergence to excited states requires more careful investigation.

There have been several other (usually brief) discussions of constraint procedures in the SCF[56-58] and MCSCF literature.[24,28,29,31,32,59] Usually these techniques do not monitor and control the individual step length amplitudes or the number of negative eigenvalues of the Hessian. Without correct step size control these procedures may move to an undesired place on the energy hypersurface that may be far from the desired stationary point. Without eigenvalue sign control, convergence may proceed to a state that is not the state of interest or even to an "unphysical" place on the energy hypersurface.

We further note that the procedure we have described above for step size and sign control,[17] as well as the mode damping procedure,[15] will in general only be invoked when far from convergence, that is, when higher order terms in Eq. (53) are important. Closer to convergence the step length amplitudes all approach zero and no constraint procedure is used.

Recently, Shepard et al.[32] have proposed and used a procedure which reduces step length amplitudes whenever a Hessian eigenvalue is smaller in magnitude than 0.1 a.u. This reduction is performed *regardless* of the magnitude of the corresponding step length amplitudes. As expected, such a constraint procedure is rather effective in destroying second-order convergence. We have never used or advocated such a constraint procedure.[13,15,17]

B. Small Hessian Eigenvalues

Small Hessian eigenvalues play a very central role for the understanding of the convergence characteristics of a Newton–Raphson MCSCF calculation. We can easily see that small eigenvalues may cause problems be reexamining Eq. (115):

$$\bar{\lambda}_i = \varepsilon_i^{-1}\bar{F}_i \qquad \text{(no sum)} \tag{115}$$

When far from convergence an $\bar{F}_i$ may be large compared to the corresponding ε_i. Thus $\bar{\lambda}_i$ may be large. Near inflection points an ε_i may be small and even of the wrong sign (of course, an eigenvalue of **G** may also be of the wrong sign because we are near the incorrect stationary point).

The eigenvalues of the configuration block ($\mathbf{A}^{CC} - \mathbf{B}^{CC}$) of the ($\mathbf{A}-\mathbf{B}$) matrix are obtained directly from Eq. (79) to be $E_m - E_0$. When a total energy difference $E_m - E_0$ is small, for example, near an avoided curve crossing, a small Hessian eigenvalue will occur, provided the configuration-orbital coupling elements are not very large. The small Hessian eigenvalue that corresponds to the total energy difference $E_m - E_0$ then of course has a predominantly configurational nature [τ of Eq. (120) is close to 1].

Small Hessian eigenvalues also may have their origin in the orbital space excitation operators due to the presence of low-lying "virtual orbitals" or due to an effect which we now discuss. Suppose that in an MCSCF calculation we have included a redundant variable that corresponds to an excitation between two completely occupied orbitals cl, cl'. The Hessian matrix elements which couple the $a_{cl}^{+}a_{cl'} - a_{cl'}^{+}a_{cl}$ excitation operator to the configuration space becomes zero for that case during a sequence of iterations since

$$a_{cl'}^{+}a_{cl'}|n\rangle = a_{cl'}^{+}a_{cl}|n\rangle = a_{cl}^{+}a_{cl'}|0\rangle = a_{cl'}^{+}a_{cl}|\rangle = 0 \tag{128}$$

The orbital part of the Hessian matrix, however, becomes a sum of gradient matrix elements. Using Eq. (128):

$$\begin{aligned}
&\langle 0|\big[(a_i^{+}a_j - a_j^{+}a_i),(a_{cl}^{+}a_{cl'} - a_{cl'}^{+}a_{cl}), H\big]|0\rangle \\
&= \tfrac{1}{2}\langle 0|\big[(a_i^{+}a_j - a_j^{+}a_i),[(a_{cl}^{+}a_{cl}{}' - a_{cl'}^{+}a_{cl}), H]\big]|0\rangle \\
&= -\tfrac{1}{2}\langle 0|\big[H,\big[a_i^{+}a_j - a_j^{+}a_i, a_{cl}^{+}a_{cl'} - a_{cl'}^{+}a_{cl}\big]\big]|0\rangle \\
&= \tfrac{1}{2}\delta_{jcl}\langle 0|\big[a_i^{+}a_{cl'} - a_{cl'}^{+}a_i, H\big]|0\rangle + \tfrac{1}{2}\delta_{icl'}\langle 0|\big[a_j^{+}a_{cl} - a_{cl}^{+}a_j, H\big]|0\rangle \\
&\quad + \tfrac{1}{2}\delta_{icl}\langle 0|\big[a_{cl'}^{+}a_j - a_j^{+}a_{cl'}, H\big]|0\rangle + \tfrac{1}{2}\delta_{jcl'}\langle 0|\big[a_{cl}^{+}a_i - a_i^{+}a_{cl}, H\big]|0\rangle
\end{aligned} \tag{129}$$

which first are zero when convergence has been reached. The Hessian matrix therefore has an eigenvalue that is approaching zero when convergence is approached and that is zero at convergence. An excitation between a completely occupied orbital and an orbital that is very close to being completely occupied gives a very similar result to the one discussed above and a small Hessian eigenvalue shows up also for such a case. These Hessian eigenvalues will be smaller the closer the partly occupied orbital is to being completely occupied. An analysis similar to the one in Eq. (129) can be carried out for excitations between empty and nearly empty orbitals. Small Hessian eigenvalues therefore are also expected to show up in such cases.

C. Step Size and Sign Controlled Nonlinear Newton–Raphson Calculations

To illustrate the convergence characteristic of the step size and sign controlled nonlinear Newton–Raphson approach we report some calculations on the $B^3\Sigma_u^-$ and $E^3\Sigma_u^-$ states of O_2. The $B^3\Sigma_u^-$ and $E^3\Sigma_u^-$ states are the two lowest states of ${}^3\Sigma_u^-$ symmetry and have an avoided crossing due to a va-

TABLE I
The Basis Set for O_2 (34 STO)

Type	Exponent
$1s$	6.83768
	9.46635
$2s$	1.67543
	2.68801
$2p$	1.65864
	3.69445
$2p_{\pm 1}$	0.30
	0.70
$3d_{0,\pm 1}$	2.0

lence-Rydberg mixing of configurations. Some Newton–Raphson calculations on the $B^3\Sigma_u^-$ and $E^3\Sigma_u^-$ states have previously been reported.[17] The complexity of the calculations varies substantially over the region of the potential energy curve. In this section and in the following sections we will often perform calculations on these states to study the efficacy of various MCSCF techniques.

The Newton–Raphson MCSCF calculations we consider use 34 STO with two diffuse sets of p orbitals. The basis set is given in Table I. The diffuse functions are included to reliably describe the Rydberg configuration. All the calculations we report except when otherwise specified will be one-step calculations, and the initial guess of orbitals is a set of grand canonical Hartree–Fock orbitals[40] with occupation $1\sigma_g^2 1\sigma_u^2 2\sigma_g^2 2\sigma_u^2 3\sigma_g^2 1\pi_u^4 1\pi_g^2$. In Section V.D a more thorough discussion of the requirements that may be imposed on the initial guess of orbitals and states is given. The calculations we consider include the four configurations in Table II. These configurations have been reported to be the dominant configurations of the CI calculations of Ref. 60.

TABLE II
Configurations for O_2 Calculations

Configuration[a]		Number of states
V_π	Core $3\sigma_g^2 1\pi_u^3 1\pi_g^3$	1
V_σ	Core $3\sigma_g^1 1\pi_u^4 1\pi_g^2 3\sigma_u^1$	2
$V_{\sigma'}$	Core $3\sigma_g^1 1\pi_u^2 1\pi_g^4 3\sigma_u^1$	2
Ry	Core $3\sigma_g^2 1\pi_u^4 1\pi_g^1 2\pi_u^1$	1

[a] Core $1\sigma_g^2 1\sigma_u^2 2\sigma_g^2 2\sigma_u^2$.

Initially we consider a $B^3\Sigma_u^-$ calculation at 2.13 a.u. and an $E^3\Sigma_u^-$ calculation at 2.10 a.u. These calculations are typical for the complexity of an average Newton–Raphson calculation on a lowest and first excited state of a given symmetry. In Tables III and IV the overall convergence characteristics of the $B^3\Sigma_u^-$ calculation at 2.13 a.u. and the $E^3\Sigma_u^-$ at 2.10 a.u. are reported. The calculations converge in six and seven iterations, respectively, to an accuracy of 10^{-10} a.u. in the total energy. The $B^3\Sigma_u^-$ state which is the lowest state of ${}^3\Sigma_u^-$ symmetry has one negative Hessian eigenvalue and three large step length amplitudes in the first iteration (iteration point 0). The step size and sign controlled algorithm thus is applied in the first iteration. In each of the second and third iterations (iteration points 1 and 2) one large step length amplitude is constrained. After the third iteration, the Newton–Raphson calculation is in the local region and the calculation converges rapidly and reliably, as expected for a second-order approach. During the entire calculation, including the initial three iterations where step size and sign control are applied, the total energy decreases rapidly and monotonically.

The $E^3\Sigma_u^-$ calculation in Table IV shows basically the same convergence characteristics as the $B^3\Sigma_u^-$ calculation in Table III. Since the $E^3\Sigma_u^-$ state is the next lowest state of ${}^3\Sigma_u^-$ symmetry, the converged Hessian should have one negative eigenvalue. No spurious negative Hessian eigenvalues were ob-

TABLE III
Convergence Characteristics of a Step Size and Sign Controlled Newton–Raphson Calculation for the $B^3\Sigma_u^-$ State of O_2 at 2.13 a.u.

Iteration point[a]	$E - E^{\text{CONV}\,b}$	$\|\mathbf{F}\|$	$\|{}^{n+1}\underline{\lambda}\| \sim \|{}^{n}\mathbf{e}\|$
0	0.0946557369	5.02×10^{-1}	$5.66\times10^{-1\,c}$
1	0.0246933274	2.18×10^{-1}	$3.17\times10^{-1\,c}$
2	0.0055035052	3.62×10^{-2}	$3.35\times10^{-1\,c}$
3	0.0009142549	3.42×10^{-2}	9.56×10^{-2}
4	0.0000433969	6.56×10^{-3}	2.65×10^{-2}
5	0.0000002359	4.30×10^{-5}	2.38×10^{-3}
6	0.0000000000	3.05×10^{-7}	1.52×10^{-5}
7	0.0000000000	$<10^{-10}$	$<10^{-8}$

[a]At iteration point n, the energy, $\mathbf{F}$, and $\mathbf{G}$ are evaluated and ${}^{n+1}\underline{\lambda}$ is determined. Thus at iteration point n, iteration $n+1$ is performed; $\|{}^{n+1}\underline{\lambda}\|$ is an approximation to $\|{}^{n}\mathbf{e}\| = \|{}^{n}\underline{\lambda} - \underline{\alpha}\|$ [see Eq. (71)].

[b]Here, $E - E^{\text{CONV}}$ is the difference in atomic units between the total energy at the present step and the converged total energy -149.3086149361 a.u.

[c]The step size and sign control algorithm[17] is applied in this iteration.

TABLE IV
Convergence Characteristics of a Step Size and Sign Controlled One-Step Newton–Raphson Calculation for the $E^3\Sigma_u^-$ State of O_2 at 2.10 a.u.

Iteration point[a]	$E - E^{\text{CONV}}$ [b]	$\|\mathbf{F}\|$	$\|^{n+1}\underline{\lambda}\| \sim \|^{n}\mathbf{e}\|$
0	0.0760320763	9.40×10^{-1}	5.21×10^{-1} [c]
1	−0.0024816821	4.24×10^{-2}	2.57×10^{-1} [c]
2	−0.0008260320	3.52×10^{-2}	4.62×10^{-1} [c]
3	0.0000947284	1.91×10^{-3}	2.76×10^{-1}
4	0.0000311601	2.06×10^{-3}	2.66×10^{-1}
5	0.0000073514	4.34×10^{-3}	6.66×10^{-2}
6	0.0000000906	2.76×10^{-4}	9.17×10^{-3}
7	0.0000000000	4.16×10^{-6}	1.03×10^{-4}
8	0.0000000000	$<10^{-10}$	—

[a]At iteration point n, the energy, $\mathbf{F}$, and $\mathbf{G}$ are evaluated and ${}^{n+1}\underline{\lambda}$ is determined. Thus, at iteration point n, iteration $n+1$ is performed; $\|^{n+1}\underline{\lambda}\|$ is an approximation to $\|^{n}e\| = \|^{n}\underline{\lambda} - \underline{\alpha}\|$ [see Eq. (71)].

[b]Here, $E - E^{\text{CONV}}$ is the difference in atomic units between the total energy at this step and the converged total energy −149.2781477108 a.u.

[c]The step size and sign control algorithm[17] is applied in this iteration.

served in this calculation since only the one desired negative Hessian eigenvalue shows up in all iterations. In the initial three iterations (iteration points 0–2) large step length amplitudes are constrained. After the third iteration the calculation is in the local region and converges rapidly and reliably. In the first iteration (iteration point 0) the total energy is above the converged total energy while the total energies of the second and third iteration (iteration points 1 and 2) are below the converged total energy. In the last iterations the total energies are again above the converged total energy. In calculations on the lowest state of a given symmetry the total energy of the reference state will always bound the converged total energy from above during the iterative procedure (the variational principle). In excited state calculations no such bounds can be established. Because we are converging to the first excited state of ${}^3\Sigma_u^-$ symmetry, fluctuations around the converged total energy may be expected during the iterative procedure. The $E^3\Sigma_u^-$ calculation shows a definite converging trend (the approximate error vector norm gets smaller except at iteration point 2) during the whole sequence of iterations. The norm of the GBT matrix slightly increases between the third and the fourth iteration points and from the fourth to the fifth iteration points, but overall a rapid decrease in the norm of the GBT vector is observed.

After having described broadly the convergence characteristics of typical Newton–Raphson calculations we continue discussing separately the global and local convergence characteristics of Newton–Raphson MCSCF calculations.

1. Global Convergence

A more complete understanding of the global convergence problem is obtained by transforming the Newton–Raphson equation to the form where the Hessian matrix is diagonal [Eq. (114)]. This form clearly demonstrates the need for using a step size and sign control algorithm in the initial couple of iterations since both very large step lengths and step sizes of the wrong sign may be encountered in the initial Newton–Raphson iterations.

As an example, in Table V we report the development of the lowest Hessian eigenvalues, the unconstrained $\bar{\lambda}_i$, constrained $\bar{\lambda}_i^c$, and the configuration mode content (τ_i) of the modes corresponding to the lowest 1–4 and 7–9 Hessian eigenvalues of the sequence of Newton–Raphson iterations in Table III ($B^3\Sigma_u^-$ at 2.13 a.u.). It is clear from Table V that one undesired negative Hessian eigenvalue shows up in the initial iteration (iteration point 0). The sign of the corresponding amplitude is changed by the step sign control algorithm. The modes corresponding to the lowest Hessian eigenvalues 2–4 are also constrained at iteration point zero. At iteration point 1 the Hessian matrix has the correct structure since it is positive definite. Only one very large step length amplitude is reduced. Iteration point 2 gives one step length amplitude of -0.5587 that is reduced to -0.3314. Iteration point 3 is in the local region and hence the calculation converges rapidly and reliably from this point on. It is clear from Table V that large step length amplitudes can be associated with small Hessian eigenvalues which may fluctuate a great deal from iteration to iteration. The larger Hessian eigenvalues show relatively smaller fluctuations during the sequence of iterations and the corresponding modes give relative small step sizes. The changes in the Hessian eigenvalues 7–9 are less than 50% from the initial to the last iteration. The corresponding step sizes are also very moderate. The larger Hessian eigenvalues show even smaller relative changes and result in very small step sizes during the sequence of iterations. When the calculation approaches the local region, the fluctuations in the smaller Hessian eigenvalues gradually become smaller.

In regions where an error term analysis is valid it is easy to see how large fluctuations in the eigenvalues can lead to large errors in $\underline{\lambda}$. From Eq. (31) we see that

$$^k e_i = \tfrac{1}{2} G_{in}^{-1}(\underline{\alpha}) K_{njk}(\underline{\alpha})\, {}^{k-1}e_j\, {}^{k-1}e_k \qquad (31)$$

$$= \tfrac{1}{2} G_{in}^{-1}(\underline{\alpha})\, \delta G_{nk}({}^k\underline{\lambda})\, {}^{k-1}e_k \qquad (130)$$

where $\delta\mathbf{G}$ is the first term in the Taylor series expansion of $\mathbf{G}$. From Eq. (31) it is obvious that second-order convergence is expected for the Newton–

TABLE V

The Lowest 1–4 and 7–9 Eigenvalues ε of the Hessian Matrix, the Unconstrained $\bar{\lambda}$, Constrained $\bar{\lambda}^c$ Step Length Amplitudes and the Configuration Mode Content τ of the Newton–Raphson Sequence of Table III ($B^3\Sigma_u^-$ of O_2 at 2.13 a.u.)

Iteration point[a]	Hessian eigenvalue number	Hessian eigenvalue ε_i	${}^{n+1}\bar{\lambda}_i$	${}^{n+1}\bar{\lambda}_i^c$	τ_i	Hessian eigenvalue number	Hessian eigenvalue ε_i	${}^{n+1}\bar{\lambda}_i$	τ_i
0	1	−0.2410	−0.1069	0.1069	0.63	7	0.4592	−0.1558	0.09
	2	0.0598	−0.7794	−0.3012	0.06	8	0.9006	−0.0277	0.95
	3	0.1840	0.6799	0.3012	0.11	9	1.2528	0.0013	0.85
	4	0.2606	0.3573	0.3012	0.03				
1	1	0.0010	−37.1987	−0.20	0.31	7	0.7874	−0.0201	0.17
	2	0.1134	−0.1884		0.40	8	1.0858	−0.0063	0.97
	3	0.2878	−0.0492		0.02	9	1.8206	0.0364	0.31
	4	0.3226	0.1226		0.07				
2	1	0.0514	−0.5587	−0.3314	0.15	7	0.8554	0.0009	0.20
	2	0.0850	−0.0243		0.40	8	1.0252	0.0012	0.05
	3	0.1824	−0.0419		0.01	9	1.1906	−0.0033	0.94
	4	0.2454	−0.0101		0.01				
3	1	0.0596	−0.0780		0.13	7	0.8062	0.0004	0.01
	2	0.1412	0.0052		0.01	8	1.2094	0.0000	0.98
	3	0.1940	−0.0181		0.14	9	1.2880	0.0106	0.10
	4	0.2250	−0.0436		0.59				
4	1	0.0484	−0.0232		0.10	7	0.8856	0.0000	0.01
	2	0.1528	−0.0029		0.04	8	1.0950	−0.0032	0.10
	3	0.1940	0.0097		0.50	9	1.2104	0.0001	0.98
	4	0.2212	0.0045		0.10				
5	1	0.0446	−0.0022		0.10	7	0.9030	0.0000	0.01
	2	0.1554	−0.0002		0.05	8	1.0428	−0.0082	0.11
	3	0.1866	−0.0008		0.51	9	1.2088	0.0000	0.98
	4	0.2228	0.0002		0.04				
6	1	0.0442	0.0000		0.10	7	0.9046	0.0000	0.01
	2	0.1556	0.0000		0.05	8	1.0536	0.0000	0.11
	3	0.1860	0.0000		0.50	9	1.2128	0.0000	0.98
	4	0.2232	0.0000						

[a]At iteration point n, the energy, **F**, and **G** are evaluated and ${}^{n+1}\lambda$ is determined. Thus at iteration point n, iteration $n+1$ is performed; $\|{}^{n+1}\underline{\lambda}\|$ is an approximation to $\|{}^{n}\mathbf{e}\| = \|{}^{n}\underline{\lambda} - \underline{\alpha}\|$ [see Eq. (71)].

Raphson approach. However, from Eq. (130) we see that if the relative changes in **G** are large, only linear convergence is expected. This is often true in intermediate regions (i.e., just prior to the local region) of the energy hypersurface.

The four configurations we are including in the O_2 calculations differ by at least two orbital replacements and the diagonal elements of the one-electron density matrix therefore become identical to the natural occupation numbers. The natural occupation numbers for the converged $B^3\Sigma_u^-$ state are $1\sigma_g(2)$, $2\sigma_g(2)$, $3\sigma_g(1.9434)$, $1\sigma_u(2)$, $2\sigma_u(2)$, $3\sigma_u(0.0566)$, $1\pi_u(1.6244)$, $2\pi_u(0.1154)$, $1\pi_g(1.2602)$. As expected from the analysis in Section V.B the eigenvector corresponding to the lowest Hessian eigenvalue has approximately 90% of the amplitude for the excitations $1\sigma_g, 2\sigma_g \rightarrow 3\sigma_g$ and $3\sigma_u \rightarrow 4\sigma_u$, $5\sigma_u$, $6\sigma_u$, $7\sigma_u$ and $2\pi_u \rightarrow 3\pi_u, 4\pi_u, 5\pi_u$ (excitations between completely occupied and nearly occupied orbitals and between nearly empty and empty orbitals). The remaining amplitude of the lowest Hessian eigenvalue is primarily in the configuration space and is predominantly on the first excited $^3\Sigma_u^-$ state since this state is the only low-lying state of $^3\Sigma_u^-$ symmetry (0.2054 a.u. above the $B^3\Sigma_u^-$).

From Table V it is further seen that the mode corresponding to the lowest Hessian eigenvalue changes character during the iterative procedure. In the initial iteration the configuration content of the mode is 0.63, whereas in the last iteration the configuration content is only 0.10. The reason for this change is that the total energy difference between the $B^3\Sigma_u^-$ state and the first excited state of $^3\Sigma_u^-$ symmetry (the nonoptimized $E^3\Sigma_u^-$ state) gradually increases during the iterative procedure because the orbitals get more and more optimized for describing the $B^3\Sigma_u^-$ state. When the $B^3\Sigma_u^-$ state becomes more optimized the energy lowers but the E state becomes less and less optimized and therefore acquires a higher total energy. The initial iteration has an energy difference between the $B^3\Sigma_u^-$ state and the unoptimized $E^3\Sigma_u^-$ state of 0.0196, while at iteration point 6 the energy difference has increased to 0.2054.

An example of a much more difficult global convergence problem is the $B^3\Sigma_u^-$ calculation at 2.05 a.u. In Table VI the global convergence characteristics of the $B^3\Sigma_u^-$ calculation at 2.05 a.u. are reported. At the initial seven iteration points the Hessian matrix is structurally incorrect. After iteration 8 (iteration point 7) the calculation is in the local region. The smallest Hessian eigenvalues show very large relative changes until the local region is reached. However, with regard to energy, the calculation is gradually and monotonically approaching the converged total energy from above. The complexity of this calculation is due to the very many small eigenvalues of the Hessian matrix. Comparing the $B^3\Sigma_u^-$ converged calculations at 2.13 and

TABLE VI
Global Convergence Characteristics of the Newton–Raphson Calculation for the $B^3\Sigma_u^-$ State of O_2 at 2.05 a.u.

Iteration point[a]	E	Hessian eigenvalue ε_i	${}^{n+1}\bar{\lambda}_i$	${}^{n+1}\bar{\lambda}_i^c$	τ_i
0	−149.200847927	−0.0282	0.7595	−0.1851	0.47
		−0.0050	−0.0249	0.0249	0.00
		0.0010	5.2503	0.2000	0.02
		0.0098	−0.3606	−0.2130	0.01
		0.0130	1.9475	0.2130	0.30
		0.0480	−0.0001		0.00
		0.0252	−0.1849		0.01
1	−149.2718590247	−0.0792	0.0783	−0.0783	0.11
		0.0070	−0.0956		0.00
		0.0121	−0.1189		0.00
		0.0164	0.0001		0.00
		0.0618	0.0404		0.01
		0.1568	−0.0127		0.28
		0.2216	−0.0050		0.00
2	−149.2780178861	−0.0012	1.0275	−0.20	0.00
		−0.0104	0.0549	−0.0549	0.00
		0.0002	0.4940	0.20	0.00
		0.0006	−0.1416		0.00
		0.0032	0.0264		0.00
		0.0150	0.0032		0.00
		0.1586	0.0034		0.28
3	−149.2785502295	−0.0182	−0.0630	0.0630	0.02
		0.0006	−7.9726	−0.20	0.00
		0.0012	0.2568	0.20	0.00
		0.0018	0.0886		0.00
		0.0044	0.0002		0.00
		0.0178	0.0024		0.00
		0.1570	0.0057		0.28
4	−149.2789667154	−0.0128	−0.2063	0.20	0.01
		0.0006	−0.1201		0.00
		0.0014	−0.0776		0.00
		0.0020	0.0475		0.00
		0.0038	−0.0303		0.00
		0.0140	0.0022		0.00
		0.1434	0.0028		0.31
5	−149.2797356270	−0.0110	−0.4600	0.20	0.00
		0.0018	0.0761		0.00
		0.0036	−0.1522		0.00
		0.0042	0.0346		0.00
		0.0126	−0.0875		0.00

TABLE VI (*Continued*)

Iteration point[a]	E	Hessian eigenvalue ε_i	${}^{n+1}\bar{\lambda}_i$	${}^{n+1}\bar{\lambda}_i^c$	τ_i
		0.0382	0.0001		0.00
		0.1388	−0.0069		0.39
6	−149.2803087224	−0.0036	0.9592	−0.20	0.05
		0.0062	0.0406		0.00
		0.0114	−0.0067		0.00
		0.0212	0.0896		0.02
		0.0636	−0.0233		0.05
		0.1248	0.0203		0.40
		0.1448	0.0000		0.00
7	−149.2809077764	0.0050	−0.4895	−0.4260	0.04
		0.0121	0.0387		0.01
		0.0198	−0.0085		0.00
		0.0275	−0.0158		0.04
		0.0449	0.0169		0.47
		0.0715	0.0259		0.05
		0.1317	0.0001		0.01
⋮					
12	−149.2813989690	0.0052	0.0000		0.00
		0.0087	0.0000		0.00
		0.0103	0.0000		0.00
		0.0307	0.0000		0.00
		0.0418	0.0000		0.64
		0.0734	0.0000		0.00
		0.1386	0.0000		0.31

[a]At iteration point n, the energy, **F**, and **G** are evaluated and ${}^{n+1}\underline{\lambda}$ is determined. Thus at iteration point n, iteration $n+1$ is performed.

2.05 a.u. shows that the smallest Hessian eigenvalues of the 2.13 a.u. calculation are about 10 times as large as the smallest Hessian eigenvalues at 2.05 a.u. The smaller Hessian eigenvalues of the 2.05 a.u. calculation occur because the natural occupation of the $3\sigma_g$ orbital (1.9964) is much closer to 2 and the natural occupation of the $3\sigma_u$ orbital (0.0036) is much closer to being zero in the 2.05 a.u. calculation than in the 2.13 a.u. calculation. As a consequence, the 2.05 a.u. calculation is a much more difficult global convergence problem than the 2.13 a.u. calculation. The gradual and monotonical decrease in the total energy of the calculation in the global region indicates that the step size and sign control algorithm works rather well.

In some cases step size and sign control algorithms may, however, show some deficiencies; for example, in the $B^3\Sigma_u^-$ calculation at 2.10 a.u. the step size and sign control procedure does not give a monotonic decrease in the total energy during the sequence of iterations. This is shown in Table VII, where we report how the lowest two total energies of a CI calculation in the MCSCF configuration space develop during the sequence of step size and sign controlled Newton–Raphson iterations. The boldface number is the total energy for the state that we are optimizing. Step size and sign control is applied in the initial 7 iterations (through iteration point 6). One negative Hessian eigenvalue shows up at iteration points 0 and 3 and mode reversal is applied in these iterations. At iteration point 3 an amplitude of -2.6826 corresponding to the negative Hessian eigenvalue -0.0030 a.u. is reversed and set equal to 0.4260. Although mode reversal certainly has to be applied in this case, it appears that a step length amplitude of less than 0.4260 would be more optimal for this case and probably would have resulted in a lower total energy in the next iteration. We would, however, point out that such a deficiency of the step size and sign control algorithm of the aforementioned character is *very* exceptional and for that reason we have not modified the step size and sign algorithm.

The global convergence characteristics of excited state calculations are very similar to the characteristics that have been observed in the lowest state calculations. One difference is that the Hessian matrix in excited state calculations should have desired negative Hessian eigenvalues. To illustrate the global convergence characteristics of an excited state calculation we report in Table VIII the development of the lowest six Hessian eigenvalues ε_i, the unconstrained $\bar{\lambda}_i$, constrainted $\bar{\lambda}_i^c$ step length amplitude, and configuration mode content τ_i of the sequence of Newton–Raphson iterations of Table IV ($E^3\Sigma_u^-$ at 2.10 a.u.). The Hessian matrix has just one negative Hessian eigenvalue in all iterations and hence no undesired negative eigenvalues show up in the calculation. The step size control algorithm is applied in the initial 3 iterations (at iteration points 0–2). It is seen that the character of the mode corresponding to the lowest Hessian eigenvalue changes character completely from initially (iteration point 0) being configurationally dominated ($\tau = 0.66$) to finally (iteration points 3–7) being of orbital nature ($\tau = 0.00$). This change is most easily understood examining the development of the lowest eigenvalues of the $\mathbf{A}^{CC} - \mathbf{B}^{CC}$ block of the Hessian matrix and assuming that the coupling with the orbital block is small. The eigenvalues of $\mathbf{A}^{CC} - \mathbf{B}^{CC}$ are the energy differences $E_n - E_0$ [see Eq. (79)], where E_0 and E_n are roots of the MCSCF CI eigenvalue problem. In Table IX we report how the lowest two roots of an MCSCF CI calculation develop during the iteration sequence of Table VIII. In all iterations the state that has been assigned $E^3\Sigma_u^-$ is the lowest root of the CI problem. In the initial iteration $E_1 - E_0$ is

TABLE VII
The Lowest CI Eigenvalues of the MCSCF Configuration List (Table II) and the Weights (CI Amplitude^2) of the Corresponding Dominant Configurations for the Newton–Raphson Sequence of the $B^3\Sigma_u^-$ State at 2.10 a.u.

Iteration point[a]	Lowest eigenvalues[b]	Configuration weight Ry[c]	Configuration weight V_π[c]
0	−149.20211563	0.830	0.156
	−149.18489141	0.169	0.751
1	**−149.25836212**	0.141	0.781
	−149.10710741	0.848	0.108
2	**−149.27099790**	0.092	0.832
	−148.97413671	0.873	0.058
3	**−149.27796839**	0.098	0.835
	−148.92061763	0.842	0.056
4	**−149.27135167**	0.087	0.840
	−148.94947506	0.871	0.053
5	**−149.27823226**	0.114	0.819
	−148.97009508	0.840	0.073
6	**−149.28499762**	0.162	0.776
	−149.02026003	0.799	0.114
7	**−149.29118188**	0.258	0.694
	−149.08785027	0.718	0.200
8	**−149.29313749**	0.403	0.560
	−149.14333597	0.581	0.336
9	**−149.29352110**	0.391	0.570
	−149.13662670	0.594	0.324
10	**−149.29352285**	0.387	0.576
	−149.13577628	0.598	0.322
11	**−149.29352285**	0.387	0.576
	−149.13579476	0.598	0.322

[a]At iteration point n, the energy, **F**, and **G** are evaluated and ${}^{n+1}\underline{\lambda}$ is determined. Thus at iteration point n, iteration $n+1$ is performed.

[b]The boldface values correspond to the $B^3\Sigma_u^-$ state. The lightface values refer to the (unoptimized) E state.

[c]Configurations Ry and V_π are defined in Table II.

TABLE VIII

The Lowest Six Hessian Eigenvalues ε, the Corresponding Unconstrained $\bar{\lambda}$, Constrained $\bar{\lambda}^c$ Step Length, and Configuration Mode Content τ of the Newton–Raphson Sequence of Table IV ($E^3\Sigma_u^-$ of O_2 at 2.10 a.u.)

Iteration point[a]	Hessian eigenvalue ε_i	${}^{n+1}\bar{\lambda}_i$	${}^{n+1}\bar{\lambda}_i^c$	τ_i
0	−0.1300	−0.0087		0.66
	0.0348	0.0173		0.01
	0.0656	0.0419		0.01
	0.0828	0.3529		0.26
	0.1356	−0.7770	−0.4260[b]	0.07
	0.2196	0.0268		0.00
1	−0.0488	0.2452	0.2098[b]	0.29
	0.0210	−0.0274		0.00
	0.0370	0.0045		0.00
	0.0980	0.0641		0.37
	0.1046	0.0075		0.00
	0.2612	0.0261		0.05
2	−0.0108	0.5294	0.4260[b]	0.03
	0.0032	−0.1387		0.00
	0.0052	−0.0886		0.00
	0.0138	0.0103		0.00
	0.0636	0.0027		0.00
	0.1120	−0.0595		0.48
3	−0.0042	0.2182		0.00
	0.0018	0.0891		0.00
	0.0024	−0.1428		0.00
	0.0110	0.0055		0.00
	0.0548	0.0007		0.00
	0.1300	−0.0167		0.39
4	−0.0028	−0.0641		0.00
	0.0010	−0.1789		0.00
	0.0016	−0.1866		0.00
	0.0058	0.0014		0.00
	0.0272	0.0015		0.00
	0.1308	−0.0036		0.38
5	−0.0056	0.0201		0.00
	0.0016	0.0397		0.00
	0.0024	0.0494		0.00
	0.0080	0.0026		0.00
	0.0376	0.0008		0.00
	0.1294	0.0013		0.39

TABLE VIII (*Continued*)

Iteration point[a]	Hessian eigenvalue ε_i	${}^{n+1}\bar{\lambda}_i$	${}^{n+1}\bar{\lambda}_i^c$	τ_i
6	−0.0050	0.0003		0.00
	0.0014	0.0065		0.00
	0.0020	0.0064		0.00
	0.0068	0.0004		0.00
	0.0324	0.0001		0.00
	0.1298	0.0001		0.39
7	−0.0050	0.0000		0.00
	0.0014	0.0001		0.00
	0.0020	0.0001		0.00
	0.0068	0.0000		0.00
	0.0322	0.0000		0.00
	0.1298	0.0000		0.39

[a]At iteration point n, the energy, **F**, and **G** are evaluated and ${}^{n+1}\lambda$ is determined. Thus at iteration point n, iteration $n+1$ is performed. $\|{}^{n+1}\bar{\lambda}\|$ is an approximation to $\|{}^{n}\mathbf{e}\| = \|{}^{n}\underline{\lambda} - \underline{\alpha}\|$ [see Eq. (71)].

[b]The step size and sign control algorithm[17] is applied to this mode in this iteration.

only 0.0173 a.u., whereas at convergence $E_1 - E_0$ has increased to 0.2526. The lowest Hessian eigenvalues therefore at convergence have a much smaller configurational content.

The $E\,^3\Sigma_u^-$ state is the first excited state of $^3\Sigma_u^-$ symmetry. It might seem a little surprising that we assign the lowest root of the MCSCF CI calculation to this state. Orbitals that are optimal for the $E\,^3\Sigma_u^-$ state (the first excited state of $^3\Sigma_u^-$ symmetry) may, however, be very far from optimal for the $B\,^3\Sigma_u^-$ state (the lowest state of $^3\Sigma_u^-$ symmetry) and the total energy of the optimized $E\,^3\Sigma_u^-$ may therefore be lower than the total energy of the $B\,^3\Sigma_u^-$ state (when nonoptimized orbitals are used to construct the $B\,^3\Sigma_u^-$ state). In the region of a potential energy curve where an avoided curve crossing occurs it is often observed that the nth state in energy of a certain symmetry is not assigned to the nth root of the MCSCF CI problem. Such a situation is often referred to as root flipping. In the present case the state we assign $E\,^3\Sigma_u^-$ has one negative Hessian eigenvalue and further the total energy of the converged state of $E\,^3\Sigma_u^-$ state is 0.01537514 a.u. above the total energy of the converged $B\,^3\Sigma_u^-$ state (see Table VII).

The global convergence of the $E\,^3\Sigma_u^-$ calculation of 2.10 a.u. is typical for an excited state calculation. Much simpler calculations may occasionally be encountered; for example, the $E\,^3\Sigma_u^-$ calculation at 2.13 a.u. is not far from being in the local region with an initial guess of grand canonical

TABLE IX
The Lowest CI Eigenvalues of the MCSCF Configuration List (Table II) and the Weights of the Corresponding Dominant Configurations for the Newton–Raphson Sequence of Table IV ($E^3\Sigma_u^-$ of O_2 at 2.10 a.u.).

Iteration point[a]	Lowest eigenvalues[b]	Configuration Weight Ry[c]	V_π[c]
0	**−149.20211563**	0.830	0.156
	−149.18489141	0.169	0.751
1	**−149.28073894**	0.840	0.153
	−149.12849771	0.153	0.737
2	**−149.27923909**	0.928	0.070
	−149.08661756	0.069	0.845
3	**−149.27814012**	0.944	0.054
	−149.03228053	0.054	0.936
4	**−149.27811808**	0.944	0.056
	−149.02553854	0.056	0.941
5	**−149.27814396**	0.941	0.057
	−149.02717088	0.058	0.939
6	**−149.27814763**	0.941	0.057
	−149.02661730	0.058	0.938
7	**−149.27814771**	0.941	0.057
	−149.02658825	0.058	0.938

[a]At iteration point n, the energy, **F**, and **G** are evaluated and $^{n+1}\underline{\lambda}$ is determined. Thus at iteration point n, iteration $n+1$ is performed.

[b]The boldface values correspond to the $E^3\Sigma_u^-$ state. The lightface values refer to the (unoptimized) B state.

[c]Configuration Ry and V_π are defined in Table II.

Hartree–Fock orbitals. In Table X we report the convergence characteristics of the $E^3\Sigma_u^-$ calculation at 2.13 a.u. The step size control algorithm is only applied in the initial iteration, where a step length of 0.63 is restricted to 0.404. The simplicity of the 2.13 a.u. calculation is caused by the fact that the numerical value of the smallest converged Hessian eigenvalue at 2.13 a.u. is 0.06 a.u. and thus is 50 times larger than the numerically smallest converged eigenvalue at 2.10 a.u. The total energy of the $E^3\Sigma_u^-$ state is also seen from Table X to fluctuate around the converged total energy.

In the two excited state $E^3\Sigma_u^-$ calculations reported so far it has not been necessary to reverse modes because the Hessian matrix had the desired number of negative eigenvalues. However, in more complicated excited state

TABLE X
Convergence Characteristics of the Step Size and Sign Controlled Newton–Raphson Sequence for the $E^3\Sigma_u^-$ State of O_2 at 2.13 a.u.

Iteration point[a]	$E - E^{CONV}$ [b]	$\|\|\mathbf{F}\|\|$	$\|\|^{n+1}\underline{\lambda}\|\| \sim \|\|^{n}\mathbf{e}\|\|$
0	0.0590242739	9.72×10^{-1}	4.62×10^{-1} [c]
1	−0.0003831946	8.36×10^{-2}	2.03×10^{-1}
2	−0.0002286547	1.36×10^{-2}	7.26×10^{-2}
3	0.0000018925	2.64×10^{-3}	5.66×10^{-3}
4	0.0000000000	1.02×10^{-5}	7.68×10^{-5}
5	0.0000000000	$<10^{-8}$	$<10^{-8}$

[a] At iteration point n, the energy, **F**, and **G** are evaluated and $^{n+1}\underline{\lambda}$ is determined. Thus, at iteration point n, iteration $n+1$ is performed. $\|^{n+1}\underline{\lambda}\|$ is an approximation to $\|^{n}\mathbf{e}\| = \|^{n}\underline{\lambda} - \underline{\alpha}\|$ [see Eq. 71)].

[b] Here, $E - E^{CONV}$ is the difference in atomic units between the total energy at this step and the converged total energy. − 149.2533729769 a.u.

[c] The step size and sign control algorithm[17] is applied in this iteration.

calculations undesired negative Hessian eigenvalues often show up. The excited state calculations reported in Ref. 17 on the $^4\Pi$ states of BO provide an example of a case in which mode reversal was needed to obtain convergence to the desired $^4\Pi$ state. In the calculation on the first excited state of $^4\Pi$ symmetry at 3.2 a.u. it was required to reverse modes in the initial 6 iterations. In the calculation on the second excited $^4\Pi$ state at 3.2 a.u. it was necessary to reverse modes in the initial 16 iterations before the Hessian matrix acquired the desired number of negative eigenvalues. However, these complications are exceptional and usual cases do not have these difficulties.[11-23]

When undesired negative eigenvalues occur in the Hessian matrix for excited state calculations, one has to face the problem of distinguishing between desired and undesired negative eigenvalues. We described in Section V.A the rules we have used. When it is required to reverse modes as many as 16 times before the Hessian matrix gets the desired number of negative eigenvalues, however, it is obvious that the procedure applied is not optimal for this case and has to be improved. We point out that the BO calculations are very complex and not typical. Our experience has been that the mode reversal technique works very well in more usual cases. A real advantage of the mode controlling technique is, of course, that it is firmly based on the theoretical characteristics of the state of interest and on assuring that no step length amplitude is extraordinarily large.

The global convergence characteristics of step size and sign controlled Newton–Raphson calculations may be summarized as follows. In the initial phase of the global convergence problem both step size and sign control have to be applied to ensure convergence to the desired state. As the calculation approaches the local region, sign control usually is not necessary since the Hessian matrix acquires the desired number of negative eigenvalues. In the last phase of the global convergence problem only step size control is normally required to bring the calculation into the local region. The complexity of the global convergence problem is closely associated with the magnitude of small Hessian eigenvalues. Generally, as the eigenvalues of the Hessian matrix become smaller, the global convergence problem becomes more complex. Furthermore, small Hessian eigenvalues show up in connection with orbitals that are very close to being completely occupied or completely unoccupied as well as when two or more MCSCF CI roots are close in energy or when the calculation is near an inflection point on the energy hypersurface.

2. *Local Convergence*

The error term analysis of a sequence of Newton–Raphson iterations in Eq. (34) is only valid in the local region. The error term which results when one Newton–Raphson iteration is carried out is given in Eq. (31). A measure of the size of the error vector is the norm of the vector given in Eq. (71). Introducing the Frobenius norm[61] $\|\cdot\|$ for a matrix $\mathbf{A}$ of dimension m and n

$$\|\mathbf{A}\| = \left(\sum_i^m \sum_j^n A_{ij}^2 \right)^{1/2} \tag{131}$$

allows us to rewrite the error term at the kth Newton–Raphson iteration point

$$\|{}^k\mathbf{e}\| \leqq \tfrac{1}{2}\|\mathbf{G}^{-1}(\alpha)\mathbf{K}(\alpha)\|\, \|{}^{k-1}\mathbf{e}\|\, \|{}^{k-1}\mathbf{e}\| \tag{132}$$

or

$$\frac{\|{}^k\mathbf{e}\|}{\|{}^{k-1}\mathbf{e}\|^2} \leqq \tfrac{1}{2}\|\mathbf{G}^{-1}(\alpha)\mathbf{K}(\alpha)\| \tag{133}$$

The ratio $\|{}^k\mathbf{e}\|/\|{}^{k-1}\mathbf{e}\|^2$ thus is smaller than or equal to the constant $\tfrac{1}{2}\|\mathbf{G}^{-1}(\alpha)\mathbf{K}(\alpha)\|$.

Let us now examine how the ratio $\|{}^k\mathbf{e}\|/\|{}^{k-1}\mathbf{e}\|^2$ develops in the local region in some of the calculations that have been described previously in detail. Initially consider the $E^3\Sigma_u^-$ calculation at 2.13 a.u. reported in Table X. At iteration points $k = 2$, 3, and 4, we obtain 1.76, 1.37, and 1.47, respectively, for the ratio $\|{}^k\mathbf{e}\|/\|{}^{k-1}\mathbf{e}\|^2$. The ratio is fairly constant during the iterative procedure. In Table XI we have reported the step length amplitudes of the modes corresponding to the lowest five Hessian eigenvalues. The step length norm corresponding to these five modes constitutes more than 99% of the full step length norm. Significant contributions originate from no more than two modes. The ratios $\|{}^k\mathbf{e}\|/\|{}^{k-1}\mathbf{e}\|^2$ are relatively constant in the local region.

The inverse Hessian eigenvalues enter directly into the error term [see Eq. (31)], and from Eq. (130) it is obvious that large fluctuations occur in the ratios $\|{}^k\mathbf{e}\|/\|{}^{k-1}\mathbf{e}\|^2$ when large fluctuations are observed in the small Hessian eigenvalues. In the $E^3\Sigma_u^-$ calculation at 2.10 a.u. of Table VIII the ratios become 3.49, 0.94, 2.07, and 1.22 for iteration points $k = 4$, 5, 6, and 7. Table VIII shows that also in this case only a few (about two or three) modes given significant contributions to the step length norm. The relative large fluctuations in the ratios of the 2.10 a.u. calculation compared to the 2.13 a.u. calculation are caused by the large fluctuations that occur in the smallest Hessian eigenvalues in the 2.10 a.u. calculation. The lowest Hessian eigenvalues at 2.10 are thus -0.0042, -0.0028, -0.0056, -0.0050, and -0.0050 for points $k = 3$, 4, 5, 6, and 7 (all in the local region), while the smallest positive eigenvalues are respectively 0.0018, 0.0010, 0.0016, 0.0014, and 0.0014. At 2.13 a.u. for points $k = 1$, 2, 3, and 4 (all in the local region) the lowest Hessian eigenvalues are -0.1238, -0.1282, -0.1312, and -0.1316, while the smallest positive eigenvalues are respectively 0.053, 0.048, 0.060, and 0.060. Hence the fluctuations in the smallest eigenvalue are relatively small for the case at 2.13 a.u. compared to the case at 2.10 a.u.

We examine a condition under which the equality sign in Eq. (133) is valid. Let us consider the basis where $\mathbf{G}(\alpha)$ is diagonal [the bar basis of Eqs. (114)–(116)] and let us assume that the mth component of the error vector $\bar{\mathbf{e}}$ dominates in both the $(k-1)$st and kth iteration point. The error vector of the kth iteration point may then be written as

$$ {}^k\bar{e}_m = \tfrac{1}{2}\varepsilon_m^{-1}\bar{K}_{mmm}\,{}^{k-1}\bar{e}_m\,{}^{k-1}\bar{e}_m \tag{134} $$

where in Eq. (134) and the following equation no summation is implied by repeated indices. Since ${}^k\bar{e}_m/{}^{k-1}\bar{e}_m^2 = \frac{1}{2}\varepsilon_m^{-1}\bar{\mathrm{K}}_{mmm}$, the equality sign of Eq. (133) is expected to be approximately valid if one of the components of the error vector $\bar{\mathbf{e}}$ dominates. Large amplitudes of ${}^{k+1}\bar{\underline{\lambda}}$ may frequently be due to the presence of small eigenvalues of $G({}^k\underline{\lambda})$. If there are several small eigenvalues ε_i which give several large $\bar{\lambda}_i$, then equality (134) is not generally

TABLE XI
The Lowest Five Hessian Eigenvalues ε_i, the Corresponding Unconstrained $\bar{\lambda}_i$, Constrained $\bar{\lambda}_i^c$ Step Length Amplitudes, and Configuration Mode Content τ_i of the Newton–Raphson Sequence of Table X ($E^3\Sigma_u^-$ of O_2 at 2.13 a.u.)

Iteration point[a]	Hessian eigenvalue ε_i	${}^{n+1}\bar{\lambda}_i$	${}^{n+1}\bar{\lambda}_i^c$	τ_i
0	−0.2052	−0.0689		0.76
	0.0316	0.6456[b]	0.3012	0.03
	0.0396	0.5580[b]	0.3012	0.03
	0.0660	0.0112		0.00
	0.1968	−0.0189		0.18
1	−0.1238	−0.1936		0.88
	0.0532	−0.0438		0.00
	0.0656	0.1287		0.01
	0.0864	0.0000		0.00
	0.2068	0.0557		0.08
2	−0.1282	−0.0656		0.87
	0.0482	−0.0087		0.00
	0.0634	−0.0273		0.01
	0.0834	0.0048		0.00
	0.1850	0.0102		0.09
3	−0.1312	0.0009		0.87
	0.0604	0.0015		0.00
	0.0808	0.0052		0.01
	0.1042	−0.0009		0.00
	0.1714	0.0010		0.09
4	−0.1316	0.00003		0.87
	0.0600	0.00000		0.00
	0.0800	0.00004		0.01
	0.1034	0.00000		0.00
	0.1718	0.00000		0.09

[a]At iteration point n, the energy, **F**, and **G** are evaluated and ${}^{n+1}\lambda$ is determined. Thus at iteration point n, iteration $n+1$ is performed; $\|{}^{n+1}\bar{\lambda}\|$ is an approximation to $\|{}^n\mathbf{e}\| = \|{}^n\underline{\lambda} - \underline{\alpha}\|$ [see Eq. (71)].

[b]The step size and sign control algorithm[17] is applied for this mode at this iteration.

expected to be valid for successive iterations (although this certainly may happen in some cases).

The development of the ratio $\|{}^{k}\mathbf{e}\|/\|{}^{k-1}\mathbf{e}\|^2$ for the $B^3\Sigma_u^-$ calculation at 2.13 a.u. (see Table V) has a fairly constant value for consecutive iterations. For iteration points $k = 4$, 5, and 6 (all in the local region) we obtain the ratios 2.90, 3.39, and 2.68, respectively. The relative small deviation in these ratios is in agreement with the fact that there is only one fairly small Hessian eigenvalue (~ 0.044) in the local region of this calculation which produces the dominant $\underline{\lambda}$ component.

The local convergence of these Newton–Raphson calculations is straightforwardly understood from the error term analysis of Section III. The ratios $\|{}^{k}\mathbf{e}\|/\|{}^{k-1}\mathbf{e}\|^2$ in the local region are approximately equal to a constant of the order 1–5. The size of the actual constant depends on the magnitude of the smallest Hessian eigenvalues.

3. *Comparison of One-Step and Two-Step Newton–Raphson Calculations*

To illustrate the convergence characteristics of the two-step Newton–Raphson calculations we report in Table XII the convergence characteristics of a two-step Newton-Raphson calculation on the $E^3\Sigma_u^-$ state at 2.10 a.u. The orbital and state rotation parameters are not determined simultaneously in the two-step approach, and a reliable measure of the total step length (and therefore also of the length of the error vector) of a given iteration is not easily available. We have for that reason only reported the difference between the total energy of a given step of the iterative procedure and the converged total energy to measure the error of a given iteration. In Table IV the convergence characteristics of the one-step Newton–Raphson approach are given. The convergence characteristics of the one- and two-step calculations are very similar, and both calculations converge to an accuracy of 10^{-10} a.u. in the total energy in seven iterations. The similar convergence characteristics of one- and two-step Newton–Raphson calculations have been observed in many other calculations.[11,13,15,17]

However, in some cases convergence difficulties have been encountered with the two-step approach where the one-step approach has converged rapidly and reliably. To illustrate this point we will examine a sequence of converged reparametrized one- and two-step calculations of the $E^3\Sigma_u^-$ state at 2.13 a.u. We have previously discussed how the four-configuration calculation of Table II can be reparameterized to a five-, six-, and seven-configuration calculation by successively adding one of the configurations core $3\sigma_g^2 1\pi_u^4 1\pi_g^1 n\pi_u$ (with $n = 3$, 4, and 5) to the four-configuration case.[17] In Table XIII we report for the converged four-, five-, six-, and seven-configuration calculations the lowest two eigenvalues of the MCSCF CI matrix

TABLE XII
Convergence Characteristics of a Two-Step Newton–Raphson Calculation for the $E^3\Sigma_u^-$ State at 2.10 a.u.

Iteration point[a]	$E - E^{\text{CONV}}$ [b]
0	0.0760320763[c]
1	−0.0039816356[c]
2	−0.0003771275[c]
3	0.0000738856
4	0.0000049646
5	−0.0000020503
6	0.0000000068
7	0.0000000000

[a]At iteration point n, the energy, **F**, and **G** are determined and ${}^{n+1}\underline{\lambda}$ is evaluated. Thus at iteration point n, iteration $n+1$ is performed.

[b]Here, $E - E^{\text{CONV}}$ is the difference in atomic units between the total energy at the present step and the converged total energy −149.2781477108 a.u.

[c]Step size control is applied in this iteration.

(column 2), the Hessian matrix (column 3), the configuration block of the Hessian matrix (column 4), and the reduced Hessian matrix (column 5). The MCSCF CI calculations of column 2 are performed using the respective MCSCF configuration lists, and the boldface MCSCF CI energy is therefore the converged MCSCF total energy. The converged MCSCF energies are the *same* for each of these reparametrized MCSCF calculations. From column 2, we see that root flipping occurs for the four- and five-configuration cases but does not occur for the six- and seven-configuration cases.

Since we are converging to the $E^3\Sigma_u^-$ state, the Hessian matrix of all the reparametrized calculations has one negative eigenvalue (see column 3). If the root flipping does not occur at one point of the iterative procedure, the reduced Hessian will have purely positive eigenvalues. Columns 4 and 5 demonstrate that the one negative eigenvalue of the Hessian matrix gets distributed to either the $\mathbf{G}^{CC}$ block (when there is no root flipping) or to the reduced Hessian (when root flipping occurs), in agreement with the derivation of Appendix D. When the root flipping occurs, the negative eigenvalue of the reduced Hessian is larger numerically than the negative eigenvalue of full Hessian, as would be expected from the analysis of Appendix D (see Fig. 1).

Table XIII demonstrates one of the difficulties which may be encountered with two-step Newton–Raphson MCSCF calculations when there are states of the same symmetry that are fairly close in energy. The five-configuration case has a negative reduced Hessian eigenvalue, -756.1 a.u. This occurs because the two lowest $^3\Sigma_u^-$ CI states (see column 2) are very close in energy. The partitioned form of the Hessian eigenvalue equation [Eq. (D.3)] then has a vertical asymptote that will be very close to the line $E = 0$ [see the plot of the multivalued function $\varepsilon(E)$ of Eq. (D.4) in Fig. 1]. Very small changes in the location of this asymptote will result in large changes of the negative eigenvalue. When a two-step calculation is carried out under such circumstances, very large fluctuations may occur in the eigenvalues of the reduced Hessian matrix, and it is not at all straightforward to devise a constraint procedure due to the behavior of these eigenvalues. In the two-step calculations reported in Table XIII, we also note that very large changes occur in the lowest eigenvalue of the reduced Hessian when the configuration list is changed. Contrary to this, the lowest eigenvalue of the Hessian matrix that is used in the one-step procedure (column 3) is very stable when the configuration list is increased. The eigenvalues of the Hessian matrix will not even fluctuate very much in the one-step procedure when the partitioning which leads to the two-step procedure is not defined (one energy difference $E_m - E_0$

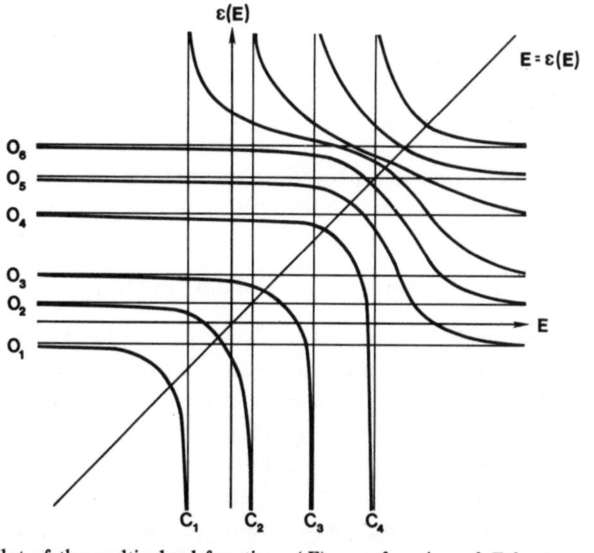

Fig. 1. A plot of the multivalued function $\varepsilon(E)$ as a function of E for a case where root flipping occurs. The branches of this function represent the eigenvalues of the matrix $\mathbf{G}^{OO} - \mathbf{G}^{OC}(\mathbf{G}^{CC} - E\mathbf{1})^{-1}\mathbf{G}^{CO}$. The eigenvalues of the Hessian matrix are represented by the interaction of these branches with the line $E = \varepsilon(E)$. The intersection of the branches with the line $E = 0$ are the eigenvalues of reduced Hessian matrix. The vertical and horizontal asymptotes are the eigenvalues of the matrices $\mathbf{G}^{CC}$ and $\mathbf{G}^{OO}$, respectively.

TABLE XIII

Converged MCSCF Results for the $E^3\Sigma_u^-$ State of O_2 at 2.13 a.u.

		CI[a] eigenvalues (a.u.)	One-step Hessian eigenvalues (a.u.)	$\mathbf{G}^{CC}$ eigenvalues (a.u.)	Two-step reduced Hessian eigenvalues (a.u.)
4 conf.[b]	Lowest	**−149.25337297**[f]	−0.13169982	0.02821812	−9.75475424
	Next lowest	−149.22515485	0.06002166	0.58710661	0.06014124
5 conf.[c]	Lowest	**−149.25337297**[f]	−0.13496382	0.00059799	−756.16741520
	Next lowest	−149.25277498	0.06002612	0.12977741	0.06015418
6 conf.[d]	Lowest	−149.26482575	−0.13541722	−0.01145278	0.06015630
	Next lowest	**−149.25337297**[f]	0.06002658	0.12611592	0.08036454
7 conf.[e]	Lowest	−149.27527358		−0.02190061	
	Next lowest	**−149.25337297**[f]		0.11310116	

[a] Configuration interaction calculation performed using the configurations defining the MCSCF problem.

[b] The first four configurations in Table II.

[c] Configurations of footnote *b* with the additional configuration (core) $3\sigma_g^2 1\pi_u^4 1\pi_g^1 3\pi_u^1$.

[d] Configurations of footnote *c* with the additional configuration (core) $3\sigma_g^2 1\pi_u^4 1\pi_g^1 4\pi_u^1$.

[e] Configurations of footnote *d* with the additional configuration (core) $3\sigma_g^2 1\pi_u^4 1\pi_g^1 5\pi_u^1$.

[f] Converged MCSCF total energy.

is equal to zero). The one-step procedure must therefore be advocated when MCSCF calculations are carried out on difficult cases, for instance, avoided crossings.

D. Initial Guess of Orbitals and States

In order to successfully carry out a Newton–Raphson sequence of iterations it is important to have an initial guess of orbitals easily available that is of such a quality that global convergence easily can be obtained with the step size and sign controlled Newton–Raphson procedure or other stable procedure. To assure convergence to the desired state, it is important to be able to obtain an initial guess of the reference state that resembles the desired state enough for the iterative procedure to converge to this state. In this section we address the problem of how to determine an initial guess of orbitals and states.

To assure convergence of a Newton–Raphson calculation it is our experience that it suffices to use as an initial guess of orbitals a set of grand canonical Hartree–Fock orbitals[40] with occupations corresponding approximately to the ones of the dominant configuration. Occupations which correspond to another configuration may in many cases also suffice;[17] for example; in the O_2, $^3\Sigma_u^-$ calculations we have reported, the V_π and the *Ry* configurations of Table II are the dominant configurations and we have used occupation numbers corresponding to the ground state $X^3\Sigma_g^-$ in the grand canonical Hartree–Fock calculation. A set of canonical restricted Hartree–Fock orbitals with $N-1$ potential virtuals[38,39] may suffice equally well as initial guess of orbitals for a Newton–Raphson MCSCF calculation.[18] We have often chosen to use a set of grand canonical Hartree–Fock orbitals as the initial guess because this set of orbitals is the easiest and most straightforward set of orbitals to determine and because the grand canonical Hartree–Fock equations are independent of the spacial and spin symmetry of the considered state, differing only in the specification of the set of occupation numbers.[40]

With an initial guess of orbitals at hand, the initial guess of the state may be obtained by selecting an appropriate root of a CI calculation in the MCSCF configuration space. The Undheim–McDonald theorem[42] is most often used for selecting this root. The Undheim–McDonald theorem tells that the nth lowest root of a certain symmetry of a CI problem is an upper bound for the nth state of that symmetry. The nth root of the MCSCF CI problem is therefore usually assigned to represent the nth state. Such an assignment suffices in most cases; an example is the $E^3\Sigma_u^-$ at 2.13 a.u. reported in Table X.

However, an assignment of the initial guess of the reference state based on the Undheim–McDonald theorem may lead to an erroneous result in some cases. An example is the $B^3\Sigma_u^-$ calculation at 2.10 a.u. reported in Table VII. Even though the $B^3\Sigma_u^-$ state is the lowest state of $^3\Sigma_u^-$ symmetry, the correct assignment corresponding to this initial guess of orbitals is the second lowest root of the MCSCF CI problem. The energy difference between the first and second root of the MCSCF CI problem is very small, and after the first iteration the state that has been assigned to the $B^3\Sigma_u^-$ state has dropped in energy to become the lowest root of the MCSCF CI problem, where it stays until convergence. The state we have assigned to be the $B^3\Sigma_u^-$ state has a dominant configuration amplitude on the V_π configuration. The amplitude of the dominant configuration changes very little and may be used when we choose the state from the MCSCF CI for a two-step calculation.

The reason why the second root of an MCSCF CI problem may be assigned as the lowest state is, of course, that orbitals which optimally describe one configuration may be very far from optimal for describing another configuration. Our initial guess of orbitals describes very poorly the dominant V_π configuration; it more optimally describes the *Ry* configuration.

When we are optimizing the $B^3\Sigma_u^-$ state the orbitals become more and more optimal for the V_π configuration, and after the first iteration the lowest root of the MCSCF CI problem has dominant weight for the V_π configuration. If we insist on using the lowest root of the initial MCSF CI problem as a guess for the $B^3\Sigma_u^-$ state and on converging to a state which had a positive definite Hessian matrix, convergence cannot be obtained. However, if we use the lowest root as an initial guess for the $E^3\Sigma_u^-$ state (which is required to have one negative Hessian eigenvalue), convergence is easily obtained as described in Table VIII. Root flipping occurs for the entire sequence of iterations in the $E^3\Sigma_u^-$ calculation at 2.10 a.u. (see Table IX). In the $B^3\Sigma_u^-$ calculation at 2.10 a.u. (Table VII) root flipping only occurs in the initial iteration. The assignment of a given root of an MCSCF CI problem to a given state may be, of course, very difficult when the Undheim–McDonald theorem cannot be used to carry out the assignment. When this is the case, a trial-and-error procedure may even have to be used before the correct assignment can be performed. As we have previously noted, since in most previous MCSCF calculations[2] no monitoring of the characteristics of the states was done, we expect that many MCSCF calculations reported in the literature are erroneous.

Occasionally when it is difficult to get a calculation to converge it may be advantageous to try to reparametrize the calculation. The philosophy of a reparametrization is to move some of the orbital parameters to the configuration space by including some additional configurations and to ensure at the same time that the total energy of the converged state does not depend on whether the reparametrization is performed or not. For example, in Ref. 17 we have shown how the O_2 MCSCF calculation which includes the four configurations of Table II may be reparameterized by including the additional configuration core $3\sigma_g^2 1\pi_u^4 1\pi_g^1 3\pi_u^1$. (See Ref. 17 and Appendix A, where a more thorough discussion can be found about this reparametrization.)

In the reported $E^3\Sigma_u^-$ calculation at 2.05 a.u. it was necessary to perform a reparametrization to get the calculation to converge. With an initial guess of grand canonical Hartree–Fock (HF) orbitals, the only "reasonable" initial CI state in the four-configuration calculation was far from having the configuration amplitudes of the converged state. The CI state was primarily valence-like (having large amplitude for the V_π configurations), with a low Rydberg weight of only 1%. The optimized state is highly correlated with a Rydberg weight of 42%. The use of grand canonical HF orbitals as an initial guess led in the four-configuration calculation to convergence to a "simple solution" without any Rydberg character at all. The convergence to this simple solution was not a result of initial spurious negative eigenvalues, since these were absent until very near the stationary point where the V_π configurations had a very large amplitude and the Ry configuration had zero amplitude for the obtained converged state. For a state that has zero amplitude on the *Ry* configuration, the operators $a^+_{n\pi_u}a_{2\pi_u}$ $(n = 3, 4, \ldots)$ become redundant. The GBT amplitude corresponding to a redundant variable is zero at convergence. In this case, the Hessian matrix also had a zero eigenvalue at convergence due to the redundant variable. Because the GBT amplitudes went faster to zero than the Hessian eigenvalue [see Eq. (129)], no warning was given that convergence was being obtained to an undesired state until right at convergence where the Hessian matrix possessed the zero eigenvalue and amplitude of zero for the *Ry* configuration. So-called simple solutions such as this one may occasionally appear in MCSCF calculations if the initial guess of configurations and orbitals is far from the desired result. This kind of solution, in which some configurations completely drop out of the calculation and some operators become redundant, is rare but not all that unusual in MCSCF calculations.[17] This case is the first one we have experienced where mode reversal was not sufficient to avoid convergence to the simple solution.

It was clear that our problems were caused by an inadequate description of the $2\pi_u$ orbital, so we decided to reparametrize the calculation to a five-configuration calculation. With the five-configuration parameterization, the initial guess for the *E* state came closer to the final configurational weights, basically because the description of the π_u space now is qualitatively more correctly treated in our initial guess. The restructuring of our parameter space led to convergence without problems. The total energy for this converged calculation with the five configurations included in the CI space is, of course, the same as the total energy obtained in a converged four-configuration calculation; however, the converged π_u orbitals in the two calculations are different. When the set of orbitals obtained in the converged five-configuration calculation was used as the initial guess of a four-configuration calculation, convergence was easily obtained to the proper stationary point with one negative eigenvalue of the Hessian and, of course, with the same total energy as in the five-configuration calculation. We obtained a weight of 55% V_π and 42% of *Ry* in the converged four-configuration calculation.

E. Conditions for Improved Iterative Methods

The previous derivation of the Newton–Raphson method as well as the convergence characteristics of actual calculations indicates that when it is possible to obtain a reasonable initial guess of the MCSCF state the constrained Newton–Raphson method is usually very stable and rapidly converging in situations where the Hessian matrix has no very small eigenvalues.[11–23, 27–37] In cases where very small Hessian eigenvalues occur, the global convergence characteristics of the method are not as satisfactory. With constraints, some 3–20 iterations may be required to get the calculation into the local region.[17] In the local region the unconstrained Newton–Raphson method converges reliably and rapidly.

The conditions that an optimization procedure should satisfy to show improved convergence properties compared to a constrained Newton–Raphson model depend on whether the method is designed for obtaining improved local or global convergence. In the local region computational efficiency to reach the stationary point is the property which decides whether a method is feasible, whereas in global convergence the stability and reliability of the method for moving to the local region of the proper stationary point when the initial guess is "far away" is the decisive factor.

Large step lengths cannot be accepted in a global method. A global method has to be designed in such a way that step sizes are so moderate that a given region is not left until the method has detected that there are no stationary points in that region. Because the global convergence problems with the step size and sign controlled Newton–Raphson algorithm may be caused by the appearance of very small Hessian eigenvalues, alternative improved global methods are required that are less sensitive to the appearance of very small Hessian eigenvalues. Such a method is obtained, for example, if the iterative function is based on a cubic expansion of the energy function. We discuss the global convergence characteristics of an iterative cubic calculation in Section VIII.

In the local region we shall show that improved computational efficiency compared to a Newton–Raphson approach may be obtained if the Hessian matrix is kept fixed during a sequence of several iterations. When a fixed Hessian series of iterations is carried out, a new energy gradient is constructed in each iteration. The difference between the new energy gradient and the one of the previous iteration contains information about the Hessian at the new point, and it may be desirable also to use this information to update the Hessian matrix during the iterative sequence.[9,21] We shall demonstrate that such Hessian update schemes constitute an efficient way for obtaining local convergence. Furthermore, Hessian updates will be shown to sometimes have attractive global convergence properties when implemented with a step size (and sign) control algorithm. In the next two sections we shall describe in more detail the convergence characteristics of improved local approaches.

VI. GENERALIZED AND FIXED HESSIAN APPROACHES

A. The Generalized Newton–Raphson Perturbative Approach

In local regions perturbation expansions can be used to solve Eq. (59). A convenient starting point for such an expansion is obtained by rearranging

Eq. (59) with $\underline{\lambda}$ equal to the stationary point $\underline{\alpha}$:

$$\alpha_k = -G_{kl}^{-1}F_l - \tfrac{1}{2}G_{kl}^{-1}K_{lmn}\alpha_m\alpha_n - \tfrac{1}{6}G_{kl}^{-1}M_{lmnp}\alpha_m\alpha_n\alpha_p + \cdots \tag{135}$$

In a given iteration (k) we may then construct an approximation ${}^k\underline{\lambda}^{(p)}$ to $\underline{\alpha}$ so that ${}^k\underline{\lambda}^{(p)}$ fulfills Eq. (135) to order p. The set of iterative functions which then is obtained is often referred to as generalized Newton–Raphson perturbative procedures.[16,18] We will now study the derivation and the convergence characteristics of these procedures. It is noted that Eq. (59) tacitly assumes an expansion around the origin (i.e., ${}^0\underline{\lambda} = \mathbf{0}$), so a nonlinear transformation is required between two iterations.

The simplest approximation to $\underline{\alpha}$ is obtained when $\underline{\alpha}$ is set equal to zero on the right-hand side of Eq. (135). We then obtain[16]

$${}^k\lambda_k^{(1)} = -G_{kl}^{-1}F_l \tag{136}$$

which is the nonlinear Newton–Raphson iterative function of Eq. (60). A second step may be carried out on Eq. (135), where $\underline{\alpha}$ on the right-hand side of Eq. (135) is from Eq. (136). Keeping all terms that are quadratic in ${}^l\underline{\lambda}^{(1)}$ gives the function

$${}^k\lambda_k^{(2)} = {}^k\lambda_k^{(1)} - \tfrac{1}{2}G_{kl}^{-1}K_{lmn}\,{}^k\lambda_m^{(1)}\,{}^k\lambda_n^{(1)} = {}^k\lambda_k^{(1)} - \tfrac{1}{2}G_{kl}^{-1}K_{lmn}G_{mi}^{-1}F_iG_{nj}^{-1}F_j \tag{137}$$

which is the Chebyshev formula.[16,62]

A general series of formulas which are consistent through still higher powers in ${}^k\underline{\lambda}^{(1)}$ may be derived through iterating on Eq. (135), keeping terms through the desired power in ${}^k\underline{\lambda}^{(1)}$. The iterative function ${}^k\underline{\lambda}^{(p)}$ is thus obtained by replacing $\underline{\alpha}$ on the right-hand side of Eq. (135) by ${}^k\underline{\lambda}^{(p-1)}$, keeping all terms through order $({}^k\underline{\lambda}^{(1)})^p$ in the power series expansion. The parameter set ${}^k\underline{\lambda}^{(\infty)}$ corresponds to the stationary point $\underline{\alpha}$ if the sequence converges.

Since ${}^k\underline{\lambda}^{(p)}$ contains all terms through order p in ${}^k\underline{\lambda}^{(1)}$, ${}^k\underline{\lambda}^{(p)} - \underline{\alpha}$ is correct up to $O[({}^k\underline{\lambda}^{(1)})^{p+1}]$. From Eq. (29) we see that ${}^k\underline{\lambda}^{(p)} - \underline{\alpha}$ is thus correct to $O({}^k\mathbf{e}^{p+1})$. In Section VIII we derive the error term of the Chebyshev formula [Eq. (137)] and report the results of some Chebyshev calculations together with the results of some other cubically convergent schemes.

B. The Linear Fixed Hessian Approach

We will now prove that formulas which show quadratic, cubic, quartic,... convergence characteristics may alternatively be derived by carrying out a sequence of fixed Hessian iterations where the $\underline{\lambda}$ parameters are determined from the equation[22]

$${}^p\lambda_i = -G_{ij}^{-1}\left(F_j + F_j({}^1\underline{\lambda}) + F_j({}^2\underline{\lambda}) + \cdots + F_j({}^{p-1}\underline{\lambda})\right) \tag{138}$$

It is assumed that the $\underline{\lambda}$ parameters at the initial iteration point are zero, and no explicit reference to this point is made in matrices evaluated at this point. The expansion in Eq. (138) replaces the calculation of the higher order derivatives at one point in Eq. (135) with the calculation of gradients at several points (e.g., at ${}^0\underline{\lambda}=0, {}^1\underline{\lambda}, {}^2\underline{\lambda}, \ldots, {}^{p-1}\underline{\lambda}$; see Appendix B). The matrix $\mathbf{F}$ is thus evaluated at points recursively defined by Eq. (138).

We will prove that when Eq. (138) is applied n times, error in ${}^n\underline{\lambda}$ will be of order $n+1$ in ${}^0\mathbf{e}$. Induction is used to prove this point. Since the initial iteration of Eq. (138) is a Newton–Raphson iteration, it is clear that the foregoing statement is true when $p=1$. For a given n it is now assumed that ${}^{n-1}\underline{\lambda}$ contains errors of order $O({}^0\mathbf{e})^n$ and it is now proved that ${}^n\underline{\lambda}$ contains errors of order $O({}^0\mathbf{e})^{n+1}$. To do this, rewrite Eq. (138) as

$$ {}^n\lambda_i = {}^{n-1}\lambda_i - G_{ij}^{-1} F_j({}^{n-1}\underline{\lambda}) \tag{139} $$

Expanding G_{ij} and $F_j({}^{n-1}\underline{\lambda})$ around the exact solution $\underline{\alpha}$ using Eq. (27) and (28) gives

$$ \begin{aligned} {}^n\lambda_i = {}^{n-1}\lambda_i &- \left(G_{rs}(\underline{\alpha}) + K_{rst}(\underline{\alpha})\left[{}^0\lambda_t - \alpha_t\right] + \cdots\right)_{ij}^{-1} \\ &\cdot\left(G_{jk}(\underline{\alpha})\left[{}^{n-1}\lambda_k - \alpha_k\right] + \tfrac{1}{2}K_{jkm}(\underline{\alpha})\left[{}^{n-1}\lambda_k - \alpha_k\right]\left[{}^{n-1}\lambda_m - \alpha_m\right] + \cdots\right) \end{aligned} \tag{140} $$

Expanding the inverse matrix of Eq. (140) then gives

$$ \begin{aligned} {}^n\lambda_i = \alpha_i &+ G_{ij}^{-1}(\underline{\alpha}) K_{jpq}\left[{}^0\lambda_p - \alpha_p\right]\left[{}^{n-1}\lambda_q - \alpha_q\right] + \cdots \\ &- \tfrac{1}{2}G_{ij}^{-1}(\underline{\alpha}) K_{jkm}(\underline{\alpha})\left[{}^{n-1}\lambda_k - \alpha_k\right]\left[{}^{n-1}\lambda_m - \alpha_m\right] + \cdots \end{aligned} \tag{141} $$

Hence ${}^n\lambda_i$ has an error vector

$$ {}^n e_i = L_{ipq}\, {}^0 e_p\, {}^{n-1} e_q \qquad \text{for} \quad n>1,\ {}^1 e_i = \tfrac{1}{2} L_{ipq}\, {}^0 e_p\, {}^0 e_q \tag{142} $$

where L_{ipq} is defined in Eq. (33). Successive use of Eq. (142) gives the error vector ${}^n\mathbf{e}$ in terms of ${}^0\mathbf{e}$

$$ {}^n e_i = L_{ipq}\, {}^0 e_p\, {}^{n-1} e_q = \tfrac{1}{2} L_{il_1k_1}\, {}^0 e_{l_1} L_{k_1l_2k_2}\, {}^0 e_{l_2} L_{k_2l_3k_3}\, {}^0 e_{l_3} \cdots L_{k_{n-1}l_nk_n}\, {}^0 e_{l_n}\, {}^0 e_{k_n} \tag{143} $$

We have thus proved that such a sequence of n fixed Hessian iterations contains errors of order $n+1$ in ${}^0\mathbf{e}$ and has the error term given in Eq. (143).

A fixed Hessian sequence may alternatively be carried out as

$$ {}^p\underline{\lambda}^L = -G_{ij}^{-1}\left(F_j + {}^1\tilde{F}_j + {}^2\tilde{F}_j + \cdots + {}^{p-1}\tilde{F}_j\right) \tag{144} $$

where ${}^{p-1}\tilde{\mathbf{F}}$ is the *GBT* vector of Eq. (54) evaluated in the orbital basis

$$ {}^{p-1}a_r^+ = \sum_s a_s^+ \left[\exp(-{}^{p-1}\kappa)\right]_{sr} \qquad (145)$$

and state basis defined through the coefficient matrix ${}^{p-1}\mathbf{C}$

$$ {}^{p-1}\mathbf{C} = {}^{0}\mathbf{C}\exp(-{}^{p-1}\mathbf{S}) \qquad (146)$$

From the previous derivation it follows straightforwardly that the sequence of fixed Hessian iterations in Eq. (144) will have the same order of convergence and the same error terms as the one of Eq. (138).[22] The advantage of using Eq. (144) is that partial derivative matrices only have to be evaluated at the point $\underline{\lambda} = \mathbf{0}$. This vector is evaluated easily (see Section III.D, Appendix B, and Appendix C).

By applying Eq. (138) or Eq. (144) cubically, quartically,..., convergent MCSCF schemes have been derived without ever explicitly constructing the third, fourth,... derivatives of the total energy.

A sequence of iterations that uses Eq. (138) or Eq. (144) results in a linear transformation of the rotational parameters $\underline{\lambda}$ between each step of the iterative procedure, as opposed to the previously used fixed Hessian approaches[17, 29, 35] that carry out a nonlinear transformation of the rotational parameters. When a sequence consists of n steps of the nonlinear fixed Hessian approach, it can be shown to have a total order of convergence of $n+1$. The proof is more involved than the proof for a linear fixed Hessian approach (see Ref. 22, where a detailed derivation is carried out).

C. Optimal Use of Fixed Hessian Approaches

We now discuss how fixed Hessian approaches most efficiently may be used in an MCSCF calculation. In the previous section it was shown that two fixed Hessian iterations, namely, a Newton–Raphson iteration followed by an iteration in which the Hessian matrix is kept fixed, have a total order of convergence of 3 with the error vector [Eq. (143)]:

$$ {}^{2}e_i = \tfrac{1}{2} L_{ijk}\,{}^{0}e_k\, L_{jmn}\,{}^{0}e_m\,{}^{0}e_n \qquad (147)$$

The error vector of the (initial) Newton–Raphson iteration [see Eq. (31)] is

$$ e_j^{\mathrm{NR}} = \tfrac{1}{2} L_{jmn}\,{}^{0}e_m\,{}^{0}e_n \qquad (148)$$

so

$$ \|\mathbf{e}^{NR}\| \leq \tfrac{1}{2}\|\mathbf{L}\,{}^{0}\mathbf{e}\|\,\|{}^{0}\mathbf{e}\| \qquad (149)$$

Using Eq. (147), Eq. (148) may be written as

$$ {}^{2}e_i = L_{ijk}\,{}^{0}e_k\, e_j^{\mathrm{NR}} \qquad (150)$$

that is

$$\|{}^2\mathbf{e}\| \leqq \|\mathbf{L}{}^0\mathbf{e}\|\,\|\mathbf{e}^{\mathrm{NR}}\| \tag{151}$$

When the Newton–Raphson iteration diverges, it is reasonable to assume that

$$\|\mathbf{e}^{\mathrm{NR}}\| > \|{}^0\mathbf{e}\| \tag{152}$$

Equations (149), (151), and (152) show that the upper bound for the length of the error vector $\|{}^2\mathbf{e}\|$ is larger than the upper bound for $\|\mathbf{e}^{\mathrm{NR}}\|$. It is therefore reasonable to apply fixed Hessian approaches only in the local region where a straightforward application of the Newton–Raphson approach (i.e., no constraints) converges.

In the local region we compare three different iterative schemes: (1) a Newton–Raphson approach, (2) a fixed Hessian approach, and (3) a combination of fixed Hessian and Newton–Raphson approaches. We initially compare the efficiency of a Newton–Raphson approach and a fixed Hessian approach. A Newton–Raphson iteration which is succeeded with $N^{\mathrm{F}}-1$ fixed Hessian iterations gives a total order of convergence of $N^{\mathrm{F}}+1$. (Note that the first step in a fixed Hessian sequence is a Newton–Raphson step.) When a Newton–Raphson iteration is succeeded by $N-1$ Newton-Raphson iterations, the total order of convergence is 2^N. The total number of fixed Hessian steps (N^{F}) which give the same order of convergence as N Newton–Raphson steps is thus

$$N^{\mathrm{F}} = 2^N - 1 \tag{153}$$

Table XIV exemplifies Eq. (153) for $N = 1, 2, \ldots, 10$. The comparison in Table XIV is solely based on order of convergence. A more accurate comparison is obtained by analyzing the corresponding error terms. The error vector obtained when carrying out N^{F} steps of the fixed Hessian series [see Eq. (143)] is

$${}^{2^N-1}e_i^{\mathrm{FH}} = \tfrac{1}{2} L_{il_1k_1} L_{k_1l_2k_2} \ldots L_{k_{(2^N-2)}l_{(2^N-1)}k_{(2^N-1)}}\, {}^0e_{l_1}\, {}^0e_{l_2} \ldots {}^0e_{l_{(2^N-1)}}\, {}^0e_{k_{(2^N-1)}} \tag{154}$$

while the error vector of the corresponding sequence of N Newton–Raphson iterations [see Eq. (34)] becomes

$${}^Ne_i^{NR} = 2^{-(2^N-1)} L^{(N)}_{ik_1\ldots k_{(2^N)}}\, {}^0e_{k_1} \ldots {}^0e_{k_{(2^N)}} \tag{155}$$

where $L^{(N)}$ is defined through the recurrence relation of Eq. (35). The error vectors in Eqs. (154) and (155) have the same order and both contain (2^N-1)

TABLE XIV
Convergence Characteristics of the Newton–Raphson versus Fixed Hessian Approaches to MCSCF

Number of Newton–Raphson iterations[a]	Equivalent number of fixed Hessian iterations[a]
1 $\left(\frac{1}{2}\right)$	1 $\left(\frac{1}{2}\right)$ (same as Newton – Raphson)
2 $\left(\frac{1}{8}\right)$	3 $\left(\frac{1}{2}\right)$
3 $\left(\frac{1}{128}\right)$	7 $\left(\frac{1}{2}\right)$
4 $\left(\frac{1}{32768}\right)$	15 $\left(\frac{1}{2}\right)$
5 $\left(\frac{1}{2.147\times 10^{9}}\right)$	31 $\left(\frac{1}{2}\right)$
⋮	⋮
10 $\left(\frac{1}{2^{1023}}\right)$	1023 $\left(\frac{1}{2}\right)$

[a] The numbers in parentheses refer to the constant multiplying the error term. See Eqs. (34) and (143).

L supermatrices connected in different ways. If we assume that the difference in the way the **L** supermatrices are coupled together does not affect the actual values of the error terms, then the difference between the error terms depends only on the constants multiplying the error terms. The error vectors satisfy then the relation

$$\|{}^{N}\mathbf{e}^{\mathrm{NR}}\| = 2^{-(2^N-2)}\|{}^{2^N-1}\mathbf{e}^{\mathrm{FH}}\| \tag{156}$$

For $N = 2$ the error term of the Newton–Raphson iteration is thus $1/4$ of the error term in the corresponding fixed Hessian series; for $N = 3$ the corresponding ratio is $1/64$; and so on. In Table XIV we have written out explicitly the error term constant for $N = 1, \ldots, 10$. When N increases, the error term of the Newton–Raphson sequence becomes smaller and smaller compared to the corresponding error terms of the fixed Hessian sequence. The error term analysis thus indicates that the number of fixed Hessian iterations of Table XIV is a lower bound for the actual number of fixed Hessian steps which have to be carried out to simulate the N Newton–Raphson steps.

The optimal use of a fixed Hessian approach is, of course, not to use a fixed Hessian procedure to exactly obtain local convergence. An uncritical use of a fixed Hessian procedure may require a very large number of fixed Hessian steps, for example if seven Newton–Raphson iterations are required to obtain convergence, a fixed Hessian approach would require at least 127 steps in which the Hessian matrix was kept fixed.

Fixed Hessian approaches are most efficiently used when combined with Newton–Raphson iterations. Two Newton–Raphson iterations may for example efficiently be replaced by three fixed Hessian iterations (see Table XIV; recall that a fixed Hessian procedure only requires evaluation of the GBT matrix at each step of the iterative procedure). In Ref. 22 it is proved that the most efficient use of fixed Hessian approaches is obtained if about one third of the total computer time (including transformation time) is spent carrying out iterations in which the Hessian matrix is kept fixed with only GBT matrix elements constructed.

D. Numerical Results

In the previous sections we have analytically examined the local convergence characteristics of Newton–Raphson and fixed Hessian iterative schemes. In this section we will report the results of some sample calculations on the $E^3\Sigma_u^-$ state of O_2 to illustrate the derivations of the previous sections. Our analytical derivations indicate that the fixed Hessian calculation may have a smaller radius of convergence than the Newton–Raphson calculation. To illustrate this point, in Table XV we report a linear fixed

TABLE XV
Convergence Characteristics of the Newton–Raphson and the Linear Fixed Hessian Approach for the $E^3\Sigma_u^-$ State of O_2 at 2.13 a.u.

Iteration point[a]	Newton–Raphson[c] $E - E^{\mathrm{CONV}\,b}$	$\|\mathbf{F}\|$	$\|{}^{n+1}\underline{\lambda} - {}^{n}\underline{\lambda}\| \sim \|{}^{n}\mathbf{e}\|$	Fixed Hessian[c] $E - E^{\mathrm{CONV}}$	$\|\mathbf{F}\|$	$\|{}^{n+1}\underline{\lambda} - {}^{n}\underline{\lambda}\| \sim \|{}^{n}$
0	0.0590242739	9.72×10^{-1}	8.72×10^{-1}	0.0590242739	9.72×10^{-1}	8.72×10^{-1}
1	0.0126959929	2.86×10^{-1}	4.02×10^{-1}	0.0126959929	2.86×10^{-1}	1.89×10^{0}
2	0.0000375222	6.00×10^{-2}	1.64×10^{-1}	0.0793910245	5.58×10^{-1}	1.71×10^{0}
3	0.0000256747	1.06×10^{-2}	1.74×10^{-2}	0.0683720648	6.18×10^{-1}	3.83×10^{0}
4	0.0000000001	1.02×10^{-4}	3.19×10^{-4}	0.2610539743	1.59×10^{0}	1.65×10^{0}
5	0.0000000000	3.90×10^{-8}	$\sim 9\times10^{-8}$	0.1923140885	1.80×10^{0}	1.23×10^{0}

[a] At iteration point n, the energy, $\mathbf{F}$, and $\mathbf{G}$ (Newton–Raphson only) are evaluated and ${}^{n+1}\underline{\lambda}$ is determin[e] Thus at iteration point n, iteration $n+1$ is performed; $\|{}^{n+1}\underline{\lambda} - {}^{n}\underline{\lambda}\|$ is an approximation to $\|{}^{n}\mathbf{e}\| = \|{}^{n}\underline{\lambda} - \underline{\alpha}\|$ [s Eq. (71)].

[b] Here, $E - E^{\mathrm{CONV}}$ denotes the difference between the total energy of the present step of the iterative pro dure and the total energy of the converged calculation. The total energy of the converged calculation – 149.2533729769 a.u.

[c] No constraints have been applied in this calculation.

Hessian [Eq. (144)] calculation and an unconstrained Newton–Raphson calculation on the $E^3\Sigma_u^-$ state of O_2 at an internuclear distance of 2.13 a.u. with our usual initial guess of grand canonical Hartree–Fock orbitals with occupations $1\sigma_g^2 1\sigma_u^2 2\sigma_g^2 2\sigma_u^2 3\sigma_g^2 1\pi_u^4 1\pi_g^2$.

The Newton–Raphson (NR) calculation in Table XV converges while the linear fixed Hessian calculation diverges. In both calculations the first iteration gives a step length norm of 0.872. A constrained NR calculation which used the same initial guess of orbitals and states has previously been reported in Table X. In the constrained NR calculations two large step length amplitudes were each reduced from 0.646 and 0.558 to 0.301 in the first iteration. The convergence rate for the NR calculations in Tables XV and X are almost identical, indicating that for this case the reduction of the step length amplitudes does not significantly affect convergence. Step sizes such as those of the initial step of the NR calculation of Table XV may actually be considered to be on the boarderline for allowed step sizes in the local region. After the first iteration both the NR calculations of Tables XV and X are definitely in the local region and no constraints are applied.

The orbitals and states obtained after constraining the step length of the first iteration in the NR calculation in Table X were used as "initial" guess of orbitals and states for the fixed Hessian series of iterations given in Table XVI. The initial iteration (the Newton–Raphson iteration) of the fixed Hessian series resulted in a step length norm of 0.203. This calculation is thus within the local region of an NR calculation. The fixed Hessian series of calculations therefore, as expected, converges. The error term of a fixed Hessian sequence of iterations [Eq. (142)] shows that the error vector decreases linearly in a fixed Hessian sequence of iterations. The norm of the approximate error vector in Table XVI is observed to decrease with approximately a factor of 0.3–0.4 in each iteration. One reason that the norm of the approximate error vector of Table XVI is not decreasing with exactly the same factor is of course that the error term analysis only is precisely correct infinitesimally close to the stationary point. Higher order terms will cause some deviations from linearity when the initial guess is not very close to being converged. The convergence behavior of Table XVI is rather typical for a fixed Hessian series when the series is initiated as soon as the local region is obtained.

In some extreme cases, the error term of a fixed Hessian series of calculations may make the fixed Hessian sequence of iterations converge extremely slowly. As an example in Table XVII we report a fixed Hessian calculation on the $E^3\Sigma_u^-$ state of O_2 at 2.10 a.u. using the orbitals and states at NR iteration point 4 of Table IV as an initial guess. Even though the fixed Hessian calculation certainly is initiated in the local region (the norm of the Newton–Raphson step is 0.266) the error terms dominate the fixed Hessian

TABLE XVI
Convergence Characteristics ($E^3\Sigma_u^-$ State of O_2 at 2.13 a.u.) of the Linear Fixed Hessian Calculation After Newton–Raphson Iteration 1 (i.e., Starting at Iteration Point 1) of Table X

Iteration point[a]	$E - E^{\text{CONV}\,b}$	$\|\mathbf{F}\|$	$\|{}^{n+1}\underline{\lambda} - {}^{n}\underline{\lambda}\| \sim \|{}^{n}\mathbf{e}\|$
0	−0.0003831946	8.36×10^{-2}	2.03×10^{-1}
1	−0.0002286547	1.36×10^{-2}	7.25×10^{-2}
2	0.0000317736	1.27×10^{-2}	3.04×10^{-2}
3	−0.0000032557	3.04×10^{-3}	1.16×10^{-2}
4	0.0000003723	2.24×10^{-3}	2.63×10^{-3}
5	−0.0000000322	1.94×10^{-4}	1.05×10^{-3}
6	0.0000000035	2.08×10^{-4}	3.18×10^{-4}
7	−0.0000000004	3.04×10^{-5}	1.25×10^{-4}
8	0.0000000000	2.40×10^{-5}	3.32×10^{-5}
9	0.0000000000	3.64×10^{-6}	1.33×10^{-5}
10	0.0000000000	2.54×10^{-6}	3.68×10^{-6}
11	0.0000000000	4.02×10^{-7}	1.49×10^{-6}
12	0.0000000000	2.80×10^{-7}	4.00×10^{-7}
13	0.0000000000	4.78×10^{-8}	1.62×10^{-7}
14	0.0000000000	3.60×10^{-8}	6.30×10^{-8}
15	0.0000000000	$<10^{-8}$	$\sim 2.5\times10^{-8}$

[a]At iteration point n, the energy and $\mathbf{F}$ are evaluated and ${}^{n+1}\underline{\lambda}$ is determined. Thus at iteration point n, iteration $n+1$ is performed; $\|{}^{n+1}\underline{\lambda} - {}^{n}\underline{\lambda}\|$ is an approximation to $\|{}^{n}\mathbf{e}\| = \|{}^{n}\underline{\lambda} - \underline{\alpha}\|$ [see Eq. (71)].

[b]Here, $E - E^{\text{CONV}}$ is the difference in atomic units between the total energy at the present step and the converged total energy. The converged total energy is −149.2533729769 a.u.

sequence of iterations to such an extent that the fixed Hessian sequence of iterations is hardly converging. If the individual step length amplitudes are analyzed in this calculation, it appears that the calculation is dominated completely by a fluctuation of a single κ element which has an initial value of −0.190 and after that becomes 0.140, −0.136, 0.118, ... Such a fluctuation is rather unusual but may occur and may make the rate of convergence of a fixed Hessian calculation extremely slow.

The error vector of the nth iteration of a fixed Hessian sequence is proportional to the error vector of the $(n-1)$st iteration with a proportionality matrix which contains the error vector of the initial iteration [see Eq. (142)]. If the fixed Hessian series therefore is initiated in a region where ${}^{0}\mathbf{e}$ is small, the error term would be smaller than if the fixed Hessian series is initiated in a region where ${}^{0}\mathbf{e}$ is large. In Table XVIII and XIX we report two fixed Hessian series of calculations on the $E^3\Sigma_u^+$ state of O_2 at 2.13 a.u., which are initiated after the second and third iteration of the Newton–Raphson calculation in Table X. Recall that Table XVI reported the fixed Hessian

TABLE XVII

Convergence Characteristics of a Linear Fixed Hessian Calculation After Newton–Raphson Iteration 4 (i.e., Starting at Iteration Point 4) of Table IV ($E^3\Sigma_u^-$ State of O_2 at 2.10 a.u.)

Iteration point[a]	$E - E^{CONV\,b}$	$\|\mathbf{F}\|$	$\|{}^{n+1}\underline{\lambda} - {}^{n}\underline{\lambda}\| \sim \|{}^{n}\mathbf{e}\|$
0	0.0000311601	2.06×10^{-3}	2.66×10^{-1}
1	0.0000073515	4.34×10^{-3}	1.73×10^{-1}
2	0.0000101563	3.30×10^{-3}	1.63×10^{-1}
3	0.0000057317	3.06×10^{-3}	1.44×10^{-1}
4	0.0000071905	2.82×10^{-3}	1.37×10^{-1}
5	0.0000046223	2.64×10^{-3}	1.27×10^{-1}
6	0.0000056164	2.54×10^{-3}	1.22×10^{-1}
7	0.0000038491	2.40×10^{-3}	1.15×10^{-1}
8	0.0000045908	2.30×10^{-3}	1.11×10^{-1}
9	0.0000032803	2.20×10^{-3}	1.06×10^{-1}
10	0.0000038580	2.12×10^{-3}	1.02×10^{-1}

[a]At iteration point n, the energy and $\mathbf{F}$ are evaluated and ${}^{n+1}\underline{\lambda}$ is determined. Thus at iteration point n, iteration $n+1$ is performed; $\|{}^{n+1}\underline{\lambda} - {}^{n}\underline{\lambda}\|$ is an approximation to $\|{}^{n}\mathbf{e}\| = \|{}^{n}\underline{\lambda} - \underline{\alpha}\|$ [see Eq. (71)].

[b]Here, $E - E^{CONV}$ is the difference in atomic units between the total energy at the present step and the converged total energy. The converged total energy is -149.2781477108 a.u.

TABLE XVIII

Convergence Characteristics ($E^3\Sigma_u^-$ State of O_2 at 2.13 a.u.) Using the Linear Fixed Hessian Approach Initiated After Iteration 2 (i.e., Starting at Iteration Point 2) of the Newton–Raphson Calculation in Table X

Iteration point[a]	$E - E^{CONV\,b}$	$\|\mathbf{F}\|$	$\|{}^{n+1}\underline{\lambda} - {}^{n}\underline{\lambda}\| \sim \|{}^{n}\mathbf{e}\|$
0	−0.0002286547	1.36×10^{-2}	7.26×10^{-2}
1	0.0000018925	2.64×10^{-3}	7.22×10^{-3}
2	0.0000000813	2.24×10^{-4}	2.09×10^{-3}
3	0.0000000051	6.72×10^{-5}	4.76×10^{-4}
4	0.0000000003	1.55×10^{-5}	1.15×10^{-4}
⋮	⋮	⋮	⋮

[a]At iteration point n, the energy and $\mathbf{F}$ are evaluated and ${}^{n+1}\underline{\lambda}$ is determined. Thus at iteration point n, iteration $n+1$ performed; $\|{}^{n+1}\underline{\lambda} - {}^{n}\underline{\lambda}\|$ is an approximation to $\|{}^{n}\mathbf{e}\| = \|{}^{n}\underline{\lambda} - \underline{\alpha}\|$ [see Eq. (71)].

[b]Here, $E - E^{CONV}$ is the difference in atomic units between the total energy at this step and the converged total energy. The converged total energy is -149.2533729769 a.u.

series initiated after the first Newton–Raphson iteration of Table X. The ratios between norms $\|^{n+1}\underline{\lambda}\|/\|^{n}\underline{\lambda}\| \sim \|^{n}\mathbf{e}\|/\|^{n-1}\mathbf{e}\|$ in Table XVI are 0.3–0.4, whereas the ratios of Table XVIII are 0.2–0.3 in qualitative agreement with the fact that the error vector $^{0}\mathbf{e}$ of the fixed Hessian sequence in Table XVIII is smaller than the error vector of the sequence of Table XVI. The $\|^{n+1}\underline{\lambda}\|/\|^{n}\underline{\lambda}\|$ ratio of Table XIX is 0.004, which is substantially smaller than the ratios of Tables XVI and XVIII, in agreement with the fact that the initial error vector $^{0}\mathbf{e}$ of the fixed Hessian sequence of iterations in Table XIX is substantially smaller than the initial error vector of the fixed Hessian sequences of Tables XVI and XVIII.

In Table XIV we have reported the number of fixed Hessian steps that are required to simulate a certain number of Newton–Raphson steps to obtain an equal total order of convergence. When error terms are as important as in the fixed Hessian calculation of Table XVII, a comparison of the total order of convergence of fixed Hessian and Newton–Raphson sequences has very little meaning. However, in less pathological cases, Table XIV gives a relatively accurate measure of the number of fixed Hessian steps that are required to simulate a certain number of NR steps. To examplify this we compare the NR sequence of iterations in Table X with the fixed Hessian series of Tables XVI, XVIII, and XIX. The fixed Hessian series of Tables XVI, XVIII, and XIX use the orbitals and states after the first, second, and third iterations of the NR series in Table X as the initial guess. Comparing Table X and XVI we see that the initial two NR iterations roughly can be replaced by three fixed Hessian iterations (one NR iteration followed by two iterations in which the Hessian matrix is kept fixed). The length of the error vec-

TABLE XIX
Convergence Characteristics of the Linear Fixed Hessian Initiated After Iteration 3 (i.e., Starting at Iteration Point 3) of the Newton–Raphson Calculation of Table X

Iteration point[a]	$E - E^{\mathrm{CONV}\,b}$	$\|\mathbf{F}\|$	$\|^{n+1}\underline{\lambda} - {}^{n}\underline{\lambda}\| \sim \|^{n}\mathbf{e}\|$
0	0.0000018925	2.64×10^{-3}	5.66×10^{-3}
1	0.0000000000	1.02×10^{-5}	5.42×10^{-5}
2	0.0000000000	1.60×10^{-7}	3.29×10^{-7}
3	0.0000000000	$<10^{-8}$	$<10^{-8}$

[a]At iteration point n, the energy and $\mathbf{F}$ are evaluated and $^{n+1}\underline{\lambda}$ is determined. Thus at iteration point n, iteration $n+1$ is performed; $\|^{n+1}\underline{\lambda} - {}^{n}\underline{\lambda}\|$ is an approximation to $\|^{n}\mathbf{e}\| = \|^{n}\underline{\lambda} - \underline{\alpha}\|$ [see Eq. (71)].

[b]Here, $E - E^{\mathrm{CONV}}$ is the difference in atomic units between the total energy at the present step and the converged total energy. The converged total energy is -149.2533729769 a.u.

tor after two NR iterations is 5.66×10^{-3}, whereas the error of three fixed Hessian steps is 1.16×10^{-2}. Three initial NR steps correspond according to Table XIV to seven fixed Hessian steps. Table X shows that three NR steps have an error vector with a norm about 7.68×10^{-5} while from Table XVI seven fixed Hessian steps have an error vector norm of approximately 1.25×10^{-4} in reasonable agreement with the predictions of Table XIV. According to Table XIV four NR iterations correspond to 15 fixed Hessian steps. Table XVI shows that the error vector after four NR iterations is less than 10^{-8} while the norm of the error vector after 15 fixed Hessian steps is about 2.5×10^{-8}. The predictions of Table XIV are thus roughly confirmed by the calculations of Tables X and XVI. A comparison of Table X and Tables XVIII and XIX shows the same trends.

In Table XIV we further reported the error term constant, that is, the constant which is a multiplicative factor on the error term. The error term constant decreases rapidly in the Newton–Raphson sequence while the constant always is $\frac{1}{2}$ in the fixed Hessian sequence. The error term analysis thus indicates that the predicted number of fixed Hessian steps is a lower limit to the actual number of fixed Hessian steps. The calculations in Tables X and XVI confirm this prediction. Because of the relation between the error term constants, the difference between the predicted and the actual number of fixed Hessian steps should increase as the number of Newton–Raphson steps increases. This behavior is not clearly observed in the calculations reported in Tables X and XVI. This may be due somewhat to our use of Eq. (71) to estimate the error.

All the fixed Hessian calculations reported so far have been carried out using Eq. (144) where a linear transformation of variables is performed in between each step of the fixed Hessian series of calculations. In Section VI.B we discussed that the linear [Eqs. (144) or (138)] and the nonlinear fixed Hessian approach have the same convergence characteristics. In Table XX we report a fixed Hessian calculation which differs from the calculation of Table XVI in that the nonlinear fixed Hessian procedure is used.[22] The result of Tables XX and XVI are in very close agreement. The agreement is so pronounced that we expect the error terms of the two calculations to be essentially the same. The error term of the nonlinear fixed Hessian approach has not yet been derived.

Thus we conclude that in the local region fixed Hessian procedures may be advantageously employed to improve the overall computational efficiency of MCSCF procedures. With these procedures only new **F** matrix elements are calculated at each iteration point. Thus these procedures may yield considerable savings over usual Newton–Raphson techniques since a smaller two-electron integral transformation is required [if, e.g., Eq. (144) is used] and a small number of matrix elements (**F** instead of both **F** and **G**) are required in each iteration. These savings, of course, may be offset by the slower con-

TABLE XX
Convergence Characteristics ($E^3\Sigma_u^-$ State of O_2 at 2.13 a.u.) of the Nonlinear Fixed Hessian Approach After Newton–Raphson Iteration 1 (Starting at Iteration Point 1) of Table X

Iteration point[a]	$E - E^{\mathrm{CONV}\,b}$	$\|\mathbf{F}\|$	$\|{}^{n+1}\underline{\lambda} - {}^{n}\underline{\lambda}\| \sim \|{}^{n}\mathbf{e}\|$
0	−0.0003831946	8.36×10^{-2}	2.03×10^{-1}
1	−0.0002286547	1.36×10^{-2}	7.25×10^{-2}
2	0.0000318517	1.31×10^{-2}	3.05×10^{-2}
3	−0.0000032488	2.98×10^{-3}	1.16×10^{-2}
4	0.0000003252	2.10×10^{-3}	2.62×10^{-3}
5	−0.0000000284	2.80×10^{-4}	1.03×10^{-3}
6	0.0000000028	1.91×10^{-4}	3.03×10^{-4}
7	−0.0000000003	2.64×10^{-5}	1.20×10^{-4}
8	0.0000000000	2.12×10^{-5}	3.07×10^{-5}

[a]At iteration point n, the energy and $\mathbf{F}$ are evaluated and ${}^{n+1}\lambda$ is determined. Thus at iteration point n, iteration $n+1$ is performed; $\|{}^{n+1}\underline{\lambda} - {}^{n}\underline{\lambda}\|$ is an approximation to $\|{}^{n}\mathbf{e}\| = \|{}^{n}\underline{\lambda} - \underline{\alpha}\|$ [see Eq. (71)].

[b]Here, $E - E^{\mathrm{CONV}}$ is the difference in atomic units between the total energy at the present step and the converged total energy. The converged total energy is − 149.2533729769 a.u.

vergence of the fixed Hessian procedures [see Eqs. (34) and (142)]. A further consideration, as shown above, is that the error vector when the fixed Hessian sequence begins should be fairly small [see Eq. (142)]. The most efficient use of fixed Hessian approaches is obtained if about $\frac{1}{3}$ of the total computer time is spent carrying out iterations in which the Hessian matrix is kept fixed with only GBT matrix elements constructed.

In the next section we examine procedures which update the initial Hessian matrix in each iteration using information contained in the gradient and step length vectors.[9, 21] These techniques may significantly speed up convergence in the local region compared to fixed Hessian procedures. Furthermore, combined with a step size and sign control algorithm, Hessian update procedures may be very useful even when far from convergence.

VII. UPDATE METHODS

Let us assume the gradient

$$\mathbf{F}(\underline{\lambda}) = \frac{\partial E(\underline{\lambda})}{\partial \underline{\lambda}} \tag{157}$$

is known at two consecutive points ${}^{k}\underline{\lambda}$ and ${}^{k+1}\underline{\lambda}$ of a sequence of iterations

$${}^{k+1}\underline{\lambda} = {}^{k}\underline{\lambda} - \mathbf{H}_k^{-1}\mathbf{F}({}^{k}\underline{\lambda}) \tag{158}$$

where $\mathbf{H}_k$ is an approximation to the Hessian matrix $\mathbf{G}$ at ${}^k\underline{\lambda}$. If $\mathbf{F}(\underline{\lambda})$ is expanded around ${}^{k+1}\underline{\lambda}$, one obtains

$$\mathbf{F}({}^k\underline{\lambda}) = \mathbf{F}({}^{k+1}\underline{\lambda}) + \mathbf{G}({}^{k+1}\underline{\lambda})({}^k\underline{\lambda} - {}^{k+1}\underline{\lambda}) + O({}^k\underline{\lambda} - {}^{k+1}\underline{\lambda})^2 \tag{159}$$

It is thus clear that $\mathbf{F}({}^k\underline{\lambda}) - \mathbf{F}({}^{k+1}\underline{\lambda})$ gives a finite difference approximation to the Hessian at ${}^{k+1}\underline{\lambda}$ multiplied with ${}^k\underline{\lambda} - {}^{k+1}\underline{\lambda}$. In the most important update methods[63] this information about the exact Hessian is built into the Hessian approximation $\mathbf{H}_{k+1}$ of the $(k+2)$nd iteration of Eq. (158). A requirement to the Hessian approximation $\mathbf{H}_{k+1}$ thus becomes that it satisfies the quasi-Newton condition[9]

$$\mathbf{H}_{k+1}[{}^k\underline{\lambda} - {}^{k+1}\underline{\lambda}] = \mathbf{F}({}^k\underline{\lambda}) - \mathbf{F}({}^{k+1}\underline{\lambda}) \tag{160}$$

which implies

$$\mathbf{H}_{k+1}[{}^k\underline{\lambda} - {}^{k+1}\underline{\lambda}] = \mathbf{G}({}^{k+1}\underline{\lambda})[{}^k\underline{\lambda} - {}^{k+1}\underline{\lambda}] + O({}^k\underline{\lambda} - {}^{k+1}\underline{\lambda})^2 \tag{161}$$

The projection of the approximate and the exact Hessian into the direction ${}^k\underline{\lambda} - {}^{k+1}\underline{\lambda}$ thus differs in second order in ${}^k\underline{\lambda} - {}^{k+1}\underline{\lambda}$.

All Hessian update methods discussed here use as the basic assumption that the quasi-Newton condition [Eq. (160)] has to be satisfied. The quasi-Newton condition contains, however, only m equations (m is the dimension of the gradient). Since the Hessian matrix has $m(m+1)/2$ unknown elements, the quasi-Newton condition therefore has to be supplemented with additional information (e.g., structural or numerical information about the exact Hessian) to determine a Hessian update. Through simulating the exact Hessian matrix numerically and structurally, Hessian update methods are partly able to obtain the superior convergence characteristics of the Newton–Raphson model without constructing the Hessian matrix explicitly.

To illustrate how the quasi-Newton condition may be supplemented with further conditions to uniquely define a Hessian update, we explicitly derive one of the most well-known update formulas, the Broyden rank-1 update.[64] The conditions which supplement the quasi-Newton condition to define some of the other well-known Hessian update methods will also be discussed, although the Broyden rank-1 formula will be the only one explicitly derived. We will thus discuss the conditions applied to determine the Powell symmetrization of the Broyden update, the Davidon–Fletcher–Powell update, and the Broyden–Fletcher–Goldfarb–Shanno update.

A. Broyden Asymmetric Rank-1 Update and Its Symmetrization

In the Broyden rank-1 update,[64] the quasi-Newton condition in Eq. (160) is supplemented with the conditions

$$\mathbf{H}_{k+1}\mathbf{P} = \mathbf{H}_k\mathbf{P} \tag{162}$$

for any vector $\mathbf{P}$ which is orthogonal to ${}^{k}\underline{\lambda} - {}^{k+1}\underline{\lambda}$:

$$\mathbf{P}^T[{}^{k}\underline{\lambda} - {}^{k+1}\underline{\lambda}] = 0 \tag{163}$$

The motivation for supplementing the quasi-Newton condition with the conditions of Eqs. (162) and (163) is that information about the exact Jacobian at ${}^{k+1}\underline{\lambda}$ only is available in the direction ${}^{k}\underline{\lambda} - {}^{k+1}\underline{\lambda}$. (A Jacobian is the first derivative of a vector function. In this case the Jacobian and the Hessian are the same.) Introducing the shorthand notation

$$\mathbf{K}_k = {}^{k+1}\underline{\lambda} - {}^{k}\underline{\lambda} \tag{164}$$

$$\mathbf{L}_k = \mathbf{F}({}^{k+1}\underline{\lambda}) - \mathbf{F}({}^{k}\underline{\lambda}) \tag{165}$$

$$\mathbf{D}_{k+1} = \mathbf{H}_{k+1} - \mathbf{H}_k \tag{166}$$

we may write Eqs. (162) and (163) as

$$\mathbf{D}_{k+1}\mathbf{P} = \mathbf{0} \tag{167}$$

for

$$\mathbf{P}^T\mathbf{K}_k = 0 \tag{168}$$

Thus, $\mathbf{D}_{k+1}$ may be written as

$$\mathbf{D}_{k+1} = \frac{\mathbf{D}'_{k+1}\mathbf{K}_k\mathbf{K}_k^T}{\mathbf{K}_k^T\mathbf{K}_k} \tag{169}$$

Here, $\mathbf{D}'_{k+1}$ is so far undetermined. When we are introducing the notation of Eqs. (164)–(166), the quasi-Newton condition in Eq. (160) gives

$$\mathbf{H}_k\mathbf{K}_k + \mathbf{D}_{k+1}\mathbf{K}_k = \mathbf{L}_k \tag{170}$$

Then, by inserting Eq. (169) into Eq. (170),

$$\mathbf{D}'_{k+1}\frac{\mathbf{K}_k\mathbf{K}_k^T}{\mathbf{K}_k^T\mathbf{K}_k}\mathbf{K}_k = \mathbf{L}_k - \mathbf{H}_k\mathbf{K}_k \tag{171}$$

we can identify uniquely $\mathbf{D}'_{k+1}\mathbf{K}_k$, and the updated approximate Hessian $\mathbf{H}_{k+1}$ may therefore be written as

$$\mathbf{H}^B_{k+1} = \mathbf{H}_k + [\mathbf{L}_k - \mathbf{H}_k\mathbf{K}_k]\frac{\mathbf{K}_k^T}{\mathbf{K}_k^T\mathbf{K}_k} \tag{172}$$

The implementation of the Broyden update [Eq. (172)] on a computer thus only requires simple matrix-vector multiplication.

The update correction to $\mathbf{H}_k$ [last term in Eq. (172)] is not a symmetric matrix. When carrying out updates on an unsymmetric Jacobian, it is acceptable to get an unsymmetric update correction [Eq. (172)], whereas in updates on a Hessian matrix one may require the update correction to be symmetric. We consider now how a symmetric analog of Eq. (172) may be determined following the derivation of Powell.[65]

Let $\mathbf{H}_k$ be a symmetric approximation to the Hessian and define $\mathbf{H}_{k+1}^{(1/2)}$ as

$$\mathbf{H}_{k+1}^{(1/2)} = \mathbf{H}_k + \frac{\mathbf{T}_k\mathbf{K}_k^T}{\mathbf{K}_k^T\mathbf{K}_k} \tag{173}$$

where

$$\mathbf{T}_k = \mathbf{L}_k - \mathbf{H}_k\mathbf{K}_k \tag{174}$$

Since $\mathbf{H}_{k+1}^{(1/2)}$ is not symmetric, Powell defined

$$\mathbf{H}_{k+1}^{(1)} = \tfrac{1}{2}\left[\mathbf{H}_{k+1}^{(1/2)} + \mathbf{H}_{k+1}^{(1/2)^T}\right] \tag{175}$$

$\mathbf{H}_{k+1}^{(1)}$ does not satisfy the quasi-Newton condition, that is,

$$\mathbf{H}_{k+1}^{(1)}\mathbf{K}_k \neq \mathbf{L}_k \tag{176}$$

so it may also be updated to satisfy the quasi-Newton condition, that is,

$$\mathbf{H}_{k+1}^{(1+1/2)} = \mathbf{H}_{k+1}^{(1)} + \frac{\mathbf{T}_k^{(1)}\mathbf{K}_k^T}{\mathbf{K}_k^T\mathbf{K}_k} \tag{177}$$

where

$$\mathbf{T}_k^{(1)} = \mathbf{L}_k - \mathbf{H}_{k+1}^{(1)}\mathbf{K}_k \tag{178}$$

Equation (177) is, however, not symmetric. A symmetrized version $\mathbf{H}_{k+1}^{(2)}$ may be determined from $\mathbf{H}_{k+1}^{(1+1/2)}$ by using an analog of Eq. (175) and so on. Powell[65] showed that the result of carrying out an infinite sequence of iterations corresponding to Eqs. (175) and (177) gives the limiting value

$$\mathbf{H}_{k+1}^{P} = \lim_{i\to\infty} \mathbf{H}_{k+1}^{(i)} = \mathbf{H}_k + \frac{1}{\mathbf{K}_k^T\mathbf{K}_k}\left[\mathbf{T}_k\mathbf{K}_k^T + \mathbf{K}_k\mathbf{T}_k^T - \frac{\mathbf{K}_k[\mathbf{T}_k^T\mathbf{K}_k]\mathbf{K}_k^T}{\mathbf{K}_k^T\mathbf{K}_k}\right] \tag{179}$$

which satisfies the quasi-Newton condition, gives a symmetric update correction, and in a sense corresponds to a symmetrized version of Eq. (172). Whereas the Broyden asymmetric update formula [Eq. (172)] is of rank 1, its symmetrized analog [Eq. (179)] is of rank 2.[9]

B. The Broyden Family (DFP and BFGS Updates)

In the numerical analysis literature the Hessian update methods have primarily been used to determine minima of functions of many variables. According to the general trends of numerical analysis, the Hessian update methods that most efficiently are able to determine function minima are the Davidon–Fletcher–Powell (DFP) and the Broyden–Fletcher–Goldfarb–Shanno (BFGS) updates (see, however, Ref. 66). The DFP and BFGS Hessian updates are members of the rank-2 Broyden update family. To understand the strengthes (and weaknesses) of the DFP and BFGS update methods, we will discuss the conditions that have to be imposed on the quasi-Newton condition to define updates that belong to the Broyden family. The conditions are as follows:

1. The updated approximate Hessian matrix has to satisfy the quasi-Newton condition of Eq. (160).
2. The updated approximate Hessian matrix has to be symmetric.
3. The updated approximate Hessian matrix has to transform under linear transformations of the variables such that the step length vector of a Newton-like method [Eq. (158)] is invariant with respect to a linear transformation of the variables.
4. When the approximate Hessian matrix is positive definite at one step of the iterative procedure, then the updated Hessian matrix must also be positive definite. Neither of the previous updates [Eqs. (172) and (179)] have this property.
5. The Hessian update has to be of at most rank 2.

The Broyden family update has one parameter, which is not determined. Special choices of this parameter gives the DFP and BFGS update formulas. To get some insight into the structure of the Broyden family updates we will now discuss in more detail the contents of conditions 1–5.

1. The importance of the quasi-Newton condition for defining Hessian update methods has already been stressed. This requirement is essential for speeding up convergence rates from the linear convergence encountered in fixed Hessian procedures. For if

$$\lim_{k \to \infty} \frac{\|(\mathbf{G}({}^{k+1}\underline{\lambda}) - \mathbf{H}_{k+1})({}^{k+1}\underline{\lambda} - {}^{k}\underline{\lambda})\|}{\|{}^{k+1}\underline{\lambda} - {}^{k}\underline{\lambda}\|} = 0, \tag{180}$$

the sequence $\{{}^k\underline{\lambda}\}$ converges superlinearily;[67] that is, $\|{}^{k+1}\underline{\lambda}-\underline{\alpha}\|/\|{}^k\underline{\lambda}-\underline{\alpha}\|$ goes toward zero.

2. The condition that the updated matrix has to be symmetric is, of course, only applicable to symmetric (Hessian) updates. When imposing the symmetry constraint on the Hessian update, we decrease the number of variables to be determined from m^2 to $m(m+1)/2$. This means, then, that more emphasis is given to the quasi-Newton condition.

3. To understand the reason for condition 3, it is instructive to examine the possible changes of the parameters ${}^{k+1}\underline{\lambda}$ that occur in a Newton–Raphson iteration

$$ {}^{k+1}\underline{\lambda} = {}^k\underline{\lambda} - \mathbf{G}^{-1}({}^k\underline{\lambda})\mathbf{F}({}^k\underline{\lambda}) \tag{181} $$

when a linear transformation is carried out among the variables

$$ \underline{\chi} = \mathbf{A}\underline{\lambda} + \mathbf{a} \tag{182} $$

The function $E(\underline{\lambda})$ may be regarded as depending on either $\underline{\chi}$ or $\underline{\lambda}$:

$$ E(\underline{\lambda}) = E(\mathbf{A}^{-1}(\underline{\chi}-\mathbf{a})) \tag{183} $$

The partial derivatives with respect to $\underline{\lambda}$ may therefore straightforwardly be related to partial derivatives with respect to $\underline{\chi}$. Consider, for example, how the gradient with respect to the variables $\underline{\lambda}$ may be related to the gradient with respect to the variables $\underline{\chi}$:

$$ F_i^{(\lambda)} = \frac{\partial E(\underline{\lambda})}{\partial \lambda_i} = (\mathbf{A}^T\mathbf{F}^{(x)})_i \tag{184} $$

Similarly, we may relate the second partial derivatives as follows:

$$ \mathbf{G}^{(\lambda)} = \mathbf{A}^T\mathbf{G}^{(x)}\mathbf{A} \tag{185} $$

Thus, in the $(k+1)$st iteration, a Newton–Raphson iteration [Eq. (181)], where $\underline{\chi}$ is considered as variable, may be written [using Eqs. (182), (184), and (185)]:

$$ \begin{aligned} {}^{k+1}\underline{\chi} &= {}^k\underline{\chi} - \left(\mathbf{G}^{(x)}({}^k\underline{\chi})\right)^{-1}\mathbf{F}^{(x)}({}^k\underline{\chi}) \\ &= \mathbf{A}^{k+1}\underline{\lambda} + \mathbf{a} \end{aligned} \tag{186} $$

From Eqs. (182) and (186) it is clear that the result of a Newton–Raphson iteration does not depend on whether the transformation is applied before or after the iteration. If we require the update methods to possess the same invariance property with respect to a linear transformation of variables as the Newton–Raphson model, a necessary and sufficient condition on the updated matrix becomes

$$\mathbf{H}_{k+1}^{(\lambda)} = \mathbf{A}^T \mathbf{H}_{k+1}^{(x)} \mathbf{A} \tag{187}$$

If we assume that the Hessian approximation of the kth iteration $\mathbf{H}_k$ possesses the invariance property, then the Hessian update matrix has to satisfy

$$\mathbf{H}_{k+1}^{(\lambda)} - \mathbf{H}_k^{(\lambda)} = \mathbf{A}^T\left(\mathbf{H}_{k+1}^{(x)} - \mathbf{H}_k^{(x)}\right)\mathbf{A} \tag{188}$$

to keep the invariance property.

4. This condition (i.e., that if the approximate Hessian matrix at one step of the iterative procedure is positive definite then the updated approximate Hessian has to be positive definite) is a useful property of an update when the update is used to determine function minima, as it ensures that the Hessian approximation will have the correct structure during the whole sequence of iterations when the structure is correct at one single step of the iterative procedure. In connection with the use of update methods for obtaining convergence in MCSCF calculations, this positive definite requirement is less important when the Hessian update method in MCSCF is used to determine excited states that have a nonpositive definite Hessian matrix.

5. The limitation that $\mathbf{H}_{k+1} - \mathbf{H}_k$ be of rank 2 complements the previously described constraints to define uniquely the Broyden family update. A rank-2 update correction may generally be written as

$$\mathbf{H}_{k+1} - \mathbf{H}_k = a\mathbf{U}\mathbf{V}^T + b\mathbf{V}\mathbf{U}^T + c\mathbf{V}\mathbf{V}^T + d\mathbf{U}\mathbf{U}^T \tag{189}$$

where a, b, c, and d are constants and $\mathbf{U}$ and $\mathbf{V}$ are vectors. Choosing the vectors $\mathbf{U}$ and $\mathbf{V}$ as

$$\mathbf{U} = \mathbf{H}_k\mathbf{K}_k; \qquad \mathbf{V} = \mathbf{L} = \mathbf{F}({}^{k+1}\underline{\lambda}) - \mathbf{F}({}^{k}\underline{\lambda}) \tag{190}$$

the Hessian update correction in Eq. (189) satisfies the invariance condition (i.e., condition 3). Conditions 1 and 2 may then be used to define the Broyden family update as

$$\mathbf{H}_{k+1} = \mathbf{H}_{k+1}^0 + a'\mathbf{W}_k\mathbf{W}_k^T \tag{191}$$

where

$$\mathbf{W}_k = \frac{\mathbf{H}_k\mathbf{K}_k}{\mathbf{K}_k^T\mathbf{H}_k\mathbf{K}_k} - \frac{\mathbf{L}_k}{\mathbf{K}_k^T\mathbf{L}_k} \tag{192}$$

and

$$\mathbf{H}^0_{k+1} = \mathbf{H}_k + \frac{\mathbf{L}_k\mathbf{L}_k^T}{\mathbf{L}_k^T\mathbf{K}_k} - \frac{\mathbf{H}_k\mathbf{K}_k\mathbf{K}_k^T\mathbf{H}_k}{\mathbf{K}_k^T\mathbf{H}_k\mathbf{K}_k} \tag{193}$$

where a' is an undefined parameter. [See Ref. 9, where a detailed derivation of Eqs. (191)–(193) is carried out.]

The conservation of positive definiteness of the approximate Hessian may be used to limit the interval of allowed a' parameters. In Ref. 9 it is shown how a positive definiteness requirement limits the interval of the parameter a' to be $[a'_{\min}, \infty]$, where $a'_{\min}$ is the largest negative number for which $\mathbf{H}^0_{k+1} + a'_{\min}\mathbf{W}_k\mathbf{W}_k^T$ is singular. It should be mentioned that to limit the interval of a' it is assumed that $\mathbf{K}_k^T\mathbf{L}_k > 0$. Which value of a' to use within the allowed interval must to a high degree rely on numerical experience. While members of the Broyden family behave identically when exact line searches are used,[68] differences are encountered with direct application of Eq. (158) (no line search). Extensive numerical tests are compiled in Ref. 69. The two most commonly used choices are the Broyden–Fletcher–Goldfarb–Shanno (BFGS) update[70]

$$a' = 0$$

$$\mathbf{H}^{\mathrm{BFGS}}_{k+1} = \mathbf{H}^0_{k+1} = \mathbf{H}_k + \frac{\mathbf{L}_k\mathbf{L}_k^T}{\mathbf{L}_k^T\mathbf{K}_k} - \frac{\mathbf{H}_k\mathbf{K}_k\mathbf{K}_k^T\mathbf{H}_k}{\mathbf{K}_k^T\mathbf{H}_k\mathbf{K}_k} \tag{194}$$

and the Davidon–Fletcher–Powell (DFP) update[71]

$$a' = a'_{\mathrm{DFP}} = \mathbf{K}_k^T\mathbf{H}_k\mathbf{K}_k$$

$$\mathbf{H}^{\mathrm{DFP}}_{k+1} = \mathbf{H}_k + \left[1 + \frac{\mathbf{K}_k^T\mathbf{H}_k\mathbf{K}_k}{\mathbf{K}_k^T\mathbf{L}_k}\right]\frac{\mathbf{L}_k\mathbf{L}_k^T}{\mathbf{L}_k^T\mathbf{K}_k} - \frac{\mathbf{H}_k\mathbf{K}_k\mathbf{L}_k^T + \mathbf{L}_k\mathbf{K}_k\mathbf{H}_k}{\mathbf{L}_k^T\mathbf{K}_k} \tag{195}$$

A rank-1 procedure can also be generated from Eq. (192):[72]

$$\mathbf{H}_{k+1} = \mathbf{H}_k + \frac{(\mathbf{L}_k - \mathbf{H}_k\mathbf{K}_k)(\mathbf{L}_k - \mathbf{H}_k\mathbf{K}_k)^T}{(\mathbf{L}_k - \mathbf{H}_k\mathbf{K}_k)^T\mathbf{K}_k} \tag{196}$$

This procedure has some disadvantages compared to, say, the DFP and BFGS updates: it does not retain positive definiteness, and it can be rather unstable. One theoretical advantage of the update (196) is that a quadratic function is minimized in at most $m+1$ iterations under reasonable conditions. Furthermore, the change is given by one vector and one scalar, which is simplifying for very large dimensions. The rank-1 update [Eq. (196)] has been used in MCSCF optimization by Eade and Robb.[73] Their implementation is, however, not optimal since they do not use the quasi-Newton condition correctly owing to neglect of changes of CI coefficients. The actual performance of the update [Eq. (196)] reported by Eade and Robb is, however, very encouraging.

C. A Variational Approach to Update Procedures

The update methods mentioned hitherto can be generated by minimizing a norm of the change $\mathbf{H}_{k+1}-\mathbf{H}_k$.[74] This point of view elucidates the relation between the different updates and thus provides help in choosing the best update for a given situation. We will therefore discuss the updates in this perspective. To do this, we define a general weighted norm

$$\|\mathbf{W}\|_{\mathbf{A}} = \|\mathbf{AWA}\| \tag{197}$$

where $\|\cdot\|$ is the Frobenius norm. Further, let $\mathbf{U}_{k+1}$ ($\mathbf{U}^{\mathrm{S}}_{k+1}$) be all matrices (symmetric matrices) that satisfy the quasi-Newton condition of iteration $k+2$.

The Broyden rank-1 update $\mathbf{H}^{\mathrm{B}}_{k+1}$ minimizes $\|\mathbf{H}'_{k+1}-\mathbf{H}_k\|_{\mathbf{I}}$, where $\mathbf{H}'_{k+1}$ belongs to $\mathbf{U}_{k+1}$. This is easily seen since for any $\mathbf{H}'_{k+1}$

$$\begin{aligned}\|\mathbf{H}^{\mathrm{B}}_{k+1}-\mathbf{H}_k\|_{\mathbf{I}} &= \left\|(\mathbf{L}_k-\mathbf{H}_k\mathbf{K}_k)\frac{\mathbf{K}_k^T}{\mathbf{K}_k^T\mathbf{K}_k}\right\|_{\mathbf{I}}\\ &= \left\|(\mathbf{H}'_{k+1}-\mathbf{H}_k)\frac{\mathbf{K}_k\mathbf{K}_k^T}{\mathbf{K}_k^T\mathbf{K}_k}\right\|_{\mathbf{I}}\\ &\leqq \|\mathbf{H}'_{k+1}-\mathbf{H}_k\|_{\mathbf{I}}\end{aligned} \tag{198}$$

where $\mathbf{I}$ represents the unit matrix (i.e., $\mathbf{A}=\mathbf{I}$). When it is essential to minimize the Frobenius norm of the change $\mathbf{H}_{k+1}-\mathbf{H}_k$, the Broyden rank-1 method seems an obvious choice. If $\|\mathbf{H}'_{k+1}-\mathbf{H}_k\|_{\mathbf{I}}$ is minimized with $\mathbf{H}'_{k+1}$ restricted to $\mathbf{U}^{\mathrm{S}}_{k+1}$, the Powell symmetrization of Broyden's rank-1 method is obtained.[74] This is shown by observing that $\|(\mathbf{H}^{P}_{k+1}-\mathbf{H}_k)\mathbf{V}\|_{\mathbf{I}}$ is smaller than $\|(\mathbf{H}'_{k+1}-\mathbf{H}_k)\mathbf{V}\|_{\mathbf{I}}$ for all choices of $\mathbf{V}$ ($\mathbf{V}$ is any arbitrary vector).

When a pattern in $\mathbf{H}_k - \mathbf{G}({}^{k+1}\underline{\lambda})$ exists, it can be advantageous to incorporate this by choosing a matrix $\mathbf{A}$ that reflects this pattern and then minimize $\|\mathbf{H}'_{k+1} - \mathbf{H}_k\|_{\mathbf{A}}$ to obtain $\mathbf{H}_{k+1}$. If $\mathbf{H}_k - \mathbf{G}({}^{k+1}\underline{\lambda})$ is expected to be proportional to $G({}^{k+1}\underline{\lambda})_{ij}$, it seems sound to weight $(\mathbf{H}_{k+1} - \mathbf{H}_k)$ so that large elements are allowed to change more than small elements. In cases where $\mathbf{G}({}^{k+1}\underline{\lambda})$ is positive definite a proper choice of $\mathbf{A}$ is then $(\mathbf{G}({}^{k+1}\underline{\lambda}))^{-1/2}$. The minimizer of $\|\mathbf{H}'_{k+1} - \mathbf{H}_k\|_{(\mathbf{G}({}^{k+1}\lambda))^{-1/2}}$ with $\mathbf{H}'_{k+1}$ restricted to $\mathbf{U}^S_{k+1}$ can be shown to be $\mathbf{H}^{\mathrm{DFP}}_{k+1}$. In the same way the minimizer with $\mathbf{H}'_{k+1}$ restricted to $\mathbf{U}^S_{k+1}$ of $\|\mathbf{H}'^{-1}_{k+1} - \mathbf{H}^{-1}_k\|_{(\mathbf{G}({}^{k+1}\lambda))^{1/2}}$ is the BFGS update. The BFGS and DFP updates thus seem appropriate when the relative errors in $\mathbf{H}_{k+1}$ (compared to $\mathbf{G}({}^{k+1}\underline{\lambda})$) are of the same magnitude and when $\mathbf{G}({}^{k+1}\underline{\lambda})$ is positive definite.

D. Inverse Updates

In the previous sections updates were carried out directly on the Hessian (Jacobian). The step length parameters ${}^{k+1}\underline{\lambda}$ were then determined from ${}^{k}\underline{\lambda}$, $\mathsf{F}({}^{k}\underline{\lambda})$ and $\mathbf{H}^{-1}_k$ as

$$ {}^{k+1}\underline{\lambda} = {}^{k}\underline{\lambda} - \mathbf{H}^{-1}_k \mathsf{F}({}^{k}\underline{\lambda}) \tag{199} $$

which corresponds to solving a set of linear inhomogenous equations

$$ \mathbf{H}_k({}^{k+1}\underline{\lambda} - {}^{k}\underline{\lambda}) = -\mathsf{F}({}^{k}\underline{\lambda}) \tag{200} $$

The solution of Eq. (199) for very large dimensions is a very time-consuming part of each iteration. It therefore becomes worthwhile to examine whether the update corrections can be carried out directly on the inverse matrix:

$$ \mathbf{H}^{-1}_{k+1} = \mathbf{H}^{-1}_k + (\mathbf{H}')^{-1} \tag{201} $$

That is, can $(\mathbf{H}')^{-1}$ be determined directly. If $\mathbf{H}^{-1}_k$ is available ${}^{k+1}\underline{\lambda}$ can be constructed directly by a multiplication of a matrix with a vector [see Eq. (199)] which is much faster than solving a set of linear inhomogeneous equations. The inverse update formulas may easily be determined from the equation

$$ \mathbf{H}_{k+1}\left(\mathbf{H}^{-1}_k + (\mathbf{H}')^{-1}\right) = \mathbf{1} \tag{202} $$

and corresponding to the update formulas of Eqs. (172), (179), (194), and

(195) we obtain

$$\left(\mathbf{H}_{k+1}^{\mathrm{B}}\right)^{-1} = \mathbf{H}_k^{-1} - \left(\mathbf{H}_k^{-1}\mathbf{L}_k - \mathbf{K}_k\right)\frac{\mathbf{K}_k^T\mathbf{H}_k^{-1}}{\mathbf{K}_k^T\mathbf{H}_k^{-1}\mathbf{L}_k} \tag{203}$$

$$\left(\mathbf{H}_{k+1}^{\mathrm{P}}\right) = \mathbf{H}_k^{-1} + \frac{1}{\mathbf{K}_k^T\mathbf{K}_k}\left\{\left(\mathbf{L}_k - \mathbf{H}_k^{-1}\mathbf{K}_k\right)\mathbf{K}_k^T + \mathbf{K}_k\left(\mathbf{L}_k^T - \mathbf{K}_k^T\mathbf{H}_k^{-1}\right) - \frac{1}{\mathbf{K}_k^T\mathbf{K}_k}\left[\mathbf{K}_k^T\mathbf{L}_k - \mathbf{K}_k^T\mathbf{H}_k^{-1}\mathbf{K}_k\right]\mathbf{K}_k\mathbf{K}_k^T\right\} \tag{204}$$

$$\left(\mathbf{H}_{k+1}^{\mathrm{BFGS}}\right)^{-1} = \mathbf{H}_k^{-1} + \left(1 + \frac{\mathbf{L}_k^T\mathbf{H}_k^{-1}\mathbf{L}_k}{\mathbf{K}_k^T\mathbf{L}_k}\right)\frac{\mathbf{K}_k\mathbf{K}_k^T}{\mathbf{K}_k^T\mathbf{L}_k} - \frac{\mathbf{K}_k\mathbf{L}_k^T\mathbf{H}_k^{-1} + \mathbf{H}_k^{-1}\mathbf{L}_k^T\mathbf{K}_k}{\mathbf{K}_k^T\mathbf{L}_k} \tag{205}$$

$$\left(\mathbf{H}_{k+1}^{\mathrm{DFP}}\right)^{-1} = \mathbf{H}_k^{-1} - \frac{\mathbf{H}_k^{-1}\mathbf{L}_k\mathbf{L}_k^T\mathbf{H}_k^{-1}}{\mathbf{L}_k^T\mathbf{H}_k^{-1}\mathbf{L}_k} + \frac{\mathbf{K}_k\mathbf{K}_k^T}{\mathbf{K}_k^T\mathbf{L}_k} \tag{206}$$

It is observed that $(\mathbf{H}_{k+1}^{\mathrm{BFGS}})^{-1}$ is obtained from $\mathbf{H}_{k+1}^{\mathrm{DFP}}$ when the substitutions

$$\mathbf{H}_k \to \mathbf{H}_k^{-1}; \qquad \mathbf{L}_k \to \mathbf{K}_k; \qquad \mathbf{K}_k \to \mathbf{L}_k \tag{207}$$

are carried out. Similarly, $\mathbf{H}_{k+1}^{\mathrm{BFGS}}$ is related to $(\mathbf{H}_{k+1}^{\mathrm{DFP}})^{-1}$ through the same substitution. The DFP and BFGS update corrections are said to be dual.[9]

E. Convergence Characteristics of Hessian Update Methods

The practical implementation of a Hessian update method consists of carrying out a sequence of iterations similar to the sequence of the linear fixed Hessian approach [Eq. (138)] but where the Hessian matrix gets updated in each iteration. If the Hessian matrix were actually constructed and used in each iteration we would perform just sequence of linear Newton–Raphson iterations [Eq. (20)]. When a sequence of N linear fixed Hessian iterations is carried out the total order of convergence in ${}^0\mathbf{e}$ is $N+1$, whereas when N linear Newton–Raphson iterations are carried out we obtain a total order of convergence of 2^N. Since the Hessian update corrections are determined to simulate the exact Hessian matrix, a sequence of N update iterations is expected to have a total order of convergence in between $N+1$ and 2^N depending on the efficiency of the update method used to simulate the exact Hessian matrix. The number of fixed Hessian iterations performed in the local region to obtain convergence is usually small (~ 10). It is therefore not realistic to expect that the Hessian updates will be very accurate and will well simulate the exact Hessian. Since the quasi-Newton condition gives information about the exact Hessian in directions corresponding to actual step lengths, it may still be expected that Hessian updates can considerably improve the structure of the part of the Hessian matrix that corresponds to variables that change significantly (i.e., that are far from converged and thereby improve considerably the convergence properties of a fixed Hessian procedure). In the next subsection we demonstrate through numerical examples that the update methods constitute a very efficient way for obtaining local convergence of an MCSCF calculation. The Hessian update methods can also be shown to converge superlinearly.[67]

The convergence problems that appear in a sequence of step size and sign controlled Newton–Raphson iterations are associated with large fluctuations in a few modes and are caused by small eigenvalues of the Hessian matrix (see Section V.B). Because the Hessian update methods update the Hessian matrix corresponding to the directions (modes) that give most convergence problems, we expect the Hessian update methods to simulate to a certain degree the global convergence characteristics of the Newton–Raphson approach. In the next subsection we return to this question and demonstrate through numerical examples how the update methods may have very promising global convergence properties.

The performance of the various update methods cannot be expected to be identical. Neither can it be expected that a single method is superior in all cases since optimization in MCSCF is carried out both to minima and saddle points. The previous subsections have provided some information about the usefulness of the different updates.

In ground state calculations the DFP and BFGS updates are appealing since they usually guarantee retention of positive definite approximations to the Hessian. However, the weighting matrix $(G({}^{k+1}\underline{\lambda}))^{-1/2}$ used in obtaining the DFP variationally does not reflect an optimal weighting, since the relative errors in the Hessian approximations are not of the same magnitude. As described in Section V the smallest eigenvalues of the Hessian may change significantly between two points of iteration whereas larger eigenvalues are more stable. So if an exact Hessian from one point is used as the starting approximation to the Hessian at another point, large elements are approximated relatively better than small elements. This is in contrast with the assumption of using a weighting matrix $\mathbf{G}({}^{k+1}\underline{\lambda})^{-1/2}$. The updates based on minimization with the neutral norm $\|\cdot\|_I$ can thus not be entirely disregarded, although they may not retain positive definite matrices.

For optimization to saddle points the DFP and BFGS updates have no theoretical advantages, and the fact that $\mathbf{H}_{k+1}^{\mathrm{P}}$ and $\mathbf{H}_{k+1}^{\mathrm{B}}$ minimizes the neutral norm $\|\cdot\|_{\mathbf{I}}$ can be of importance. Further, the denominators in $\mathbf{H}_{k+1}^{\mathrm{P}}$ and $\mathbf{H}_{k+1}^{\mathrm{B}}$ [Eqs. (172) and (179)] are the square of the step length, which remains strictly positive. The denominators in $\mathbf{H}_{k+1}^{\mathrm{BFGS}}$ and $\mathbf{H}^{\mathrm{DFP}}$ [Eqs. (194) and (195)] contain $\mathbf{L}_k^T\mathbf{K}_k$ and $\mathbf{K}_k^T\mathbf{H}_k\mathbf{K}_k$. They can become near-singular. Therefore, it is expected that the Broyden rank-1 and its symmetrized form can be used with advantage when converging to a saddle point.

F. Numerical Results

To illustrate how update methods work in actual calculations we will now describe calculations carried out on the $B^3\Sigma_u^-$ state at 2.13 a.u. and on the $E^3\Sigma_u^-$ state at 2.13 and 2.10 a.u. To examine the methods for ground state calculations we initially consider the $B^3\Sigma_u^-$ calculation at 2.13 a.u.

In Table XXI we report the result of a Broyden rank-1 calculation. The initial Hessian for both the update and the fixed Hessian approaches is from after iteration 2 (at iteration point 2) of the Newton–Raphson calculation in Table III. Updates begin in the subsequent iteration. The update sequence converges, while a series of linear fixed Hessian iterations with the same initial guess of orbitals diverges. This indicates the importance of updating or recalculating the Hessian matrix during the iterative procedure. Thus, the region of convergence is extended by the update of the Hessian matrix.

To investigate the region of convergence in more detail we describe some update calculations that are initiated at previous steps of the Newton–Raphson series of iterations in Table III. Step size and sign control were carried out in these steps of the Newton–Raphson calculation. Step size and sign control were implemented in the update calculations in a way similar to that of the Newton–Raphson approach (see Section V). Step size and sign control is usually carried out on symmetric matrices. The Broyden rank-1 update gives an unsymmetric Hessian approximation, so the update calculations we initially report for obtaining global convergence will be using the Powell symmetrization of the Broyden rank-1 update (the BFGS and DFP update methods will be discussed later).

A Powell update calculation is reported in Table XXII, which uses an initial Hessian calculated after the first iteration (at iteration point 1) of Table III. The Powell sequence of iterations in Table XXII converges. In the initializing Newton–Raphson iteration a step length amplitude of 37.2 was constrained, and in the subsequent Powell iteration a step length amplitude was constrained. The Powell sequence of iterations fluctuates in the initial six iterations, and the total energy at iteration point 5 is farther away from the converged total energy than the total energy at iteration point 2. During the iterations where fluctuations in the total energy occur, the approximation to the Hessian matrix gradually improves. After iteration point 5 the step sizes are accurate enough to ensure rapid convergence. The gradual improvement of the Hessian approximation is reflected in a stabilization of the smallest eigenvalues of the Hessian approximation. To illustrate this point,

TABLE XXI

Fixed Hessian and Broyden Rank-1 Update Which Use an Initial **G** from Iteration Point 2 of Table III ($B^3\Sigma_u^-$ of O_2 at 2.13 a.u.)

Iteration point[b]	Broyden rank-1 update[a]			Fixed Hessian[a]		
	$E - E^{CONV c}$	$\|\mathbf{F}\|$	$\|^{n+1}\underline{\lambda} - {}^{n}\underline{\lambda}\| \sim \|^{n}\mathbf{e}\|$	$E - E^{CONV c}$	$\|\mathbf{F}\|$	$\|^{n+1}\underline{\lambda} - {}^{n}\underline{\lambda}\| \sim \|^{n}\mathbf{e}\|$
0	0.0055035052	3.62×10^{-2}	$3.35 \times 10^{-1\,d}$	0.005503502	3.62×10^{-2}	$3.35 \times 10^{-1\,d}$
1	0.0009142549	3.42×10^{-2}	1.73×10^{-1}	0.0009142549	3.42×10^{-2}	2.72×10^{-1}
2	0.0008941631	2.58×10^{-2}	7.96×10^{-2}	0.0031811111	3.54×10^{-2}	4.34×10^{-1}
3	0.0000767315	1.03×10^{-2}	2.96×10^{-2}	0.0055891388	7.96×10^{-2}	1.03
4	0.0000036257	1.57×10^{-3}	6.04×10^{-3}	0.0388891594	2.64×10^{-1}	2.33
5	0.0000000457	2.20×10^{-4}	8.71×10^{-4}	0.1468562383	1.56×10^{0}	4.42×10^{-1}
6	0.0000000064	5.88×10^{-5}	4.75×10^{-4}			
7	0.0000000000	4.41×10^{-6}	3.26×10^{-5}			

[a] The first iteration listed is the Newton–Raphson iteration. Updates begin at iteration point 1. The fixed Hessian approach uses **G** from iteration point 0.

[b] At iteration point n, the energy and **F** are evaluated and $^{n+1}\underline{\lambda}$ is determined. Thus at iteration point n, iteration $n+1$ is performed; $\|^{n+1}\underline{\lambda} - {}^{n}\underline{\lambda}\|$ is an approximation to $\|^{n}\mathbf{e}\| = \|^{n}\underline{\lambda} - \underline{\alpha}\|$ [see Eq. (71)].

[c] Here, $E - E^{CONV}$ is the difference in atomic units between the total energy at the present step and the converged total energy -149.3086149361 a.u.

[d] The step size and sign algorithm[17] is applied in this iteration.

TABLE XXII

Powell Update Which Uses an Initial **G** from Iteration Point 1 of Table III ($B^3\Sigma_u^-$ of O_2 at 2.13 a.u.).

Iteration point[a]	$E - E^{CONV}$[c]	$\|\mathbf{F}\|$	$\|{}^{n+1}\underline{\lambda} - {}^{n}\underline{\lambda}\| \sim \|{}^{n}\mathbf{e}\|$
0[b]	0.0246933274	2.18×10^{-1}	3.17×10^{-1}[d]
1	0.0055035052	3.62×10^{-2}	2.44×10^{-1}[d]
2	0.0006525723	2.40×10^{-2}	1.80×10^{-1}
3	0.0003031638	2.98×10^{-2}	4.10×10^{-2}
4	0.0000577782	7.78×10^{-3}	2.27×10^{-1}
5	0.0013706643	2.60×10^{-2}	2.04×10^{-1}
6	0.0000027061	1.84×10^{-3}	7.64×10^{-3}
7	0.0000004323	1.11×10^{-3}	1.96×10^{-3}
8	0.0000000258	1.35×10^{-4}	6.14×10^{-4}
9	0.0000000865	6.56×10^{-5}	3.52×10^{-4}
10	0.0000000008	2.42×10^{-5}	5.85×10^{-5}
11	0.0000000002	9.06×10^{-6}	3.96×10^{-5}
12	0.0000000000	4.00×10^{-6}	4.31×10^{-5}

[a]At iteration point n, the energy and **F** are evaluated and ${}^{n+1}\underline{\lambda}$ is determined. Thus at iteration point n, iteration $n+1$ is performed; $\|{}^{n+1}\underline{\lambda} - {}^{n}\underline{\lambda}\|$ is an approximation to $\|{}^{n}\mathbf{e}\| = \|{}^{n}\underline{\lambda} - \alpha\|$ [see Eq. (71)].

[b]Initializing constrained Newton–Raphson iteration. Updates start at iteration point 1.

[c]Here, $E - E^{CONV}$ is the difference in atomic units between the total energy at the present step and the converged total energy -149.3086149361 a.u.

[d]The step size and sign algorithm[17] is applied in this iteration.

in Table XXIII we report the development of the lowest four eigenvalues of the approximation Hessian matrix during the sequence of iterations of Table XXII. The four lowest eigenvalues of the exact Hessian matrix at the converged point are 0.0442, 0.1556, 0.1860, and 0.2232 (see Table V).

It may be seen in Table XXIII that the lowest eigenvalues fluctuate a little in the initial iterations. As a general trend, however, the smallest eigenvalues of the approximate Hessian matrix are rapidly improved during the iterative procedure, and close agreement is observed between the four lowest eigenvalues of the converged approximate Hessian matrix and the exact Hessian matrix. The gradual decrease in $E - E^{CONV}$, which is interrupted at iteration point 5, can be ascribed to the fact that the approximate Hessian matrix at iteration point 4 has the smallest eigenvalue (0.0082). In all previous and subsequent iterations the smallest eigenvalue of the approximate Hessian is 5–10 times as large as the smallest eigenvalue at iteration point 4. This small eigenvalue at iteration point 4 disappears after the Hessian update at iteration point 5 is carried out.

TABLE XXIII
Lowest Four Eigenvalues of the Hessian Update Matrix in the Powell Update Sequence of Iterations in Table XXII ($B^3\Sigma_u^-$ of O_2 at 2.13 a.u.)

Iteration point[a]	Lowest eigenvalues			
0	0.0010	0.1134	0.2878	0.3226
1	0.0240	0.0862	0.2860	0.3262
2	0.0506	0.1056	0.2860	0.3256
3	0.0448	0.1012	0.2848	0.3272
4	0.0082	0.1204	0.2776	0.3302
5	0.0522	0.1364	0.2754	0.3302
6	0.0548	0.1336	0.2756	0.3302
7	0.0382	0.1780	0.2751	0.3302
8	0.0578	0.1760	0.2710	0.3306
9	0.0434	0.1756	0.2560	0.3314
10	0.0544	0.1780	0.2500	0.3308
11	0.0884	0.1541	0.1798	0.3362
12	0.0390	0.1622	0.1798	0.3340
13	0.0434	0.1622	0.1812	0.3334

[a]At iteration point n, the energy and **F** are evaluated and ${}^{n+1}\underline{\lambda}$ is determined. Thus at iteration point n, iteration $n+1$ is performed.

The convergence characteristics of a Hessian update (as well as a Newton–Raphson) MCSCF calculation are very strongly affected by the modes corresponding to the small eigenvalues of the Hessian matrix. The small eigenvalues of the Hessian matrix get substantially improved when updates are carried out. The larger eigenvalues are not affected much by the update; for example, the lowest seven to nine eigenvalues at iteration point 0 are 0.7874, 1.0858, 1.8206, respectively, whereas at iteration point 13 they are 0.8022, 1.1780, and 1.8284. The corresponding converged Hessian eigenvalues (nonlinear transformation) are 0.9046, 1.0536, and 1.2128 (see Table V).

Thus the Powell update is able to converge from a region where it is necessary to restrict step sizes in both the initializing Newton–Raphson iteration and in the consecutive sequence of Powell update iterations. When the Powell sequence of iterations are started out at one previous iteration of the Newton–Raphson procedure (the initial **G** is from iteration point 0 of Table III, and updates start on the subsequent iteration), the Powell sequence of iterations is found not to converge. The initializing Hessian matrix is structurally incorrect at this step of the Newton–Raphson sequence, and in the current implementation the Powell sequence of iterations is not able to improve the structure of the Hessian matrix enough to make the Powell sequence of iterations converge. (See Ref. 21 for a more thorough discussion of this matter.)

The BFGS and DFP methods have also been applied, using as an initial **G** the Newton-Raphson calculation at iteration point 1 of Table III. The BFGS and DFP calculations are reported in Ref. 21 and have similar convergence characteristics as those obtained for the Powell calculation of Table XXII. The DFP and BFGS calculations converge in 10 and 12 iterations, respectively, whereas the Powell sequence of iterations converges in 12 iterations. Calculations which further investigate the global convergence characteristics of BFGS and DFP updates on the lowest state of a given symmetry are currently being carried out.[75]

After having discussed the region of convergence of update methods in ground state calculations, we now carry out a corresponding investigation for excited states. To do so, we describe Powell and Broyden rank-1 update calculations on the $E^3\Sigma_u^-$ state of O_2 at the internuclear distances of 2.13 and 2.10 a.u. The Newton–Raphson calculation at 2.13 a.u. is a very simple MCSCF calculation and requires only five Newton–Raphson iterations to converge (see Table X). The Newton–Raphson calculation at 2.10 a.u. is more complicated and requires eight iterations to converge (see Table IV). The difference in the complexity of the two calculations can be ascribed to the fact that the converged Hessian matrix at 2.13 a.u. has an eigenvalue at convergence of 6.0×10^{-2} (see Table XI) whereas the smallest Hessian eigenvalue (in magnitude) at convergence is 1.4×10^{-3} at 2.10 a.u. (see Table VIII).

In Table XXIV we report the Broyden rank-1 and the fixed Hessian calculation at 2.13 a.u. The fixed Hessian sequence of iterations diverges, while the Broyden rank-1 calculation converges. The Broyden rank-1 calculation converges in 11 iterations, while the Newton–Raphson iteration converges in 5 iterations. The convergence characteristics of the Broyden calculation of Table XXIV show great similarity with the Powell calculation of Table XXII. In the initial 6 Broyden iterations in Table XXIV the calculation does not show a definite convergence trend. At iteration point 3, $E - E^{\mathrm{CONV}}$ is 10 times as large as the corresponding value of iteration point 1. The energy difference $E - E^{\mathrm{CONV}}$ at iteration point 6 is also of the same magnitude as the value at iteration point 1. After iteration point 6, the Broyden calculation converges rapidly and monotonically. During the initial sequence of iterations the approximative Hessian matrix gains more and more similarity with the exact Hessian matrix. After iteration point 6, steps which bring the calculation to converge rapidly can be taken. Since the fixed Hessian calculation diverges, the region of convergence is thus extended by updating the Hessian matrix.

To further examine the extension in the region of convergence we now describe some calculations on the $E^3\Sigma_u^-$ state of O_2 at the internuclear distance of 2.10 a.u. The Newton–Raphson calculation is given in Table IV. Step sizes were controlled in the initial three Newton–Raphson iterations (see Table VIII). The Broyden rank-1 update calculations initiated using **G** at Newton-Raphson iteration points 4 and 3 of Table IV are reported in Tables XXV and XXVI. The Broyden rank-1 calculation that has been started out closest to convergence (Table XXV) converges rapidly and monotonically. When the Broyden calculations have been started farther from convergence (Table XXVI), $E - E^{\mathrm{CONV}}$ very slowly improves during the initial iterations and then converges rapidly and monotonically. A Powell update sequence has also been initiated using **G** from Newton–Raphson iteration point 2 of Table IV. Here, the Powell update calculation is not converging, and it appears that the region of convergence of the Powell updates are smaller for excited states than for ground state calculations—in particular when the converged Hessian matrix has very small eigenvalues.

In order to compare the region of convergence of the Broyden rank-1 update methods with the region of convergence of the DFP and BFGS update methods when applied on excited states, the calculations on the $E^3\Sigma_u^-$ state at 2.10 a.u. were repeated using BFGS and DFP updates. The Broyden rank-1 update calculation converges when the update is initialized using **G**

TABLE XXIV

Convergence Characteristics of the Broyden Rank-1 Update and Linear Fixed Hessian Calculations of the $E^3\Sigma_u^-$ State of O_2 at 2.13 a.u. Using an Initial **G** from Iteration Point 0 of Table X

Iteration	Broyden rank-1			Fixed Hessian		
point[a]	$E - E^{\mathrm{CONV}\,c}$	$\lVert\mathbf{F}\rVert$	$\lVert{}^{n+1}\underline{\lambda} - {}^{n}\underline{\lambda}\rVert \sim \lVert{}^{n}\underline{\mathbf{e}}\rVert$	$E - E^{\mathrm{CONV}\,c}$	$\lVert\mathbf{F}\rVert$	$\lVert{}^{n+1}\underline{\lambda} - {}^{n}\underline{\lambda}\rVert \sim \lVert{}^{n}\underline{\mathbf{e}}\rVert$
0[b]	0.0590242739	9.72×10^{-1}	4.62×10^{-1}[b]	0.0590242739	9.72×10^{-2}	4.62×10^{-1}[d]
1	−0.0003831946	8.36×10^{-2}	1.37×10^{-1}	−0.0003831946	8.36×10^{-2}	1.45×10^{-1}
2	−0.0002434388	4.66×10^{-2}	2.92×10^{-1}	0.0000353020	5.16×10^{-2}	3.50×10^{-1}
3	0.0038464235	1.33×10^{-1}	2.18×10^{-1}	0.0068589871	1.90×10^{-1}	1.34
4	0.0002185597	5.06×10^{-2}	1.41×10^{-1}	0.0522554397	4.08×10^{-1}	1.34
5	0.0001156611	2.04×10^{-2}	1.97×10^{-1}	0.0400059992	5.10×10^{-1}	3.64
6	0.0001917947	3.04×10^{-2}	1.29×10^{-1}	0.2462884409	1.57×10^{0}	5.32
7	0.0000104537	4.62×10^{-2}	1.02×10^{-2}			
8	0.0000025132	2.96×10^{-2}	8.69×10^{-3}			
9	0.0000014093	1.95×10^{-3}	4.26×10^{-3}			
10	0.0000000010	8.38×10^{-5}	2.41×10^{-4}			
11	0.0000000000	4.64×10^{-5}	9.19×10^{-5}			

[a]At iteration point n, the energy and **F** are evaluated and ${}^{n+1}\underline{\lambda}$ is determined. Thus at iteration point n, iteration $n+1$ is performed. $\lVert{}^{n+1}\underline{\lambda} - {}^{n}\underline{\lambda}\rVert$ is an approximation to $\lVert{}^{n}\mathbf{e}\rVert = \lVert{}^{n}\underline{\lambda} - \underline{\alpha}\rVert$ [see Eq. (71)].

[b]Initializing constrained Newton–Raphson iteration. Updates start at iteration point 1.

[c]Here, $E - E^{\mathrm{CONV}}$ is the difference in atomic units between the total energy at this step and the converged total energy −149.2533729769 a.u.

[d]The step size and sign control algorithm[17] is applied in this iteration.

TABLE XXV

Broyden Rank-1 Update After Newton–Raphson Iteration Point 4 of Table IV ($E^3\Sigma_u^-$ of O_2 at 2.10 a.u.): Initial **G** is from Iteration Point 4 of Table IV

Iteration point[a]	$E - E^{\text{CONV}}$ [c]	$\|\mathbf{F}\|$	$\|{}^{n+1}\underline{\lambda} - {}^{n}\underline{\lambda}\| \sim \|{}^{n}\mathbf{e}\|$
0[b]	0.0000311601	2.06×10^{-3}	2.66×10^{-1}
1	0.0000073514	4.34×10^{-3}	1.06×10^{-1}
2	0.0000010598	8.14×10^{-4}	2.79×10^{-2}
3	0.0000000067	2.82×10^{-4}	5.76×10^{-3}
4	0.0000000005	1.17×10^{-4}	1.98×10^{-3}
5	0.0000000000	1.97×10^{-5}	2.79×10^{-4}

[a]At iteration point n, the energy and **F** are evaluated and ${}^{n+1}\underline{\lambda}$ is determined. Thus at iteration point n, iteration $n+1$ is performed; $\|{}^{n+1}\underline{\lambda} - {}^{n}\underline{\lambda}\|$ is an approximation to $\|{}^{n}\mathbf{e}\| = \|{}^{n}\underline{\lambda} - \underline{\alpha}\|$ [see Eq. (71)].

[b]Initializing Newton–Raphson iteration. Updates start at iteration point 1.

[c]Here, $E - E^{\text{CONV}}$ is the difference in atomic units between the total energy at this step and the converged total energy -149.2781477108 a.u.

TABLE XXVI

Broyden Rank-1 Update After Newton–Raphson Iteration Point 3 of Table IV ($E^3\Sigma_u^-$ of O_2 at 2.10 a.u.): Initial **G** is from Iteration Point 3 of Table IV

Iteration point[a]	$E - E^{\text{CONV}}$ [c]	$\|\mathbf{F}\|$	$\|{}^{n+1}\underline{\lambda} - {}^{n}\underline{\lambda}\| \sim \|{}^{n}\mathbf{e}\|$
0[b]	0.0000947284	1.91×10^{-1}	2.76×10^{-1}
1	0.0000311601	2.06×10^{-2}	2.84×10^{-1}
2	0.0000141971	1.82×10^{-2}	1.98×10^{-1}
3	0.0000034206	2.76×10^{-3}	5.55×10^{-2}
4	0.0000002208	3.60×10^{-4}	9.39×10^{-3}
5	0.0000000165	2.56×10^{-4}	4.47×10^{-3}
6	0.0000000084	9.42×10^{-5}	3.45×10^{-3}

[a]At iteration point n, the energy and **F** are evaluated and ${}^{n+1}\underline{\lambda}$ is determined. Thus at iteration point n, iteration $n+1$ is performed; $\|{}^{n+1}\underline{\lambda} - {}^{n}\underline{\lambda}\|$ is an approximation to $\|{}^{n}\mathbf{e}\| = \|{}^{n}\underline{\lambda} - \underline{\alpha}\|$ [see Eq. (71)].

[b]Initializing Newton–Raphson iteration. Updates start at iteration point 1.

[c]Here, $E - E^{\text{CONV}}$ is the difference in atomic units between the total energy at this step and the converged total energy -149.2781477108 a.u.

TABLE XXVII
BFGS and DFP Update After Newton–Raphson Iteration Point 4 of Table IV ($E^3\Sigma_u^-$ of O_2 at 2.10 a.u.): Initial **G** is from Iteration Point 4 of Table IV

Iteration	BFGS			DFP		
point[a]	$E - E^{\text{CONV}\,c}$	$\|\mathbf{F}\|$	$\|{}^{n+1}\underline{\lambda} - {}^{n}\underline{\lambda}\| \sim \|{}^{n}\mathbf{e}\|$	$E - E^{\text{CONV}\,c}$	$\|\mathbf{F}\|$	$\|{}^{n+1}\underline{\lambda} - {}^{n}\underline{\lambda}\| \sim \|{}^{n}\mathbf{e}\|$
0[b]	0.0000311601	2.06×10^{-3}	2.66×10^{-1}	0.0000311601	2.06×10^{-3}	2.66×10^{-1}
1	0.0000073514	4.34×10^{-3}	9.62×10^{-2}	0.0000073514	4.34×10^{-3}	9.61×10^{-2}
2	0.0000004867	7.34×10^{-4}	2.06×10^{-2}	0.0000004674	6.30×10^{-4}	2.00×10^{-2}
3	0.0000000085	4.12×10^{-4}	1.31×10^{-2}	0.0000000006	3.82×10^{-4}	1.93×10^{-2}
4	0.0000000651	7.34×10^{-4}	9.62×10^{-3}	0.0000001973	1.26×10^{-3}	1.69×10^{-2}
5	0.0000000017	3.50×10^{-5}	8.22×10^{-4}	0.0000000000	5.72×10^{-6}	2.87×10^{-4}
6	0.0000000002	1.88×10^{-5}	5.58×10^{-4}			
7	0.0000000000	3.16×10^{-6}	6.36×10^{-5}			

[a]At iteration point n, the energy and **F** are evaluated and ${}^{n+1}\underline{\lambda}$ is determined. Thus at iteration point n, iteration $n+1$ is performed; $\|{}^{n+1}\underline{\lambda} - {}^{n}\underline{\lambda}\|$ is an approximation to $\|{}^{n}\mathbf{e}\| = \|{}^{n}\underline{\lambda} - \underline{\alpha}\|$ [see Eq. (71)].
[b]Initializing Newton–Raphson iteration. Updates start at iteration point 1.
[c]Here, $E - E^{\text{CONV}}$ is the difference in atomic units between the total energy at the present step and the converged total energy -149.2781477108 a.u.

from the Newton–Raphson iteration points 4 or 3 (Tables XXV and XXVI) of the Newton–Raphson calculation of Table IV. In Table XXVII we report the DFP and BFGS update calculations initiated using **G** from iteration point 4 of Table IV. The Broyden rank-1 (Table XXV) and the DFP calculations converge in five iterations, whereas the BFGS calculation uses seven iterations to converge. All three update calculations converge monotonically and with about the same convergence characteristics. In Table XXVIII we report the DFP and BFGS update calculations initiated after Newton–Raphson iteration point 3 of Table IV. Both the DFP and BFGS calculations diverge. The Broyden rank-1 calculation was reported in Table XXVI and converges in seven iterations. The Broyden rank-1 update approach thus has a larger region of convergence in these excited state calculations than do the DFP and BFGS update approaches.

To understand why Broyden rank-1 update converges when the DFP and BFGS methods diverge, we analyze the Hessian updates that result in the three update methods. In the Broyden rank-1 update calculation in Table XXVI Hessian update corrections are in general very small. In the initial Broyden rank-1 update iteration the largest change in an approximate Hessian element is 3.16×10^{-3}, corresponding to a Hessian element that is changed from -1.78×10^{-2} to -1.46×10^{-2}. The exact Hessian element (when a nonlinear transformation of variables is carried out) is -1.16×10^{-2}. The Broyden rank-1 update thus significantly improves the agreement be-

TABLE XXVIII

BFGS and DFP Updates After Newton–Raphson Iteration Point 3 of Table IV ($E^3\Sigma_u^-$ of O_2 at 2.10 a.u.): Initial **G** is from Iteration Point 3 of Table IV

Iteration	BFGS			DFP		
point[a]	$E - E^{\mathrm{CONV}\,c}$	$\|\mathbf{F}\|$	$\|{}^{n+1}\underline{\lambda} - {}^{n}\underline{\lambda}\| \sim \|{}^{n}\mathbf{e}\|$	$E - E^{\mathrm{CONV}\,c}$	$\|\mathbf{F}\|$	$\|{}^{n+1}\underline{\lambda} - {}^{n}\underline{\lambda}\| \sim \|{}^{n}\mathbf{e}\|$
0[b]	0.0000947284	1.91×10^{-1}	2.76×10^{-1}	0.0000947284	1.91×10^{-1}	2.76×10^{-1}
1	0.0000311601	2.06×10^{-2}	2.85×10^{-1}[d]	0.0000311601	2.06×10^{-2}	2.85×10^{-1}[d]
2	0.0000993762	2.78×10^{-2}	2.04×10^{-1}[d]	0.0000934239	2.78×10^{-2}	2.02×10^{-1}[d]
3	0.0000140309	4.22×10^{-2}	2.16×10^{-1}[d]	0.0000407694	9.44×10^{-3}	2.00×10^{-1}[d]
4	0.0002170358	6.20×10^{-2}	2.40×10^{-1}[d]	0.0000902366	2.72×10^{-2}	2.01×10^{-1}[d]
5	0.0001324723	6.60×10^{-2}	3.30×10^{-1}[d]	0.0000858022	4.34×10^{-2}	1.95×10^{-1}
6	0.0000614033	8.96×10^{-2}	2.86×10^{-1}[d]	0.0000964965	2.82×10^{-2}	3.64×10^{-2}
7	0.0017914279	1.08×10^{-1}	3.51×10^{-1}[d]	0.0000880822	2.64×10^{-2}	2.00×10^{-1}[d]
8	0.0024049328	1.28×10^{-1}	5.79×10^{-1}[d]			

[a]At iteration point n, the energy and **F** are evaluated and ${}^{n+1}\underline{\lambda}$ is determined. Thus at iteration point n, iteration $n+1$ is performed; $\|{}^{n+1}\underline{\lambda} - {}^{n}\underline{\lambda}\|$ is an approximation to $\|{}^{n}\mathbf{e}\| = \|{}^{n}\underline{\lambda} - \underline{\alpha}\|$ [see Eq. (71)].

[b]Initializing Newton–Raphson iteration. Updates start at iteration point 1.

[c]Here, $E - E^{\mathrm{CONV}}$ is the difference in atomic units between the total energy at the present step and the converged total energy -149.2781477108 a.u.

[d]The step size and sign control algorithm[17] is applied in this iteration.

tween the approximate Hessian matrix and the exact Hessian matrix. The Hessian updates can be justified further through the information about the exact Hessian that is contained in the quasi-Newton condition. The approximate Hessian element changes from -1.78×10^{-2} to -1.48×10^{-2} corresponds to a change of a variable ($\underline{\lambda}$ element change) of 0.23. The aforementioned trends are found in all the consecutive iterations. The success of the Broyden rank-1 update method can thus be ascribed to the fact that relative small changes occur in the Hessian updates and that these small changes are directly related to information about the exact Hessian matrix that is available in the quasi-Newton condition.

In the first BFGS iterations of Table XXVIII (the iterations using an initial **G**(0) of Newton–Raphson iteration point 3 of Table IV) very large changes occur in the approximate Hessian. For example a diagonal element of the Hessian matrix is changed from 2.08 to -34.8. The corresponding exact Hessian element is 2.50. Fourteen elements are changed by more than 2.0. These changes cannot at all be justified on the basis of information contained in the quasi-Newton condition. The aforementioned change of 36.8 is related to a change in a variable (${}^{n}\underline{\lambda}$ element change) of 0.005. The large changes in the approximate Hessian can be traced back to very small denominators in the BFGS updates [see Eq. (194)]. The denominators $\mathbf{F}^T\mathbf{K}$ and

$\mathbf{K}^T\mathbf{HK}$ of the BFGS update are -3.8×10^{-6} and 3.5×10^{-5}, respectively, in the initial update iteration of Table XXVIII. The wild changes of the Hessian approximation result in a undesired negative Hessian eigenvalue of -48. The same behavior continues during the update sequence of iterations, and the Hessian approximation fast loses any connection to the exact Hessian matrix. Therefore the BFGS update calculation diverges.

In the DFP update sequence of iterations we expect a behavior analogous to the one observed in the BFGS update calculation, since the denominators of a DFP update [see Eq. (195)] are the same as the denominators of a BFSG update. As an example of this behavior we note that in the first DFP update a diagonal element of the approximate Hessian is changed from 2.08 to 354, where the exact Hessian element is 2.50. We thus conclude that when updates are applied on excited states, the Broyden rank-1 and the Powell updates have larger regions of convergence than BFGS and DFP updates. This conclusion is contrary to the experience of numerical analysis. We should point out here that the experience of numerical analysis is obtained basically in connection with function minimization. The foregoing conclusions are obtained when optimizations are carried out to saddle points, and our conclusions therefore may differ from those in numerical analysis.

An investigation of the efficiency of the update methods can be obtained by comparing Broyden rank-1 and Powell update calculations with Newton–Raphson calculations. We consider the calculations on the $B^3\Sigma_u^-$ state at 2.13 a.u. In Tables XXI and XXII the Broyden rank-1 and Powell update calculations are reported initiated after Newton–Raphson iteration points 2 and 1, respectively, of Table III. Using the energy difference $E - E^{\mathrm{CONV}}$ as a measure of the distance to the desired stationary point, in Table XXIX we have tabulated the number of update iterations of Tables XXI and XXII that are equivalent to a given number of Newton–Raphson iterations of Table III. The actual numbers in Table XXIX have been determined by requiring that the update give an energy approximation which is at least as good as the one of the equivalent number of Newton–Raphson iterations. When the update calculation is started out relatively early (after Newton–Raphson iteration point 1 of Table III), the update calculation fluctuates in the initial update iterations and a large number of update iterations (five) is required to reach a point that is at least as good as the one obtained by carrying out one additional Newton–Raphson iteration. After the fluctuations in the initial update iterations, the update calculation converges rapidly and monotonically and some two or three update iterations are in general found to simulate one Newton–Raphson iteration.

If the update calculation is started out after Newton–Raphson iteration point 2 (initial $\mathbf{G}$ from iteration point 2), the update calculation converges rapidly and monotonically and some two or three update iterations simulate

TABLE XXIX
The Number of Powell Update Iterations That Is Equivalent to a Given Number of Newton–Raphson Iterations ($B^3\Sigma_u^-$ of O_2 at 2.13 a.u.)

Powell update starts after Newton–Raphson iteration point 1 of Table III		Powell update starts after Newton–Raphson iteration point 2 of Table III	
Number of Newton–Raphson iterations	Equivalent number of Powell updates	Number of Newton–Raphson iterations	Equivalent number of Powell updates
1	5	1	3
2	5	2	4
3	7	3	6
4	11		

one Newton–Raphson iteration. A similar analysis may be performed on the $E^3\Sigma_u^-$ calculation at 2.10 a.u. for the Broyden rank-1 update calculations (Table XXVI) initiated after Newton–Raphson iteration point 3 of Table IV. This calculation gives a result similar to that of the $B^3\Sigma_u^-$ calculation. If the update is started close to convergence, the update results indicate that the number of update iterations required to converge is about twice the number required in the corresponding Newton–Raphson sequence. If the update is started earlier and fluctuations occur in the initial update iterations, the ratio may be as large as 3.

The computational work involved in carrying out an update iteration is very similar to the work involved in a fixed Hessian iteration. The convergence characteristics of fixed Hessian calculations have previously been described (Section VI). In the local region the Hessian updates do not significantly improve the convergence rate in the initial few iterations compared to a fixed Hessian approach. However, after these few iterations, Hessian updates become extremely important for obtaining efficient and rapid convergence. It is also of importance to note that the region of convergence of Hessian update methods is larger than the region of convergence of a fixed Hessian approach. Hessian update methods therefore generally can be used much more reliably and efficiently than fixed Hessian approaches to obtain local convergence of an MCSCF calculation. The Hessian update methods even have some promising global convergence properties.

G. Other Applications of Update Methods

Update methods have only recently been introduced in wavefunction optimization.[21, 73] We anticipate that they will also ease the optimization of many types of wavefunctions besides MCSCF wavefunctions.

The methods can be modified so that one part of the Hessian is calculated exactly and other parts are approximated through update procedures. For example, this variant can be used to approximate coupling elements (the block $\mathbf{A}^{CO} - \mathbf{B}^{CO}$) while the other parts of the Hessian are calculated exactly. Another conceivable use is to calculate exactly the part of the Hessian that can be constructed from integrals with one free index, while approximating other parts of the Hessian. Investigations of the usefulness of new update procedures[76, 77] of, say, the Oren–Lauenberg[77] type may also be of interest.

VIII. CUBIC CONTRIBUTIONS IN MCSCF OPTIMIZATION

A. Theory

When the energy expansion in Eq. (53) in truncated after terms which are quadratic in $\underline{\lambda}$, the nonlinear Newton–Raphson iterative function [Eq. (60)] is straightforwardly derived. If a cubic expansion of the energy function is considered, the cubic iterative function is determined

$$F_i + G_{ij}\lambda_j + \tfrac{1}{2}K_{ijk}\lambda_j\lambda_k = 0 \tag{208}$$

When the Hessian matrix contains small eigenvalues, a serious slowdown of the global convergence of a Newton–Raphson calculation may be observed[9, 15, 17, 18, 22] (see Section V). Such a deficiency may not show up when an iterative cubic function is used. The cubic term may dominate the modes that cause problems in the Newton–Raphson approach. To clarify this point, we transform Eq. (208) to the basis where the Hessian matrix is diagonal [Eq. (114)]. We then obtain

$$\bar{F}_i + \varepsilon_{ij}\bar{\lambda}_j + \tfrac{1}{2}\bar{C}_{ij}\bar{\lambda}_j = 0 \tag{209}$$

where

$$\bar{\mathbf{C}} = \mathbf{U}^{+}\mathbf{K}\underline{\lambda}\mathbf{U} \tag{210}$$

and hence

$$\bar{\lambda}_i = -(\boldsymbol{\varepsilon} + \tfrac{1}{2}\bar{\mathbf{C}})_{ij}^{-1}\bar{F}_j \tag{211}$$

Equations (115) and (211) have the same structure and differ formally only in the inverse matrix where the $\boldsymbol{\varepsilon}$ matrix of Eq. (115) is changed to $\boldsymbol{\varepsilon} + \frac{1}{2}\bar{\mathbf{C}}$. Peculiarities in the convergence behavior that occur in the Newton–Raphson method due to small Hessian eigenvalues (e.g., near an inflection point) may be of less importance when Eq. (211) is used as an iterative function, since the term $\frac{1}{2}\bar{\mathbf{C}}$ is often of such magnitude that it will dominate the very small $\boldsymbol{\varepsilon}$ elements. The modes of the Newton–Raphson approach that are dominated by the small eigenvalues of $\mathbf{G}$ may therefore be expected to be controlled by the term $\frac{1}{2}\overline{\mathbf{C}}$ in Eq. (211), and the large step length amplitudes should be much less troublesome with the use of an iterative cubic MCSCF.

A solution to Eq. (208) may only be determined in terms of an iterative procedure, the solution of which then will satisfy Eq. (208). Each determination of the step length vector $\underline{\lambda}$ which solves Eq. (208) is called a *macro-iteration*. When the cubic term dominates certain modes, a solution to Eq. (208) cannot be determined perturbatively.

We demonstrate now how a Newton–Raphson technique may be used to solve Eq. (208). We define a vector function

$$f_i(\underline{\lambda}) = F_i + G_{ij}\lambda_j + \tfrac{1}{2}K_{ijk}\lambda_j\lambda_k \tag{212}$$

and attempt to determine iteratively a set of $\underline{\lambda}$ parameters such that the vector $\mathbf{f}(\underline{\lambda})$ becomes equal to zero. Carrying out a Taylor expansion of $\mathbf{f}(\underline{\lambda})$ about the point ${}^0\underline{\lambda}$ gives

$$f_i(\underline{\lambda}) = f_i({}^0\underline{\lambda}) + f'_{ij}({}^0\underline{\lambda})(\lambda_j - {}^0\lambda_j) + \cdots \tag{213}$$

The first derivative matrix is determined from Eq. (212) to be

$$f'_{ij}(\underline{\lambda}) = G_{ij} + K_{ijk}\lambda_k \tag{214}$$

When a Newton–Raphson technique is used to obtain a solution to Eq. (208), only the linear terms in the Taylor expansion are kept and $\mathbf{f}(\underline{\lambda})$ is set equal to zero:

$$f_i({}^0\underline{\lambda}) + f'_{ij}({}^0\underline{\lambda})(\lambda_j - {}^0\lambda_j) = 0 \tag{215}$$

Thus the Newton–Raphson step length formally becomes[20]

$$\Delta^0\lambda_j = \lambda_j - {}^0\lambda_j = -f'({}^0\underline{\lambda})^{-1}_{ij} f_j({}^0\underline{\lambda}) \tag{216}$$

The step length correction to ${}^0\underline{\lambda}$ is thus obtained by solving Eq. (216). An updated set of $\underline{\lambda}$ parameters is determined as

$${}^1\underline{\lambda} = {}^0\underline{\lambda} + \Delta^0\underline{\lambda} \tag{217}$$

These ${}^1\underline{\lambda}$ parameters can then be used for the next application of Eq. (216) [i.e., ${}^1\underline{\lambda}$ replaces ${}^0\underline{\lambda}$ in Eqs. (212) and (214)], and the iterative sequence is continued until the correction term $\Delta\underline{\lambda}$ is smaller than a certain tolerance.

Each single iteration of such an iterative procedure [Eqs. (216) and (217)] will be referred to as a *micro-iteration*. A converged series of micro-iterations thus gives a set of $\underline{\lambda}$ parameters that satisfies Eq. (208). When a Newton–Raphson sequence of micro-iterations is performed, the step length corrections of a micro-iteration will, of course, show second-order convergence.

When the cubic iterative function of Eq. (208) is applied far from convergence, the cubic term cannot usually be considered as a small correction to the second-order function for the modes corresponding to the small eigenvalues of the Hessian matrix. If a Newton–Raphson (unconstrained) MCSCF step [Eq. (60)] is used as initial guess for a series of micro-iterations [Eqs. (216) and (217)], such a series of iterations may diverge. To determine a solution of the cubic iterative function in Eq. (208) when far from convergence, we have used the Newton–Raphson MCSCF step size and sign controlled rotational parameters[17] (see Section V) as an initial guess of the series of micro-iterations and have obtained convergence in the sequence of micro-iterations in few iterations. However, in the calculations we discuss in the next subsection (VIII.B), the *final* $\underline{\lambda}$ from the series of micro-iterations for a certain macro-iteration is unconstrained even though the initial guess is the constrained Newton–Raphson $\underline{\lambda}$.

We expect that the iterative cubic approach is particularly effective and reliable when far from convergence (the global convergence problem) since one more term is included in the expansion of the energy [Eq. (208)]. This is, in fact, borne out by the results to be presented in Section VIII.B. Hence, often the iterative cubic approach does not require that constraints are applied far from convergence. In that region one iterative cubic iteration can replace two to five constrained Newton–Raphson iterations. In addition, we show that iterative cubic step length amplitudes are often mimicked fairly well by the mode-controlled Newton–Raphson values.[17] The constraint values for both mode damping[15] and mode controlling[17] have been determined by a somewhat "trial-and-error" process (often euphemistically called "numerical experience"), and this agreement is very encouraging.[20]

When the cubic iterative functions of Eq. (208) are used in the local region, a solution to Eq. (208) may be determined perturbatively. Such a procedure is easily described by rearranging Eq. (208) as

$$\lambda_i = -G_{ij}^{-1}F_j - \tfrac{1}{2}G_{il}^{-1}K_{ljk}\lambda_j\lambda_k \tag{218}$$

A solution to Eq. (218) is then obtained by carrying out a sequence of iterations in which we initially set $\underline{\lambda}$ equal to zero on the right-hand side of Eq. (218). We then obtain the set of $\underline{\lambda}$ values of a Newton–Raphson MCSCF iteration [Eq. (60)]. The Newton–Raphson set of parameters are then inserted back into the right-hand side of Eq. (218), and the Chebyshev formula of Eq. (137) is determined. The iterative procedure may be continued until a self-consistent set of $\underline{\lambda}$ parameters are determined, which then constitutes a solution to Eq. (208). We should point out that this sequence of iterations will only converge linearly.

To understand the local convergence characteristics of the cubic iterative function we will now derive the error term of the cubic iterative function. To do so we insert the Chebyshev formula into the right-hand side of Eq. (218), keeping terms through third order in the Newton–Raphson step length:

$$\lambda_i = {}^{\mathrm{NR}}\lambda_i - \tfrac{1}{2} G_{ij}^{-1} K_{jkl}\, {}^{\mathrm{NR}}\lambda_k\, {}^{\mathrm{NR}}\lambda_l + \tfrac{1}{2} G_{ij}^{-1} K_{jkl}\, {}^{\mathrm{NR}}\lambda_k G_{lp}^{-1} K_{pqr}\, {}^{\mathrm{NR}}\lambda_q\, {}^{\mathrm{NR}}\lambda_r + O\left(({}^{\mathrm{NR}}\lambda)^4\right) \tag{219}$$

where ${}^{\mathrm{NR}}\underline{\lambda}$ is the Newton–Raphson step length of Eq. (60). The error term of one Newton–Raphson iteration has previously been derived [see, e.g., Eq. (31)]. To understand the structure of the error term of the cubic iterative function it is necessary to determine the error term of the Newton–Raphson iteration to one higher order and therefore carry out the expansions in Eqs. (27) and (28) to one higher order in

$${}^{0}e = {}^{0}\lambda - \alpha \tag{220}$$

(the error vector for the initial guess of orbitals and states for the iterative cubic iteration). We obtain

$${}^{\mathrm{NR}}\lambda_i = \alpha_i + \tfrac{1}{2} G_{ij}^{-1}(\underline{\alpha}) K_{jmp}(\underline{\alpha})\, {}^{0}e_m\, {}^{0}e_p + \tfrac{1}{3} G_{ij}^{-1}(\underline{\alpha}) M_{jmpq}(\underline{\alpha})\, {}^{0}e_m\, {}^{0}e_p\, {}^{0}e_q - \tfrac{1}{2} G_{ij}^{-1}(\underline{\alpha}) K_{jrm}(\underline{\alpha})\, {}^{0}e_m G_{rk}^{-1}(\underline{\alpha}) K_{kpq}(\underline{\alpha})\, {}^{0}e_p\, {}^{0}e_q + O\left({}^{0}e^4\right) \tag{221}$$

The $\mathbf{K}$ supermatrix may also be expanded around the stationary point $\underline{\alpha}$:

$$K_{ijk}({}^{0}\underline{\lambda}) = K_{ijk}(\underline{\alpha}) + M_{ijkl}(\underline{\alpha})\left[{}^{0}\lambda_l - \alpha_l\right] + \cdots \tag{222}$$

Hence

$$-\tfrac{1}{2}\left(G({}^{0}\underline{\lambda})\right)_{ij}^{-1} K_{jkl}({}^{0}\underline{\lambda}) = -\tfrac{1}{2} G_{ij}^{-1}(\underline{\alpha}) K_{jkl}(\underline{\alpha}) - \tfrac{1}{2} G_{ij}^{-1}(\underline{\alpha}) M_{ijkl}\, {}^{0}e_l + \tfrac{1}{2} G_{ij}^{-1}(\underline{\alpha}) K_{jmp}(\underline{\alpha})\, {}^{0}e_m G_{pq}^{-1}(\underline{\alpha}) K_{qkl}(\underline{\alpha}) + \cdots \tag{223}$$

Inserting Eqs. (221) and (223) into Eq. (219) and retaining only terms through third order in ${}^0\mathbf{e}$ gives the error term of one iteration of the cubic iterative function:

$$e_i^{IC} = \lambda_i - \alpha_i = -\tfrac{1}{6} G_{ij}^{-1}(\underline{\alpha}) M_{jmpq}(\underline{\alpha})\, {}^0e_m\, {}^0e_p\, {}^0e_q \tag{224}$$

Equation (224) explicitly demonstrates third-order convergence of the iterative cubic MCSCF procedure.

We will now show that if an iterative procedure is used where only the initial Newton–Raphson and one additional micro-iteration of Eq. (216) are performed, then the errors that are introduced by not converging the sequence of micro-iterations will be of fourth order in the error vector ${}^0\mathbf{e}$. The set of rotational parameters obtained when a Newton–Raphson and a micro-iteration of Eq. (216) is carried out becomes

$${}^1\lambda_i = {}^{NR}\lambda_i - (f'({}^{NR}\underline{\lambda}))_{ij} f({}^{NR}\underline{\lambda})_j \tag{225}$$

where

$${}^{NR}\underline{\lambda} = -\mathbf{G}^{-1}\mathbf{F} \tag{226}$$

Equation (213) and (214) may be inserted into Eq. (225)

$${}^1\lambda_i = {}^{NR}\lambda_i - \left(G_{pq} + K_{pqr}\, {}^{NR}\lambda_r\right)_{ij}^{-1} \left(F_j + G_{jk}\, {}^{NR}\lambda_k + \tfrac{1}{2} K_{jkl}\, {}^{NR}\lambda_k\, {}^{NR}\lambda_l\right) \tag{227}$$

and expansion of the inverse matrix in Eq. (227) gives, when terms are kept through third order in ${}^{NR}\underline{\lambda}$,

$$\begin{aligned} {}^1\lambda_i = {}^{NR}\lambda_i &- \tfrac{1}{2} G_{ij}^{-1} K_{jkl}\, {}^{NR}\lambda_k\, {}^{NR}\lambda_l \\ &+ \tfrac{1}{2} G_{ij}^{-1} K_{jkl}\, {}^{NR}\lambda_k G_{lp}^{-1} K_{pqr}\, {}^{NR}\lambda_p\, {}^{NR}\lambda_r + O\left(({}^{NR}\lambda)^4\right) \end{aligned} \tag{228}$$

Subtracting Eq. (228) from Eq. (219) (the iterative function of the cubic iterative procedure through third order in ${}^{NR}\underline{\lambda}$) and denoting $\underline{\lambda}$ of Eq. (219) ${}^{IC}\lambda$ then gives

$${}^{IC}\underline{\lambda} - {}^1\underline{\lambda} = O\left(({}^{NR}\lambda)^4\right) \tag{229}$$

which shows that the errors that are introduced by not carrying out the sequence of micro-iterations to convergence are of fourth order in the error

The error vector of one iteration of the perturbative (Chebyshev) cubic MCSCF technique may be obtained by only keeping the first two terms in Eq. (219). Denoting the error term of the Chebyshev formula $\mathbf{e}^{CB}$ we obtain
The error vector of one iteration of the perturbative (Chebyshev) cubic MCSCF technique may be obtained by only keeping the first two terms in Eq. (219). Denoting the error term of the Chebyshev formula $\mathbf{e}^{CB}$ we obtain

$$e_i^{CB} = e_i^{IC} + \tfrac{1}{2} G_{ij}^{-1}(\underline{\alpha}) K_{jkl}(\underline{\alpha})^0 e_k G_{lp}^{-1}(\underline{\alpha}) K_{pqr}(\underline{\alpha})^0 e_q {}^0 e_r \tag{230}$$

A cubic calculation may also be carried out using two steps of a fixed Hessian approach. Subsequently, a new **G** is constructed, two fixed Hessian steps performed, and so on. We will refer to such a calculation as a recursive cubic calculation. In Eq. (143) we demonstrated that the error term of a recursive cubic iteration is

$$e_i^{RC} = \tfrac{1}{2} L_{ijk} {}^0 e_k L_{jpq} {}^0 e_p {}^0 e_q \tag{231}$$

$$= \tfrac{1}{2} G_{ij}^{-1}(\underline{\alpha}) K_{jk1}(\underline{\alpha})^0 e_k G_{1p}^{-1}(\underline{\alpha}) K_{pqr}(\underline{\alpha})^0 e_p {}^0 e_r \tag{232}$$

We note that

$$e_i^{CB} = e_i^{IC} + e_i^{RC} \tag{233}$$

and that, hence, if either the iterative cubic or the recursive cubic procedures diverge locally, it is expected that the perturbative cubic (Chebyshev) will in general also diverge. In Section VI.C we demonstrated that the radius of convergence of a recursive cubic calculation is generally smaller than the radius of convergence for a Newton–Raphson calculation. Both the Chebyshev and the recursive cubic approaches therefore are useful only in the local region of an MCSCF calculation. In Table XXX we show the total

TABLE XXX
Total Order of Convergence for a Sequence of Second- and Third-Order Iterations

Iteration	Second-order MCSCF	Third-order MCSCF
1	2	3
2	4	9
3	8	27
4	16	81
5	32	243

local order of convergence of a cubic procedure versus a second-order procedure.

The only previous analysis of cubic contributions was by Yaffe and Goddard.[24] However, in their analysis, they did not include coupling with the CI states. Coupling terms enter in second order (the NR equation) and are particularly important when converging to excited states.[13-15] Calculations that neglect coupling terms therefore cannot be expected to demonstrate either third-order or second-order convergence in general. The analysis by Yaffe and Goddard was primarily for the Chebyshev procedure. Hence, applicability without additional constraints is limited to the local region. Yaffe and Goddard performed SCF and GVB calculations using the Chebyshev procedure.

In Appendix E we discuss the explicit construction of the cubic contributions for both the iterative and Chebyshev cubic procedures. By explicitly constructing third-derivative terms multiplied by a vector, efficient procedures have been developed.[16] In addition, large amounts of computer storage are not required since a matrix (and not a supermatrix) is calculated and stored. In Appendix F we discuss the various two-electron integral transformations required for Newton–Raphson, fixed Hessian, and cubic procedures. A new transformation technique is presented in Section IX.A and Appendix F that is extremely efficient for the transformations of the integrals required for the cubic procedures. In Section IX we discuss the overall efficiency of Newton–Raphson and cubic procedures.

B. Numerical Results

To demonstrate the global convergence characteristics of the iterative cubic approach and the local convergence properties of the iterative, the perturbative, and the recursive cubic approaches we first describe an O_2 calculation on the $E\,^3\Sigma_u^-$ state at 2.16 a.u.[20] In Table XXXI the results of carrying out an iterative cubic calculation are reported. For a comparison, the results of the corresponding step size and sign controlled Newton–Raphson calculation are also given in Table XXXI. The Newton–Raphson calculation converges in six iterations and is thus of average complexity.

We first consider the global convergence problem. The initial Newton–Raphson iteration gives a step length of 1.95, which the step size control automatically constrains to 0.72. The iterative cubic step length is 0.58. In the Newton–Raphson calculation there is one constraint applied which reduces a step length amplitude from 1.86 to 0.43. The corresponding step length amplitude with the (unconstrained) iterative cubic procedure is also 0.43. It is, of course, somewhat fortuitous that this amplitude is constrained with the mode-controlled Newton–Raphson approach to be exactly the value from the iterative cubic MCSCF. However, we have observed that the constrained

TABLE XXXI
Convergence Characteristics of the Iterative Cubic and a Newton–Raphson Calculation of the $E^3\Sigma_u^-$ State of O_2 at 2.16 a.u.

Iteration point[a]	Iterative Cubic $E - E^{\text{CONV}\,b}$ (a.u.)	$\|\mathbf{F}\|$	$\|{}^{n+1}\underline{\lambda}\| \sim \|{}^{n}\mathbf{e}\|$	Newton–Raphson $E - E^{\text{CONV}\,b}$ (a.u.)	$\|\mathbf{F}\|$	$\|{}^{n+1}\underline{\lambda}\| \sim \|{}^{n}\mathbf{e}\|$
0	0.0614068393	1.04×10^{0}	5.78×10^{-1}	0.0614068393	1.04×10^{0}	1.95×10^{0} (7.21×10^{-1}
1	−0.0004083693	1.26×10^{-1}	1.68×10^{-1}	0.0028461916	1.22×10^{-1}	4.74×10^{-1}
2	−0.0000158313	5.38×10^{-3}	1.58×10^{-2}	−0.0001981711	6.54×10^{-2}	2.62×10^{-1}
3	0.0000000000	6.64×10^{-6}	2.61×10^{-5}	0.0000485658	1.50×10^{-2}	2.55×10^{-2}
4	0.0	$<10^{-8}$	$<10^{-8}$	0.0000000086	2.26×10^{-4}	5.39×10^{-4}
5				0.0000000000	1.30×10^{-6}	1.32×10^{-7}
6				0.0	$<10^{-8}$	$<10^{-8}$

[a]At iteration point n, the energy and all derivative matrices are evaluated and ${}^{n+1}\underline{\lambda}$ is determined. Th at iteration point n iteration $n+1$ is performed; $\|{}^{n+1}\underline{\lambda}\|$ is an approximation to $\|{}^{n}\mathbf{e}\| = \|{}^{n}\underline{\lambda} - \underline{\alpha}\|$ [see Eq. (7

[b]Here, $E - E^{\text{CONV}}$ denotes the difference between the total energy of the considered step of the iterat procedure minus the total energy of the converged calculation. The total energy of the converged calculati is −149.2537057620 a.u.

[c]This is the value of the norm after one constraint with the mode controlling procedure[17] is used to duce the size of one step length amplitude.

values of the step length amplitudes of the step size and sign controlled Newton–Raphson approach are often very close to the iterative cubic values.

It is seen that the norms of the error vectors in the cubic iterative procedure are decreasing approximately cubically. The ratios $\|{}^{n+1}\underline{\lambda}\|/\|{}^{n}\underline{\lambda}\|^3$ (note that ${}^{n+1}\underline{\lambda}$ is determined at iteration point n) for $n = 1$, 2, and 3 are 0.87, 3.48, and 6.26, respectively, and hence are fairly constant, demonstrating the third-order convergence characteristics. In the Newton–Raphson sequence of iterations the ratios $\|{}^{n+1}\underline{\lambda}\|/\|{}^{n}\underline{\lambda}\|^2$ for $n = 2$, 3, 4, and 5 (1.17, 0.37, 0.83, and 0.45) are also fairly constant as expected of a second-order procedure.

In all cases, the series of micro-iterations which is required to obtain a solution to the cubic iterative function of Eq. (208) is carried out using the Newton–Raphson technique of Eq. (216). If a Newton–Raphson (unconstrained) MCSCF step [Eq. (60)] is used as initial guess for the series, in many cases such a series is found not to converge. To obtain a solution of the cubic iterative function when far from convergence (e.g., in the first macro-iteration) requires the Newton–Raphson step size and sign controlled parameters as an initial guess of the sequence. With such an initial guess, the sequence of micro-iterations converges in few iterations. An example of the convergence of a sequence of micro-iterations is given in Table XXXII. Step size control is only applied in the initial micro-iteration of the first macro-iteration. Since we are using a Newton–Raphson technique to determine the step lengths of the micro-iterations, these step lengths, as expected, are diminishing quadratically.

The error vector obtained after the first iterative cubic iteration is carried out is $\sim 1.68\times 10^{-1}$. When two Newton–Raphson steps are carried out the error vector norm is $\sim 2.62\times 10^{-1}$, and when three NR steps are performed the error vector norm is $\sim 2.55\times 10^{-2}$ (see Table XXXI). The initial macro-iteration of the iterative cubic approach thus is able to replace some number between 2 and 3 Newton–Raphson iterations. It is seen from Table XXXII that the norm of the error of the second micro-iteration in the first macro-iteration is 2.37×10^{-2}. If the second micro-iteration is discarded, it is seen from Table XXXII that the error after the first macro-iteration is $\sim 1.92\times 10^{-1}$ $(1.68\times 10^{-1}+2.37\times 10^{-2})$, which is *smaller* than the error after the second Newton–Raphson iteration (2.62×10^{-1}). Thus a first macro-iteration consisting of an initial step size and sign controlled Newton–Rapshon iteration and one micro-iteration has a smaller approximate error vector norm than the first two Newton–Raphson iterations.

The first derivative of the function $\mathbf{f}(\underline{\lambda})$ given in Eq. (212) is

$$f'_{ij}(\underline{\lambda}) = G_{ij} + K_{ijk}\lambda_k \tag{234}$$

Equation (234) is a first-order approximation to the Hessian matrix (using a linear transformation of variables) at the point defined by the step length parameters $\underline{\lambda}$ of the micro-iteration; that is,

$$G_{ij}(\underline{\lambda}) = G_{ij} + K_{ijk}\lambda_k + O(\underline{\lambda})^2 \tag{235}$$

TABLE XXXII
The Norm of the Correction to the Step Length $||\lambda||$ of Each Micro-iteration for the Cubic Iterative Approach of Table XXXI

Micro-iteration	Macro-iteration 1	2	3	4
0^a	7.21×10^{-1}	1.68×10^{-1}	1.58×10^{-2}	2.61×10^{-5}
1	3.14×10^{-1}	6.71×10^{-2}	1.64×10^{-4}	5.47×10^{-10}
2	2.37×10^{-2}	2.82×10^{-2}		

[a]Micro-iteration 0 gives the step length of the step size and sign controlled Newton–Raphson iteration used as the initial guess for solving Eq. (208).

The stability of the cubic iterative procedure may be understood by examining how the lowest eigenvalues of $\mathbf{f}'(\underline{\lambda})$ develop during a sequence of micro-iterations. In Table XXXIII we report the lowest four eigenvalues of $\mathbf{f}'(\underline{\lambda})$ in the sequence of micro-iterations of the first macro-iteration of Table XXXI. The eigenvalues of the micro-iteration point 0 denote the Hessian eigenvalues [$\underline{\lambda} = \mathbf{0}$ in Eq. (214)]. It may be seen that a large change occurs in the small eigenvalues during the iterative sequence. The lowest eigenvalues of the Hessian matrix at macro-iteration 2 are -0.13904, 0.03028, 0.03536, and 0.05078, in farily good agreement with the converged eigenvalues of the micro-iteration sequence of Table XXXIII. $\mathbf{f}'(\underline{\lambda})$ is a first approximation to the Hessian matrix of macro-iteration 2 and therefore also contains one negative eigenvalue.

The unconstrained Newton-Raphson and also the perturbative cubic and recursive cubic approaches [see Eq. (151)] diverge for this case.[20] Only the iterative cubic approach and, of course, the step size and sign controlled Newton–Raphson approach have attractive global convergence properties. The details of constrained Newton–Raphson procedures are usually determined by numerical experience. Thus, when far from convergence, the iterative cubic approach is preferable. One iteration of the iterative cubic approach can replace two to five constrained Newton–Raphson steps when far from convergence. In Section IX, Appendices E and F, and Ref. 20, we demonstrate that the iterative cubic procedure is computationally of equal efficiency when applied far from convergence compared to a step sign and size controlled Newton-Raphson approach. The iterative cubic approach therefore has very attractive global convergence characteristics. More calculations, however, have to be carried out on difficult MCSCF cases before a full understanding of the global convergence characteristics of the iterative cubic approach is obtained.

TABLE XXXIII

Lowest Four Eigenvalues (a.u.) of $\mathbf{f}'(\underline{\lambda})$ in Eq. (214) for the Initial (Macro-iteration 1) Series of Micro-iterations of Table XXXI

	Lowest four eigenvalues of $\mathbf{f}'(\underline{\lambda})$			
Micro-iteration	1	2	3	4
0	−0.21250	0.00348	0.00788	0.01594
1	−0.19264	0.02984	0.04206	0.05584
2	−0.24320	0.03058	0.04344	0.04768
⋮				
Converged	−0.26002	0.02778	0.03800	0.04356

To illustrate the local convergence properties of cubically convergent approaches we also have carried out a recursive cubic and a perturbative cubic calculation on the $E^3\Sigma_u^-$ state at 2.16 a.u., using the orbitals and states obtained from the first iterative cubic macro-iteration as an initial guess. These calculations are reported in Table XXXIV together with the corresponding iterative cubic and the Newton–Raphson calculation results. Since the "initial guess" of orbitals and states is in the local region, the cubic approaches all show third-order convergence characteristics and as expected converge faster than the (unconstrained) Newton–Raphson calculation. As expected from Table XXX, approximately one iteration is saved in the local region by using cubic rather than quadratic MCSCF approaches. However, substantial differences are observed in the convergence rate of the three cubic approaches. This difference is caused by the different error terms of the three cubic approaches. The perturbative cubic approach has both the error term of the iterative cubic and the recursive cubic approaches [see Eq. (233)] and converges slowest of the three cubic procedures. The iterative cubic approach converges faster than the recursive cubic approach. An explanation may be that the error term of the iterative cubic and the recursive cubic approaches contains, respectively, one and two inverse Hessian matrices and that the small eigenvalues of the Hessian matrix dominate the error term.

To further demonstrate the local convergence properties of these cubic procedures, we report calculations on the $E^3\Sigma_u^-$ state of O_2 at 2.2 a.u. The basis set used in these calculations is given in Table XXXV, and the configurations are given in Table II. The initial guess for the Newton–Raphson part of these calculations is a set of grand canonical Hartree–Fock orbitals with occupation $1\sigma_g^2\, 1\sigma_u^2\, 2\sigma_g^2\, 2\sigma_u^2\, 1\pi_u^{3.4}\, 3\sigma_g^{1.7}\, 1\pi_g^{0.6}\, 3\sigma_u^{0.3}$. The initial set of orbitals used in these MCSCF calculations was obtained at iteration point 4 of a sign and size controlled Newton–Raphson calculation for the $E^3\Sigma_u^-$ state (the first point where no constraints are required). Hence, the reported calculations are in the local region.

In Table XXXVI we show the results of the Newton–Raphson, iterative cubic, Chebyshev (perturbative) cubic, and recursive cubic MCSCF procedures. For these calculations convergence is obtained for all the cubic procedures in three iterations. The Newton–Raphson approach requires four iterations. As expected (see Table XXX) one iteration is saved in the local region by use of a cubic procedure.

For this case the recursive cubic is converging slightly faster than the other cubic procedures. The smallest (in magnitude) Hessian eigenvalue at iteration point 0 of 0.052 is relatively large, and only relatively modest fluctuations are observed in the smallest Hessian eigenvalues during the sequence of iterations. The ratios $\|{}^{n+1}\underline{\lambda}\|/\|{}^{n}\underline{\lambda}\|^3$ for the cubic procedures (see Table

TABLE XXXIV

Convergence Characteristics of the Iterative Cubic, Perturbative (Chebyshev) Cubic, a.u., Starting with Orbitals Obtained After One Macro-iteration of the Iterative Cubic

	Iterative cubic			Perturbative cubic		
Iteration point[a]	$E - E^{\text{CONV}\,b}$ (a.u.)	$\lVert\mathbf{F}\rVert$	$\lVert{}^{n+1}\underline{\lambda}\rVert \sim \lVert{}^{n}\mathbf{e}\rVert$	$E - E^{\text{CONV}\,b}$ (a.u.)	$\lVert\mathbf{F}\rVert$	$\lVert{}^{n+1}\underline{\lambda}\rVert \sim \lVert{}^{n}\mathbf{e}\rVert$
0	−0.0004083693	1.26×10^{-1}	1.68×10^{-1}	−0.0004083693	1.26×10^{-1}	1.73×10^{-1}
1	−0.0000158313	5.38×10^{-3}	1.65×10^{-2}	−0.0001005968	1.24×10^{-2}	4.07×10^{-2}
2	0.0000000000	6.64×10^{-6}	2.81×10^{-5}	0.0000000055	6.88×10^{-5}	1.97×10^{-4}
3	0.0	$<10^{-8}$	$<10^{-8}$	0.0000000000	$<10^{-8}$	1.30×10^{-8}
4				0.0	$<10^{-8}$	$<10^{-8}$

[a]At iteration point n, the energy and all derivative matrices are evaluated and ${}^{n+1}\underline{\lambda}$ is determined. Thus, at iteration point n, iteration $n+1$ is performed; $\lVert{}^{n+1}\underline{\lambda}\rVert$ is an approximation to $\lVert{}^{n}\mathbf{e}\rVert = \lVert{}^{n}\underline{\lambda} - \underline{\alpha}\rVert$ [see Eq. (71)].

XXXVI) are almost constant. As explained previously, we expect

$$\lVert{}^{n+1}\lambda\rVert \leqq K\lVert{}^{n}\lambda\rVert^{3} \tag{236}$$

Thus these calculations demonstrate what might be called "classic" behavior; that is; $K \sim 1$ and the equality approximately holds in Eq. (236).

All the cubic procedures efficiently bring the MCSCF calculation to convergence when the "initial guess" of orbitals and states is in the local region. The three cubic approaches considered thus all have attractive local convergence properties. For further discussions of the efficiency of these procedures see Section IX, Appendixes E and F, and Ref. 20.

C. Summary and Conclusions

We have derived three cubic approaches—the iterative cubic, the perturbative cubic, and the recursive cubic:

1. The iterative cubic approach determines the set of rotational parameters from a cubic energy function. A solution of the cubic iterative function is defined in terms of an iterative solution, and we describe how a Newton–Raphson technique may be used to determine such a solution.
2. In the perturbative (Chebyshev) cubic approach the rotational parameters are determined such as to include the first-order perturbation correction to a Newton–Raphson set of rotational parameters.
3. The rotational parameters of the recursive cubic approach are obtained by carrying out a two-point fixed Hessian sequence of iterations. Such a sequence of iterations has previously been shown to have cubic convergence characteristics.

Recursive Cubic, and Newton–Raphson MCSCF for the $E^3\Sigma_u^-$ State of O_2 at 2.16 MCSCF

Recursive cubic			Newton–Rapshon		
$E - E^{CONV\,b}$ (a.u.)	$\|\|F\|\|$	$\|\|^{n+1}\underline{\lambda}\|\| \sim \|\|^{n}e\|\|$	$E - E^{CONV\,b}$ (a.u.)	$\|\|F\|\|$	$\|\|^{n+1}\underline{\lambda}\|\| \sim \|\|^{n}e\|\|$
−0.0004083693	1.26×10^{-1}	1.68×10^{-1}	−0.0004083693	1.26×10^{-1}	1.68×10^{-1}
0.0000220099	6.48×10^{-3}	3.65×10^{-2}	0.0000811349	1.12×10^{-2}	4.43×10^{-2}
0.0000000000	1.80×10^{-5}	6.12×10^{-5}	0.0000000111	6.72×10^{-4}	2.57×10^{-3}
0.0	$<10^{-8}$	$<10^{-8}$	0.0000000000	1.89×10^{-6}	4.01×10^{-6}
			0.0	$<10^{-8}$	$<10^{-8}$

[b]Here, $E - E^{CONV}$ denotes the difference between the total energy of the considered step of the iterative procedure minus the total energy of the converged calculation. The total energy of the converged calculation is − 149.2537057620 a.u.

An error term analysis has been carried out for these three cubic approaches. To demonstrate the cubic convergence of the three cubic models, calculations were carried out with an "initial guess" of orbital and states in the local region. Differences in the rate of convergence of the three cubic procedures can be ascribed to the different structure of the error terms of the three approaches. All three cubic procedures have attractive local convergence properties.

We also carried out calculations with the three cubic procedures when the initial guess of orbitals was far from convergence. Only the iterative cubic approach has attractive global convergence characteristics. The iterative cubic procedure by itself (with no constraints) is able to correctly bring a calculation into the local region even when the Hessian matrix of the initial iteration has spurious negative eigenvalues. Large step lengths occur in a

TABLE XXXV
26 STO O_2 Basis Set ($R_{AB} = 2.2$ a.u.)

Function	Exponent
1s	7.65781
2s	2.68801
	1.67543
	0.9
2p	3.69445
	1.65864
	0.9

TABLE XXXVI

Convergence Characteristics in the Local Region of the Iterative Cubic, Perturbative (Chebyshev)

	Iterative Cubic				Perturbative Cubic			
Iteration point[a]	$E - E^{\text{CONV}\,b}$ (a.u.)	$\|\mathbf{F}\|$	$\|{}^{n+1}\underline{\lambda}\| \sim \|{}^{n}\mathbf{e}\|$	$\frac{\|{}^{n+1}\underline{\lambda}\|}{\|{}^{n}\underline{\lambda}\|^3}$	$E - E^{\text{CONV}\,b}$ (a.u.)	$\|\mathbf{F}\|$	$\|{}^{n+1}\underline{\lambda}\| \sim \|{}^{n}\mathbf{e}\|$	$\frac{\|{}^{n+1}\underline{\lambda}\|}{\|{}^{n}\underline{\lambda}\|^3}$
0	0.0017425033	4.26×10^{-2}	3.14×10^{-1}		0.0017425033	4.26×10^{-2}	2.69×10^{-1}	
1	0.0000144510	3.20×10^{-3}	2.59×10^{-2}	0.84	0.0000088059	6.08×10^{-3}	2.10×10^{-2}	1.08
2	0.0000000000	2.88×10^{-6}	2.35×10^{-5}	1.35	0.0000000000	3.16×10^{-6}	7.21×10^{-6}	0.78
3	0.0000000000	$<10^{-10}$	$<10^{-10}$		0.0000000000	$<10^{-10}$	$<10^{-10}$	
4								

[a]At iteration point n, the energy and all derivative matrices are evaluated and ${}^{n+1}\underline{\lambda}$ is determined. Thus, at iteration point n, iteration $n+1$ is performed; $\|{}^{n+1}\underline{\lambda}\|$ is an approximation to $\|{}^{n}\mathbf{e}\| = \|{}^{n}\underline{\lambda} - \underline{\alpha}\|$ [see Eq. (71)].

Newton–Raphson MCSCF approach because the Hessian matrix has small eigenvalues. When the iterative function is based on a third-order expansion of the energy function, the convergence of a sequence of iterations is not to the same extent dominated by the small eigenvalues of the Hessian matrix. Large step lengths of the Newton–Raphson procedure often are artifacts of truncating the energy function after second-order terms. In all our calculations, step length amplitudes from the iterative cubic procedure have been found to be very moderate.

When far from convergence (the global problem), the cubic iterative procedure therefore may constitute a real alternative to constrained Newton–Raphson MCSCF. The cubic calculations, however, indicate that the step size and sign controlled Newton–Raphson approaches often approximate fairly well the steps of an iterative cubic procedure.

IX. EFFECTIVE IMPLEMENTATION AND COMBINATION OF MCSCF PROCEDURES

In this review we have made no attempt to discuss in depth other MCSCF approaches such as Fock operator or super-CI procedures. We also have heretofore somewhat emphasized our own recent work. This, of course, is because we are most familiar with our own contributions and because we have performed similar MCSCF calculations using constrained and unconstrained Newton–Raphson, fixed Hessian, updated Hessian, and cubic procedures so that the nature of MCSCF convergence may be viewed as a more coherent whole.

Cubic, Recursive Cubic, and the Newton-Raphson MCSCF for the $E^3\Sigma_u^-$ State of O_2 at 2.20 a.u.

Recursive Cubic				Newton–Raphson			
$E - E^{CONV\,b}$ (a.u.)	$\|\mathbf{F}\|$	$\|^{n+1}\underline{\lambda}\| - \|^{n}\mathbf{e}\|$	$\dfrac{\|^{n+1}\underline{\lambda}\|}{\|^{n}\underline{\lambda}\|^3}$	$E - E^{CONV\,b}$ (a.u.)	$\|\mathbf{F}\|$	$\|^{n+1}\underline{\lambda}\| - \|^{n}\mathbf{e}\|$	$\dfrac{\|^{n+1}\underline{\lambda}\|}{\|^{n}\underline{\lambda}\|^2}$
0.0017425033	4.26×10^{-2}	2.19×10^{-1}		0.0017425033	4.26×10^{-2}	2.19×10^{-1}	
0.0000077516	4.82×10^{-3}	2.39×10^{-2}	2.28	0.0000683846	8.24×10^{-3}	7.14×10^{-2}	1.49
0.0000000000	2.30×10^{-6}	7.94×10^{-6}	0.58	0.0000001390	1.30×10^{-3}	1.62×10^{-3}	0.32
0.0000000000	$<10^{-10}$	$<10^{-10}$		0.0000000000	9.60×10^{-7}	2.71×10^{-6}	1.03
				0.0000000000	$<10^{-10}$	$<10^{-10}$	

[b]Here, $E - E^{CONV}$ denotes the difference between the total energy at iteration point n minus the total energy of the converged calculation. The total energy of the converged calculation is −149.9712386378 a.u. Note that these calculations use the 26 STO basis set of Table XXXV. The initial guess is the set of orbitals from after iteration 4 of a constrained Newton–Raphson calculation.[17]

For super-CI and Fock operator approaches we, in general, refer the interested reader to the literature.[1,2] The super-CI approach[2,78–80] does not contain all second-order terms. It is expected that these contributions are particularly important for difficult cases. In the local region, super-CI converges only linearly. Fock operator procedures also are usually formulated without all second order contributions.[1,2] In recent work by Das[33] and Werner and Meyer,[29–30] complete second order Fock operator procedures have been formulated and studied.

Unitary exponential operators have been used in nuclear physics[52,53,81] for many years. The first attempts to use unitary exponential operators in MCSCF in chemical physics were by Levy[82] and Yaffe and Goddard.[24] Yaffe and Goddard[24] also developed both approximate second-order and approximate Chebyshev (third-order) techniques for the SCF and the MCSCF methods. They performed several SCF and GVB calculations with their approximate second and third order methods. For the MCSCF case Yaffe and Goddard did not include coupling with the MCSCF CI space, so their calculations were not fully second or third order. This can be critical, particularly for convergence to excited states. Further early work on second-order SCF optimization has been done by Douady et al.[25] (submission date, July 25, 1977; acceptance, July, 27 1977).

The first theoretical treatment of the general second-order MCSCF case using unitary exponential operators including coupling was by Dalgaard and Jørgensen.[10] They formulated the two-step MCSCF approach in the language of second quantization. Yeager and Jørgensen[11] and Dalgaard[12] made further theoretical developments on the second-order procedures. Dalgaard derived the one-step second-order procedure in the Brillouin formulation and presented some one-step calculations using the model of Pariser and Parr and of Pople.[83] The Brillouin formulation results with an unsymmetric "Hessian" matrix. Yeager and Jørgensen derived the one-step second-order procedure in the energy formulation and demonstrated relations between the one- and the two-step second-order procedures (some of which are described in Section III.C). Yeager and Jørgensen performed the first calculations with both the one- and the two-step second-order techniques.

At the 1979 Sanibel symposia D. Hopper and C. C. J. Roothaan lectured on new MCSCF procedures.[26] They commented there that their presentation was largely a reformulation in first quantized (wavefunction) form of the earlier work of Dalgaard and Jørgensen.[10] In addition, they derived the one-step MCSCF procedure. Roothaan, Detrich, and Hopper subsequently submitted a manuscript on their developments.[26] To date they have reported no calculational results with their procedures. At the same 1979 Sanibel meeting Yeager presented work based on the already submitted paper of Yeager and Jørgensen.[11] Yeager discussed the theoretical development of both the one-step and the two-step MCSCF procedures as well as calculational results.

Since 1979 there have been several outstanding contributions in second-order MCSCF. It is impossible to adequately discuss them all in detail. Except for certain specific examples, we refer interested readers to the original literature.[2, 10–37, 84–85]

The previous sections discussed primarily theoretical aspects of the iterative procedures. The present section complements this discussion with descriptions of how the procedures can be further modified so computational efforts are minimized. We first discuss (Section IX.A) how the procedures can be modified further in order to simplify integral transformations. Simplifications of the handling of the CI space are then indicated in Section IX.B. In Section IX.C we link different procedures together. It is shown that knowledge at one point can make computations at other points more effective. We then summarize the convergence characteristics of the methods mentioned and outline the methods' relative complexity. On the basis of this, efficient combinations of MCSCF procedures are discussed. In Section IX.D, we discuss two other ways of including higher order terms.

A. Simplifications of Transformations

Internal rotations in the subspace of completely occupied orbitals or internal rotations in the subspace of completely unoccupied orbitals do not change the total energy of the wavefunction. We shall now study a technique where transformations in these spaces are used to simplify two-electron transformations.

To do this, we introduce a formal partitioning of the set of properly symmetrized basis orbitals, b_k^+. Three subsets, I, A, and S, of basis orbitals are defined so subset I matches the set of completely occupied (inactive) molecular orbitals, subset A matches the set of partly occupied (active) molecular orbitals, and subset S matches the set of completely unoccupied (secondary) molecular orbitals. Two sets of orbitals are said to match if for all symmetry types the number of orbitals of a given symmetry in one set equals the number of orbitals of the same symmetry in the other set. In this section basis orbitals from set I and inactive molecular orbitals are denoted $i, j, k\ldots$; basis orbitals from set A and active molecular orbitals are denoted a, b, c; and basis orbitals from set S and secondary molecular orbitals are denoted s, t, u, v. General indices are denoted o, p, q, r. The matrix $\mathbf{C}$ which expands $\{a_k^+\}$ in terms of $\{b_k^+\}$ is partitioned thus:

$$\begin{Bmatrix} \mathbf{a}_{\mathrm{I}}^+ \\ \hline \mathbf{a}_{\mathrm{A}}^+ \\ \hline \mathbf{a}_{\mathrm{S}}^+ \end{Bmatrix} = \left\{\begin{array}{c|c|c} \mathbf{C}^{\mathrm{II}} & \mathbf{C}^{\mathrm{IA}} & \mathbf{C}^{\mathrm{IS}} \\ \hline \mathbf{C}^{\mathrm{AI}} & \mathbf{C}^{\mathrm{AA}} & \mathbf{C}^{\mathrm{AS}} \\ \hline \mathbf{C}^{\mathrm{SI}} & \mathbf{C}^{\mathrm{SA}} & \mathbf{C}^{\mathrm{SS}} \end{array}\right\} \begin{Bmatrix} \mathbf{b}_{\mathrm{I}}^+ \\ \hline \mathbf{b}_{\mathrm{A}}^+ \\ \hline \mathbf{b}_{\mathrm{S}}^+ \end{Bmatrix} \tag{237}$$

The energy truncated to second order is a function of the parameters $\underline{\lambda}$

$$E(\underline{\lambda}) = E_0 + \underline{\lambda}^T \mathbf{F} + \tfrac{1}{2}\underline{\lambda}^T \mathbf{G} \underline{\lambda}. \tag{238}$$

Introducing a new set of variables $\underline{\lambda}'$,

$$\underline{\lambda}' = \mathbf{P}'^{-1}\underline{\lambda} \tag{239}$$

where $\mathbf{P}'^{-1}$ is nonsingular, the terms in $E(\underline{\lambda})$ are transformed to

$$E(\underline{\lambda}') = E_0 + \underline{\lambda}'^{\mathrm{T}} \mathbf{P}'^{\mathrm{T}} \mathbf{F} + \tfrac{1}{2}\underline{\lambda}'^{\mathrm{T}} (\mathbf{P}'^{\mathrm{T}} \mathbf{G} \mathbf{P}') \underline{\lambda}' \tag{240}$$

The set of $\underline{\lambda}$ parameters that correspond to the stationary point of Eq. (238) may be determined directly from Eq. (238). Alternatively this set of $\underline{\lambda}$ parameters may be evaluated through determining the set of $\underline{\lambda}'$ parameters that make Eq. (240) stationary and then using the transformation in Eq. (239).

We will now investigate the possibility of choosing $\mathbf{P}'$ so $\mathbf{P}'^{\mathrm{T}}\mathbf{F}$ and $\mathbf{P}'^{\mathrm{T}}\mathbf{G}\mathbf{P}'$ can be calculated easier than $\mathbf{F}$ and $\mathbf{G}$. We are only interested in orbital rotations, so $\mathbf{P}'$ (the partition of $\mathbf{P}'$ corresponding to orbital and state parameters) is confined to be of the form

$$\mathbf{P}' = \left\{\begin{array}{c|c} \boldsymbol{\Pi} & \mathbf{0} \\ \hline \mathbf{0} & \mathbf{1} \end{array}\right\} \tag{241}$$

The matrix $\mathbf{\Pi}$ may be defined as a direct product matrix

$$\Pi_{pp',qq'} = P_{pq}P_{p'q'} \tag{242}$$

where pp' (or qq') is an orbital excitation index. The matrix $\mathbf{P}$ may be furthermore required to be a block diagonal with blocks dimensioned as the partitioning used in Eq. (237):

$$\mathbf{P} = \left\{\begin{array}{c|c|c} \mathbf{P}^{\mathrm{II}} & \mathbf{0} & \mathbf{0} \\ \hline \mathbf{0} & \mathbf{P}^{\mathrm{AA}} & \mathbf{0} \\ \hline \mathbf{0} & \mathbf{0} & \mathbf{P}^{\mathrm{SS}} \end{array}\right\} \tag{243}$$

We are now ready to study elements of $\mathbf{P}'^{\mathrm{T}}\mathbf{G}\mathbf{P}'$. The element $(\mathbf{P}'^{\mathrm{T}}\mathbf{G}\mathbf{P}')_{si,tj}$ becomes (see Appendix C for further notation):

$$\begin{aligned}(\mathbf{P}'^{\mathrm{T}}\mathbf{G}\mathbf{P}')_{si,tj} &= \mathbf{P}'_{uk,si}G_{uk,vl}\mathbf{P}'_{vl,tj} \\ &= P_{ki}P_{us}\{4[4(uk|vl)-(uv|kl)-(ul|vk)] \\ &\quad +4\delta_{kl}(D^{\mathrm{I}}_{uv}+D^{\mathrm{A}}_{uv})-4\delta_{uv}(D^{\mathrm{I}}_{kl}+D^{\mathrm{A}}_{kl})\}P_{lj}P_{vt}\end{aligned} \tag{244}$$

Introducing the notation

$$\tilde{a}^+_r = P_{r'r}a^+_{r'} \tag{245}$$

and corresponding integrals and **D** elements

$$\begin{aligned}\tilde{D}_{pq} &= P_{p'p}P_{q'q}D_{p'q'} \\ (\tilde{o}\tilde{p}|\tilde{q}\tilde{r}) &= P_{o'o}P_{p'p}P_{q'q}P_{r'r}(o'p'|q'r')\end{aligned} \tag{246}$$

one obtains

$$\begin{aligned}(\mathbf{P}'^{T}\mathbf{G}'\mathbf{P}')_{si,tj} &= 4[4(\tilde{i}\tilde{s}|\tilde{j}\tilde{t})-(\tilde{s}\tilde{t}|\tilde{i}\tilde{j})-(\tilde{s}\tilde{j}|\tilde{t}\tilde{i})] \\ &\quad +4(\mathbf{P}^{\mathrm{T}}\mathbf{P})_{ij}(\tilde{D}^{\mathrm{I}}_{st}+\tilde{D}^{\mathrm{A}}_{st})-4(\mathbf{P}^{\mathrm{T}}\mathbf{P})_{st}(\tilde{D}^{\mathrm{I}}_{ij}+\tilde{D}^{\mathrm{A}}_{ij})\end{aligned} \tag{247}$$

Equation (247) shows that $(\mathbf{P}'^{\mathrm{T}}\mathbf{G}\mathbf{P}')_{si,tj}$ has the same structure as $G_{si,tj}$, but with transformed integrals and **D** elements. Furthermore, δ_{qr} has been replaced with $(\mathbf{P}^{\mathrm{T}}\mathbf{P})_{qr}$. The construction of $(\mathbf{P}'^{\mathrm{T}}\mathbf{G}\mathbf{P}')_{si,tj}$ from $(\tilde{o}\tilde{p}|\tilde{q}\tilde{r})$ and $\tilde{\mathbf{D}}$ is thus basically identical to the construction of $G_{si,tj}$ from $(op|qr)$ and **D**.

An element $(\mathbf{P}'^{\mathrm{T}}\mathbf{G}\mathbf{P}')_{ai,bj}$ becomes

$$
\begin{aligned}
(\mathbf{P}'^{\mathrm{T}}\mathbf{G}\mathbf{P}')_{ai,bj} = {} & 4\left[\rho^{(2)}_{\tilde{b}\tilde{a},cd}(cd|\tilde{i}\tilde{j}) + 2\rho^{(2)}_{\tilde{b}d,\tilde{a}c}(c\tilde{i}|d\tilde{j})\right] \\
& + 2\left[\mathbf{P}^{\mathrm{T}}_{ac} - (\mathbf{P}^{\mathrm{T}}\boldsymbol{\rho}^{(1)})_{ac}\right]\left[4(c\tilde{i}|\tilde{b}\tilde{j}) - (\tilde{b}\tilde{i}|c\tilde{j}) - (\tilde{b}\tilde{i}|\tilde{i}\tilde{j})\right] \\
& + 2\left[\mathbf{P}^{\mathrm{T}}_{bc} - (\mathbf{P}^{\mathrm{T}}\boldsymbol{\rho}^{(1)})_{bc}\right]\left[4(c\tilde{j}|\tilde{a}\tilde{i}) - (\tilde{a}\tilde{j}|c\tilde{i}) - (\tilde{a}c|\tilde{i}\tilde{j})\right] \\
& + 2(\mathbf{P}^{\mathrm{T}}\boldsymbol{\rho}^{(1)}\mathbf{P})_{ab}\tilde{D}^{\mathrm{I}}_{ij} + 2(\mathbf{P}^{\mathrm{T}}\mathbf{P})_{ij}\left(2\tilde{D}^{\mathrm{I}}_{ab} + 2\tilde{D}^{\mathrm{A}}_{ab} - \tilde{D}_{ba}\right) \\
& - 4(\mathbf{P}^{\mathrm{T}}\mathbf{P})_{ab}\left(\tilde{D}^{\mathrm{I}}_{Ij} + \tilde{D}^{\mathrm{A}}_{ij}\right)
\end{aligned} \tag{248}
$$

It is seen that the formula for $(\mathbf{P}'^{\mathrm{T}}\mathbf{G}\mathbf{P}')_{ai,bj}$ is not, in general, obtained from the similar formula for $G_{ai,bj}$ by replacing $(op|qr)$, D_{qr}, δ_{qr} with $(\tilde{o}\tilde{p}|\tilde{q}\tilde{r})$, $\tilde{D}_{qr}, (\mathbf{P}^{\mathrm{T}}\mathbf{P})_{qr}$. In order to avoid complications, $\mathbf{P}^{\mathrm{AA}}$ is set equal to $\mathbf{1}$. The transformation to orbitals of class A can be simplified by other means, but for reasons of clarity we will not discuss this here.

With $\mathbf{P}^{\mathrm{AA}}$ as the unit matrix, all elements of $\mathbf{P}'^{\mathrm{T}}\mathbf{F}$ and $\mathbf{P}'^{\mathrm{T}}\mathbf{G}\mathbf{P}'$ are reported in Appendix C. All elements of $\mathbf{P}'^{\mathrm{T}}\mathbf{G}\mathbf{P}'$ and $\mathbf{P}'^{\mathrm{T}}\mathbf{F}$ are seen to correspond to elements of $\mathbf{G}$ and $\mathbf{F}$ with the changes

$$
\begin{aligned}
(op|qr) &\to (\tilde{o}\tilde{p}|\tilde{q}\tilde{r}) \\
D_{qr} &\to \tilde{D}_{qr} \\
\delta_{qr} &\to (\mathbf{P}^{\mathrm{T}}\mathbf{P})_{qr}
\end{aligned} \tag{249}
$$

This is due to the absence of indices corresponding to completely occupied and completely unoccupied orbitals in two-electron density matrices in the formulas for $\mathbf{F}$ and $\mathbf{G}$ (see Appendix C).

So far we have only introduced a new set of variables $\underline{\lambda}'$ [Eq. (239)] and used this set of variables to rewrite the total energy expression as in Eq. (240). Corresponding to this new set of variables, formulas have been derived for the "gradient" $\mathbf{P}'^{\mathrm{T}}\mathbf{F}$ and the "Hessian" $\mathbf{P}'^{\mathrm{T}}\mathbf{G}\mathbf{P}'$. We will now show that with an appropriate choice of the matrix $\mathbf{P}$ in Eq. (243) the two-electron integral transformations that are required for constructing $\mathbf{P}'^{\mathrm{T}}\mathbf{G}\mathbf{P}$ and $\mathbf{P}'^{\mathrm{T}}\mathbf{F}$ are simpler than those required for constructing $\mathbf{G}$ and $\mathbf{F}$ by themselves. To calculate $\mathbf{P}'^{\mathrm{T}}\mathbf{G}\mathbf{P}'$ and $\mathbf{P}'^{\mathrm{T}}\mathbf{F}$, integrals must be transformed to the tilde ($\tilde{}$) basis [Eq. (237)]. The corresponding orbital expansion is

$$
\begin{Bmatrix} \tilde{\mathbf{a}}^{+}_{\mathrm{I}} \\ \hline \tilde{\mathbf{a}}^{+}_{\mathrm{A}} \\ \hline \tilde{\mathbf{a}}^{+}_{\mathrm{S}} \end{Bmatrix} = \begin{Bmatrix} \mathbf{P}^{\mathrm{II}} & \mathbf{0} & \mathbf{0} \\ \mathbf{0} & \mathbf{1} & \mathbf{0} \\ \mathbf{0} & \mathbf{0} & \mathbf{P}^{\mathrm{SS}} \end{Bmatrix} \begin{Bmatrix} \mathbf{C}^{\mathrm{II}} & \mathbf{C}^{\mathrm{IA}} & \mathbf{C}^{\mathrm{IS}} \\ \mathbf{C}^{\mathrm{AI}} & \mathbf{C}^{\mathrm{AA}} & \mathbf{C}^{\mathrm{AS}} \\ \mathbf{C}^{\mathrm{SI}} & \mathbf{C}^{\mathrm{SA}} & \mathbf{C}^{\mathrm{SS}} \end{Bmatrix} \begin{Bmatrix} \mathbf{b}^{+}_{\mathrm{I}} \\ \hline \mathbf{b}^{+}_{\mathrm{A}} \\ \hline \mathbf{b}^{+}_{\mathrm{S}} \end{Bmatrix} \tag{250}
$$

A choice that simplifies the construction of $(\tilde{o}\tilde{p}|\tilde{q}\tilde{r})$ is

$$\begin{aligned} \mathbf{P}^{\mathrm{II}} &= \mathbf{C}^{\mathrm{II}^{-1}} \\ \mathbf{P}^{\mathrm{SS}} &= \mathbf{C}^{\mathrm{SS}^{-1}} \end{aligned} \tag{251}$$

Then

$$\begin{Bmatrix} \tilde{\mathbf{a}}_{\mathrm{I}}^{+} \\ \hline \tilde{\mathbf{a}}_{\mathrm{A}}^{+} \\ \hline \tilde{\mathbf{a}}_{\mathrm{S}}^{+} \end{Bmatrix} = \left\{ \begin{array}{c|c|c} \mathbf{1} & \mathbf{C}'^{\mathrm{IA}} & \mathbf{C}'^{\mathrm{IS}} \\ \hline \mathbf{C}'^{\mathrm{AI}} & \mathbf{C}'^{\mathrm{AA}} & \mathbf{C}'^{\mathrm{AS}} \\ \hline \mathbf{C}'^{\mathrm{SI}} & \mathbf{C}'^{\mathrm{SA}} & \mathbf{1} \end{array} \right\} \begin{Bmatrix} \mathbf{b}_{\mathrm{I}}^{+} \\ \hline \mathbf{b}_{\mathrm{A}}^{+} \\ \hline \mathbf{b}_{\mathrm{S}}^{+} \end{Bmatrix} \tag{252}$$

The set of orbitals $\{\tilde{a}_k^+\}$ are nonorthogonal with the choice of the matrix $\mathbf{P}$ in Eq. (251). The basis $\{\tilde{a}_k^+\}$ is, however, only used to construct the modified two-electron integrals (etc.) that are required in evaluating $\mathbf{P}'^{\mathrm{T}}\mathbf{G}\mathbf{P}'$ [e.g., see Eq. (247) and Appendix C] and $\mathbf{P}'^{\mathrm{T}}\mathsf{F}$, so that the nonorthogonality of the $(\tilde{\ })$ orbitals is of no consequence. The two-electron integrals in $\mathbf{P}'^{\mathrm{T}}\mathbf{G}\mathbf{P}'$ and $\mathbf{P}'^{\mathrm{T}}\mathsf{F}$ are easier to evaluate than the ones required for constructing $\mathbf{G}$ and $\mathbf{F}$ since the transformation matrix $\mathbf{C}'$ that is obtained through the choice of $\mathbf{P}$ in Eq. (251) reduces the transformations needed in second-order procedures. This is because the usual largest block $\mathbf{C}^{\mathrm{SS}}$ now becomes a unit matrix, $\mathbf{C}'^{\mathrm{SS}} = \mathbf{1}$. The advantages of introducing the reduced transformations are discussed in detail in Appendix F.

The additional work required to be able to perform these reduced transformations are inversion and multiplication of matrices [see Eqs. (251) and (239)]. Due to the direct product character of $\mathbf{P}'$, only matrices with the dimensions equal to the number of orbitals are involved. These additional matrix operations are therefore of negligible complexity.

The simplified reduced matrix Eq. (252) can also be used in connection with gradient based methods and one-point cubic methods. In gradient calculations iterations are of the type

$$^{i+1}\underline{\lambda} = {}^{i}\underline{\lambda} - \mathbf{H}^{-1}{}_{i}\mathsf{F}({}^{i}\underline{\lambda}) \tag{253}$$

By using the reduced matrix $\mathbf{C}'$, the vector $\mathbf{P}'^{\mathrm{T}}\mathsf{F}$ is calculated. Then F is calculated by multiplying $\mathbf{P}'^{\mathrm{T}}\mathsf{F}$ with $(\mathbf{P}'^{\mathrm{T}})^{-1}$, which again is simple because of the direct product nature of $\mathbf{P}'$.

The use of the reduced expansion matrix [Eq. (251)] in the iterative cubic procedure is less obvious. The variables in the cubic energy expansion

$$E(\underline{\lambda}) = E_0 + \underline{\lambda}^{\mathrm{T}}\mathsf{F} + \tfrac{1}{2}\underline{\lambda}^{\mathrm{T}}\mathbf{G}\underline{\lambda} + \tfrac{1}{6}\mathbf{K}\underline{\lambda}\underline{\lambda}\underline{\lambda} \tag{254}$$

are transformed according to Eq. (239)

$$\mathbf{E}(\underline{\lambda}') = E_0 + \underline{\lambda}'^{\mathrm{T}}(\mathbf{P}'^{\mathrm{T}}\mathbf{F}) + \tfrac{1}{2}\underline{\lambda}'^{\mathrm{T}}(\mathbf{P}'^{\mathrm{T}}\mathbf{G}\mathbf{P}')\underline{\lambda}' + \tfrac{1}{6}\mathbf{K}(\mathbf{P}'\underline{\lambda}')(\mathbf{P}'\underline{\lambda}')(\mathbf{P}'\underline{\lambda}') \quad (255)$$

This cubic approximation can be minimized with respect to $\underline{\lambda}'$. In Appendix E a method to construct the terms necessary for minimizing Eq. (243) is outlined. The cubic contributions $\mathbf{K}\underline{\lambda}$ in Eq. (254) are introduced by calculating a "Hessian" with transformed integrals and density matrices. The transformed integrals are of the type

$$\begin{aligned}(op|\bar{q}r) = {} & \kappa_{oo'}(o'p|qr) + \kappa_{pp'}(op'|qr) \\ & + \kappa_{qq'}(op|q'r) + \kappa_{rr'}(op|qr') \end{aligned} \quad (256)$$

An analysis very similar to the analysis in Appendix E can be carried out for the modified cubic function in Eq. (255). The cubic terms are now introduced by calculation of a "Hessian" of the type $\mathbf{P}'^{\mathrm{T}}\mathbf{K}(\mathbf{P}'\underline{\lambda}')\mathbf{P}'$. Instead of transformed integrals of the type of Eq. (256), the transformed integrals become

$$\begin{aligned}(\tilde{o}\tilde{p}|\bar{\tilde{q}}\tilde{r}) = {} & (\tilde{o}'\tilde{p}|\tilde{q}\tilde{r})(\mathbf{P}^{\mathrm{T}}\mathbf{P}\kappa')_{oo'} + (\tilde{o}\tilde{p}'|\tilde{q}\tilde{r})(\mathbf{P}^{\mathrm{T}}\mathbf{P}\kappa')_{pp'} \\ & + (\tilde{o}\tilde{p}|\tilde{q}'\tilde{r})(\mathbf{P}^{\mathrm{T}}\mathbf{P}\kappa')_{qq'} + (\tilde{o}\tilde{p}|\tilde{q}\tilde{r}')(\mathbf{P}^{\mathrm{T}}\mathbf{P}\kappa')_{rr'} \end{aligned} \quad (257)$$

From Eq. (255) it is easily seen that only integrals in the tilde ($\tilde{\ }$) basis are needed. The corresponding transformations are discussed in Appendix F.

The aforementioned modifications can result in numerical instabilities. These can be eliminated by choosing $\mathbf{P}^{\mathrm{II}}$ and $\mathbf{P}^{\mathrm{SS}}$ so that $\mathbf{P}^{\mathrm{II}}\mathbf{C}^{\mathrm{II}}$ and $\mathbf{P}^{\mathrm{SS}}\mathbf{C}^{\mathrm{SS}}$ are general diagonal matrices instead of unit matrices. The reduction of transformation times is not affected by this.

Thus, by redefining our MCSCF approaches in terms of the tilde ($\tilde{\ }$) basis set above, we are able to significantly reduce transformation times (see Appendix F). Some small additional work is required with this technique, that is, matrix inversion [Eq. (251)] and multiplication [to obtain $\underline{\lambda}$ from Eq. (239)]. These two matrix multiplications are $\propto N^3$ and hence, in general, will not significantly increase computer times.

B. Effective Treatment of Large CI Expansions

When the number n_{CI} of variables in the CI expansion of the MCSCF state is large, it is important to have effective algorithms to calculate terms connected with changes of the CI expansion. The purpose of this section is to discuss such algorithms. We first show how a basis for the CI space orthogonal to $|0\rangle$ can be defined through $n_{\mathrm{CI}} - 1$ variables. This parametrization is then used to construct the $\mathbf{G}^{CO}$ and $\mathbf{G}^{CC}$ blocks of the Hessian in a simple manner. We then discuss a direct second order MCSCF method of use when the complete Hessian cannot be constructed and stored in central memory.

1. *Basis for the Orthogonal Complement States to* $|0\rangle$

In the ansatz [Eq. (9)] for the MCSCF wavefunction a basis for the space $\{|\Phi_g\rangle\}/\{|0\rangle\}$ is needed. To define such a basis let $|\Phi_0\rangle$ be an arbitrary configuration state function which has non-zero overlap with $|0\rangle$. The CI expansion of $|0\rangle$ is

$$|0\rangle = C_{00}|\Phi_0\rangle + \sum_{g\neq 0} C_{g0}|\Phi_g\rangle \tag{258}$$

A set of parameters $S'_g(g \neq 0)$ is defined so

$$e^{i\hat{S}'}|\Phi_0\rangle = \exp i\left[i\sum_{g\neq 0} S'_g(\langle|\Phi_g\rangle\langle\Phi_0| - |\Phi_0\rangle\langle\Phi_g|)\right]|\Phi_0\rangle = |0\rangle \tag{259}$$

Since[12]

$$\exp\left[-\left(S'_g|\Phi_g\rangle\langle\Phi_0| - |\Phi_0\rangle\langle\Phi_g|\rangle\right)\right]|\Phi_0\rangle = \cos d|\Phi_0\rangle - \frac{\sin d}{d}S'_g|\Phi_g\rangle \tag{260}$$

where

$$d = \sqrt{\sum_{g\neq 0} S'^2_g} \tag{261}$$

and the parameters S'_g can be chosen as

$$S'_g = \frac{-C_{g0}d}{\sin d}, \qquad d = \cos^{-1}C_{00} \tag{262}$$

With this choice, Eq. (259) is fulfilled and the subspace $\{|k\rangle\}$ orthogonal to $|0\rangle$ may be constructed

$$|k\rangle = \exp\left[-S'_g\left(|\Phi_g\rangle\langle\Phi_0| - |\Phi_0\rangle\langle\Phi_g|\right)\right]|\Phi_k\rangle, \qquad k \neq 0 \tag{263}$$

By expanding the exponential operator in Eq. (263) one obtains[12]

$$|k\rangle = |\Phi_k\rangle + \frac{\sin d}{d}S'_k|\Phi_0\rangle - S'_k\left(\frac{\cos d - 1}{d^2}\right)|*\rangle \tag{264}$$

where

$$|*\rangle = \sum_{g \neq 0} S'_g |\Phi_g\rangle \tag{265}$$

The simple expansion Eq. (264) is due to the rank-2 structure of $\hat{S}'$.

The $\mathbf{G}^{CC}$ matrix

$$G^{CC}_{kl} = 2\langle k|H|l\rangle - 2\delta_{kl}E_0 \tag{266}$$

requires the transformation

$$\langle k|H|l\rangle = C_{gk}\langle \Phi_g|H|\Phi_{g'}\rangle C_{g'l} \tag{267}$$

The direct transformation Eq. (267) requires, in general, a number of multiplications proportional to n^3_{CI}, where n_{CI} is the dimension (e.g., number of configuration state functions) of the MCSCF CI space. Such a transformation can be extremely time consuming, for example, where $n_{CI} \approx 10^3$. The transformation in Eq. (267) is, however, not an n^3_{CI} algorithm when $|k\rangle$ is defined as in Eq. (264). Then we have

$$\begin{aligned}\langle k|H|l\rangle = {} & \langle \Phi_k|H|\Phi_l\rangle + \frac{\sin^2 d}{d^2} S'_k S'_l \langle \Phi_0|H|\Phi_0\rangle \\ & + \frac{\sin d}{d}(S'_l\langle \Phi_k|H|\Phi_0\rangle + S'_k\langle \Phi_0|H|\Phi_l\rangle) \\ & - \frac{(\cos d - 1)}{d^2}(S'_l\langle \Phi_k|H|*\rangle + S'_k\langle *|H|\Phi_l\rangle) \\ & - \frac{(\cos d - 1)\sin d}{d^3}(S'_k S'_l)\langle *|H|\Phi_0\rangle \\ & + \frac{(\cos d - 1)^2}{d^4}(S'_k S'_l)\langle *|H|*\rangle\end{aligned} \tag{268}$$

where the Einstein summation convention is not used. The basis Eq. (263) thus enables us to construct $\langle k|H|l\rangle$ from $\langle \Phi_g|H|\Phi_l\rangle$ with a number of multiplications proportional to n^2_{CI}. The construction of $\mathbf{G}^{CO}$ requires a trans-formation from $\langle 0|[\mathbf{Q}^+, H]|\Phi_g\rangle$. This transformation is also simplified significantly by defining $|k\rangle$ as in Eq. (263).

Other ways of bypassing what was thought to be a n^3_{CI} bottleneck in constructing $\langle k|H|l\rangle$ have been introduced by other workers.[2, 30, 84] In these methods the complete set of CSFs $\{|\Phi_g\rangle\}$ is used to span the CI manifold.

(The "Hessian" thereby obtained is related to the augmented Hessian[86] by similarity transformations.) This introduces a redundant variable which has been counteracted in different ways. Werner and Meyer[29, 30] have introduced the normalization condition as a constraint via a Lagrange multiplier. Lengsfield and Liu[84] have eliminated the extra degree of freedom by introducing projection operators. The extra degree of freedom can also be eliminated by using the methods developed for the augmented Hessians since the "Hessian" obtained by Lengsfield and Liu[84] actually is an augmented Hessian.

2. *Direct MCSCF*

If the size of the CI expansion is very large, $\mathbf{G}^{\mathrm{CO}}$ and $\mathbf{G}^{\mathrm{CC}}$ cannot conveniently be stored in central memory, while $\mathbf{G}^{\mathrm{OO}}$ usually is of manageable dimensions. In this case a direct second-order MCSCF procedure (i.e., a method that solves a set of Newton–Raphson equations without setting up the Hessian) is of interest. The Newton–Raphson equations [Eq. (72)] can be solved iteratively in several ways. Many techniques for the iterative solution of linear equations have been developed.[2, 87–89] The Newton–Raphson equations can also be solved by the update procedures discussed in Section VII. In either case it is necessary in each step to calculate a vector of the type

$$\mathbf{f}(\underline{\lambda}) = \mathbf{F} + \mathbf{G}\underline{\lambda} \tag{269}$$

The vector $\mathbf{G}\underline{\lambda}$ is

$$\begin{pmatrix} \mathbf{G}^{\mathrm{OO}} & \mathbf{G}^{\mathrm{OC}} \\ \mathbf{G}^{\mathrm{CO}} & \mathbf{G}^{\mathrm{CC}} \end{pmatrix} \begin{pmatrix} \underline{\kappa} \\ \mathbf{S} \end{pmatrix}$$
$$= \begin{pmatrix} \mathbf{G}^{\mathrm{OO}}\underline{\kappa} + \langle \delta 0 | [\mathbf{Q}^+ - \mathbf{Q}, H] | 0 \rangle + \langle 0 | [\mathbf{Q}^+ - \mathbf{Q}, H] | \delta 0 \rangle \\ \langle 0 | [\mathbf{R}^+ - \mathbf{R}, [(Q_i^+ - Q_i)\kappa_i, H]] | 0 \rangle + \langle \delta 0 | [\mathbf{R}^+ - \mathbf{R}, H] | 0 \rangle + \langle 0 | [\mathbf{R}^+ - \mathbf{R}, H] | \delta 0 \rangle \end{pmatrix} \tag{270}$$

where

$$|\delta 0\rangle = - S_{n0} |n\rangle \tag{271}$$

If $\mathbf{G}^{\mathrm{OO}}$ cannot be stored either, one obtains

$$\begin{pmatrix} \mathbf{G}^{\mathrm{OO}} & \mathbf{G}^{\mathrm{OC}} \\ \mathbf{G}^{\mathrm{CO}} & \mathbf{G}^{\mathrm{CC}} \end{pmatrix} \begin{pmatrix} \underline{\kappa} \\ \mathbf{S} \end{pmatrix}$$
$$= \begin{pmatrix} \langle 0 | [\mathbf{Q}^+ - \mathbf{Q}, [(Q_i^+ - Q_i)\kappa_i, H]] | 0 \rangle + \langle \delta 0 | [\mathbf{Q}^+ - \mathbf{Q}, H] | 0 \rangle + \langle 0 | [\mathbf{Q}^+ - \mathbf{Q}, H] | \delta 0 \rangle \\ \langle 0 | [\mathbf{R}^+ - \mathbf{R}, [(Q_i^+ - Q_i)\kappa_i, H]] | 0 \rangle + \langle \delta 0 | [\mathbf{R}^+ - \mathbf{R}, H] | 0 \rangle + \langle 0 | [\mathbf{R}^+ - \mathbf{R}, H] | \delta 0 \rangle \end{pmatrix} \tag{272}$$

The operator $[(Q_i^+ - Q_i)\kappa_i, H]$ corresponds to H with replaced integrals

$$h_{rs} \to \tilde{h}_{rs} = \kappa_{rp} h_{ps} + \kappa_{sp} h_{rp}$$
$$(rs|tu) \to (rs\tilde{|}tu) = \kappa_{rp}(ps|tu) + \kappa_{sp}(rp|tu) + \kappa_{tp}(rs|pu) + \kappa_{up}(rs|tp) \tag{273}$$

If Eq. (270) is used only integrals $(rs\tilde{|}tu)$ with all four indices corresponding to occupied orbitals are needed. These integrals can be constructed from $(rs|tu)$ by a transformation that requires about $\frac{1}{2}n^4N$ multiplications (n is the number of occupied orbitals and N is the total number of orbitals). If Eq. (272) is used, then integrals $(rs\tilde{|}tu)$ with one unrestricted index are needed. These integrals can be constructed from $(rs|tu)$ in about $\frac{3}{2}N^2n^3$ multiplications.

Introducing the notation

$$H' = H + [\kappa, H]$$
$$|0'\rangle = |0\rangle + |\delta 0\rangle \tag{274}$$

one obtains to first order in $\underline{\lambda}$

$$\mathbf{F} + \begin{pmatrix} \mathbf{G}^{\mathrm{OO}} & \mathbf{G}^{\mathrm{OC}} \\ \mathbf{G}^{\mathrm{CO}} & \mathbf{G}^{\mathrm{CC}} \end{pmatrix} \begin{pmatrix} \underline{\kappa} \\ \mathbf{S} \end{pmatrix} = \begin{pmatrix} \mathbf{G}^{\mathrm{OO}}\underline{\kappa} + \langle 0'|[\mathbf{Q}^+ - \mathbf{Q}, H]|0'\rangle \\ \langle 0'|[\mathbf{R}^+ - \mathbf{R}, H']|0'\rangle \end{pmatrix} \tag{275}$$

or alternatively

$$\mathbf{F} + \begin{pmatrix} \mathbf{G}^{\mathrm{OO}} & \mathbf{G}^{\mathrm{OC}} \\ \mathbf{G}^{\mathrm{CO}} & \mathbf{G}^{\mathrm{CC}} \end{pmatrix} \begin{pmatrix} \underline{\kappa} \\ \mathbf{S} \end{pmatrix} = \begin{pmatrix} \langle 0'|[\mathbf{Q}^+ - \mathbf{Q}, H']|0'\rangle \\ \langle 0'|[\mathbf{R}^+ - \mathbf{R}, H']|0'\rangle \end{pmatrix} \tag{276}$$

From Eqs. (275) and (276) it is seen that $\mathbf{F} + \mathbf{G}\underline{\lambda}$ can be calculated with a gradient routine with small modifications. The hamiltonian H is (partly in Eq. (275)) replaced by H' and $|0\rangle$ is replaced by $|0'\rangle$. The replacement of $|0\rangle$ with $|0'\rangle$ corresponds to the optimization of the CI coefficients. The orbital optimization corresponds to the replacement of H with H'.

If a usual direct CI iteration[88,89] is performed H' and $\langle 0'|[\mathbf{Q}^+ - \mathbf{Q}, H']|0'\rangle$ are not calculated. The construction of $\langle 0'|[\mathbf{Q}^+ - \mathbf{Q}, H]|0'\rangle$ requires about Nn^4 multiplications. The extra work for a direct second-order MCSCF iteration compared to a direct CI iteration is thus about $\frac{3}{2}Nn^4$ operations if Eq. (275) is used. A direct second-order MCSCF iteration is thus not expected to be significantly more complicated than a direct CI iteration.

The direct second-order MCSCF is, however, initiated by a two-electron transformation and [if Eq. (275) is used] a construction of $\mathbf{G}^{\mathrm{OO}}$. Furthermore, convergence in direct second-order MCSCF results in a second-order approximation to a stationary point, whereas convergence in direct CI calculations gives the stationary point.

C. Combination and Overview of Methods

In this section we show how different methods can be linked together to give an effective and sophisticated polyalgorithm. We first outline how second- or third-order information at one point can be used together with first-order information at other points. A summary of the methods is then given, and convergence characteristics are compared. By weighing the convergence characteristics of a given method with the complexity of a method, relative efficiencies are then obtained. This information is used finally to sketch the optimal combination of the procedures.

1. Combination of Gradient-Based Methods with Second-Order Methods

The fixed Hessian method (Section VI.B) combines second-order information of one point with first-order information of other points. This yields an algorithm that is effective only if a few iterations with a fixed Hessian are carried out before the Hessian is reevaluated.

It was noted in Section VII.E that the first two iterations of a fixed Hessian sequence in the local region seem as effective as the corresponding first two update iterations. When the initial Hessian approximation is an exact Hessian of a point in the local region, the imposition of the quasi-Newton condition does not usually improve the Hessian approximation in the first few update iterations. Since

$$\mathbf{G}({}^{k}\underline{\lambda})({}^{k+1}\underline{\lambda}-{}^{k}\underline{\lambda})=\mathbf{G}({}^{k+1}\underline{\lambda})({}^{k+1}\underline{\lambda}-{}^{k}\underline{\lambda})+O({}^{k+1}\underline{\lambda}-{}^{k}\underline{\lambda})^{2} \tag{277}$$

it is seen that using $\mathbf{G}({}^{k}\underline{\lambda})$ as a Hessian approximation, $\mathbf{H}_{k+1}$ is consistent with Eq. (161). Since this equation is the rationale behind the quasi-Newton condition Eq. (160), it is obvious that fixed Hessian methods and update methods based on Eq. (160) often behave similarly in the first iteration where the Hessian is updated.

In order to obtain new information about $\mathbf{G}({}^{k+1}\underline{\lambda})$ from $\mathbf{F}({}^{k+1}\underline{\lambda})$, when iteration $k+1$ is a Newton–Raphson iteration, $\mathbf{F}({}^{k+1}\underline{\lambda})$ and $\mathbf{G}({}^{k+1}\underline{\lambda})({}^{k+1}\underline{\lambda}-{}^{k}\underline{\lambda})$ are expanded through second order

$$\begin{aligned}\mathbf{F}({}^{k+1}\underline{\lambda})&=\mathbf{F}({}^{k}\underline{\lambda})+\mathbf{G}({}^{k}\underline{\lambda})({}^{k+1}\underline{\lambda}-{}^{k}\underline{\lambda})+\tfrac{1}{2}\mathbf{K}({}^{k}\underline{\lambda})({}^{k+1}\underline{\lambda}-{}^{k}\underline{\lambda})({}^{k+1}\underline{\lambda}-{}^{k}\underline{\lambda})\\ \mathbf{G}({}^{k+1}\underline{\lambda})({}^{k+1}\underline{\lambda}-{}^{k}\underline{\lambda})&=\mathbf{G}({}^{k}\underline{\lambda})({}^{k+1}\underline{\lambda}-{}^{k}\underline{\lambda})+\mathbf{K}({}^{k}\underline{\lambda})({}^{k+1}\underline{\lambda}-{}^{k}\underline{\lambda})({}^{k+1}\underline{\lambda}-{}^{k}\underline{\lambda})\end{aligned} \tag{278}$$

Thus, the exact Hessian $\mathbf{G}({}^{k+1}\underline{\lambda})$ satisfies through second order

$$\mathbf{G}({}^{k+1}\underline{\lambda})({}^{k+1}\underline{\lambda}-{}^{k}\underline{\lambda}) = -\mathbf{G}({}^{k}\underline{\lambda})({}^{k+1}\underline{\lambda}-{}^{k}\underline{\lambda})+2\mathbf{F}({}^{k+1}\underline{\lambda})-2\mathbf{F}({}^{k}\underline{\lambda}) \tag{279}$$

A condition for an Hessian approximation $\mathbf{H}_{k+1}$, when the exact Hessian $\mathbf{G}({}^{k}\underline{\lambda})$ is known, is thus

$$\mathbf{H}_{k+1}({}^{k+1}\underline{\lambda}-{}^{k}\underline{\lambda}) = -\mathbf{G}({}^{k}\underline{\lambda})({}^{k+1}\underline{\lambda}-{}^{k}\underline{\lambda})+2\mathbf{F}({}^{k+1}\underline{\lambda})-2\mathbf{F}({}^{k}\underline{\lambda}) \tag{280}$$

If iteration $k+1$ is an unmodified Newton–Raphson iteration [Eq. (20)], Eq. (280) becomes

$$\mathbf{H}_{k+1}({}^{k+1}\underline{\lambda}-{}^{k}\underline{\lambda}) = 2\mathbf{F}({}^{k+1}\underline{\lambda})-\mathbf{F}({}^{k}\underline{\lambda}) \tag{281}$$

The close resemblance between Eqs. (281) and (160) is noted.

If an update procedure based on Eqs. (280) and (281) is carried out, the corresponding approximate Hessian $\mathbf{H}_{k+1}$ fulfills

$$[\mathbf{H}_{k+1}-\mathbf{G}({}^{k+1}\underline{\lambda})]({}^{k+1}\underline{\lambda}-{}^{k}\underline{\lambda}) = O({}^{k+1}\underline{\lambda}-{}^{k}\underline{\lambda})^3 \tag{282}$$

We thus expect that updates based on Eqs. (280) and (281) will converge faster than the corresponding fixed Hessian iteration, even in the initial iterations.

The update condition [Eq. (280) or (281)] must be supplemented with other conditions. Based on the results discussed in Section VII.F, we suggest that $\mathbf{H}_{k+1}$ can be chosen so that $\|\mathbf{H}_{k+1}-\mathbf{G}({}^{k}\underline{\lambda})\|$ is minimal.

2. *Combination of Gradient-Based Methods with the Iterative Cubic Method*

If iteration k is an iterative cubic iteration at ${}^{k-1}\underline{\lambda}$, we obtain a first-order approximation $\mathbf{G}'_k$ to the exact Hessian at the resulting iteration point ${}^{k}\underline{\lambda}$:

$$\begin{aligned}\mathbf{G}({}^{k}\underline{\lambda}) &= \mathbf{G}'_k + O({}^{k}\underline{\lambda}-{}^{k-1}\underline{\lambda})^2\\ &= \mathbf{G}({}^{k-1}\underline{\lambda})+\mathbf{K}({}^{k-1}\underline{\lambda})({}^{k}\underline{\lambda}'-{}^{k-1}\underline{\lambda})+O({}^{k}\underline{\lambda}-{}^{k-1}\underline{\lambda})^2\end{aligned} \tag{283}$$

Typically ${}^{k}\underline{\lambda}'$ is the last point in the sequence of micro-iterations of macro-iteration k at which the first-order Hessian [Eq. (235)] is calculated. The reason we allow ${}^{k}\underline{\lambda}'$ to differ from ${}^{k}\underline{\lambda}$ is that the first-order Hessian [Eq. (235)] is usually not evaluated at the final point ${}^{k}\underline{\lambda}$. In Eq. (283) ${}^{k}\underline{\lambda}'$ is at least a first-order approximation to ${}^{k}\underline{\lambda}$ in the following sense

$${}^{k}\underline{\lambda}' = {}^{k}\underline{\lambda}+\underline{\delta}, \qquad \|\underline{\delta}\| = C\|{}^{k-1}\underline{\lambda}-\underline{\alpha}\|^2 \tag{284}$$

where $\underline{\alpha}$ is the stationary point of interest. Later we will prove a method which requires that ${}^{k}\underline{\lambda}' - {}^{k}\underline{\lambda}$ have third-order (rather than second-order) errors, that is, ${}^{k}\underline{\lambda}'$ is then closer to ${}^{k}\underline{\lambda}$.

We can now perform a sequence of "modified fixed Hessian" iterations

$$\begin{aligned} {}^{k}\underline{\lambda}^{\text{MFH}} &= {}^{k}\underline{\lambda} \\ {}^{k+i}\underline{\lambda}^{\text{MFH}} &= {}^{k+i-1}\underline{\lambda}^{\text{MFH}} - \mathbf{G}_k'^{-1}\mathbf{F}({}^{k+i-1}\underline{\lambda}^{\text{MFH}}), \qquad i = 1,2,\ldots,n \end{aligned} \tag{285}$$

The method defined by Eq. (285) uses third-order information at ${}^{k-1}\underline{\lambda}$. We now show that this increases the convergence rate compared to usual fixed Hessian methods (Section VI.B).

The Newton–Raphson iteration at ${}^{k}\underline{\lambda}$ is

$${}^{k+1}\underline{\lambda}^{\text{NR}} = {}^{k}\underline{\lambda} - (\mathbf{G}({}^{k}\underline{\lambda}))^{-1}\mathbf{F}({}^{k}\underline{\lambda}) \tag{286}$$

A procedure consisting of one iterative cubic iteration and one Newton–Raphson iteration has order 6 $(3 \cdot 2)$:

$$\|{}^{k+1}\underline{\lambda}^{\text{NR}} - \underline{\alpha}\| = \|{}^{k-1}\underline{\lambda} - \underline{\alpha}\|^{6} \tag{287}$$

Since iteration k is cubic, the resulting gradient $\mathbf{F}({}^{k}\underline{\lambda})$ is of third order in $\|{}^{k-1}\underline{\lambda} - \underline{\alpha}\|$. We then have

$$\begin{aligned} \|{}^{k+1}\underline{\lambda}^{\text{MFH}} - \underline{\alpha}\| &= \|{}^{k}\underline{\lambda} - \left[\mathbf{G}({}^{k}\underline{\lambda}) + O({}^{k-1}\underline{\lambda} - \underline{\alpha})^{2}\right]^{-1}\mathbf{F}({}^{k}\underline{\lambda}) - \underline{\alpha}\| \\ &= \|{}^{k+1}\underline{\lambda}^{\text{NR}} + O({}^{k-1}\underline{\lambda} - \underline{\alpha})^{5} - \underline{\alpha}\| \\ &= O({}^{k-1}\underline{\lambda} - \underline{\alpha})^{5} \end{aligned} \tag{288}$$

The procedure consisting of one iterative cubic iteration and one modified fixed Hessian iteration is a fifth-order procedure.

The foregoing result can be generalized inductively. For an arbitrary value of i in Eq. (285), one obtains

$$\|{}^{k+i}\underline{\lambda}^{\text{MFH}} - \underline{\alpha}\| = C\|{}^{k-1}\underline{\lambda} - \underline{\alpha}\|^{3+2i} \tag{289}$$

By calculating $1,2,3,\ldots$ gradients after one iterative cubic iteration, Eq. (285) gives a procedure of order $5,7,9,\ldots$.

The values of n in Eq. (285) corresponding to an optimal combination of the two iteration types involved can easily be found. The optimal number of gradient evaluations per cubic iteration is obtained when approximately 30%

of computer processing time is used to construct gradients.[16] It is thus not advisable to keep the same "Hessian" $\mathbf{G}'_k$ for very many iterations.

The fast convergence obtained with the algorithm Eq. (285) is due to the good approximation $\mathbf{G}'_k$ one has to $\mathbf{G}({}^k\underline{\lambda})$. It is easily shown that $\mathbf{G}'_k$ is a Hessian approximation that satisfies Eqs. (161) and (282) to relevant order. If one wants to obtain new information about $\mathbf{G}({}^k\underline{\lambda})$ from $\mathsf{F}({}^k\underline{\lambda})$ and a first-order Hessian [Eq. (235)], update conditions Eqs. (161) and (282) thus can not be used. We now derive an update condition that gives new information about $\mathbf{G}({}^k\underline{\lambda})$ from Eq. (235) and $\mathsf{F}({}^k\underline{\lambda})$. Such a procedure should usually give faster convergence than the modified fixed Hessian procedure hitherto discussed. $\mathsf{F}({}^k\underline{\lambda})$ and $\mathbf{G}({}^k\underline{\lambda})({}^k\underline{\lambda}-{}^{k-1}\underline{\lambda})$ are expanded to third order:

$$\mathsf{F}({}^k\underline{\lambda})=\frac{1}{3!}\mathbf{M}({}^{k-1}\underline{\lambda})({}^k\underline{\lambda}-{}^{k-1}\underline{\lambda})({}^k\underline{\lambda}-{}^{k-1}\underline{\lambda})({}^k\underline{\lambda}-{}^{k-1}\underline{\lambda})+\cdots$$

$$\mathbf{G}({}^k\underline{\lambda})({}^k\underline{\lambda}-{}^{k-1}\underline{\lambda})=\mathbf{G}''_k({}^k\underline{\lambda}-{}^{k-1}\underline{\lambda})+\frac{1}{2}\mathbf{M}({}^k\underline{\lambda}-{}^{k-1}\underline{\lambda})({}^k\underline{\lambda}-{}^{k-1}\underline{\lambda})({}^k\underline{\lambda}-{}^{k-1}\underline{\lambda})+\cdots \tag{290}$$

In the expansion of $\mathsf{F}({}^k\underline{\lambda})$ we used the fact that iteration k is an iterative cubic iteration, so the first three terms in the expansion of F [Eq. (59)] vanish. In Eq. (290)

$$\mathbf{G}''_k=\mathbf{G}({}^{k-1}\underline{\lambda})+\mathbf{K}({}^{k-1}\underline{\lambda})({}^k\underline{\lambda}-{}^{k-1}\underline{\lambda})+O({}^k\underline{\lambda}-{}^{k-1}\underline{\lambda})^3 \tag{291}$$

contrary to the previous requirement [Eq. (284)]. A third-order condition to $\mathbf{G}({}^k\underline{\lambda})$ is thus

$$\mathbf{G}({}^k\underline{\lambda})({}^k\underline{\lambda}-{}^{k-1}\underline{\lambda})=\mathbf{G}''({}^k\underline{\lambda}-{}^{k-1}\underline{\lambda})+3\mathsf{F}({}^k\underline{\lambda}) \tag{292}$$

When $\mathbf{G}''_k$ is known, a condition for the Hessian approximation in the following iteration is thus

$$\mathbf{H}_k({}^k\underline{\lambda}-{}^{k-1}\underline{\lambda})=\mathbf{G}''_k({}^k\underline{\lambda}-{}^{k-1}\underline{\lambda})+3\mathsf{F}({}^k\underline{\lambda}) \tag{293}$$

If $\mathbf{H}_k$ is in accord with Eq. (293), it satisfies the equation

$$[\mathbf{H}_k-\mathbf{G}({}^k\underline{\lambda})]({}^k\underline{\lambda}-{}^{k-1}\underline{\lambda})=\mathbf{0}+O({}^k\underline{\lambda}-{}^{k-1}\underline{\lambda})^4 \tag{294}$$

When the new error ${}^k\underline{\lambda}-\underline{\alpha}$ is nearly parallel with ${}^k\underline{\lambda}-{}^{k-1}\underline{\lambda}$, an iteration with this choice of $\mathbf{H}_k$ must be very similar to a Newton–Raphson iteration. If the error ${}^k\underline{\lambda}-\underline{\alpha}$ is nearly orthogonal to ${}^k\underline{\lambda}-{}^{k-1}\underline{\lambda}$, the use of update condition Eq. (293) does not improve the rate of convergence significantly compared to Eq. (285) with $n = 1$.

The third-order condition Eq. (293) must be supplemented with other requirements in order to define $\mathbf{H}_k$ unambiguously. Based on experience with conventional update methods (Section VII), we suggest that $\mathbf{H}_k$ should be chosen so the norm of $\mathbf{H}_k - \mathbf{G}_k''$ is minimal.

3. *Overview of Methods and Optimal Combinations of Methods*

The convergence characteristics of the iterative cubic method, the step controlled Newton–Raphson method, update methods, and fixed Hessian methods are summarized in Table XXXVII. The typical regions of convergence and the nonlocal and local convergence rates are described for each method. For clarity, in this section we do not include analyses of other cubic procedures or "infinite order" techniques. Of course, it is straightforward to extend the analysis of this section to the other cubic procedures.

The iterative cubic method requires the smallest number of iterations to converge. The small number of iterations required with this method is partly gained at the expense of increased complexity of a single iterative cubic iteration. In order to study relative efficiencies of different methods, one must study the amount of work involved in a single iteration of a given method. The basis for this analysis is given in Table XXXVIII. In Table XXXVIII the most time-consuming parts in an iteration of the different methods are stated.

Table XXXVII stresses the need for different methods for local and global convergence. Only the iterative cubic method and the step controlled Newton–Raphson method can reliably be used to bring the iteration point to a point relatively close to the stationary point of interest. While the nonlocal part of the optimization can be reliably performed by just two methods, all four methods mentioned in Tables XXXVII and XXXVIII can, in principle, be used for the local optimization. We thus discuss separately the optimal methods for the nonlocal and the local optimization.

Typically some two or three constrained Newton–Raphson iterations are required to move the iteration point into the local region when the initial point is defined by a CI eigenvector with Hartree–Fock orbitals. Compared to this, only one iterative cubic iteration is typically required to bring the iteration point into the local region. It was demonstrated in Section VIII that usually only one micro-iteration is needed to follow the initial Newton–Raphson iteration in an iteration cubic iteration. This number of micro-iterations is sufficient when two Newton–Raphson iterations bring the calcula-

TABLE XXXVII
Summary of Convergence Characteristics of Different Optimization Methods

	Method			
Characteristics	Step size and sign controlled (constrained) Newton–Raphson	Iterative cubic	Update procedures	Fixed Hessian
Region of convergence	Almost always converges even when the Hessian has an incorrect structure	Usually converges even when the Hessian has an incorrect structure. May not need constraints	Usually converges from a region where Hessian has correct number of negative eigenvalues	Converges only from a point relatively close to stationary point
Effectiveness of method when far from convergence compared to Newton–Raphson	Unconstrained Newton–Raphson often converges very slowly or diverges. Properly constrained NR approach usually converges	One iterative cubic iteration approximates typically 2–5 constrained Newton–Raphson iterations	3–5 update iterations approximate typically 1 constrained Newton–Raphson iteration	Fixed Hessian techniques diverge if Newton–Raphson diverges
Local convergence	Second order convergence (in region with no constraints)	Third-order convergence	Superlinear convergence	Linear convergence
Effectiveness of method in local region compared to (unconstrained) Newton–Raphson	Second-order convergence in local region (no constraints applied)	One iterative cubic iteration approximates 1.3–1.6 Newton–Raphson iterations	2 update iterations approximate typically 1 Newton–Raphson iteration	$2^N - 1$ fixed Hessian iterations approximate typically N Newton–Raphson iterations

TABLE XXXVIII
Summary of Requirements of an Iteration in Different Optimization Methods

Parts of an Iteration	Method			
	Step controlled Newton–Raphson	Iterative Cubic	Update	Fixed Hessian
Type of transformation (see Appendix F for explanation of notation)	T_2	$T_3 + T_\mu$	T_1	T_1
Number of constructions of Hessian	1	2	0	0
Number of constructions of gradient	1	1	1	1
Number of triangularizations used for solving linear equations	1	2	0	0

tion to a region with pronounced local character. This assumption about the number of micro-iterations is tacitly used in Table XXXVIII.

We now compare these two nonlocal procedures: two Newton–Raphson iterations and one iterative cubic iteration consisting of the above "1 + 1" micro-iterations. One such iterative cubic iteration and two Newton–Raphson iterations is seen from Table XXXVIII to have different requirements only with respect to transformations. The iterative cubic iteration requires a T_3 transformation plus a T_μ transformation. Two Newton–Raphson iterations require two T_2 transformations. We refer to Appendix F for more information about the different transformations. From Tables F.1 and F.4 it is seen that the number of multiplications in $T_3 + T_\mu$ is about $0.75–0.86N^5$ when the number of occupied orbitals is half the total number of orbitals N. When $\frac{1}{5}$ of the set of orbitals corresponds to occupied orbitals the corresponding number of multiplications is about $0.23–0.24N^5$. For two T_2 transformations; one obtains operation counts $1.10N^5$ and $0.28N^5$ for the ratios $\frac{1}{2}$ and $\frac{1}{5}$, respectively. The number of operations in the transformations of an iterative cubic iteration is thus about 30% lower than the number of operations for the transformations of two Newton–Raphson iterations.

If it is of importance to reduce the integral transformations in the MCSCF optimization, the iterative cubic may be a real alternative to step controlled Newton–Raphson method for the global optimization. The usefulness of the iterative cubic method is furthermore increased by the use of subsequent "modified fixed Hessian" iterations [Eq. (285)] or "third-order updates" corresponding to Eq. (293). The difference in efficacy between the two global methods is not great, and the step controlled Newton–Raphson method may be preferred as a global method because it is conceptually simpler and in many ways more flexible.

However, as we pointed out in Section VII, the constraints imposed on Newton–Raphson step-length amplitudes are usually imposed based somewhat on numerical experience. With the iterative cubic procedure our experience to date has indicated that with reasonable starting orbitals and CI coefficients often no constraints are necessary. Hence, particularly in very difficult cases, the iterative cubic approach may be useful since it may, in fact, replace several additional constrained Newton–Raphson iterations.

The optimal combination of optimization methods in the local region is not finally settled. All four methods in Tables XXXVII and XXXVIII will now be compared, and it will be shown that at least three of the methods discussed in Tables XXXVII and XXXVIII can be ingredients of a very fast local algorithm.

Our numerical experience indicates an iterative cubic iteration is typically equivalent to 1.3–1.6 Newton–Raphson iterations in the local region. One Newton–Raphson iteration is typically equivalent to two conventional update iterations. The number of fixed Hessian iterations corresponding to one Newton–Raphson depends on how "old" is the Hessian in use. The first two fixed Hessian iterations after a Newton–Raphson iteration correspond roughly to one Newton–Raphson iteration. A significantly larger number of fixed Hessian iterations are required to simulate more Newton–Raphson iterations as discussed in Section VI.C.

From Tables F.2 and F.3 in Appendix F it is seen that the two T_1 transformations required in two update or fixed Hessian iterations are more time consuming than one T_2 transformation which is required in one Newton–Raphson iteration. From Tables E.3 and E.4 (Appendix E) it is seen that the transformations $T_3 + T_\mu$ required in an iterative cubic iteration have a higher operation count than the 1.3 Newton–Raphson T_2 transformations required in the similar 1.3 Newton–Raphson iterations. In the local region the Newton–Raphson method thus is seen to require the fewest number of multiplications in the required integral transformation if one cubic iterative replaces 1.3 Newton–Raphson iterations. If one iterative cubic iteration replaces 1.6 Newton–Raphson iterations, then the transformations in the iterative cubic method appear slightly more efficient.

The differences in requirements of the integral transformations are not very large, and so other parts of an iteration must be studied in order to get reasonable assignments of the various methods' local efficiency. A conventional Newton–Raphson iteration requires the construction and triangularization of an Hessian. These are absent in fixed Hessian and update methods, whereas they are required two times in an iterative cubic iteration. The iterative cubic method is thus often not an efficient local method, and we leave it out in the following discussion. The fixed Hessian and update methods are effective local methods. The slightly increased transformation times in these methods are usually outweighed by the absence of Hessian constructions and triangularizations.

The update methods are usually more efficient than fixed Hessian methods. If update methods are used in the local region it will often be efficient to use these methods alone to obtain convergence. Only in cases where slow convergence is observed should the Hessian approximation be reset by a new Newton–Raphson calculation.

Update methods usually are superior to the fixed Hessian methods. However, even the most stable update method can encounter instabilities in extreme cases. In these situations the fixed Hessian method is preferred, since the behavior of the method is determined by the initial Newton–Raphson iteration. The Newton–Raphson method is very stable in the local region.

The optimal number of fixed Hessian iterations per Newton–Raphson corresponds to the use of about 30% of cpu time on fixed Hessian iterations (transformations included) (see Section VI.C). Since a fixed Hessian iteration usually is at least three times faster than a Newton–Raphson iteration, the Newton–Raphson method should not be used alone in the local region. The procedure with a large number of fixed Hessian iterations is also not an effective choice.

The only remaining feature is how and when one should go from a global method to a local method. We now discuss the case where the iterative cubic method is the global method in use. It is our experience that the transition to a local method can be performed after the iterative cubic method converges to a point where the first-order corrected Hessian [Eq. (235)] has the correct structure, that is, the correct number of negative eigenvalues. The transition to a local method can therefore usually be done after one iterative cubic iteration. We then propose the use of a modified fixed Hessian iteration [Eq. (285)] or a third-order update iteration [Eq. (293)] as the initial local iteration. These methods are the only methods that reuse the third-order information obtained in an iterative cubic iteration.

D. Higher (Infinite) Order Procedures

In Section VIII it was demonstrated that an algorithm that optimizes a cubic energy approximation [Eq. (208)] is very stable. Other methods that include certain cubic, quartic,… terms in $\underline{\kappa}$ have been developed.[19,29,30,90] We will now describe these. Since these methods' basic aim is to improve the orbital optimization, we will focus on this and for simplicity delete the coupling terms connected to the state optimization.

Werner and Meyer[29,30] have introduced a method in which the energy is expanded through second order in $e^{i\kappa}-\mathbf{1}$:

$$\begin{aligned}
E(\kappa) &= \langle 0|e^{-i\hat{\kappa}}He^{i\hat{\kappa}}|0\rangle \\
&= h_{rs}\langle 0|e^{-i\hat{\kappa}}a^{+}_{r\sigma}e^{i\hat{\kappa}}e^{-i\hat{\kappa}}a_{s\sigma}e^{i\hat{\kappa}}|0\rangle \\
&\quad + \tfrac{1}{2}(pq|rs)\langle 0|e^{-\hat{\kappa}}a^{+}_{p\sigma}e^{i\hat{\kappa}}e^{-i\hat{\kappa}}a^{+}_{r\sigma'}e^{i\hat{\kappa}}e^{-i\hat{\kappa}}a_{s\sigma'}e^{i\hat{\kappa}}e^{-i\hat{\kappa}}a_{q\sigma}e^{i\kappa}|0\rangle \\
&\simeq E^{\mathrm{WM}} = E_0 + 2(e^{i\kappa}-\mathbf{1})_{rk}h_{rs}\rho^{(1)}_{sk} + 2(e^{i\kappa}-\mathbf{1})_{pk}(pq|rs)\rho^{(2)}_{kq,rs} \\
&\quad + (e^{i\kappa}-\mathbf{1})_{rk}(e^{i\kappa}-\mathbf{1})_{sl}\rho^{(1)}_{kl}h_{rs} + 2(e^{i\kappa}-\mathbf{1})_{pk}(e^{i\kappa}-\mathbf{1})_{rl}\rho^{(2)}_{kq,ls}(pq|rs) \\
&\quad + (e^{i\kappa}-\mathbf{1})_{pk}(e^{i\kappa}-\mathbf{1})_{ql}\rho^{(2)}_{kl,rs}(pq|rs)
\end{aligned} \tag{295}$$

In Eq. (295) the usual nonrelativistic Hamiltonian is assumed, and $\rho^{(1)}$ and $\rho^{(2)}$ are the one-electron density matrix and symmetric two-electron density matrix.

Werner and Meyer express E^{WM} [Eq. (295)] by introducing the following matrices

$$\begin{aligned}
F^{pq}_{rs} &= \rho^{(1)}_{pq}h_{rs} + \rho^{(2)}_{pq,kl}(kl|rs) \\
G^{pq}_{rs} &= F^{pq}_{rs} + 2\rho^{(2)}_{pk,ql}(kr|ls),
\end{aligned} \tag{296}$$

where p and q correspond to occupied orbitals. By denoting

$$(e^{i\kappa}-\mathbf{1})_{kl} = \Delta_{kl}$$

Eq. (295) becomes

$$E^{\mathrm{WM}} = E_0 + 2F^{ps}_{rs}\Delta_{rp} + G^{pq}_{rs}\Delta_{rp}\Delta_{sq} \tag{297}$$

The energy approximation Eq. (297) includes terms cubic, quartic, ... in κ due to the expansion of Δ:

$$\Delta = i\kappa - \frac{1}{2}\kappa^2 - \frac{i}{6}\kappa^3 + \frac{1}{24}\kappa^4 + \cdots \tag{298}$$

However, there are also cubic, quartic, ... terms in Eq. (295) which are missing. Since the basic variable κ occurs as $e^{i\kappa}$, E^{WM} is periodical in κ. The approximate energy E^{WM} therefore always has a minimum. The second-order energy functional Eq. (19) defining the uncontrolled Newton–Raphson iteration does not always have a minimum, but the restricted Newton–Raphson problem [Eq. (110)] always does.

In a given iteration the $\mathbf{F}^{rs}$ and $\mathbf{G}^{rs}$ matrices are constructed and a "point" $\boldsymbol{\kappa}$ is found where E^{WM} is stationary; that is,

$$\frac{\partial E^{\mathrm{WM}}}{\partial \kappa_{ij}} = 2F_{rs}^{ps}\frac{\partial \Delta_{rp}}{\partial \kappa_{ij}} + 2G_{rs}^{pq}\frac{\partial \Delta_{rp}}{\partial \kappa_{ij}}\Delta_{sq} = 0 \tag{299}$$

The derivatives of $\boldsymbol{\Delta}$ occurring in Eq. (299) can be expressed by the generators $\mathbf{E}_{kl}$ of the orthogonal matrix group.[91] The only nonzero elements of $\mathbf{E}_{kl}$ are

$$(\mathbf{E}_{kl})_{kl} = 1, \qquad (\mathbf{E}_{kl})_{lk} = -1 \tag{300}$$

By performing algebraic manipulations very similar to those in Appendix B, one obtains

$$\frac{\partial}{\partial \kappa_{ij}}(e^{i\boldsymbol{\kappa}} - \mathbf{1}) = e^{i\boldsymbol{\kappa}}\left(i\mathbf{E}_{ij} - \frac{1}{2}\left[\boldsymbol{\kappa}, E_{ij}\right] - \frac{i}{3!}\left[\boldsymbol{\kappa}, \left[\boldsymbol{\kappa}, \mathbf{E}_{ij}\right]\right] + \cdots\right) \tag{301}$$

In Eq. (301) commutators between matrices are introduced. The exact derivative Eq. (301) can easily be obtained by diagonalizing $\boldsymbol{\kappa}$ in analogy with derivatives in Appendix B. However, since E^{WM} has errors starting in third order, Eq. (301) does not need to be evaluated to high accuracy. Direct use of the first three or four terms of Eq. (301) therefore gives a satisfactory approximation to the derivative of $\boldsymbol{\Delta}$. Werner and Meyer solve Eq. (299) with approximate derivatives of $\boldsymbol{\Delta}$ by a modified Gauss–Seidel procedure.[29] They report that satisfactory convergence is obtained if different "tricks" to improve convergence are used.

After having outlined the basic ideas of the method proposed by Werner and Meyer, we now discuss how the construction of $\mathbf{F}^{rs}$ and $\mathbf{G}^{rs}$ [Eq. (296)] can be replaced by the construction of matrices that are easier to construct.

The construction of the $\mathbf{F}^{rs}$ and $\mathbf{G}^{rs}$ matrices [Eq. (296)] requires a two-electron integral transformation equal to the transformation required by other second-order procedures. In Section IX.A it was shown that the two-electron integral transformation used in the Newton–Raphson method can be simplified. A similar simplification in the summations over secondary orbitals can be obtained in connection with the method of Werner and Meyer. Introduce the tilde (˜) basis as in Eq. (250) with

$$\mathbf{P}^{\mathrm{II}} = \mathbf{1}, \qquad \mathbf{P}^{\mathrm{SS}} = \mathbf{C}^{\mathrm{SS}-1} \tag{302}$$

and perform the corresponding integral transformation. This gives rise to modified **F** and **G** matrices

$$\tilde{F}^{pq}_{rs} = \rho^{(1)}_{pq}\tilde{h}_{rs} + \rho^{(2)}_{pq,kl}(\tilde{k}\tilde{l}|\tilde{r}\tilde{s})$$
$$\tilde{G}^{pq}_{rs} = \tilde{F}^{pq}_{rs} + 2\rho^{(2)}_{pk,ql}(\tilde{k}\tilde{r}|\tilde{l}\tilde{s}) \tag{303}$$

Since the upper indices in $\tilde{F}$ and $\tilde{G}$ correspond to occupied orbitals [see Eq. (296)], which are the same in the molecular orbital basis and in the tilde basis, one obtains

$$F^{pq}_{rs} = \tilde{F}^{pq}_{kl}(\mathbf{P}^{-1})_{kr}(\mathbf{P}^{-1})_{ls}$$
$$G^{pq}_{rs} = \tilde{G}^{pq}_{kl}(\mathbf{P}^{-1})_{kr}(\mathbf{P}^{-1})_{ls} \tag{304}$$

where

$$\mathbf{P} = \left(\begin{array}{c|c|c} \mathbf{1} & \mathbf{0} & \mathbf{0} \\ \hline \mathbf{0} & \mathbf{1} & \mathbf{0} \\ \hline \mathbf{0} & \mathbf{0} & \mathbf{P}^{\mathrm{SS}} \end{array}\right) \tag{305}$$

If Eq. (304) is introduced in the energy approximation [Eq. (297)], one obtains

$$E^{\mathrm{WM}} = E_0 + 2\tilde{F}^{ps}_{mn}(\mathbf{P}^{-1})_{mr}(\mathbf{P}^{-1})_{ns}\Delta_{rp} + \tilde{G}^{pq}_{mn}(\mathbf{P}^{-1})_{mr}(\mathbf{P}^{-1})_{ns}\Delta_{rp}\Delta_{sq} \tag{306}$$

In the first summation in Eq. (306) $(\mathbf{P}^{-1})_{ns}$ equals δ_{sn}, since s corresponds to an occupied orbital (s occurs as upper index in $\tilde{F}^{ps}_{mn}$). Equation (306) then becomes

$$E^{\mathrm{WM}} = E_0 + 2\tilde{F}^{ps}_{ms}\tilde{\Delta}_{mp} + \tilde{G}^{pq}_{mn}\tilde{\Delta}_{mp}\tilde{\Delta}_{nq} \tag{307}$$

where

$$\tilde{\Delta}_{lk} = (\mathbf{P}^{-1}\Delta)_{lk} \tag{308}$$

By introducing a new set of variables [Eq. (308)], integrals are thus only needed in the tilde basis. As discussed in Section IX.A and Appendix F, this simplifies transformations significantly.

The method of Werner and Meyer is thus not significantly more involved than the Newton–Raphson method. Higher order terms are included. However, it is not obvious that the cubic and higher order terms included are the most essential cubic and higher order terms, so theoretical predications about the method are not easy to make.

The reported numerical experience[29,30] with the Werner–Meyer method is now summarized. In the nonlocal region, Werner and Meyer report improved convergence can be sometimes obtained by using their procedure without coupling. However, similar improvements are sometimes observed in the Newton–Raphson technique by not including coupling for the first few iterations.[13-15] Several iterations with the Werner–Meyer procedure are usually required to enter the local region.[29,30] This is in contrast with the iterative cubic method discussed in Section VII. In the local region, the inclusion of Werner and Meyer's higher order terms are reported to often slow convergence.[29-30]

Line search methods can exactly optimize a function in a subspace defined by a single variable. These methods are very stable, and only minor modifications have to be introduced in order to guarantee global convergence. We now discuss a method proposed by Igawa et al.[19,90] that optimizes the exact function in a subspace defined by *several* variables.

A diagonal $\hat{\kappa}$ operator is defined

$$\hat{\kappa}^d = \sum_A \gamma_A \hat{n}_A, \qquad \hat{n}_A = \sum_\sigma a^+_{A\sigma} a_{A\sigma} \quad \text{(no Einstein summation)} \tag{309}$$

If the basis in which $\hat{\kappa}$ is assumed diagonal were the usual molecular orbital basis, the *Ansatz* [Eq. (309)] introduces only a phase factor in the trial function. By choosing another basis, that is, another set of number operators $\{\hat{n}_A\}$, nontrivial variations are introduced by Eq. (309). The maximal dimension of the space

$$|0^d\rangle = e^{i\hat{\kappa}^d}|0\rangle \tag{310}$$

can be shown[19] to be

$$K = \min\{N, 2n\} \tag{311}$$

where N is the total number of orbitals and n is the number of occupied orbitals. Since K is usually smaller than the dimension of the complete orbital optimization manifold, the *Ansatz* Eq. (310) is only a subspace of the complete MCSCF manifold.

The energy corresponding to the ansatz Eq. (310) is

$$\begin{aligned} E^d &= \langle 0|e^{-i\hat{\kappa}^d} H e^{i\hat{\kappa}^d}|0\rangle \\ &= \langle 0|H|0\rangle + \langle 0|e^{i\hat{\kappa}^d}\left[e^{-i\hat{\kappa}^d}, H\right]|0\rangle \end{aligned} \tag{312}$$

It will now be shown that a compact expression for the energy Eq. (312) can be derived for an arbitrary $\hat{\kappa}^d$ [Eq. (309)]. Since the number operators commute, the exponential operator in Eq. (310) can be factorized

$$e^{i\Sigma_A\gamma_A\hat{n}_A} = \prod_{A=1,K} e^{i\gamma_A\hat{n}_A} \quad \text{(no Einstein summation)} \tag{313}$$

The relation

$$e^{i\gamma_r\hat{n}_r} = (1-\hat{n}_r) + e^{i\gamma_r}\hat{n}_r \quad \text{(no sum)} \tag{314}$$

is easily derived. By using Eqs. (313) and (314) and commutators for fermions, the following expressions are derived:

$$e^{-i\Sigma_B\gamma_B\hat{n}_B}\left[\sum_\sigma a^+_{r\sigma}a_{s\sigma}, e^{i\Sigma_A\gamma_A\hat{n}_A}\right] = (e^{-i(\gamma_r-\gamma_s)}-1)\sum_\sigma a^+_{r\sigma}a_{s\sigma}$$

$$\text{(no Einstein summation)} \tag{315a}$$

$$e^{-i\Sigma_B\gamma_B\hat{n}_B}\left[\sum_{\sigma,\sigma'} a^+_{r\sigma}a^+_{s\sigma'}a_{t\sigma'}a_{u\sigma}, e^{i\Sigma_A\gamma_A\hat{n}_A}\right]$$
$$= (e^{-i(\gamma_r+\gamma_s-\gamma_t-\gamma_u)}-1)\sum_{\sigma,\sigma'} a^+_{r\sigma}a^+_{s\sigma'}a_{t\sigma'}a_{u\sigma} \quad \text{(no Einstein summation)}$$
$$\tag{315b}$$

The exact energy (Eq. (312)) with the nonrelativistic Born–Oppenheimer Hamiltonian expressed in the basis where $\hat{\kappa}^d$ is diagonal becomes

$$E^d = \sum_{A,B}^{K} h_{AB}e^{-i(\gamma_A-\gamma_B)}\rho^{(1)}_{AB}$$
$$+\tfrac{1}{2}\sum_{A,B,C,D}^{K}(AB|CD)e^{-i(\gamma_A+\gamma_C-\gamma_B-\gamma_D)}\rho^{(2)}_{AB,CD} \tag{316}$$

With the conventional MCSCF *Ansatz* [Eq. (11)] the energy dependence of orbital rotations is

$$E = \langle 0|e^{-i\hat{\kappa}}He^{i\hat{\kappa}}|0\rangle$$
$$= h_{rs}(e^{i\kappa})_{r'r}(e^{i\kappa})_{s's}\rho^{(1)}_{r's'}$$
$$+\tfrac{1}{2}(rs|tu)(e^{i\kappa})_{r'r}(e^{i\kappa})_{s's}(e^{i\kappa})_{t't}(e^{i\kappa})_{u'u}\rho^{(2)}_{r't',s'u'} \tag{317}$$

While the general energy expression, Eq. (317), involves eightfold summations for $\hat{\kappa} \neq \hat{0}$, a nonzero $\hat{\kappa}^d$ in Eq. (316) only introduces fourfold summations and phasefactors. The exact energy [Eq. (316)] for an arbitrary operator $\hat{\kappa}^d$ [Eq. (309)] is thus easily obtained. The same is true for derivatives with respect to the γ parameters. The exact energy can thus easily be optimized in the subspace defined by Eqs. (309) and (310) once the Hamiltonian is known in this basis. In the method of Igawa et al.[19] a given iteration first defines a basis where $\hat{\kappa}^{\mathrm{d}}$ is diagonal, then the transformation of integrals to this basis is performed, and finally the exact energy is optimized in this space.

The basis in which $\hat{\kappa}^d$ is diagonal is chosen as follows. First an usual MCSCF iteration with, for example, the constrained Newton–Raphson method or the conjugate gradient technique is performed. This gives a reference operator $\hat{\kappa}^{\mathrm{ref}}$.

$$\hat{\kappa}^{\mathrm{ref}} = \kappa_{rs}^{\mathrm{ref}} a_{r\sigma}^{+} a_{s\sigma} \tag{318}$$

The matrix $\boldsymbol{\kappa}^{\mathrm{ref}}$ defined by Eq. (318) is then diagonalized

$$\boldsymbol{\kappa}^{\mathrm{ref}} \mathbf{U} = i \mathbf{U} \boldsymbol{\gamma}^{\mathrm{ref}} \tag{319}$$

The eigenvectors of $\boldsymbol{\kappa}^{\mathrm{ref}}$ are then used to define the basis in which $\hat{\kappa}^d$ is constructed

$$a_A^{+} = a_r^{+} U_{rA} \tag{320}$$

where $\{a_r^{+}\}$ denotes the molecular orbital basis. The $\hat{\kappa}^d$ [Eq. (312)] is then defined as

$$\hat{\kappa}^d = \left(\Delta\gamma_A + \gamma_A^{\mathrm{ref}}\right) a_{A\sigma}^{+} a_{A\sigma} = \gamma_A \hat{n}_A \tag{321}$$

A subspace around the reference point is thus constructed.

The step lengths are defined only by the γ parameters, and these are optimized exactly (i.e., no constraints on the exact γ's are required). It is thus expected that the method is particularly effective in the nonlocal region, where most other methods need an empirically based constraint scheme. It is not likely that the exact optimization in the subspace will improve local convergence, just as line search is not advisable locally.[9]

Experience with the method[19] stresses these points. No constraints for the γ's are needed in the nonlocal region in order to obtain convergence. The actual rate of convergence in the nonlocal region depends often on the selected subspace. If the eigenvectors determining the subspace are obtained from an undamped Newton–Raphson iteration, convergence is often slow but steady. If a $\boldsymbol{\kappa}$ matrix from a constrained Newton–Raphson iteration is used to define the subspace, convergence is faster. The exact optimization in a subspace can then reduce the total number of iterations in the nonlocal region by approximately 1 or more. In the local region, the exact optimization in a subspace does not improve convergence.

The extra computational work due to the exact optimization in a subspace is the extra transformation, which is required to construct the Hamiltonian in the "diagonal basis." This transformation required about $(K/2)N^4$ operations. Since K is usually equal to $2n$ [Eq. (311)], the operation count for this transformation is about nN^4. This transformation has a higher multiplication count than the reduced transformations needed in a second-order iteration (Section IX.A).

This infinite order method is very time consuming. However, the exact optimization of the energy in a subspace makes the method extremely stable and reliable. The method is thus expected to be useful when the energy function changes so abruptly that exact information about the energy at many points in a subspace is needed in order to properly define a step.

Conventional line search in the nonlocal region has been investigated with promising results.[9] Comparison studies between the optimization in a subspace as discussed and more conventional line search procedures are currently under way and will be reported soon.[92]

X. SUMMARY AND CONCLUSIONS

Current theoretical research in MCSCF primarily involves the study of proper characterization of a multiconfigurational state and the development of efficient and reliable optimization procedures. Until recent developments with second- and higher order MCSCF schemes, it was often very difficult to converge to the correct state of interest.[1,2] Previously, the characteristics of a multiconfigurational state were usually not properly examined, and many reported calculations undoubtedly converged to an undesired stationary point on the energy hypersurface. Since we have already discussed at length, particularly in Section IX, effective implementation and combination of methods, in this section we will only briefly restate a few main points.

We have discussed in Section IV some desired characteristics of an MCSCF state. If converging to the nth state in energy of a certain symmetry, these include the following:

1. The Hessian has $n-1$ negative eigenvalues.
2. The MCTDHF calculation using the MCSCF state as a reference state is stable. In the symmetry block of the MCSCF reference state the MCTDHF should have $n-1$ negative energy differences.
3. In the CI using the MCSCF configuration state functions, the state of interest is the nth state.

Since the MCSCF state includes a finite (usually small) number of configurations and the orbitals are usually expanded using a finite basis set, some (or, in rare cases, all) of these criteria may not be fulfilled for a converged MCSCF state that represents the nth state—in which case it is usually most important to fulfull the criteria that are of particular relevance to subsequent calculations which use the MCSCF state and orbitals. As has been demonstrated through both theory and calculations, characteristic 3 (above) is the one most often not fulfilled and in many cases is not essential (e.g., especially since reparameterization may cause a converged MCSCF state to change from being the $(n-1)$st to the nth CI state).

The MCSCF convergence problem differs significantly in the local and the nonlocal regions. When far from convergence (the global convergence problem) it is most important for a technique to move a calculation to a region on the energy hypersurface close to the proper stationary point. Either constrained Newton–Raphson or iterative cubic methods may be used for the global convergence problem. Constrained Newton–Raphson approaches usually seem to work well.[13,15,18] Recent calculations with the Fletcher constraint algorithm which guarantees convergence for the lowest state of a certain symmetry are extremely promising.[55] The iterative cubic technique is apparently both efficient and reliable.[20] Iterative cubic calculations have also demonstrated the validity of parameters chosen by numerical experience for the mode-controlling technique. "Infinite order" techniques are interesting and appear useful.[19,29,30] However, exactly what is occurring theoretically and calculationally with these is much less clear (e.g., it is not at all certain that the most important variables are taken to infinite order, and calculations are quently performed without coupling). Constraints are still usually required with these procedures when far from convergence. Augmented Hessian approaches[31,32,86] often also require constraints when far from convergence and currently do not appear to offer any significant advantages over adequately and properly constrained Newton–Raphson approaches.[18]

When in the local region, the Hessian has the proper number of negative eigenvalues and Newton–Raphson step length amplitudes are not large. Convergence with the Newton–Raphson approach is rapid and reliable. In this region computational efficiency is most important. Current theory and numerical experience indicates that an update procedure such as the Broyden rank-1 method should be used.[21,75] A combination of fixed Hessian and Newton–Raphson approaches is also useful, provided no more than about 30% of the CPU time (including transformation time) is used on fixed Hessian approaches.[22] Infinite order approaches are not, in general, advocated in the local region.

Often wavefunction optimization approaches have been based almost entirely on numerical experience. Occasionally this has led to disasterous consequences. We and others have recently explicitly demonstrated, in this paper and elsewhere, how both detailed theoretical and computational studies should be combined in order to properly design the most efficient MCSCF procedures which reliably converge to the correct stationary point on the energy hypersurface.

APPENDIX A: REDUNDANT VARIABLES

For real orbitals, a general unitary transformation of the reference state $|0\rangle$ may be written as

$$|\tilde{0}\rangle = \exp i\hat{\kappa} \exp i\hat{S}|0\rangle \tag{A.1}$$

where $\hat{\kappa}$ and $\hat{S}$ are defined in Eqs. (5) and (7). The excitation operators $\{a_r^+ a_s - a_r^+ a_s\}$ in $\hat{\kappa}$ and $\{|k\rangle\langle 0| - |0\rangle\langle k|\}$ in $\hat{S}$ may span a basis with linearly dependencies. When optimizing a state or when evaluating the response of an MCSCF state to an external one-electron perturbation, such linear dependencies must be eliminated.

The generalized Brillouin's theorem (GBT)[10-12,78]

$$\langle 0|[a_r^+ a_s, H]|0\rangle = 0 \tag{A.2}$$

$$\langle 0|[|k\rangle\langle 0|, H]|0\rangle = \langle 0|H|k\rangle = 0 \tag{A.3}$$

is derived by considering first-order variations of in the reference state $|0\rangle$. This variational space is embedded in the space spanned by

$$\{|0\rangle; |k\rangle; (a_r^+ a_s - a_s^+ a_r)|0\rangle, \quad r > s\} = \{|j\rangle; (a_r^+ a_s - a_s^+ a_r)|0\rangle, \quad r > s\} \tag{A.4}$$

Because the orbitals are real, Eq. (A.2) may be written as

$$\langle 0|H(a_r^+ a_s - a_s^+ a_r)|0\rangle = 0 \tag{A.5}$$

Operators $a_{cl}^+ a_{cl'}$ or $a_p^+ a_{p'}$, where here cl, cl' refer to orbitals doubly occupied in all MCSCF CI configurations and p, p' refer to unoccupied orbitals, trivially fulfill the generalized Brillouin's theorem at all points on the energy hypersurface; that is,

$$\langle 0|[a_{cl}^+ a_{cl'}, H]|0\rangle = 0 \tag{A.6}$$

$$\langle 0|[a_p^+ a_{p'}, H]|0\rangle = 0 \tag{A.7}$$

These operators are trivially redundant and can be easily eliminated immediately from an MCSCF calculation.

It is more complicated to eliminate operators due to linear dependencies in the space spanned by the sets $\{a_r^+ a_s - a_s^+ a_r\}$ and $\{|k\rangle\langle 0| - |0\rangle\langle k|\}$ [Eq.

(A.4)]. A linear dependency in the set (A.4) may be determined if

$$\begin{aligned}(a_x^+ a_y - a_y^+ a_x)|0\rangle &= |0\rangle\langle 0|a_x^+ a_y - a_y^+ a_x|0\rangle + \sum_{k\neq 0} |k\rangle\langle k|a_x^+ a_y - a_y^+ a_x|0\rangle \\ &+ \sum_{(r,s)\neq(x,y)} (a_r^+ a_s - a_s^+ a_r)|0\rangle \\ &\times \langle 0|(a_r^+ a_s - a_s^+ a_r)(a_x^+ a_y - a_y^+ a_x)|0\rangle \end{aligned} \tag{A.8}$$

These operators, $a_x^+ a_y - a_y^+ a_x$ in Eq. (A.8), are thus redundant in the sense that, if the GBT in Eqs. (A.2) and (A.3) is satisfied for the nonredundant set of operators (i.e., without $a_x^+ a_y - a_y^+ a_x$), the GBT will be automatically satisfied for the redundant set of operators (i.e., the set which includes $a_x^+ a_y - a_y^+ a_x$). The operators $a_{cl}^+ a_{cl'} - a_{cl'}^+ a_{cl}$ and $a_p^+ a_{p'} - a_{p'}^+ a_p$ in Eqs. (A.6) and (A.7) trivially fulfill Eq. (A.8). However, there are other operators which may also fulfill Eq. (A.8) and should not be included in the calculations.

The states $\{|0\rangle, |k\rangle\} = \{|j\rangle\}$ are related to the configuration space $\{|\phi_g\rangle\}$ through a unitary transformation. Hence, Eq. (A.8) may be reexpressed as

$$\begin{aligned}(a_x^+ a_y - a_y^+ a_x)|0\rangle &= \sum_g |\phi_g\rangle\langle \phi_g|a_x^+ a_y - a_y^+ a_x|0\rangle \\ &+ \sum_{(r,s)\neq(x,y)} (a_r^+ a_s - a_s^+ a_r)|0\rangle \\ &\times \langle 0|(a_r^+ a_s - a_s^+ a_r)(a_x^+ a_y - a_y^+ a_x)|0\rangle \end{aligned} \tag{A.9}$$

If the last term in Eq. (A.9) is neglected, we determine only linear dependices between the operator $a_x^+ a_y - a_y^+ a_x$ and $\{|0\rangle, |k\rangle\}$. Only if both terms in Eq. (A.9) are included are all redundant operators of an MCSCF calculation eliminated. In Ref. 17 examples are given where the last term in Eq. (A.9) is required in order to eliminate all redundant variables of the MCSCF calculation.

An operator that fulfills Eq. (A.8) usually gives a zero eigenvalue only at a stationary point. This can easily be demonstrated since

$$\begin{aligned}&\langle 0|[a_t^+ a_u - a_u^+ a_t, H, a_x^+ a_y - a_y^+ a_x]|0\rangle \\ &= \tfrac{1}{2}\langle 0|[[a_t^+ a_u - a_u^+ a_t, H], a_x^+ a_y - a_y^+ a_x]|0\rangle \\ &\quad + \tfrac{1}{2}\langle 0|[a_t^+ a_u - a_u^+ a_t, [H, a_x^+ a_y - a_y^+ a_x]]|0\rangle \end{aligned} \tag{A.10}$$

$$\begin{aligned}&= \langle 0|[[a_t^+ a_u - a_u^+ a_t, H], a_x^+ a_y - a_y^+ a_x]|0\rangle \\ &\quad - \tfrac{1}{2}\langle 0|[[a_t^+ a_u - a_u^+ a_t, a_x^+ a_y - a_y^+ a_x], H]|0\rangle \\ &= 2\langle 0|[a_t^+ a_u - a_u^+ a_t, H](a_x^+ a_y - a_y^+ a_x)|0\rangle \\ &\quad - (\delta_{ux}\langle 0|[a_t^+ a_y, H]|0\rangle - \delta_{yt}\langle 0|[a_x^+ a_u, H]|0\rangle \\ &\quad - \delta_{xt}\langle 0|[a_u^+ a_y, H]|0\rangle + \delta_{yu}\langle 0|[a_x^+ a_t, H]|0\rangle \end{aligned} \tag{A.11}$$

and

$$\langle 0|\left[|k\rangle\langle 0| - |0\rangle\langle k|, H, a_x^+ a_y - a_y^+ a_x\right]|0\rangle$$

$$= -2\langle 0|\left[H, a_x^+ a_y - a_y^+ a_x\right]|k\rangle \tag{A.12}$$

$$= -2\langle 0|H\left(a_x^+ a_y - a_y^+ a_x\right)|k\rangle + 2\langle 0|\left(a_x^+ a_y - a_y^+ a_x\right)H|k\rangle \tag{A.13}$$

At a stationary point (i.e., where the GBT, Eq. (A.2), is fulfilled) the last four terms in Eq. (A.11) are zero. If we assume that Eq. (A.8) is valid when $|0\rangle$ is replaced by $|j\rangle$ it can easily be seen that at a stationary point all the Hessian elements involving $a_x^+ a_y - a_y^+ a_x$ are composed of linear combinations of Hessian elements which do not involve $a_x^+ a_y - a_y^+ a_x$. Hence in this case, if the redundant operator $a_x^+ a_y - a_y^+ a_x$ is included in the operator set, a zero eigenvalue of the Hessian results. This is generally true *only* at convergence. An example of this behavior is given in Section V.B. If the redundant operator is included in the set, serious convergence problems may result since, for example, the Hessian eigenvalues may approach zero more rapidly than do the corresponding redundant GBT amplitudes and large step length amplitudes.

APPENDIX B: THE FIRST PARTIAL DERIVATIVE OF THE TOTAL ENERGY

We consider in the following how the first partial derivative (the gradient) of the energy

$$E\left({}^0\underline{\kappa} + \underline{\kappa}, {}^0\mathbf{S} + \mathbf{S}\right) = \langle 0|\exp(-i\hat{S})\exp(-i\hat{\kappa})H\exp(i\hat{\kappa})\exp(i\hat{S})|0\rangle \tag{B.1}$$

where

$$\hat{\kappa} = i\sum_{r>s}\left({}^0\kappa_{rs} + \kappa_{rs}\right)\left(a_r^+ a_s - a_s^+ a_r\right) \tag{B.2}$$

$$\hat{S} = i\sum_{k\neq 0}\left({}^0S_{k0} + S_{k0}\right)\left(|k\rangle\langle 0| - |0\rangle\langle k|\right) \tag{B.3}$$

may be evaluated at an arbitrary point $({}^0\underline{\kappa}, {}^0\mathbf{S})$. Such a derivative becomes important when deriving MCSCF iterative approaches that carry out a linear transformation among the variables between each step of the iterative procedure. Initially we consider how the differential $\exp[-(\hat{\lambda} + d\hat{\lambda})] -$

$\exp(-\hat{\lambda})$, where $\hat{\lambda}$ is an arbitrary operator, may be evaluated:

$$\begin{aligned}\exp[-(\hat{\lambda}+d\hat{\lambda})]-\exp(-\hat{\lambda}) &= \exp(-\hat{\lambda})[\exp(\hat{\lambda})\exp(-(\hat{\lambda}+d\hat{\lambda})]^{-1})\\ &= \exp(-\hat{\lambda})\exp(\hat{\lambda}z)\exp[-(\hat{\lambda}+d\hat{\lambda})z]|_{z=1}\\ &\quad -\exp(\hat{\lambda}z)\exp[-(\hat{\lambda}+d\hat{\lambda})z]|_{z=0}\\ &= \exp(-\hat{\lambda})\int_0^1 dz\,\frac{d}{dz}[\exp(\hat{\lambda}z)\exp(-(\hat{\lambda}+d\hat{\lambda})z)]\\ &= \exp(-\hat{\lambda})\int_0^1 dz\,\exp(\hat{\lambda}z)(\hat{\lambda}-\hat{\lambda}-d\hat{\lambda})\exp[-(\hat{\lambda}+d\hat{\lambda})z] \end{aligned} \tag{B.4}$$

Since we wish to determine the differential $d(\exp(-\hat{\lambda}))$, all terms which are not linear in $d\hat{\lambda}$ may be neglected. We get

$$d(\exp(-\hat{\lambda})) = -\exp(-\hat{\lambda})\int_0^1 dz\,\exp(\hat{\lambda}z)\,d\hat{\lambda}\exp(-\hat{\lambda}z) \tag{B.5}$$

Defining a superoperator $\hat{\hat{\lambda}}$ corresponding to $\hat{\lambda}$

$$\hat{\hat{\lambda}} = [\hat{\lambda}, f] \tag{B.6}$$

(the $\hat{\hat{}}$ notes the superoperator) for an arbitrary operator f gives

$$\begin{aligned} d(\exp(-\hat{\lambda})) &= -\exp(-\hat{\lambda})\int_0^1 dz\left(\sum_{n=0}^{\infty}\frac{(z)^n}{n!}\hat{\hat{\lambda}}^n\,d\lambda\right)\\ &= -\exp(-\hat{\lambda})\sum_{n=0}^{\infty}\frac{1}{(n+1)!}\hat{\hat{\lambda}}^n\,d\hat{\lambda} \end{aligned} \tag{B.7}$$

If $\hat{\lambda}$ is a sum of certain operators O_i with expansion coefficients L_i

$$\hat{\lambda} = \sum_i ({}^0L_i + L_i)O_i$$

we have from Eq. (B.7) that

$$\frac{\partial\exp(-\hat{\lambda})}{\partial L_i} = -\exp(-\hat{\lambda})\left(\sum_n\frac{1}{(n+1)!}\hat{\hat{\lambda}}^n O_i\right) \tag{B.8}$$

The relation

$$\frac{\partial\exp(\hat{\lambda})}{\partial L_i} = \left(\sum_n\frac{1}{(n+1)!}\hat{\hat{\lambda}}^n O_i\right)\exp(\hat{\lambda}) \tag{B.9}$$

may be derived in an equivalent way.

B.1. Derivative with Respect to Orbital Parameters

To determine the partial derivative of the total energy with respect to the orbital parameters, it becomes convenient to introduce the notation

$$ {}^0\hat{\kappa} = i \sum_{r>s} {}^0\kappa_{rs}(a_r^+ a_s - a_s^+ a_r) \tag{B.10}$$

$$ {}^0\hat{S} = i \sum_{k \neq 0} {}^0S_{k0}(|k\rangle\langle 0| - |0\rangle\langle k|) \tag{B.11}$$

$$ {}^0\hat{\hat{\kappa}} a_r^+ a_s = \left[{}^0\hat{\kappa}, a_r^+ a_s\right] \tag{B.12}$$

We then obtain

$$ F({}^0\underline{\lambda})_{rs} = \left. \frac{\partial E({}^0\underline{\kappa} - \underline{\kappa}, {}^0\mathbf{S} + \mathbf{S})}{\partial \kappa_{rs}} \right|_{\substack{\mathbf{S}=0 \\ \underline{\kappa}=0}} $$

$$ = \langle 0| \exp(-i{}^0\hat{S}) \left(\sum_n \frac{(-i)^n}{(n+1)!} \left({}^0\hat{\hat{\kappa}}^n (a_r^+ a_s - a_s^+ a_r)\right) \right) $$

$$ \times \exp(-i{}^0\hat{\kappa}) H \exp(i{}^0\hat{\kappa}) \exp(i{}^0\hat{S})|0\rangle $$

$$ - \langle 0| \exp(-i{}^0\hat{S}) \exp(-i{}^0\hat{\kappa}) H \exp(i{}^0\hat{\kappa}) $$

$$ \times \left(\sum_n \frac{(-i)^n}{(n+1)!} \left({}^0\hat{\hat{\kappa}}^n (a_r^+ a_s - a_s^+ a_r)\right) \right) \exp(i{}^0\hat{S})|0\rangle $$

$$ = \langle \tilde{0}| \left[\sum_n \frac{(-i)^n}{(n+1)!} \left({}^0\tilde{\hat{\hat{\kappa}}}^n (\tilde{a}_r^+ \tilde{a}_s - \tilde{a}_s^+ \tilde{a}_r)\right), H \right] |\tilde{0}\rangle $$

$$ = 2\langle \tilde{0}| \left[\sum_n \frac{(-i)^n}{(n+1)!} \left({}^0\tilde{\hat{\hat{\kappa}}}^n \tilde{a}_r^+ \tilde{a}_s\right), H \right] |\tilde{0}\rangle \tag{B.13}$$

where

$$ \tilde{a}_r^+ = \exp(i{}^0\hat{\kappa}) a_r^+ \exp(-i{}^0\hat{\kappa}) \tag{B.14}$$

$$ |\tilde{0}\rangle = \exp(i{}^0\hat{\kappa}) \exp(i{}^0\hat{S})|0\rangle \tag{B.15}$$

Since

$$ (-i)^n {}^0\tilde{\hat{\hat{\kappa}}}^n \tilde{a}_r^+ \tilde{a}_s = \sum_{pq} C_{rs,pq}^n \tilde{a}_p^+ \tilde{a}_q \tag{B.16}$$

we get

$$ F({}^0\underline{\lambda})_{rs} = 2 \sum_n \frac{1}{(n+1)!} \sum_{pq} C_{rs,pq}^n \langle \tilde{0}| \left[\tilde{a}_p^+ \tilde{a}_q, H\right] |\tilde{0}\rangle \tag{B.17}$$

and the energy gradient elements that correspond to a linear transformation among the variables thus become a sum of the ordinary gradient elements $2\langle\tilde{0}|[\tilde{a}_p^+\tilde{a}_q, H]|\tilde{0}\rangle$. The evaluation of $F(^0\underline{\lambda})_{rs}$ through determining $C^n_{rs,pq}$ is very cumbersome if high accuracy is required, and we describe in the following how Eq. (B.16) may be rewritten to allow an easy high-accuracy evaluation.

Since the matrix $i^0\boldsymbol{\kappa}$ is hermitian, it may be diagonalized of a unitary matrix

$$i^0\boldsymbol{\kappa} = \mathbf{U}\boldsymbol{\varepsilon}\mathbf{U}^+ \tag{B.18}$$

where $\boldsymbol{\varepsilon}$ is a real diagonal matrix. The $^0\tilde{\hat{\kappa}}$ may therefore be rewritten as

$$^0\tilde{\hat{\kappa}} = \sum \varepsilon_p \bar{\hat{n}}_p \tag{B.19}$$

where

$$\bar{\hat{n}}_p = \bar{a}_p^+\bar{a}_p \tag{B.20}$$

$$\bar{a}_p^+ = \sum_j \tilde{a}_j^+ (\mathbf{U})_{jp} \tag{B.21}$$

$$\bar{a}_p = \sum_j \tilde{a}_j (\mathbf{U}^+)_{pj} \tag{B.22}$$

Transforming all operators in Eq. (B.13) into the overbar ($\bar{}$) basis gives

$$F(^0\underline{\lambda})_{rs} = 2\sum_n \frac{(-i)^n}{(n+1)!}\langle\tilde{0}|\left[\sum_{\substack{k\\pq}}\left(\left(\varepsilon_k\bar{\hat{n}}_k\right)^n \bar{a}_p^+\bar{a}_q\right), H\right]|\tilde{0}\rangle U_{pr}^+ U_{sq} \tag{B.23}$$

Since

$$\left(\sum_k \varepsilon_k \bar{\hat{n}}_k\right)\left(\bar{a}_p^+\bar{a}_q\right) = (\varepsilon_p - \varepsilon_q)\bar{a}_p^+\bar{a}_q \tag{B.24}$$

we may rewrite Eq. (B.23) as

$$\begin{aligned} F(^0\underline{\lambda})_{rs} &= 2\sum_{\substack{n\\pq}} \frac{(-i)^n}{(n+1)!}(\varepsilon_p - \varepsilon_q)^n \langle\tilde{0}|\left[\bar{a}_p^+\bar{a}_q, H\right]|\tilde{0}\rangle U_{pr}^+ U_{sq} \\ &= \sum_{pq} g(p,q)\bar{f}(p,q) U_{pr}^+ U_{sq} \end{aligned} \tag{B.25}$$

where

$$g(p,q)=\sum_n \frac{(-i)^n}{(n+1)!}(\varepsilon_p-\varepsilon_q)^n \tag{B.26}$$

$$\bar{f}(p,q)=2\langle\tilde{0}|[\bar{a}_p^+\bar{a}_q, H]|\tilde{0}\rangle \tag{B.27}$$

If $\varepsilon_p \neq \varepsilon_q$, then $g(p,q)$ may be written as

$$\frac{i}{\varepsilon_p-\varepsilon_q}\sum_{n=0}^{\infty}\frac{(-i)^{n+1}}{(n+1)!}(\varepsilon_p-\varepsilon_p)^{n+1}=\frac{i}{\varepsilon_p-\varepsilon_q}\left[\exp\{-i(\varepsilon_p-\varepsilon_q)\}-1\right] \tag{B.28}$$

Equation (B.28) is numerically unstable for small values of $\varepsilon_p-\varepsilon_q$, and it then becomes advantageous to use Eq. (B.25) directly to evaluate $f(p,q)$. If, for example, Eq. (B.25) is applied when $\varepsilon_p-\varepsilon_q<10^{-5}$, only the first few terms in the series in Eq. (B.25) are sufficient to give $g(p,q)$ to machine accuracy. The gradient elements in an iterative procedure based on both carrying out a linear and nonlinear transformation of the variables require evaluation of the elements $\langle\tilde{0}|[\tilde{a}_r^+\tilde{a}_s, H]|\tilde{0}\rangle$. The additional work required for constructing the gradient elements corresponding to a linear transformation is the diagonalization of ${}^0\boldsymbol{\kappa}$ and the matrix multiplications in Eq. (B.25).

B.2. Derivatives with Respect to State Parameters

The partial derivative of the total energy in Eq. (B.1) with respect to state parameters becomes using Eqs. (B.8), (B.9), and (B.10):

$$\begin{aligned}F({}^0\underline{\lambda})_n &= \left.\frac{\partial E({}^0\underline{\kappa}+\underline{\kappa}, {}^0\mathbf{S}+\mathbf{S})}{\partial S_{n0}}\right|_{\substack{\mathbf{S}=\mathbf{0}\\ \underline{\kappa}=\mathbf{0}}} \\ &= \langle 0|\left[\sum_m \frac{(-i)^m}{(m+1)!}\,{}^0\hat{\hat{S}}^m R_{n0}\right]\exp(-{}^0i\hat{S}) \\ &\quad\times\exp(-{}^0i\hat{\kappa})H\exp({}^0i\hat{\kappa})\exp({}^0i\hat{S})|0\rangle \\ &\quad-\langle 0|\exp(-{}^0iS)\exp(-{}^0i\hat{\kappa})H\exp({}^0i\hat{\kappa}) \\ &\quad\times\exp({}^0i\hat{S})\left[\sum_m \frac{(-i)^m}{(m+1)!}\,{}^0\hat{\hat{S}}^m R_{n0}\right]|0\rangle \\ &= \langle\tilde{0}|\left[\sum \frac{(-i)^m}{(m+1)!}\,{}^0\tilde{\hat{\hat{S}}}^m\tilde{R}_{n0}, H\right]|\tilde{0}\rangle\end{aligned} \tag{B.29}$$

where $|\tilde{0}\rangle$ is defined in Eq. (B.15) and the superoperator ${}^0\tilde{\hat{\hat{S}}}$ is defined through the relations

$$ {}^{0}\hat{\tilde{S}}f = \left[{}^{0}\tilde{S}, f\right], \qquad {}^{0}\tilde{S} = i\sum_{m} {}^{0}S_{m0}\tilde{R}_{m0} \tag{B.30}$$

$$ \tilde{R}_{m0} = |\tilde{m}\rangle\langle\tilde{0}| - |\tilde{0}\rangle\langle\tilde{m}| \tag{B.31}$$

The most straightforward way of evaluating Eq. (B.29) would be to reuse the method of the previous section and thus to diagonalize ${}^{0}\mathbf{S}$. If the number of the CI states is up in the range of 100–1000, this part of the iterative procedure may be computationally rather demanding. Because of the simple rank-2 structure of ${}^{0}\mathbf{S}$, an easier method may be developed.

Let us consider the series

$$ \sum_{m} \frac{(-i)^{m}}{(m+1)!}\,{}^{0}\hat{\tilde{S}}^{m}\tilde{R}_{n0} = \tilde{R}_{n0} - \frac{i}{2}\left[{}^{0}\tilde{S}, \tilde{R}_{n0}\right] + \frac{1}{3!}\left[{}^{0}\tilde{S}, \left[{}^{0}\tilde{S}, \tilde{R}_{n0}\right]\right] + \cdots \tag{B.32}$$

By denoting $(|\tilde{n}\rangle\langle\tilde{m}| - |\tilde{m}\rangle\langle\tilde{n}|)$ with $\tilde{R}_{nm}$ we may express the second and third terms of Eq. (B.32) as

$$ {}^{0}\hat{\tilde{S}}\tilde{R}_{n0} = \left[{}^{0}\tilde{S}, \tilde{R}_{n0}\right] = \sum i\,{}^{0}S_{m0}\tilde{R}_{nm} \tag{B.33}$$

$$ {}^{0}\hat{\tilde{S}}^{2}\tilde{R}_{n0} = \left[{}^{0}\tilde{S}, \left[{}^{0}\tilde{S}, \tilde{R}_{n0}\right]\right] = i\,{}^{0}S_{n0}\,{}^{0}\tilde{S} + P\tilde{R}_{n0}; \qquad P = \sum_{k} S_{k0}^{2} \tag{B.34}$$

Using Eq. (B.31), we obtain

$$ {}^{0}\hat{\tilde{S}}^{3}\tilde{R}_{n0} = \left[{}^{0}\tilde{S}, \left[{}^{0}\tilde{S}, \left[{}^{0}\tilde{S}, \tilde{R}_{n0}\right]\right]\right] = P\left[{}^{0}\tilde{S}, \tilde{R}_{n0}\right] = P\,{}^{0}\hat{\tilde{S}}\tilde{R}_{n0} \tag{B.35}$$

and Eqs. (B.33) and (B.34) may be generalized

$$ {}^{0}\hat{\tilde{S}}^{2n}\tilde{R}_{n0} = (P)^{n-1}\left(i\,{}^{0}S_{n0}\,{}^{0}\tilde{S} + P\tilde{R}_{n0}\right) \tag{B.36}$$

$$ {}^{0}\hat{\tilde{S}}^{2n+1}\tilde{R}_{n0} = i(P)^{n}\sum_{m} {}^{0}S_{m0}\tilde{R}_{nm} \tag{B.37}$$

We thus have

$$ \begin{aligned} \sum_{m=0}^{\infty} \frac{(-i)^{m}}{(m+1)!}\,{}^{0}\hat{\tilde{S}}^{m}\tilde{R}_{n0} &= \left(1 - \frac{P}{3!} + \frac{P^{2}}{5!} - \frac{P^{3}}{7!} + \cdots\right)\tilde{R}_{n0} \\ &\quad + \left(\frac{1}{3!} - \frac{P}{5!} + \frac{P^{2}}{7!} + \cdots\right){}^{0}S_{n0}\left(\sum_{m} {}^{0}S_{m0}\tilde{R}_{m0}\right) \\ &\quad + \left(\frac{1}{2} - \frac{P}{4!} + \frac{P^{2}}{6!} + \cdots\right)\sum_{m} {}^{0}S_{m0}\tilde{R}_{nm} \end{aligned} \tag{B.38}$$

Denoting the series

$$\psi = 1 - \frac{P}{3!} + \frac{P^2}{5!} - \frac{P^3}{7!} + \cdots \tag{B.39}$$

$$\phi = \frac{1}{3!} - \frac{P}{5!} + \frac{P^2}{7!} - \cdots \tag{B.40}$$

we may write the gradient elements $F({}^0\underline{\lambda})_n$ as

$$F({}^0\underline{\lambda})_n = \psi\langle\tilde{0}|[\tilde{R}_{n0}, H]|\tilde{0}\rangle + \phi\,{}^0S_{n0}\sum_m {}^0S_{m0}\langle\tilde{0}|[\tilde{R}_{m0}, H]|\tilde{0}\rangle \tag{B.41}$$

The calculation of ψ and ϕ is very simple. The additional work required for constructing the gradient elements in the configuration space corresponding to a linear transformation among the variables is thus the summations in the last term of Eq. (B.41), which are extremely simple to carry out even for very large configuration spaces.

APPENDIX C: FORMULAS FOR (P′ᵀF(0)) AND (P′ᵀG(0)P′)

To simplify the computation of gradient-like and Hessian-like elements, we divide orbitals in three classes: class i, consisting of all inactive (completely occupied) orbitals; class a, consisting of all active (partly occupied) orbitals; and class s, consisting of all secondary (completely unoccupied) orbitals. Orbitals from class i are denoted $i, j, k, \ldots$; orbitals from class a are denoted $a, b, c, \ldots$; orbitals from class s are denoted $s, t, u, \ldots$. General indices are denoted o, p, q, r.

Four types of orbital excitations exist:

$$\begin{aligned} &a_a^+ a_i - a_i^+ a_a \\ &a_s^+ a_i - a_i^+ a_s \\ &a_a^+ a_b - a_b^+ a_a \\ &a_s^+ a_a - a_a^+ a_s \end{aligned} \tag{C.1}$$

By taking the specific nature of an orbital excitation into account, several simplifications can be obtained. The number of commutators in a given expression can be increased for operators of the type $a_s^+ a_i - a_i^+ a_s$:

$$\langle 0|[a_s^+ a_i, H]|0\rangle = \langle 0|[[a_s^+, H], a_i]_+|0\rangle \tag{C.2}$$

By introducing the anticommutator an extra delta function is introduced, so the summations are reduced. The occurrence of an inactive orbital in the one-

and two-electron density matrices allows simplifications:

$$\begin{aligned}\langle 0|a_{i\sigma}^{+}a_{r\sigma}|0\rangle &= 2\delta_{i,r}\\ \langle 0|a_{i\sigma}^{+}a_{o\sigma'}^{+}a_{p\sigma'}a_{r\sigma}|0\rangle &= 2\delta_{i,r}\langle 0|a_{o\sigma'}^{+}a_{p\sigma'}|0\rangle - \delta_{i,p}\langle 0|a_{o\sigma'}^{+}a_{r\sigma'}|0\rangle\end{aligned} \tag{C.3}$$

In Eq. (C.3) we have explicitly written spin indices. As pointed out by Siegbahn et al.,[35] density matrices are thus only needed for indices corresponding to active orbitals.

The construction of relevant formulas are also eased by the use of the following relations:

$$\begin{aligned}\langle 0|[T_i^+ - T_i, H]|0\rangle &= 2\langle 0|[T_i^+, H]|0\rangle\\ \langle 0|[T_i^+ - T_i, T_j^+ - T_j, H]|0\rangle &= 2\langle 0|[T_i^+, T_j^+ - T_j, H]|0\rangle\\ \langle 0|[T_i^+, T_j^+ - T_j, H]|0\rangle &= \langle 0|[T_i^+, [T_j^+ - T_j, H]]|0\rangle\\ &\quad + \tfrac{1}{2}\langle 0|[[T_i^+, T_j^+ - T_j], H]|0\rangle\end{aligned} \tag{C.4}$$

The computations are further facilitated with the use of *symmetric* two-electron density matrices

$$\rho_{ad,bc}^{(2)} = \tfrac{1}{2}\langle 0|a_{a\sigma}^{+}a_{b\sigma'}^{+}a_{c\sigma'}a_{d\sigma}|0\rangle + \tfrac{1}{2}\langle 0|a_{a\sigma}^{+}a_{c\sigma'}^{+}a_{b\sigma'}a_{d\sigma}|0\rangle \tag{C.5}$$

and one-electron density matrices

$$\rho_{ab}^{(1)} = \langle 0|a_{a\sigma}^{+}a_{b\sigma}|0\rangle \tag{C.6}$$

A combination of the aforementioned relations together with Eqs. (83) and (84) allows reasonable, easy derivations of the following. Siegbahn et al.[35] give formulas for the gradient $\mathbf{F(0)}$ and $\mathbf{G(0)}$. We list here the formulas for $\mathbf{P'}^{T}\mathbf{F(0)}$ and $\mathbf{P'}^{T}\mathbf{G(0)P'}$, where $\mathbf{P'}$ is a direct-product matrix:

$$\mathbf{P'} = \left(\begin{array}{c|c}\mathbf{P}^{OO} & \mathbf{P}^{OC}\\ \hline \mathbf{P}^{CO} & \mathbf{P}^{CC}\end{array}\right)\left(\begin{array}{c|c}\mathbf{P}^{OO} & \mathbf{P}^{OC}\\ \hline \mathbf{P}^{CO} & \mathbf{P}^{CC}\end{array}\right) = \left(\begin{array}{c|c}\mathbf{P} & \mathbf{0}\\ \hline \mathbf{0} & \mathbf{1}\end{array}\right)\left(\begin{array}{c|c}\mathbf{P} & \mathbf{0}\\ \hline \mathbf{0} & \mathbf{1}\end{array}\right) \tag{C.7}$$

In Eq. (C.7) the indices have been partitioned in an orbital excitation part (O), and a state excitation part (C). The matrix $\mathbf{P}$ has the form

$$\left(\begin{array}{c|c|c}\mathbf{P}^{II} & \mathbf{0} & \mathbf{0}\\ \hline \mathbf{0} & \mathbf{1} & \mathbf{0}\\ \hline \mathbf{0} & \mathbf{0} & \mathbf{P}^{SS}\end{array}\right) \tag{C.8}$$

where the orbital space has been partitioned in its usual three parts. Construction of $\mathbf{P}'^{\mathrm{T}}\mathbf{F}(\mathbf{0})$ and $\mathbf{P}'^{\mathrm{T}}\mathbf{G}(\mathbf{0})\mathbf{P}'$ is required when the reduced transformations discussed in Section IX.A are used. To compact expressions we introduce the following:

$$
\begin{aligned}
(\tilde{o}\tilde{p}\,|\,\tilde{q}\tilde{r}) &= P_{o'o}P_{p'p}P_{q'q}P_{r'r}(o'p'\,|\,q'r') \\
\tilde{h}_{pq} &= P_{p'p}P_{q'q}h_{p'q'}
\end{aligned}
\tag{C.9}
$$

and modified Fock matrices

$$
\begin{aligned}
\tilde{D}_{ap} &= \rho^{(1)}_{ab}\tilde{D}^{\mathrm{I}}_{pb} + \rho^{(2)}_{ab,cd}(\tilde{p}\tilde{b}\,|\,\tilde{c}\tilde{d}) \\
\tilde{D}_{ip} &= 2(\tilde{D}^{\mathrm{I}}_{ip} + \tilde{D}^{\mathrm{A}}_{ip}) \\
\tilde{D}^{\mathrm{I}}_{pq} &= \tilde{h}_{pq} + [2(\tilde{p}\tilde{q}\,|\,ii) - (\tilde{p}i\,|\,\tilde{q}i)] \\
\tilde{D}^{\mathrm{A}}_{pq} &= \rho^{(1)}_{ab}[2(\tilde{p}\tilde{q}\,|\,\tilde{a}\tilde{b}) - \tfrac{1}{2}(\tilde{p}\tilde{a}\,|\,\tilde{q}\tilde{b})]
\end{aligned}
\tag{C.10}
$$

The two-electron indices in $\tilde{D}^{\mathrm{I}}$ in Eq. (C.10) involves two different sets of orbitals. This does not introduce complications since $\tilde{D}^{\mathrm{I}}$ usually is constructed directly from integrals over basis orbitals. The formulas below are equivalent to the formulas presented by Siegbahn et al. for $\mathbf{P} = \mathbf{1}$.

C.1. Formulas for $\mathbf{P}'^{\mathrm{T}}\mathbf{F}(0)$

Five types of elements $\mathbf{P}'^{\mathrm{T}}\mathbf{F}(0)$ exist:

$$
\begin{aligned}
(\mathbf{P}'^{\mathrm{T}}\mathbf{F}(0))_{ai} &= P'_{(bj)(ai)}\langle 0|[a^{+}_{b\sigma}a_{j\sigma} - a^{+}_{j\sigma}a_{b\sigma}, H]|0\rangle \\
&= P_{ji}\langle 0|[a^{+}_{a\sigma}a_{j\sigma} - a^{+}_{j\sigma}a_{a\sigma}, H]|0\rangle \\
&= 2(\tilde{D}_{ai} - \tilde{D}_{ia}) \\
&= 2\rho^{(1)}_{ac}D^{\mathrm{I}}_{ac} + 2\rho^{(2)}_{ac,de}(\tilde{i}c\,|\,de) - 4(\tilde{D}^{\mathrm{I}}_{ia} + \tilde{D}^{\mathrm{A}}_{ia})
\end{aligned}
\tag{C.11a}
$$

$$
\begin{aligned}
(\mathbf{P}'^{\mathrm{T}}\mathbf{F}(0))_{si} &= 2(\tilde{D}_{is} - \tilde{D}_{si}) \\
&= -4(\tilde{D}^{\mathrm{I}}_{is} + \tilde{D}^{\mathrm{A}}_{is})
\end{aligned}
\tag{C.11b}
$$

$$
\begin{aligned}
(\mathbf{P}'^{\mathrm{T}}\mathbf{F}(0))_{ab} = &-2\rho^{(1)}_{bc}\tilde{D}^{\mathrm{I}}_{ac} + 2\rho^{(1)}_{ac}\tilde{D}^{\mathrm{I}}_{bc} \\
&-2\rho^{(2)}_{bc,de}(\tilde{a}\tilde{c}\,|\,\tilde{d}\tilde{e}) \\
&+2\rho^{(2)}_{ac,de}(\tilde{b}\tilde{c}\,|\,\tilde{d}\tilde{e})
\end{aligned}
\tag{C.11c}
$$

$$
(\mathbf{P}'^{\mathrm{T}}\mathbf{F}(0))_{sa} = -2\rho^{(1)}_{ab}\tilde{D}^{\mathrm{I}}_{sb} - 2\rho^{(2)}_{ab,cd}(\tilde{s}\tilde{b}\,|\,\tilde{c}\tilde{d})
\tag{C.11d}
$$

The "derivative" with respect to a state variable is

$$(\mathbf{P}'^{\mathrm{T}}\mathbf{F}(0))_n = \langle 0|[|n\rangle\langle 0| - |0\rangle\langle n|, H]|0\rangle = -2\langle 0|H|n\rangle \quad \text{(C.11e)}$$

The transformed gradient ($\mathbf{P}'\mathbf{F}(\mathbf{0})$) can thus be calculated from $\tilde{D}^{\mathrm{I}}$, $\tilde{D}^{\mathrm{A}}$ and two-electron integrals $(\tilde{a}\tilde{b}|\tilde{c}\tilde{p})$. This is the basis for the reduced T_1 transformation discussed in Section IX.A.

C.2. Formulas for ($\mathbf{P}'^{\mathrm{T}}\mathbf{G}(0)\mathbf{P}'$)

There are 10 formulas for the orbital–orbital Hessian, four formulas for the orbital–state part of the Hessian, and one formula for the state–state part of the Hessian:

$$\begin{aligned}(\mathbf{P}'^{\mathrm{T}}\mathbf{G}(0)\mathbf{P}')_{ai,bj} = {} & 2\left[\rho^{(2)}_{ba,cd}(\tilde{c}\tilde{d}|\tilde{i}\tilde{j}) + 2\rho^{(2)}_{bd,ca}(\tilde{c}\tilde{i}|\tilde{d}\tilde{j})\right] \\ & + 2(\delta_{ac} - \rho^{(1)}_{ac})(4(\tilde{c}\tilde{i}|\tilde{b}\tilde{j}) - (\tilde{b}\tilde{i}|\tilde{c}\tilde{j}) - (\tilde{b}\tilde{c}|\tilde{i}\tilde{j})) \\ & + 2(\delta_{bc} - \rho^{(1)}_{bc})(4(\tilde{c}\tilde{i}|\tilde{a}\tilde{i}) - (\tilde{a}\tilde{j}|\tilde{c}\tilde{i}) - (\tilde{a}\tilde{c}|\tilde{i}\tilde{j})) \\ & + 2\rho^{(1)}_{ab}\tilde{D}^{\mathrm{I}}_{ab} + 2(\mathbf{P}^{\mathrm{T}}\mathbf{P})_{ij}(2\tilde{D}^{\mathrm{I}}_{ab} + 2\tilde{D}^{\mathrm{A}}_{ab} - \tilde{D}_{ba}) \\ & - 4\delta_{ab}(\tilde{D}^{\mathrm{I}}_{ij} + \tilde{D}^{\mathrm{A}}_{ij}) \end{aligned} \quad \text{(C.12a)}$$

$$\begin{aligned}(\mathbf{P}'^{\mathrm{T}}\mathbf{G}(0)\mathbf{P}')_{ai,sj} = {} & 2(2\delta_{ab} - \rho^{(1)}_{ab})(4(\tilde{s}\tilde{j}|\tilde{b}\tilde{i}) - (\tilde{s}\tilde{b}|\tilde{i}\tilde{j}) - (\tilde{s}\tilde{i}|\tilde{b}\tilde{j})) \\ & + 4(\mathbf{P}^{\mathrm{T}}\mathbf{P})_{ij}(\tilde{D}^{\mathrm{I}}_{sa} + \tilde{D}^{\mathrm{A}}_{sa}) - (\mathbf{P}^{\mathrm{T}}\mathbf{P})_{ij}\tilde{D}_{as} \end{aligned} \quad \text{(C.12b)}$$

$$\begin{aligned}(\mathbf{P}'^{\mathrm{T}}\mathbf{G}(0)\mathbf{P}')_{ai,bc} = {} & 4\left[(\tilde{d}\tilde{i}|\tilde{c}\tilde{e})\rho^{(2)}_{ad,be} - (\tilde{d}\tilde{i}|\tilde{b}\tilde{e})\rho^{(2)}_{ad,ce}\right] \\ & + 2\left[(\tilde{c}\tilde{i}|\tilde{d}\tilde{e})\rho^{(2)}_{ab,de} - (\tilde{b}\tilde{i}|\tilde{d}\tilde{e})\rho^{(2)}_{ac,de}\right] \\ & - 2\rho^{(1)}_{bd}\left[4(\tilde{a}\tilde{i}|\tilde{c}\tilde{d}) - (\tilde{a}\tilde{d}|\tilde{c}\tilde{i}) - (\tilde{a}\tilde{c}|\tilde{d}\tilde{i})\right] \\ & + 2\rho^{(1)}_{cd}\left[4(\tilde{a}\tilde{i}|\tilde{b}\tilde{d}) - (\tilde{a}\tilde{d}|\tilde{b}\tilde{i}) - (\tilde{a}\tilde{b}|\tilde{d}\tilde{i})\right] \\ & + 2\rho^{(1)}_{ab}\tilde{D}^{\mathrm{I}}_{ic} - 2\delta_{ab}\tilde{D}_{ci} - 2\rho^{(1)}_{ac}\tilde{D}_{ib} + 2\delta_{ac}\tilde{D}_{bi} \end{aligned} \quad \text{(C.12c)}$$

$$\begin{aligned}(\mathbf{P}'^{\mathrm{T}}\mathbf{G}(0)\mathbf{P}')_{ai,sb} = {} & -2\left[\rho^{(2)}_{ab,cd}(\tilde{s}\tilde{i}|\tilde{c}\tilde{d}) + 2\rho^{(2)}_{ac,bd}(\tilde{s}\tilde{d}|\tilde{c}\tilde{i})\right] \\ & + 2\rho^{(1)}_{bc}(4(\tilde{s}\tilde{c}|\tilde{a}\tilde{i}) - (\tilde{s}\tilde{i}|\tilde{a}\tilde{c}) - (\tilde{s}\tilde{a}|\tilde{c}\tilde{i})) \\ & - 2\rho^{(1)}_{ab}\tilde{D}^{\mathrm{I}}_{si} + 2\delta_{ab}(\tilde{D}^{\mathrm{I}}_{si} + \tilde{D}^{\mathrm{A}}_{si}) \end{aligned} \quad \text{(C.12d)}$$

$$
\begin{aligned}
(\mathbf{P}'^{\mathrm{T}}\mathbf{G}(0)\mathbf{P}')_{ab,cd} = {} & 4\big[(\tilde{a}\tilde{e}|\tilde{c}\tilde{f})\rho^{(2)}_{be,df} + (\tilde{b}\tilde{e}|\tilde{d}\tilde{f})\rho^{(2)}_{ae,cf} \\
& - (\tilde{b}\tilde{e}|\tilde{c}\tilde{f})\rho^{(2)}_{ae,df} - (\tilde{a}\tilde{e}|\tilde{d}\tilde{f})\rho^{(2)}_{be,df}\big] \\
& + 2\big[(\tilde{a}\tilde{c}|\tilde{e}\tilde{f})\rho^{(2)}_{bd,ef} + (\tilde{b}\tilde{d}|\tilde{e}\tilde{f})\rho^{(2)}_{ac,ef} \\
& - (\tilde{a}\tilde{d}|\tilde{e}\tilde{f})\rho^{(2)}_{bc,ef} - (\tilde{b}\tilde{c}|\tilde{e}\tilde{f})\rho^{(2)}_{ad,ef}\big] \\
& + 2\big[\rho^{(1)}_{bd}\tilde{D}^{\mathrm{I}}_{ac} + \rho^{(1)}_{ac}\tilde{D}^{\mathrm{I}}_{bd} - \rho^{(1)}_{ad}\tilde{D}^{\mathrm{I}}_{bc} - \rho^{(1)}_{bc}\tilde{D}^{\mathrm{I}}_{ad}\big] \\
& + 2\big[\delta_{ad}\tilde{D}_{cb} + \delta_{bc}\tilde{D}_{da} - \delta_{bd}\tilde{D}_{ca} - \delta_{ac}\tilde{D}_{db}\big]
\end{aligned}
\tag{C.12e}
$$

$$
\begin{aligned}
(P'^{\mathrm{T}}G(0)P')_{ab,sc} = {} & 4\big[(\tilde{a}\tilde{d}|\tilde{e}\tilde{s})\rho^{(2)}_{bd,ce} - (\tilde{b}\tilde{d}|\tilde{e}\tilde{s})\rho^{(2)}_{ad,ce}\big] \\
& + 2\big[(\tilde{a}\tilde{s}|\tilde{d}\tilde{e})\rho^{(2)}_{bc,de} - (\tilde{b}\tilde{s}|\tilde{d}\tilde{e})\rho^{(2)}_{ac,de}\big] \\
& + 2\rho^{(1)}_{bc}\tilde{D}^{\mathrm{I}}_{sa} - 2\rho^{(1)}_{ac}\tilde{D}^{\mathrm{I}}_{sb}
\end{aligned}
\tag{C.12f}
$$

$$
\begin{aligned}
(\mathbf{P}'^{\mathrm{T}}\mathbf{G}(0)\mathbf{P}')_{si,tj} = {} & 4(4(\tilde{s}\tilde{i}|\tilde{t}\tilde{j}) - (\tilde{s}\tilde{t}|\tilde{i}\tilde{j}) - (\tilde{s}\tilde{j}|\tilde{i}\tilde{t})) \\
& + 4(\mathbf{P}^{\mathrm{T}}\mathbf{P})_{ij}(\tilde{D}^{\mathrm{I}}_{st} + \tilde{D}^{\mathrm{A}}_{st}) \\
& - 4(\mathbf{P}^{\mathrm{T}}\mathbf{P})_{st}(\tilde{D}^{\mathrm{I}}_{ij} + \tilde{D}^{\mathrm{A}}_{ij})
\end{aligned}
\tag{C.12g}
$$

$$
\begin{aligned}
(\mathbf{P}'^{\mathrm{T}}\mathbf{G}(0)\mathbf{P}')_{si,ab} = {} & 2\rho^{(1)}_{bc}\big[4(\tilde{a}\tilde{c}|\tilde{i}\tilde{s}) - (\tilde{a}\tilde{i}|\tilde{c}\tilde{s}) - (\tilde{a}\tilde{s}|\tilde{c}\tilde{i})\big] \\
& - 2\rho^{(1)}_{ac}\big[4(\tilde{b}\tilde{c}|\tilde{i}\tilde{s}) - (\tilde{b}\tilde{i}|\tilde{c}\tilde{s}) - (\tilde{b}\tilde{s}|\tilde{c}\tilde{i})\big]
\end{aligned}
\tag{C.12h}
$$

$$
\begin{aligned}
(\mathbf{P}'^{\mathrm{T}}\mathbf{G}(0)\mathbf{P}')_{si,ta} = {} & 2\rho^{(1)}_{ab}(4(\tilde{s}\tilde{i}|\tilde{t}\tilde{b}) - (\tilde{s}\tilde{b}|\tilde{t}\tilde{i}) - (\tilde{s}\tilde{t}|\tilde{b}\tilde{i})) \\
& - 2(\mathbf{P}^{\mathrm{T}}\mathbf{P})_{st}(2\tilde{D}^{\mathrm{I}}_{ai} + 2\tilde{D}^{\mathrm{A}}_{ai} - \tilde{D}_{ai})
\end{aligned}
\tag{C.12i}
$$

$$
\begin{aligned}
(\mathbf{P}'^{\mathrm{T}}\mathbf{G}(0)\mathbf{P}')_{sa,tb} = {} & 2(\rho^{(2)}_{ab,cd}(\tilde{s}\tilde{t}|\tilde{c}\tilde{d}) + 2\rho^{(2)}_{ad,bc}(\tilde{s}\tilde{d}|\tilde{t}\tilde{c})) \\
& + 2\rho^{(1)}_{ab}\tilde{D}^{\mathrm{I}}_{st} - 2(\mathbf{P}^{\mathrm{T}}\mathbf{P})_{st}\tilde{D}_{ba}
\end{aligned}
\tag{C.12j}
$$

Coupling-elements have the form

$$
(\mathbf{P}'^{\mathrm{T}}\mathbf{G}(0))_{pq,n} = -2P_{p'p}P_{q'q}\langle 0|\big[a^{+}_{p'\sigma}a_{q'\sigma} - a^{+}_{q'\sigma}a_{p'\sigma}, H\big]|n\rangle \tag{C.12k}
$$

These elements have the same structure as $\mathbf{P}'^{\mathrm{T}}\mathbf{F}(0)$ [Eq. (C.11)], but regular density matrices have been replaced by transition density matrices. The four types of coupling elements can thus be obtained from Eq. (C.11e) by replacing *symmetric* density matrices with symmetric transition density matrices.

The state–state part of $\mathbf{P}'^{\mathrm{T}}\mathbf{G}(0)\mathbf{P}'$ becomes

$$\mathbf{G}(0)_{nm} = +2\langle n|H|m\rangle - 2\delta_{nm}\langle 0|H|0\rangle \tag{C.121}$$

These elements can be calculated from integrals over active orbitals and $\mathbf{D}^{\mathrm{I}}$.

It has thus been demonstrated that $\mathbf{P}'^{\mathrm{T}}\mathbf{G}(0)\mathbf{P}'$, where $\mathbf{P}'$ has the structure indicated in Eqs. (C.7) and (C.8), can be calculated from integrals in the tilde basis and a one-electron operator. This is the basis for the reduced integral transformations T_2 and T_2' discussed in Section IX.A.

APPENDIX D: EIGENVALUES OF THE FULL HESSIAN AND THE REDUCED HESSIAN MATRIX

One desirable feature of the nth excited state of a given symmetry is that the Hessian matrix for that state has n negative eigenvalues.[13, 15, 16] In the one-step second-order approach, the Hessian matrix

$$\begin{pmatrix} \mathbf{G}^{\mathrm{OO}} & \mathbf{G}^{\mathrm{OC}} \\ \mathbf{G}^{\mathrm{CO}} & \mathbf{G}^{\mathrm{CC}} \end{pmatrix}$$

is part of the iterative function [see Eqs. (60), (75), and (76)] and the number of negative eigenvalues of the Hessian matrix may therefore be used to direct the calculation to the desired stationary point. In the two-step second order procedure, it is the reduced Hessian matrix $\mathbf{G}^{\mathrm{OO}} - \mathbf{G}^{\mathrm{OC}}(\mathbf{G}^{\mathrm{CC}})^{-1}\mathbf{G}^{\mathrm{CO}}$ and the configuration block $\mathbf{G}^{\mathrm{CC}}$ that appear explicitly in the iterative function [see Eq. (80)]. Therefore in a two-step calculation it becomes important, in order to direct the calculation to the desired stationary point, to know how the negative eigenvalues of the nth excited state Hessian matrix get distributed into the reduced Hessian matrix and the configuration interaction matrix. We show in this appendix that if the Hessian matrix has n negative eigenvalues, and $\mathbf{G}^{\mathrm{CC}}$ contains m negative eigenvalues, then the reduced Hessian matrix must contain $n - m$ negative eigenvalues. We use multidimensional partitioning technique to show this point.[41] For the case where there are n total negative eigenvalues, all in the $\mathbf{G}^{\mathrm{CC}}$ block of the Hessian matrix, Shepard et al.[32] have previously proved, using multidimensional partitioning theory, that the reduced Hessian matrix is positive definite.

The eigenvalue problem for the Hessian matrix

$$\begin{pmatrix} \mathbf{G}^{\mathrm{OO}} & \mathbf{G}^{\mathrm{OO}} \\ \mathbf{G}^{\mathrm{CO}} & \mathbf{G}^{\mathrm{CC}} \end{pmatrix} \begin{pmatrix} \mathbf{U}^{\mathrm{O}} \\ \mathbf{U}^{\mathrm{O}} \end{pmatrix} = \begin{pmatrix} \mathbf{U}^{\mathrm{O}} \\ \mathbf{U}^{\mathrm{C}} \end{pmatrix} E \tag{D.1}$$

transforms, using partitioning theory, to an equation of the reduced dimension

$$\left(\mathbf{G}^{\mathrm{OO}}+\mathbf{G}^{\mathrm{OC}}(E\mathbf{1}-\mathbf{G}^{\mathrm{CC}})^{-1}\mathbf{G}^{\mathrm{CO}}\right)\mathbf{U}^{\mathrm{O}}=\mathbf{U}^{\mathrm{O}}E \tag{D.2}$$

If we introduce the matrix function

$$\mathbf{L}(E)=\mathbf{G}^{\mathrm{OO}}+\mathbf{G}^{\mathrm{OC}}(E\mathbf{1}-\mathbf{G}^{\mathrm{CC}})^{-1}\mathbf{G}^{\mathrm{CO}} \tag{D.3}$$

and define the multivalued function $\underline{\varepsilon}(E)$ to contain the eigenvalues of $\mathbf{L}(E)$, then from the partitioned form of the Hessian eigenvalue equation [Eq. (D.3)] it is clear that the eigenvalues of the Hessian matrix occur at the values of $\mathbf{L}(E)$ for which

$$E=\underline{\varepsilon}(E) \tag{D.4}$$

In Fig. 1 (see p. 61) we have plotted the multivalued function $\underline{\varepsilon}(E)$ as a function of E for the case where the configuration space $\mathbf{G}^{\mathrm{CC}}$ has a dimension of 4 and the orbital excitation space a dimension of 6. Generalization to arbitrary dimensions is straightforward. The branches of $\underline{\varepsilon}(E)$ have horizontal asymptotes at the eigenvalues of $\mathbf{G}^{\mathrm{OO}}$ denoted $\mathrm{O}_1, \mathrm{O}_2, \ldots, \mathrm{O}_6$ and vertical asymptotes at the eigenvalues of $\mathbf{G}^{\mathrm{CC}}$ denoted C_1, C_2, C_3, and C_4. Each branch of $\underline{\varepsilon}(E)$ is a nonincreasing function of E, and except when the matrices involved have a very special structure the individual branches of $\underline{\varepsilon}(E)$ satisfy a noncrossing rule.[41] The eigenvalues of the reduced Hessian matrix are values of $\underline{\varepsilon}(E)$ for $E=0$. A comprehensive discussion of the structural characteristics of a multidimensional function $\underline{\varepsilon}(E)$ has been given by Löwdin.[41] In Fig. 1 the Hessian matrix has two negative eigenvalues [points of intersection with $E=\underline{\varepsilon}(E)$] and $\mathbf{G}^{\mathrm{CC}}$ has one negative eigenvalue. We thus assume that we are converging to the second excited state of a given symmetry and that root flipping occurs. From the figure it is clear that the one negative eigenvalue of $\mathbf{G}^{\mathrm{CC}}$ introduces one negative eigenvalue in the Hessian matrix. As a result of the root flipping the second branch of $\underline{\varepsilon}(E)$, which results in the second negative eigenvalue of the Hessian matrix, has a positive asymptote (C_2). Because of its crossing with the line $E=\underline{\varepsilon}(E)$ at the negative E value, this branch must cross the line at $E=0$ for a negative function value of $\underline{\varepsilon}(0)$ of more negative value (larger magnitude) than the crossing at $E=\underline{\varepsilon}(E)$. The reduced Hessian therefore has one negative eigenvalue.

A generalization of this result to the case where the Hessian has n negative eigenvalues and $\mathbf{G}^{\mathrm{CC}}$ has m negative eigenvalues is simple and straightforwardly yields the conclusion that the reduced Hessian has $n-m$ negative eigenvalues. We should mention that of course $\mathbf{G}^{\mathrm{OO}}$ should have no more than n negative eigenvalues.

APPENDIX E: CONSTRUCTION OF CUBIC CONTRIBUTIONS FOR THE ITERATIVE CUBIC AND PERTURBATIVE (CHEBYSHEV) CUBIC PROCEDURES

E.1. The Cubic Formulas

In this appendix we describe in detail how to calculate the cubic contributions in Eqs. (137) and (208).[16]

The direct calculation of the cubic supermatrix K_{ijk} in Eq. (15) when ${}^0\underline{\lambda} = \mathbf{0}$ is difficult for several reasons. First, there are approximately $N^3 \cdot n^3$ terms, where N is the total number of orbitals and n is the number of occupied orbitals. Second is the fact that all of the density matrices $\langle n | a_i^+ a_j^+ a_k a_l | m \rangle$, where $|n\rangle$ and $|m\rangle$ are the MCSCF CI states, are required. Third is the fact that the calculation of $K_{ijk}\lambda_k$ requires about $N^3 n^3$ multiplications. We develop here an alternative method based on the fact that cubic terms always occur as $K_{ijk} v_k$, where $\mathbf{v}$ is a vector and, hence, we calculate $K_{ijk} v_k$ directly. In Eq. (208) $\mathbf{v}$ is the vector $\underline{\lambda}$, and in Eq. (137) $\mathbf{v}$ is ${}^{\mathrm{NR}}\underline{\lambda}$.

Writing out $K_{ijk}\lambda_k$ explicitly, we get

$$K_{ijk}\lambda_k = \langle 0 | \left[T_i^+ - T_i, T_j^+ - T_j, T_k^+ - T_k, H \right] | 0 \rangle \lambda_k \tag{E.1}$$

where $\mathbf{T}^+$ and $\underline{\lambda}$ are defined in Eqs. (52) and (51), respectively. Introducing the shorthand notation

$$T_i^+ - T_i = \begin{pmatrix} Q_i^+ - Q_i \\ R_i^+ - R_i \end{pmatrix} = X_i = \begin{pmatrix} U_i \\ V_i \end{pmatrix} \tag{E.2}$$

we may write Eq. (E.1) as

$$K_{ijk}\lambda_k = \langle 0 | \left[X_i, X_j, X_k, H \right] | 0 \rangle \lambda_k \tag{E.3}$$

We will evaluate Eq. (E.3) by considering the four possible cases: (1) X_i and X_j both refer to orbital excitation operators; (2) X_i refers to an orbital excitation operator and X_j to a state excitation operator; (3) X_i refers to a state excitation operator and X_j to an orbital excitation operator; and (4) both X_i and X_j refer to state excitation operators.

Case (1): i, j both orbital indices. Equation (E.1) may now be written as

$$K_{ijk}\lambda_k = \langle 0 | \left[U_i, U_j, U_k, H \right] | 0 \rangle \kappa_k + \langle 0 | \left[V_k, \left[U_i, U_j, H \right] \right] | 0 \rangle S_{k0} \tag{E.4}$$

The first term in Eq. (E.4) may be written as

$$\tfrac{1}{6}\{\langle 0|[U_i,[U_j,[U_k,H]]]|0\rangle + \langle 0|[U_i,[U_k,[U_j,H]]]|0\rangle + \langle 0|[U_j,[U_i,[U_k,H]]]|0\rangle + \langle 0|[U_j,[U_k,[U_i,H]]]|0\rangle + \langle 0|[U_k,[U_i,[U_j,H]]]|0\rangle + \langle 0|[U_k,[U_j,[U_i,H]]]|0\rangle\}\kappa_k \quad \text{(E.5)}$$

If we commute the operators in Eq. (E.5) such that in terms of the type $\langle 0|[U,[U,[U,\mathrm{H}]]]|0\rangle$, U_k becomes commuted with H, that is, $[U_k, H]$, we obtain

$$\begin{aligned}\langle 0|[U_i,U_j,U_k,H]|0\rangle\kappa_k = {} & \tfrac{1}{2}\langle 0|[U_i,[U_j,[U_k,H]]]|0\rangle\kappa_k \\ & + \tfrac{1}{2}\langle 0|[U_j,[U_i,[U_k,H]]]|0\rangle\kappa_k \\ & + \tfrac{1}{3}\langle 0|[U_i,[[U_k,U_j],H]]|0\rangle\kappa_k \\ & + \tfrac{1}{3}\langle 0|[U_j,[[U_k,U_i],H]]|0\rangle\kappa_k \\ & + \tfrac{1}{6}\langle 0|[[U_k,U_i],[U_j,H]]|0\rangle\kappa_k \\ & + \tfrac{1}{6}\langle 0|[[U_k,U_j],[U_i,H]]|0\rangle\kappa_k \end{aligned} \quad \text{(E.6)}$$

Denoting

$$[\hat{\kappa}, H] = \tilde{H},$$

and commuting terms in Eq. (E.6), we obtain

$$\begin{aligned}\langle 0|[U_i,U_j,U_k,H]|0\rangle\kappa_k = {} & \langle 0|[U_i,U_j,\tilde{H}]|0\rangle + \tfrac{1}{2}\langle 0|[[U_k,U_j],[U_i,H]]|0\rangle\kappa_k \\ & + \tfrac{1}{2}\langle 0|[[U_k,U_i],[U_j,H]]|0\rangle\kappa_k \\ & + \tfrac{1}{2}\langle 0|[[U_k,U_j],[U_i,H]]|0\rangle\kappa_k \\ & + \tfrac{1}{6}\langle 0|[[U_i,[U_k,U_j]],H]|0\rangle\kappa_k \\ & + \tfrac{1}{6}\langle 0|[[U_j,[U_k,U_i]],H]|0\rangle\kappa_k \end{aligned} \quad \text{(E.7)}$$

It may readily be seen that the term $\langle 0|[U_i, U_j, \tilde{H}]|0\rangle$ corresponds to a Hessian element with a modified Hamiltonian H, which we now evaluate: H has the form

$$H = \sum_{rs} h_{rs} a_r^+ a_s + \tfrac{1}{2}\sum_{rstu} \langle rs|ut\rangle a_r^+ a_s^+ a_t a_u$$

and

$$\tilde{H} = [\hat{\kappa}, H] = \sum_{rs} \tilde{h}_{rs} a_r^+ a_s + \tfrac{1}{2} \sum_{rstu} \langle rs\tilde{|}ut\rangle a_r^+ a_s^+ a_t a_u \tag{E.8}$$

where

$$\tilde{h}_{rs} = \sum_p (h_{rp}\kappa_{sp} + h_{ps}\kappa_{rp})$$

$$\langle rs\tilde{|}tu\rangle = \sum_p (\kappa_{rp}\langle ps|tu\rangle + \kappa_{sp}\langle rp|tu\rangle + \kappa_{tp}\langle rs|pu\rangle + \kappa_{up}\langle rs|tp\rangle) \tag{E.9}$$

The modified Hamiltonian $\tilde{H}$ is thus identical to H with transformed integrals. However, the integral transformation is a *one-index* transformation and may therefore easily be carried out. The residual terms of Eq. (E.7) are just straightforward Hessian and GBT matrix elements (i.e., the commutator of two one-body operators is a sum of one-body operators). We also point out that

$$\langle 0|[[U_k, U_j], [U_i, H]]|0\rangle\kappa_k \tag{E.10}$$

contains some terms of the form

$$\langle 0|[a_{cl}^+ a_{cl'}, [U_i, H]]|0\rangle \quad \text{and} \quad \langle 0|[a_{un}^+ a_{un'}, [U_i, H]]|0\rangle \tag{E.11}$$

where cl and cl' refer to orbitals that are completely occupied and un and un' refer to orbitals which are completely unoccupied. These terms, however, become zero as $a_{cl}^+ a_{cl'}$ or $a_{un}^+ a_{un'}$ operates directly on the reference state. This is why the commutators $[U_k, U_j]$ are chosen to be the outer part of the commutator operator string.

To simplify the second term in Eq. (E.4), it is convenient to introduce the notation

$$|\delta 0\rangle = -\sum S_{n0} R_n^+ |0\rangle = -\sum_n S_{n0}|n\rangle \tag{E.12}$$

With this notation we may rewrite the second term of Eq. (E.4) as

$$\langle 0|[V_n, [U_i, U_j, H]]|0\rangle S_{n0} = \langle 0|[U_i, U_j, H]|\delta 0\rangle + \langle \delta 0|[U_i, U_j, H]|0\rangle \tag{E.13}$$

Equation (E.13) thus corresponds to a Hessian element with modified density matrix elements

$$\langle 0|a_r^+ a_s|0\rangle \rightarrow \langle 0|a_r^+ a_s|\delta 0\rangle + \langle \delta 0|a_r^+ a_s|0\rangle$$
$$\langle 0|a_r^+ a_s^+ a_t a_u|0\rangle \rightarrow \langle \delta 0|a_r^+ a_s^+ a_t a_u|0\rangle + \langle 0|a_r^+ a_s^+ a_t a_u|\delta 0\rangle \quad \text{(E.14)}$$

We have thus shown how the cubic term in Eq. (E.4) may be efficiently evaluated.

Case (2) i orbital index, j state index. In this case we may express the cubic term in Eq. (E.3) as

$$K_{ijk}\lambda_k = \langle 0|\left[V_j,[U_i,U_k,H]\right]|0\rangle\kappa_k + \langle 0|\left[V_j,V_k,[U_i,H]\right]|0\rangle S_{k0} \quad \text{(E.15)}$$

We consider initially the first term of Eq. (E.15)

$$\langle 0|\left[V_j,[U_i,U_k,H]\right]|0\rangle\kappa_k = \langle 0|\left[V_j,[U_i,[U_k,H]]\right]|0\rangle\kappa_k$$
$$+\tfrac{1}{2}\langle 0|\left[V_j,[[U_k,U_i],H]\right]|0\rangle\kappa_k \quad \text{(E.16)}$$

The first term on the right-hand side of Eq. (E.16) is a Hessian element with the modified Hamiltonian

$$\langle 0|\left[V_j,[U_i,\tilde{H}]\right]|0\rangle \quad \text{(E.17)}$$

whereas the second term in Eq. (E.16) is a sum of standard Hessian elements.

Denoting

$$|\delta j\rangle = S_{j0}|0\rangle \quad \text{(E.18)}$$

we may express the second term in Eq. (E.15) as

$$\langle 0|\left[V_j,V_k,[U_i,H]\right]|0\rangle S_{k0} = -2\langle \delta j|[U_i,H]|0\rangle - 2\langle 0|[U_i,H]|\delta 0\rangle \quad \text{(E.19)}$$

Eq. (E.19) describes coupling elements of the Hessian with modified density matrices

$$\langle j|a_r^+ a_s|0\rangle \rightarrow \langle \delta j|a_r^+ a_s|0\rangle + \langle 0|a_r^+ a_s|\delta 0\rangle$$
$$\langle j|a_r^+ a_s^+ a_t a_u|0\rangle \rightarrow \langle \delta j|a_r^+ a_s^+ a_t a_u|0\rangle + \langle j|a_r^+ a_s^+ a_t a_u|\delta 0\rangle \quad \text{(E.20)}$$

We have now shown that Eq. (E.15) may be rewritten as two modified coupling elements of the Hessian matrix plus an additional sum of ordinary Hessian coupling elements.

Case (3): i state index, j orbital index. Since $K_{ijk}\lambda_k = K_{jik}\lambda_k$, this term may be straightforwardly evaluated from Case (2), above.

Case (4): i, j both state indices. The cubic term in Eq. (E.3) may now be written as

$$K_{ijk}\lambda_k = \langle 0|\left[V_i, V_j, [U_k, H]\right]|0\rangle\kappa_k + \langle 0|\left[V_i, V_j, V_k, H\right]|0\rangle S_{k0} \quad \text{(E.21)}$$

The first term in Eq. (E.21) may straightforwardly be expressed as

$$2\langle i|\tilde{H}|j\rangle - 2\delta_{ij}\langle 0|\tilde{H}|0\rangle \quad \text{(E.22)}$$

which corresponds to configuration interaction matrix elements of the modified Hamiltonian of Eq. (E.8). The second term in Eq. (E.21) may be written as

$$\begin{aligned}\langle 0|\left[V_i, V_j, V_k, H\right]|0\rangle S_{k0} = \tfrac{1}{6}\{&\langle 0|\left[V_i, \left[V_j[V_k, H]\right]\right]|0\rangle \\ &+ \langle 0|\left[V_i, \left[V_k, \left[V_j, H\right]\right]\right]|0\rangle \\ &+ \langle 0|\left[V_j, \left[V_i, \left[V_k, H\right]\right]\right]|0\rangle \\ &+ \langle 0|\left[V_j, \left[V_k, \left[V_i, H\right]\right]\right]|0\rangle \\ &+ \langle 0|\left[V_k, \left[V_i, \left[V_j, H\right]\right]\right]|0\rangle \\ &+ \langle 0|\left[V_k, \left[V_j, \left[V_i, H\right]\right]\right]|0\rangle\}S_{k0} \quad \text{(E.23)}\end{aligned}$$

Expanding the V operators gives

$$\langle 0|\left[V_i, V_j, V_k, H\right]|0\rangle S_{k0} = \tfrac{8}{3}S_{i0}\langle 0|H|j\rangle + \tfrac{8}{3}S_{j0}\langle i|H|0\rangle + \tfrac{8}{3}\delta_{ij}S_{k0}\langle k|H|0\rangle \quad \text{(E.24)}$$

such that the term in Eq. (E.24) is just a sum of standard CI matrix elements. Of course, in the two-step MCSCF procedure, Eq. (E.24) is identically zero. Equation (E.21) can thus be written as a sum of standard and modified CI matrix elements.

E.2. Construction of the Cubic Terms for in the Iterative Cubic Procedure

In the previous section it was shown that the matrix elements of $K_{ijk}\lambda_k$ could be expressed as a sum of three terms. The first term corresponds to a

Hessian with the "Hamiltonian" $[\kappa, H]$, and normal density elements, the second term is a "Hessian" with changed density elements and the hamiltonian H. The third term is a sum of Hessian elements and gradient elements. A straightforward construction of $\mathbf{K}\underline{\lambda}$ requires thus the construction of two Hessians for each choice of $\underline{\lambda}$. It is, however, possible to construct $\mathbf{G}+\mathbf{K}\underline{\lambda}$ to the required accuracy (first order in $\underline{\lambda}$) by only one Hessian construction. To do this, introduce

$$\begin{aligned} |0'\rangle &= |0\rangle + |\delta 0\rangle \\ |j'\rangle &= |j\rangle + |\delta j\rangle \\ H' &= H + [\kappa, H]. \end{aligned} \tag{E.25}$$

If second-order terms are neglected, the four types of elements $G_{ij} + K_{ijk}\lambda_k$ separately become

$$\begin{aligned} &1.\quad \langle 0'|[Q_i^+ - Q_i, Q_j^+ - Q_j, H']|0'\rangle + G'_{ij} \\ &2.\quad -2\langle j'|[Q_i^+ - Q_i, H']|0'\rangle + G'_{ij} \\ &3.\quad -2\langle i'|[Q_j^+ - Q_j, H']|0'\rangle + G'_{ij} \\ &4.\quad 2\langle i'|H'|j'\rangle - 2\delta_{ij}\langle 0'|H|0'\rangle + G'_{ij} \end{aligned} \tag{E.26}$$

where G'_{ij} denotes the sum of Hessian and gradient elements.

In order to construct $(G_{ij} + K_{ijk}\lambda_k)$, the first-order transformed hamiltonian H' must be constructed. An algorithm based directly on Eq. (E.9) requires

$$N'_\mu = \left(\tfrac{3}{2}r^2 - 2r^4 + r^5\right)N^5 \tag{E.27a}$$

multiplications where r is n/N. The operation count in Eq. (E.27a) is based on the assumption that the only $\boldsymbol{\kappa}$ elements which are identically zero occur when two indices correspond to unoccupied orbitals.[20] Often a very large fraction of the zero $\boldsymbol{\kappa}$ elements occur when both indices correspond to occupied orbitals. In this case the operation count is about[20]

$$N''_\mu = \left(\tfrac{3}{2}r^2 - \tfrac{3}{2}r^3 - \tfrac{1}{2}r^4 + \tfrac{1}{2}r^5\right)N^5 \tag{E.27b}$$

The timings for carrying out this microtransformation are discussed in Appendix F. The microtransformation is not the most time-consuming step in the setting up $(G_{ij} + K_{ijk}\lambda_k)$.

Construction of $\mathbf{G}'$ from $\mathbf{G}$ and $\mathbf{F}$ are not time consuming. The operation count in construction of the orbital-orbital part of $\mathbf{G}'$ is proportional to r^3N^5.

E.3. Calculation of the Cubic Terms in the Perturbative Cubic Procedure

In the perturbative cubic approach [Eq. (137)] only a vector $K_{ijk}{}^{\mathrm{NR}}\lambda_i{}^{\mathrm{NR}}\lambda_k$ is needed. The vector $K_{ijk}{}^{\mathrm{NR}}\lambda_i{}^{\mathrm{NR}}\lambda_k$ can be calculated through a gradient routine, by simple extensions of the ideas developed previously.

APPENDIX F: TWO-ELECTRON INTEGRAL TRANSFORMATIONS IN MCSCF

Transformations of two-electron integrals are the most time-consuming part of an MCSCF iteration when a small or medium sized CI expansion is used. Even when a thousand symmetrized configuration states are included, the transformation occupies a large fraction of an iteration.[35, 84] The iteration procedures discussed in Sections III, V, VI, VII, and VIII require different transformations. An understanding of each of the transformations' relative complexity is therefore required for combining the methods optimally. In the following we first outline the complexity of usual transformations.[79, 93–95] We then show how the techniques discussed in Section IX.A simplify the transformations. The number of multiplications directly programmed (the operation count) is used to measure the complexity of a given algorithm. This index is easy to calculate, but it is not a precise indicator of relative CPU times due to neglect of differences at machine language level and the like. Only the terms in the operation count that are of fifth order are included.

F.1. Traditional Two-Electron Integral Transformations

In fixed Hessian methods and update methods only gradients are calculated. This involves integrals (see Appendix C and Section III.D) with three indices confined to occupied (both completely and partially) orbitals. The transformation T_1 required to construct these integrals can be depicted as

$$T_1\colon\quad (\alpha\beta|\gamma\delta) \rightarrow (ph_3|h_2h_1) \tag{F.1}$$

In Eq. (F.1) and in the following molecular orbitals in general are denoted r, s, t, u and basis orbitals are denoted by $\alpha, \beta, \gamma, \ldots$. Fully or partly occupied molecular orbitals are indexed as $h_1, h_2, \ldots,$ and completely unoccupied orbitals are indexed as $p_1, p_2, \ldots$. Simplifications can sometimes be obtained if indices corresponding to occupied orbitals are divided into indices corresponding to inactive orbitals and indices corresponding to active orbitals.[35] However, for simplicity we will neglect this reduction.

The transformation T_1 can be straightforwardly carried out as indicated in Ref. 20. The operation count N_1 for T_1 is readily obtained

$$N_1 = \left(\frac{r}{2} + \frac{r^2}{4} + r^3 - \frac{3}{8}r^4\right)N^5 \tag{F.2}$$

In Eq. (F.2) r is n/N, where n is the number of occupied (both fully and partially) orbitals and N is the total number of orbitals. A straightforward T_1 transformation is shown in Fig. 2.

In second-order procedures the Hessian is also required. Explicit formulas for "Hessian"-like elements are given in Appendix C and Section III.D.

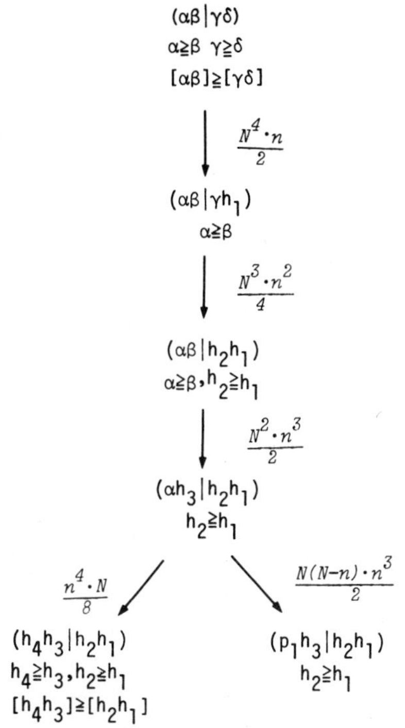

Fig. 2. Outline of a straightforward two-electron integral transformation, T_1, used in gradient-based methods. Multiplication counts for each step is indicated. Here, N is the total number of orbitals, and n is the number of occupied orbitals. The canonical index $[\alpha\beta]$ equals $\max(\alpha,\beta)\times[\max(\alpha,\beta)+1]/2+\min(\alpha,\beta)$.

From Appendix C and Section III.D it may be seen that the following integrals are needed:

$$\begin{aligned} &(rs|h_2h_1) \\ &(rh_2|sh_1) \end{aligned} \tag{F.3}$$

The transformations that set up the sets of integrals indicated in Eq. (F.3) have been studied by Ruedenberg et al.[79] They propose two algorithms T_2 and T_2' with operation counts N_2 and N_2':

$$N_2 = \left(r + \frac{3}{2}r^2 - \frac{11}{6}r^3 + \frac{3}{8}r^4\right)N^5 \tag{F.4}$$

$$N_2' = \left(\frac{r}{2} + \frac{5}{2}r^2 - \frac{11}{6}r^3 + \frac{3}{8}r^4\right)N^5 \tag{F.5}$$

Ruedenberg et al.[79] argue that the smaller operation count in T_2' is often off-

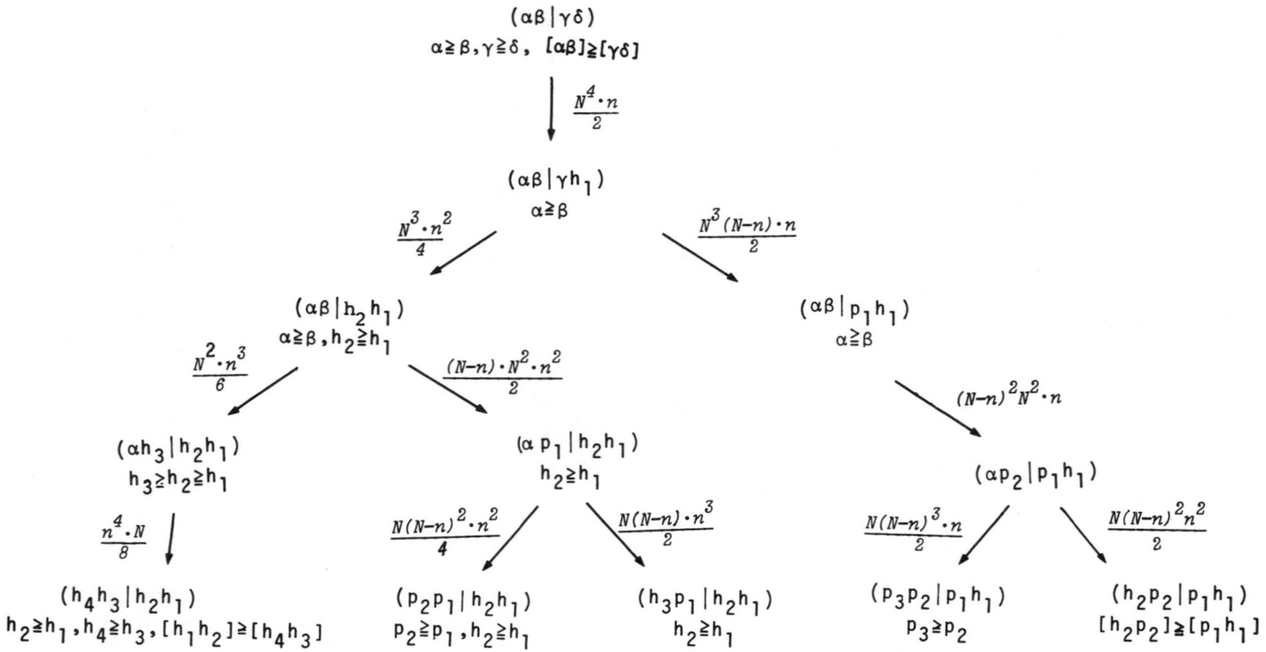

Fig. 3. Outline of a straightforward two-electron integral transformation, T_3, used in one-point cubic methods. See the legend to Fig. 2 for further explanation.

set by increased I/O processing. As a result of this they implemented T_2 in the ALIS program package.

Two very different transformations are involved in the one-point cubic procedures (iterative and perturbative cubic). Initially a transformation T_3 is required

$$T_3: \quad (\alpha\beta|\gamma\delta) \to (rs|th_1) \tag{F.6}$$

A straightforward T_3 algorithm is sketched in Fig. 3. The corresponding operation count N_3 is

$$N_3 = \left(\tfrac{5}{2}r - \tfrac{5}{2}r^2 + \tfrac{7}{6}r^3 - \tfrac{1}{8}r^4\right)N^5 \tag{F.7}$$

As shown in Appendix E a micro-transformation T_μ is also required in these cubic procedures. The corresponding operation count N_μ has the limits (see Appendix E)

$$N'_\mu = \left(\tfrac{3}{2}r^2 - 2r^4 + r^5\right)N^5$$
$$N''_\mu = \left(\tfrac{3}{2}r^2 - \tfrac{3}{2}r^3 - \tfrac{1}{2}r^4 + \tfrac{1}{2}r^5\right)N^5 \tag{F.8}$$

Although none of the procedures discussed requires a complete two-electron transformation, it is of interest to consider such a transformation for purposes of comparison. A very efficient method for the complete transformation, T_4, is suggested by Elbert.[95] The operation count is

$$N_4 = \tfrac{25}{24}N^5 \tag{F.9}$$

Values of N_1, N_2, N'_2, N_3, N'_μ, and N''_μ are shown in Table F.1 for the typical values $r = \frac{1}{2}, \frac{1}{3}, \frac{1}{4}, \frac{1}{5}$. The operation counts N_2, N_3, N_4 go toward the same limit as $r \to 1$, while N_1 becomes greater than N_4 in this limit. This can be traced back to the third step of T_1 in Fig. 2. N'_2 also becomes greater than N_4 as $r \to 1$. Table F.1 indicates that T_3 is somewhat more involved than $T'_2(T_2)$ and

TABLE F.1
Operation Counts for Conventional Transformations[a]

r	N_1	N_2	N'_2	N_3	N'_μ	N''_μ
$\frac{1}{2}$	0.41	0.67	0.67	0.76	0.28	0.17
$\frac{1}{3}$	0.23	0.44	0.38	0.60	0.15	0.11
$\frac{1}{4}$	0.15	0.32	0.25	0.49	0.09	0.07
$\frac{1}{5}$	0.12	0.25	0.19	0.41	0.06	0.05

[a] In units of N^5

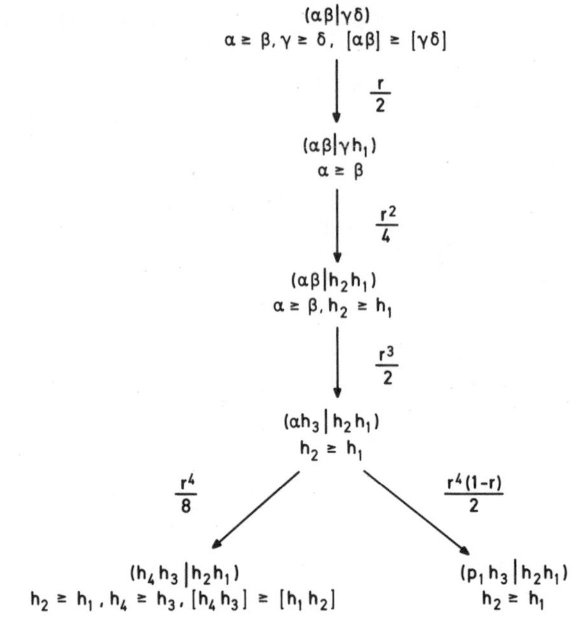

Fig. 4. Outline of the reduced two-electron transformation, T_1. Multiplication counts ($\times N^5$) corresponding to the simplified expansion matrix [Eq. (252)] are shown; r is n/N. See the legend to Fig. 2 for further explanation.

that $T_2'(T_2)$ is more involved than T_1. This is due to the increased number of integrals with indices corresponding to unoccupied orbitals.

F.2. Simplified Transformations

In Section IX.A a method to reduce the number of multiplications in the transformations was outlined. The method changed the expansion matrix **C** so that an unoccupied molecular orbital was spanned by only $n+1$ (instead of N) basis orbitals. Reductions in the expansion of a completely occupied orbital were also made, but this will not concern us now. We shall discuss here how these simplifications affect the operation counts of various transformations.

We first discuss how the gradient transformation T_1 is affected by the reduction of the matrix **C**′ [Eq. (252)]. The reduced T_1 transformation is summarized in Fig. 4. By summing the operation counts for the individual steps, the total number of multiplications for the reduced T_1 is

$$\tilde{N}_1 = \left(\frac{r}{2} + \frac{r^2}{4} + \frac{r^3}{2} + \frac{5}{8}r^4 - \frac{r^5}{2}\right)N^5 \tag{F.9a}$$

TABLE F.2
Comparison Between Operation Counts for Normal (N_1) and Reduced ($\tilde{N}_1$) Two-Electron Integral Transformation Required for Gradient Procedures[a]

r	N_1	$\tilde{N}_1$
$\frac{1}{2}$	0.41	0.40
$\frac{1}{3}$	0.23	0.22
$\frac{1}{4}$	0.15	0.15
$\frac{1}{5}$	0.12	0.11

[a]In units of N^5.

In Table F.2, $\tilde{N}_1$ is compared to N_1 for our usual choices of r. It is seen that $\tilde{N}_1$ is not significantly smaller than N_1. This transformation consists primarily of summing over occupied orbitals, so the small difference between N_1 and $\tilde{N}_1$ is not surprising.

For second-order procedures T_2 and T_2' transformations can be used to set up the necessary $(\tilde{r}\tilde{s}|\tilde{t}\tilde{u})$. From Fig. 5 it is seen that an application of T_2 in

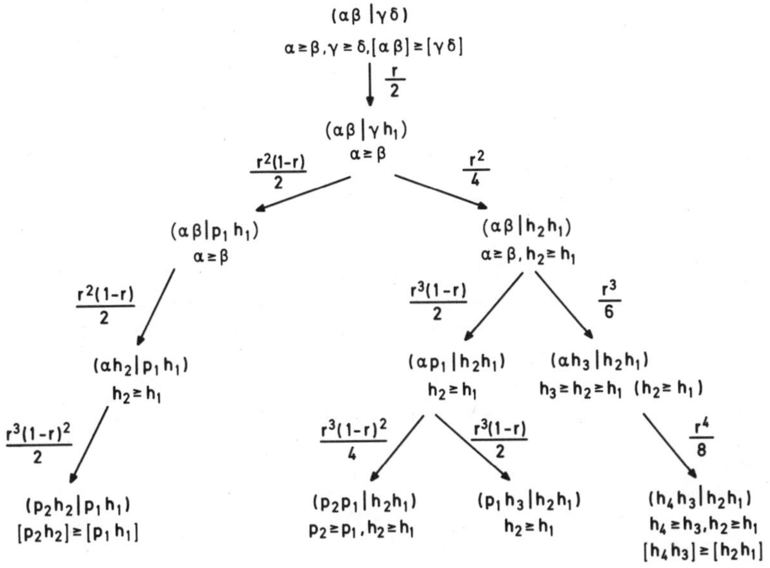

Fig. 5. Outline of the reduced two-electron transformations, $\tilde{T}_2$. See the legends to Figs. 2 and 4 for further explanation.

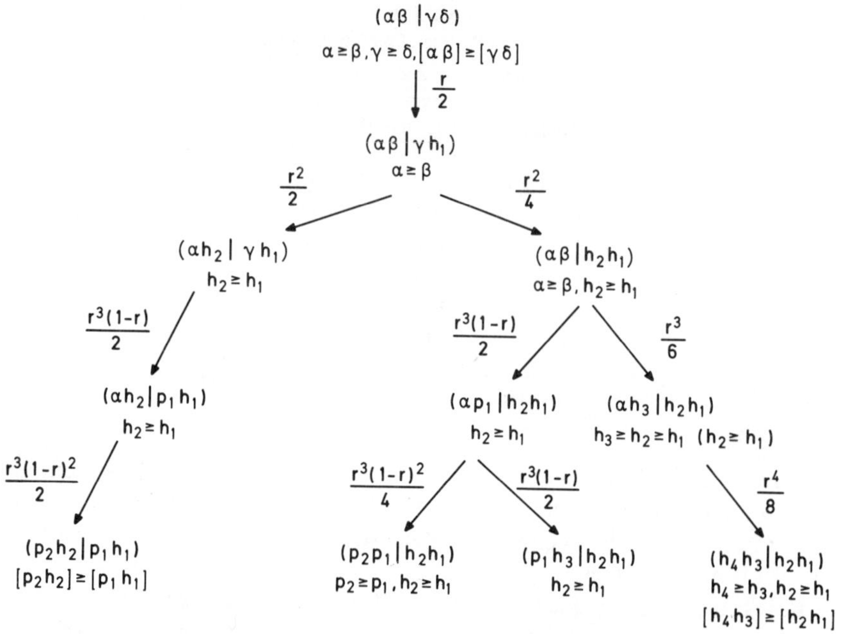

Fig. 6. Outline of the reduced two-electron integral transformations, $\tilde{T}_2'$. See the legends to Figs. 2 and 4 for further explanation.

TABLE F.3

Operation Counts for Reduced ($\tilde{N}_2$ and $\tilde{N}_2'$) and Normal (N_2 and N_2') Two-Electron Integral Transformations Required for Second-Order MCSCF Procedures[a]

r	N_2	$\tilde{N}_2$[b]	N_2'	$\tilde{N}_2'$[b]
$\frac{1}{2}$	0.67	0.55	0.67	0.58
$\frac{1}{3}$	0.44	0.31	0.38	0.31
$\frac{1}{4}$	0.32	0.21	0.25	0.20
$\frac{1}{5}$	0.25	0.15	0.19	0.14

[a] In units of N^5.
[b] From Eqs. (F.10) and (F.11).

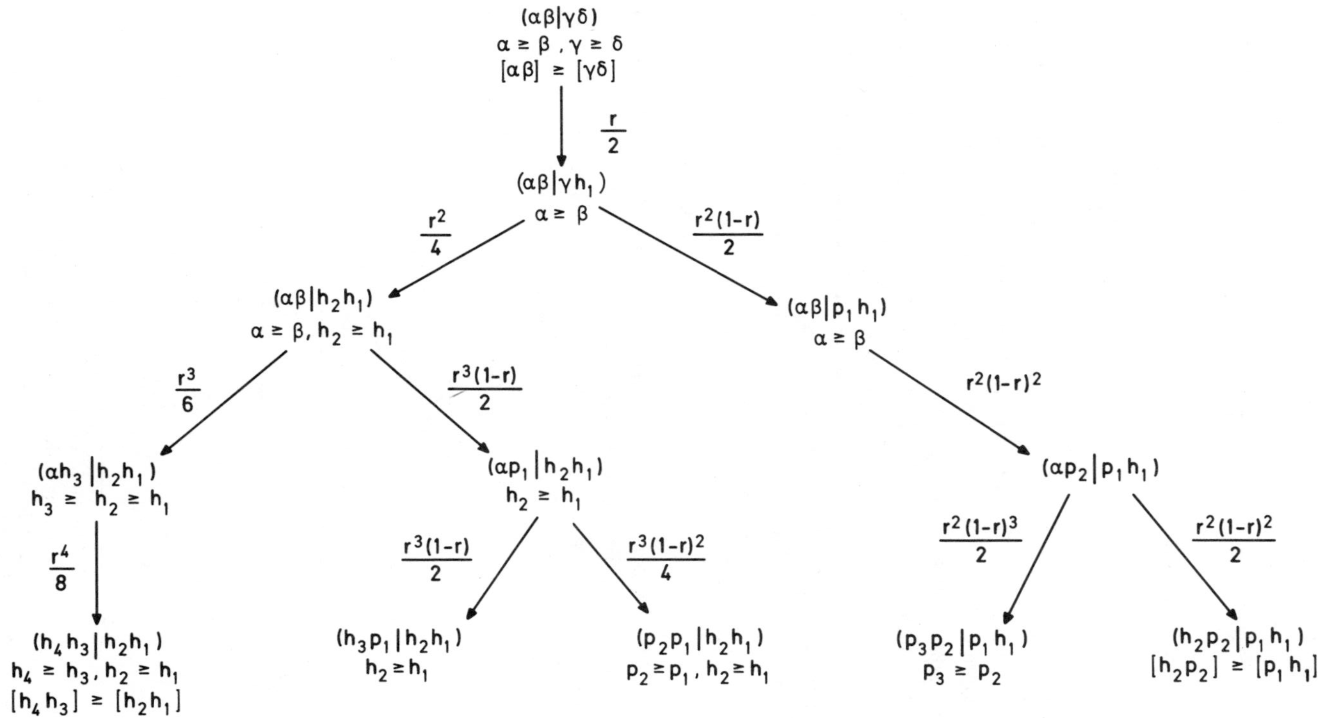

Fig. 7. Outline of the reduced two-electron transformations, $\tilde{T}_3$. See the legends to Figs. 2 and 4 for further explanation.

connection with $\mathbf{C}'$ gives an operation count

$$\tilde{N}_2 = \left(\frac{r}{2} + \frac{5}{4}r^2 + \frac{11}{12}r^3 - \frac{19}{8}r^4 + \frac{3}{4}r^5\right)N^5 \tag{F.10}$$

When T_2' is used the modified operation count is

$$\tilde{N}_2' = \left(\frac{r}{2} + \frac{3}{4}r^2 + \frac{29}{12}r^3 - \frac{23}{8}r^4 + \frac{3}{4}r^5\right)N^5 \tag{F.11}$$

The operation counts $\tilde{N}_2$ and $\tilde{N}_2'$ are reported in Table F.3. It may be seen that operation counts for T_2 are especially decreased. The greater change in operation counts of T_2 is due to the larger part of the T_2 transformation that consists of transforming to indices corresponding to unoccupied orbitals.

The major part of the original transformation T_3 [Eq. (F.6)] corresponds to summing over expansions of unoccupied orbitals. Thus, the simplified expansion matrix will decrease the operation count very significantly for T_3. The operation counts for the individual steps of the reduced T_3 transformation are given in Fig. 7. The reduced total operation count is

$$\tilde{N}_3 = \left(\frac{r}{2} + \frac{11}{4}r^2 - \frac{43}{12}r^3 + \frac{13}{8}r^4 - \frac{1}{4}r^5\right)N^5 \tag{F.12}$$

Values of $\tilde{N}_3$ are compared to N_3 for different values of r in Table F.4. As expected, very significant reductions are encountered; for example, the reduced operation count $\tilde{N}_3$ is about 50% of the operation count of N_3 for $r = \frac{1}{4}$ or $r = \frac{1}{5}$. By comparing Table F.4 and Table F.1 it is furthermore seen

TABLE F.4
Operation Counts for Normal (N_3) and Simplified ($\tilde{N}_3$) Two-Electron Integral Transformations Needed for Cubic MCSCF Procedures[a]

r	N_3	$\tilde{N}_3$
$\frac{1}{2}$	0.76	0.58
$\frac{1}{3}$	0.60	0.36
$\frac{1}{4}$	0.49	0.25
$\frac{1}{5}$	0.41	0.18

[a] In units of N^5. Here, $r = n/N$, where n is the number of occupied orbitals and N is the total number of orbitals.

that the operation counts for the reduced cubic transformation are below operations counts for the original T_2 transformations.

While the operation counts N_1, N_2, N_2', N_3 differ significantly for small values of r, the reduced operation counts $\tilde{N}_1, \tilde{N}_2, \tilde{N}_2', \tilde{N}_3$ are very similar for small values of r. For $r = \frac{1}{5}$ it is seen that $\tilde{N}_1$ is 0.11, $\tilde{N}_2$ is 0.15, and $\tilde{N}_3$ is 0.18. Although it is not advisable to use these numbers to make precise quantitative predictions about relative computer processing times, they do indicate that the reduced T_3 and T_2 transformations can be carried out with computer times of the same magnitude as the reduced T_1 transformation.

Acknowledgments

We would especially like to thank Esper Dalgaard and Jan Linderberg for their many helpful comments and suggestions; Diane Lynch, Jeff Nichols, Preben Albertsen, and Peter Swanstrøm for theoretical and computational assistance; and Birgitte Buus (Aarhus) and Sandy Seligman (Texas A&M) for scientific editing and typing.

We also appreciate receiving copies of many MCSCF articles prior to publication, including preprints from E. Dalgaard, D. G. Hopper, J. Detrich, C. C. J. Roothaan, B. Lengsfield, R. Shepard, I. Shavitt, J. Simons, H. – J. Werner, W. Meyer, C. Nellin, F. A. Matsen, G. Bacskay, B. Brooks, W. Laidig, H. F. Schaefer, N. Kosugi, H. Kuroda, G. Das, B. Liu, J. van Lenthe, and D. Yarkony.

J.O., P.J. and D.L.Y. acknowledge research support from the National Science Foundation (Grant CHE-8023352) and the Robert A. Welch Foundation (Grant A-770). J.O. also acknowledges support from the Danish Natural Science Research Council, and D.L.Y. has received travel assistance from NATO. (Grant RG 193.80)

References

1. D. R. Hartree, W. Hartree, and B. Swirles, *Phil. Trans. Roy. Soc.* (London) **A238**, 229 (1939); G. Das and A. C. Wahl, *J. Chem. Phys.* **44**, 87 (1966); A. Veillard, *Theoret. Chim. Acta* **4**, 22 (1966); J. Hinze and C. C. J. Roothaan, *Progr. Theoret. Phys. Suppl.* **40**, 37 (1967); see, for example, A. C. Wahl and G. Das, in *Modern Theoretical Chemistry*, H. J. Schaefer III, ed., Plenum Press, New York, 1977, and references therein.
2. See, for example, *Recent Developments and Applications of Multiconfiguration Hartree–Fock Methods*, Proceedings of the National Resource for Computational Chemistry Workshop held at Texas A&M University, July 1980, M. Dupuis, ed.
3. U. Fock, *Z. Physik* **61**, 126 (1930); J. C. Slater, *Phys. Rev.* **35**, 210 (1930).
4. G. G. Hall, *Proc. Roy. Soc.* (London) **A205**, 541 (1951); G. G. Hall, *Proc. Roy. Soc.* (London) **A208**, 328; C. C. J. Roothaan, *Rev. Mod. Phys.* **23**, 69 (1951); R. K. Nesbet, *Proc. Roy. Soc.* (London) **A230**, 312 (1955); C. C. J. Roothaan, *Rev. Mod. Phys.* **32**, 179 (1960); S. Huzinaga, *Phys. Rev.* **120**, 866 (1960); S. Huzinaga, *Phys. Rev.* **122**, 131 (1961); F. W. Birss and S. Fraga, *J. Chem. Phys.* **38**, 2252 (1963); F. W. Birss and S. Fraga, *J. Chem. Phys.* **40**, 3203, 3207, 3212 (1964); C. C. J. Roothaan and P. S. Bagus, *Methods Comp. Phys.* **2**, 47 (1963); W. J. Hunt, T. H. Dunning, and W. A. Goddard, *Chem. Phys. Lett.* **3**, 609 (1969).
5. T. A. Koopmans, *Physica* **1**, 104 (1933).

6. Yale Babylonian Collection Tablet No. 7289; see also Otto Neugebauer and A. Sachs, *Mathematical Cuneiform Texts*, American Oriental Society, New Haven, Conn., 1945.
7. A. Ralston and P. Rabinowitz, *A First Course in Numerical Analysis*, 2nd ed., McGraw-Hill, New York, 1978.
8. J. F. Traub, *Iterative Methods for the Solution of Equations*, Prentice-Hall, Englewood Cliffs, N.J., 1964.
9. R. Fletcher, *Practical Methods of Optimization*, Vol. 1, Wiley, New York, 1980.
10. E. Dalgaard and P. Jørgensen, *J. Chem. Phys.* **69**, 3833 (1978).
11. D. L. Yeager and P. Jørgensen, *J. Chem. Phys.* **71**, 755 (1979); D. L. Yeager, 1979 Sanibel Symposia.
12. E. Dalgaard, *Chem. Phys. Lett.* **65**, 559 (1979).
13. D. L. Yeager and P. Jørgensen, *Mol. Phys.* **39**, 587 (1980).
14. P. Jørgensen, P. Albertsen, and D. L. Yeager, *J. Chem. Phys.* **72**, 6466 (1980); **73**, 5408 (1980).
15. D. L. Yeager, P. Albertsen, and P. Jørgensen, *J. Chem. Phys.* **73**, 2811 (1980).
16. P. Jørgensen, J. Olsen, and D. L. Yeager, *J. Chem. Phys.* **75**, 5802 (1981).
17. J. Olsen, P. Jørgensen, and D. L. Yeager, *J. Chem. Phys.* **76**, 527 (1982).
18. D. L. Yeager, D. Lynch, J. Nichols, P. Jørgensen, and J. Olsen, *J. Phys. Chem.* **86**, 2140 (1982).
19. A. Igawa, D. L. Yeager, and H. Fukutome, *J. Chem. Phys.* **76**, 5388 (1982).
20. J. Olsen, P. Jørgensen, and D. L. Yeager, *J. Chem. Phys.* **77**, 356 (1982).
21. J. Olsen and P. Jørgensen, *J. Chem. Phys.* **77**, 6109 (1982).
22. J. Olsen, P. Jørgensen, and D. L. Yeager, "Second and Higher Order Convergence in Linear and Non-linear Multiconfigurational Hartree–Fock Theory," *Int. J. Quantum Chem.* (in press).
23. P. Jørgensen and J. Simons, *Second Quantization-Based Methods in Quantum Chemistry*, Academic Press, New York, 1981.
24. L. G. Yaffe and W. A. Goddard III, *Phys. Rev.* **A13**, 1682 (1976).
25. L. G. J. Douady, J. Ellinger, R. Subra, and B. Levy, *J. Chem. Phys.* **72**, 1452 (1980).
26. C. C. J. Roothaan, J. Detrich, and D. G. Hopper, *Int. J. Quantum Chem.* **S13**, 93 (1979); D. G. Hopper and C. C. J. Roothaan, 1979 Sanibel Symposia.
27. P. Siegbahn, A. Heiberg, B. Roos, and B. Levy, *Phys. Scr.* **21**, 323 (1980).
28. B. Lengsfield III, *J. Chem. Phys.* **73**, 382 (1980).
29. H.-J. Werner and W. Meyer, *J. Chem. Phys.* **73**, 2342 (1980).
30. H.-J. Werner and W. Meyer, *J. Chem. Phys.* **74**, 5794 (1981).
31. R. Shepard and J. Simons, *Int. J. Quantum Chem.* **S14**, 211 (1980).
32. R. Shepard, I. Shavitt, and J. Simons, *J. Chem. Phys.* **76**, 543 (1982).
33. G. Das, *J. Chem. Phys.* **74**, 5775 (1981).
34. F. A. Matsen and C. J. Nelin, *Int. J. Quantum Chem.* **20**, 861 (1981).
35. P. Siegbahn, J. Almlöf, A. Heiberg, and B. Roos, *J. Chem. Phys.* **74**, 2384 (1981).
36. A. U. Nemukhin, J. Almlöf, and A. Heiberg, *Chem. Phys.* **57**, 197 (1981); J. Almlöf, A. U. Nemukhin, and A. Heiberg, *Int. J. Quantum Chem.* **88**, 655 (1981).

37. C. W. Bauschlicher Jr., P. S. Bagus, D. R. Yarkony, and B. Lengsfield, *J. Chem. Phys.* **74**, 3965 (1981).

38. W. J. Hunt and W. A. Goddard III, *Chem. Phys. Lett.* **3**, 414 (1969).

39. S. Iwata and K. Morokuma, *Theoret. Chim. Acta* **33**, 4972 (1974).

40. S. F. Abdulnur, J. Linderberg, Y. Öhrn, and P. W. Thulstrup, *Phys. Rev.* **A6**, 889 (1972).

41. P. O. Löwdin, in *Perturbation Theory and Its Application in Quantum Mechanics*, C. H. Wilcox, ed., Wiley, New York, 1966.

42. E. A. Hylleraas and B. Undheim, *Z. Phys.* **65**, 759 (1930); J. K. L. MacDonald, *Phys. Rev.* **43**, 830 (1933).

43. D. L. Yeager and P. Jørgensen, *Chem. Phys. Lett.* **65**, 77 (1979).

44. E. Dalgaard, *J. Chem. Phys.* **72**, 816 (1980).

45. P. Albertsen, P. Jorgensen, and D. L. Yeager, *Mol. Phys.* **41**, 409 (1980).

46. P. Albertsen, P. Jørgensen, and D. L. Yeager, *Int. J. Quantum Chem.* **14S**, 249 (1980).

47. P. Albertsen, P. Jørgensen, and D. Yeager, *Chem. Phys. Lett.* **76**, 354 (1980).

48. D. L. Yeager, J. Olsen, and P. Jørgensen, *Int. J. Quantum Chem.* **15S**, 151 (1981).

49. J. Nichols and D. L. Yeager, *Chem. Phys. Lett.* **84**, 77 (1981).

50. D. Lynch, M. F. Herman, and D. L. Yeager, *Chem. Phys.* **64**, 69 (1982).

51. A. Banerjee, J. W. Kenney, and J. Simons, *Int. J. Quantum Chem.* **16**, 1209 (1980).

52. D. J. Thouless, *Nucl. Phys.* **21**, 255.

53. D. J. Rowe, *Nuclear Collective Motion Models and Theory*, Methuen, London, 1970.

54. J. Linderberg and Y. Öhrn, *Propagators in Quantum Chemistry*, Academic Press, New York, 1973.

55a. P. Jørgensen, P. Swanstrøm, and D. Yeager, *J. Chem. Phys.* **78**, 347 (1983).

55b. J. Golab, D. Yeager, and P. Jørgensen, "Proper Characterization of MCSCF Stationary Points," *Chem. Phys.* (submitted for publication).

56. V. R. Saunders and I. H. Hillier, *Int. J. Quantum Chem.* **7**, 699 (1973).

57. M. F. Guest and V. R. Saunders, *Mol. Phys.* **28**, 819 (1974).

58. J. E. Grabenstetter and F. Grein, *Mol. Phys.* **31**, 1469 (1976).

59. V. A. Kuprievich and O. V. Shamko, *Int. J. Quantum Chem.* **9**, 1009 (1975).

60. H. Tatewaki, K. Tanaka, F. Sasaki, S. Obara, K. Ohno, and M. Yoshimine, *Int. J. Quantum Chem.* **15**, 533 (1979).

61. G. W. Stewart, *Introduction to Matrix Computations*, Academic Press, New York, 1973.

62. C. E. Fröberg, *Introduction to Numerical Analysis*, 2nd ed., Addison-Wesley, Reading, Mass., 1969.

63. F. E. Dennis and F. F. Moré, *SIAM Rev.* **19**, 46 (1977); E. Spedicto, in *Towards Global Optimization*, L. C. W. Dixon and G. P. Szegö, eds., North-Holland Publs., Amsterdam, 1978.

64. C. G. Broyden, *Math. Comp.* **19**, 577 (1965).

65. M. J. D. Powell, *Math. Progr.* **1**, 26 (1971).

66. H. Kleinmichel, *Num. Math.* **38**, 219 (1981); E. Spedicto, in *Towards Global Optimization*, L. C. W. Dixon and G. P. Szegö, eds., North-Holland Publs., Amsterdam, 1978, 209.

67. J. E. Dennis and J. J. Moré, *Math. Comp.* **28**, 549 (1974).

68. L. C. W. Dixon, *Opt. Theoret. Appl.* **10**, 34 (1972).

69. R. W. H. Sargent and P. I. Sebastian, in *Numerical Methods for Non-linear Optimization*, F. A. Lootsma, ed., Academic Press, New York, 1972; D. M. Himmelblau, ibid.

70. C. G. Broyden, *J. Inst. Math. Appl.* **6** 70, 222 (1970); R. Fletcher, *Comp. J.* **13**, 317 (1970); D. Goldfarb, *Math. Comp.* **24**, 23 (1970); D. F. Shanno, ibid. **24**, 647 (1970).

71. W. Davidon, AEC Research and Development Rep. ANL 5990, Argonne National Lab., Argonne, Ill., 1959; R. Fletcher and M. Powell; *Comp. J.* **6**, 163 (1963).

72. C. G. Broyden, *Math. Comp.* **21**, 368 (1967); B. A. Murtagh and R. W. H. Sargent, in *Optimization*, R. Fletcher, ed., Academic Press, New York 1969.

73. R. H. A. Eade and M. A. Robb, *Chem. Phys. Lett.* **83**, 362 (1981).

74. J. E. Dennis and R. R. Schnabel, *SIAM Rev.* **21**, 443 (1979), and references therein.

75. D. Simmons, J. Golab, J. Olsen, and D. Yeager, *Chem. Phys. Lett.* (to be submitted).

76. W. C. Davidon, *Math. Progr.* **9**, 1 (1975).

77. S. S. Oren, *Math. Progr.* **7**, 351 (1974).

78. B. Levy and G. Berthier, *Int. J. Quantum Chem.* **2**, 307 (1968).

79. See, for example, K. Ruedenberg, L. M. Cheung, and S. T. Elbert, *Int. J. Quantum Chem.* **16**, 1069 (1979), and references therein.

80. F. Grein and T. C. Chang, *Chem. Phys. Lett.* **1**, 44 (1971); T. C. Chang and W. H. E. Schwarz, *Theoret. Chim. Acta* **44**, 45 (1977); F. Grein and A. Banerjee, *Int. J. Quantum Chem.* **59**, 147 (1975); A. Banerjee and F. Grein, *J. Chem. Phys.* **66**, 1054 (1977).

81. D. J. Rowe, *Nucl. Phys.* **A107**, 99 (1968).

82. B. Levy, *Int. J. Quantum Chem.* **4**, 297 (1970); B. Levy, *Chem. Phys. Lett.* **4**, 17 (1969).

83. R. Pariser and R. G. Parr, *J. Chem. Phys.* **21**, 446 (1953); **21**, 767 (1953); J. A. Pople, *Trans. Faraday Soc.* **42**, 1357 (1953).

84. B. Lengsfield III and B. Liu, *J. Chem. Phys.* **75**, 478 (1981).

85. G. B. Bacskay, *Chem. Phys.* **61**, 385 (1981); G. B. Bacskay, *Chem. Phys.* **65**, 383 (1982).

86. D. Yarkony, *Chem. Phys. Lett.* **77**, 634 (1981).

87. R. S. Varga, *Matrix Iterative Analysis*, Prentice-Hall, Englewood Cliffs, N.J., 1962.

88. *Numerical Algorithms in Chemistry: Algebraic Methods*, Proceedings of a Workshop of the National Resource for Computational Chemistry held at Lawrence–Berkeley Laboratory (1978).

89. B. Roos, P. Siegbahn, in *Methods of Electronic Structure*, H. F. Schaefer III, ed., Plenum Press, New York, 1977.

90. A. Igawa and H. Fukutome, *Progr. Theoret. Phys.* **54**, 1266 (1975).

91. F. Paldus, in *Theoretical Chemistry: Advances and Perspectives*, H. Eyring and D. Hendersen, eds. Vol. 2, Academic Press, New York, 1976.

92. D. Simmons, D. Lynch, and D. Yeager (to be submitted).

93. C. F. Bender, *J. Comp. Phys.* **9**, 547 (1972).

94. I. Shavitt, in *Methods in Electronic Structure Theory*, H. F. Schaefer III, ed., Plenum Press, 1977.

95. S. Elbert, in *Numerical Algorithms in Chemistry: Algebraic Methods*, Proceedings of a Workshop of the National Resource for Computation in Chemistry held at Lawrence–Berkeley Laboratory, 1978.

TWO-PHOTON SPECTROSCOPY OF PERTURBED BENZENES

LIONEL GOODMAN AND RICHARD P. RAVA*

Department of Chemistry
Rutgers University
New Brunswick, New Jersey

CONTENTS

I. INTRODUCTION

By midcentury many of the regularities in the optical (ultraviolet) spectra of aromatic molecules had become clear. The striking similarities between spectra of dissimilar aromatic hydrocarbons prompted Platt[1] to classify the

*Graduate School Fellow 1979–1982.

spectra in terms of states (L_b, L_a, $B_{a,b}$, $C_{a,b}$, etc.) of a parent $4n+2$ atom cyclic polyene. Platt's classification led Moffitt[2] to demonstrate that the forbidden transitions to the L_b and L_a states derive intensity from coupling to the very intense (doubly degenerate) $B_{a,b}$ transition, found frequently in the far ultraviolet. The important feature of Moffitt's work is that it demonstrated that inductive electrostatic perturbations (such as produced by fluorine substitution) couples B to L_b much more strongly than to L_a, but the coupling caused by cross-linking perturbations (such as occurs in naphthalene) is just reversed. Further, it led to an understanding of the great vibronic activity in L_a transitions and concomitant weakness of L_b. In Moffitt's ideas an "odd" vibration [like ν_6 in benzene (Wilson numbering[3])] is effective in coupling L_a to $B_{a,b}$, but not in coupling L_b.

In a parallel development Murrell and Longuet-Higgins,[4] while introducing the idea of charge-transfer (CT) states in discussing conjugative electrostatic perturbations (such as produced by methyl and amino substitutions), showed how the benzene transitions change in intensity and shift in frequency in a characteristic manner, depending on the nature (inductive or conjugative), the number, and the position of the substituents. Final proof for coupling of the L_b and L_a transitions to $B_{a,b}$ was obtained by polysubstitution intensities. The theoretical rule is $1:1:1:4:0$ for (1):(1,2):(1,3):(1,4):(1,2,3) orientations of identical substituents around benzene.[4-7] If L_b obtained intensity directly from CT transitions derived from methyl substitution, for example, the equivalent rule would be $\sim 0:1:1:\sim 0:1$. The former predictions are in excellent accord with the known spectra of substituted benzenes and have established to a great extent our sense of understanding of aromatic molecule spectra.

Extensive attention has been given to the molecular spectra of aromatic compounds excited by the simultaneous absorption of two photons (e.g., 73 references are cited in Friedrich and McClain's 1980 review[8] of two-photon (TP) molecular electronic spectroscopy). These spectra are excited by visible laser photons, generally in the green-yellow region from 5000 to 6000 Å into excited states near 2500–3000 Å. Though it has only been a little over a decade since the first reports of moderate resolution TP molecular spectra, and the data are still in evolution, remarkable similarities and regularities are already apparent. Attention in this chapter will be directed toward the systematics of TP spectra of perturbed benzenes, both isotopic and electrostatic. Enough is now known to understand the mechanisms and relationships involved in the TP spectroscopy of the perturbed benzene $^1B_{2u}$ state.

Excellent comprehensive articles on TP spectroscopy in general include those by McClain[9] in 1974, Mahr[10] in 1975, and McClain and Harris[11] in 1977. The TP selection rules derive from the three basic terms in the TP transition amplitude:

$$S = S^0 + S^s + S^a \tag{1}$$

where S^0 is a (symmetric) scalar term that requires transitions between states of identical symmetry; S^s is a symmetric tensor term that follows selection rules appropriate for electric quadrupole transitions; and S^a is an antisymmetric tensor term that obeys selection rules appropriate for magnetic dipole transitions but is zero in a single laser experiment that employs photons of identical frequency and polarization (i.e., as in all spectra being discussed in this chapter). This should be compared to the one-photon (OP) transition amplitude which follows selection rules appropriate for electric dipole transitions. The interest in the study of TP spectra arises because only transitions between states of the same parity are allowed in TP spectroscopy whereas the opposite is true in OP spectroscopy. In addition, the interplay of the three terms in Eq. (1) for similar (e.g., isotopically substituted) molecules has no counterpart in OP spectroscopy.

Perturbation effects on the benzene states and normal OP spectroscopy of aromatic molecules have been discussed in Refs. 1, 2, and 5–7 and in two very thorough articles by Albrecht.[12] Benzene holds a special position as the testing ground for molecular theories. Anticipation arises that careful examination of the TP spectra of perturbed benzenes will not only prove rewarding in developing the theory of TP spectra of aromatic molecules but also in furthering understanding of their electronic structure. In the following sections we will discuss the effects of different perturbations on the benzene TP spectrum and attempt to establish an understanding of TP spectra of aromatic molecules that is analogous to our understanding of normal optical spectra.

II. TWO-PHOTON SPECTRUM OF THE PARENT $^1B_{2u} \leftarrow {}^1A_{1g}$ BENZENE TRANSITION

The C_6H_6 TP spectrum is parity forbidden for the $^1B_{2u} \leftarrow {}^1A_{1g}$ transition, that is, no origin band is observed. The results of analyses of well-resolved spectra obtained in the vapor phase by different techniques,[13–19] as shown in Fig. 1 by fluorescence excitation, as well as in crystals[20–23] at low temperatures, indicate that the forbidden character is induced by three vibrations, all of u parity. The skeletal mode $\nu'_{14}(b_{2u})$ is dominant, and the b_{2u} vibrational normal coordinate Q_{14} requires vibronic coupling of the final B_{2u} state (f) to an A_{1g} state (k) (i.e., $\Gamma(f)\times\Gamma(k)\times\Gamma(Q_{14}) = B_{2u}\times A_{1g}\times b_{2u} = A_{1g}$).

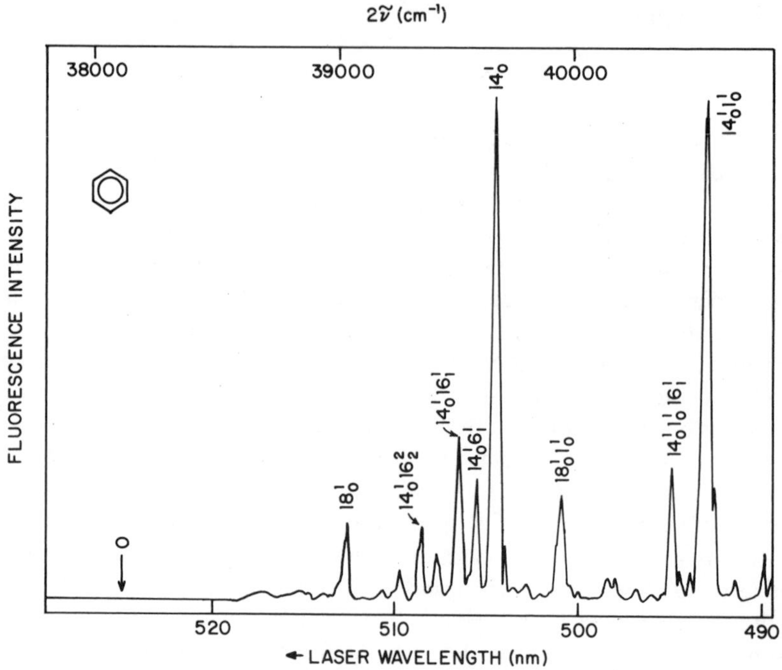

Fig. 1. Fundamental region of the normalized two-photon fluorescence excitation spectrum of the $^1B_{2u} \leftarrow {}^1A_{1g}$ transition in benzene vapor; *O* designates the missing origin band.

It is thought that the primary perturbing A_{1g} state is the ground state,[13] and this is partially confirmed by calculations[24] and is discussed further in Section VII. The in-plane hydrogen wagging e_{1u} mode ν'_{18} is also active, although considerably less so than ν'_{14}; ν'_{17}, an out-of-plane e_{2u} mixed skeletal and hydrogen wagging mode, is still less active; and ν''_{15}, a b_{2u} mode, is active only in the hot band spectrum.

With the exception of b_{1u}, modes of all symmetry classes that satisfy symmetry requirements for vibronic coupling are active. The b_{1u} modes couple to A_{2g} states ($B_{2u} \times A_{2g} \times b_{1u} = A_{2g}$), but these states can only generate identity forbidden transitions,[13] which are not observable in a single laser experiment.

III. MASS PERTURBATIONS

Two types of spectroscopic isotope effects should be distinguished: the frequency and the intensity changes. The first are due to diagonal kinetic energy perturbations and are routinely observed upon replacing a hydrogen atom by deuterium. Intensity changes are more subtle, being caused by the effect of isotopic substitution on the forms of the normal modes of vibration, due to off-diagonal kinetic energy perturbations.

There is an important difference between the active modes appearing in the TP and OP spectra of the ${}^1B_{2u} \leftarrow {}^1A_{1g}$ benzene transition: the prominent TP coupling mode ν_{18} possesses substantial hydrogen character; on the other hand, the only OP coupling mode, ν_6 (shown in Fig. 2), is primarily skeletal. The way is thus paved for deuterium isotope effects to appear in the TP spectrum which are not possible in OP.

A. Harmonic Mode Scrambling

The discussion will concentrate on comparison of the hot band spectra (transitions arising from a level with one vibrational quantum excited in the ground state to the zero level of the excited state) and cold band spectra (transitions arising from the zero level of the ground state to a level with one quantum of an excited state vibration) of symmetrically deuterated *s*-$C_6H_3D_3$. These studies have been extended to and verified by many other isotopically substituted benzene spectra.[24-28a]

The isotopic perturbation affects the kinetic energy of the molecule because of the change in mass but has a negligible effect on the potential energy.[29, 30] The result is that the normal vibrations of the unsubstituted molecule may become scrambled in the labeled molecule, as well as undergo frequency shifts. Since the normal coordinates of C_6H_6 are well known for the ground state and form a complete set, we may expand the normal modes of the isotopic benzene in terms of those for C_6H_6. This expansion may be written

$$Q_i^{D''} = \sum_{j \in \Gamma(D)} a_{ij} Q_j^{0''} \tag{2}$$

where $Q_i^{0''}$ and $Q_i^{D''}$ are the ground state normal modes for the h_6 and labeled molecule, respectively; $\Gamma(D)$ is the symmetry of the modes in the labeled molecule point group, the sum being over one symmetry class; and a_{ij} is the transformation matrix. This matrix is given for *s*-$C_6H_3D_3$ in Table I for the e' and a_2' symmetry classes as determined from the calculations of Albrecht[31] using the force field of Whiffen.[32] The matrix indicates that large mixing of the e' normal coordinates $Q_{18}^{0''}$ and $Q_9^{0''}$ to form $Q_{18}^{D''}$ and $Q_9^{D''}$ in *s*-$C_6H_3D_3$, as well as the a_2' normal coordinates $Q_3^{0''}$ and $Q_{14}^{0''}$, will take place. Examination of Table I shows that the 1101 and 1257 cm^{-1} modes have major parentage in Q_9^0 and Q_3^0, respectively. These modes have been historically designated ν_9 and ν_3 in the heavy molecules despite the strong mode scrambling.[31, 35] Since both Q_{18}^0 and Q_{14}^0 are active in the C_6H_6 TP spectrum, there

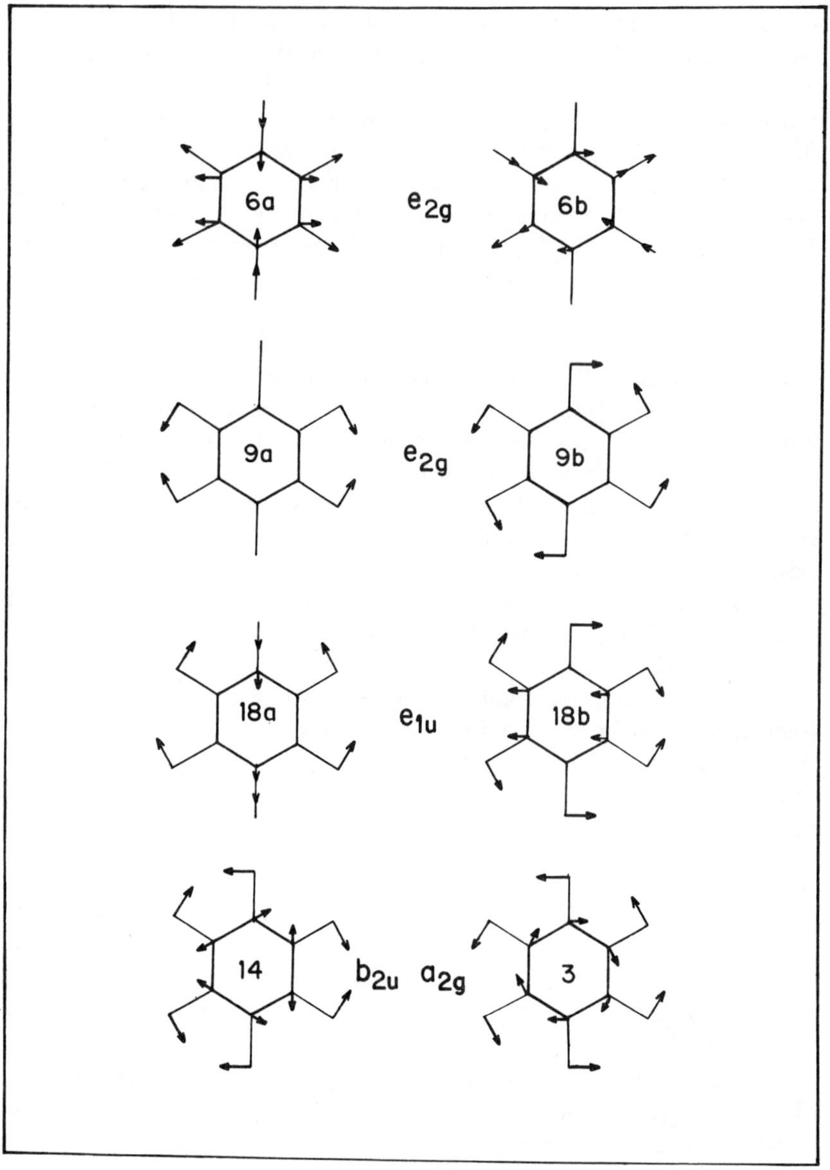

Fig. 2. Zero-point displacement diagrams [H motions are weighted by $(m_H/m_C)^{1/2}$] for some ground state normal modes active in one- and two-photon spectra of isotopic benzenes.

TABLE I
Harmonic Mode Scrambling Matrices for Ground State Normal Modes of s-$C_6H_3D_3$[a]

e'^{c}: Mode	$\nu(\text{cm}^{-1})^{b}$	Q_6^0	Q_7^0	Q_8^0	Q_9^0	Q_{18}^0	Q_{19}^0	Q_{20}^0
Q_6^D	592	0.97	−0.01	0.02	0.06	0.09	−0.03	0.01
Q_7^D	2274	−0.06	−0.62	−0.13	0.04	−0.07	0.09	0.59
Q_8^D	1575	−0.03	−0.05	0.98	−0.05	−0.07	0.09	0.06
$Q_9^{D\,d}$	1101	0.00	−0.03	−0.01	−0.72	0.66	0.13	0.03
$Q_{18}^{D\,d}$	833	−0.18	0.00	0.07	0.50	0.67	−0.18	0.00
Q_{19}^D	1412	0.00	−0.03	−0.07	0.31	0.12	0.93	0.03
Q_{20}^D	3053	0.00	0.70	0.00	0.00	0.00	0.00	0.72

$a_2'^{c}$ Mode	$\nu(\text{cm}^{-1})^{b}$	Q_3^0	Q_{14}^0	Q_{15}^0
Q_3^D	1257	0.62	−0.55	0.55
Q_{14}^D	1318	0.59	0.80	0.12
Q_{15}^D	909	−0.39	0.20	0.75

[a]The tabulated values are the a_i of Eq. 2.
[b]Observed frequencies are from Ref. 31.
[c]D_{3h} symmetry designation.
[d]These modes are strongly scrambled. Frequency shifts in the isotopic benzenes indicate that it is more useful to label the 833 cm^{-1} frequency mode as Q_{18}^D and that at 1101 cm^{-1} as Q_9^D.

is a definite and clear prediction that new bands should appear in the spectrum of s-$C_6H_3D_3$ derived from ν_3 and ν_9 in C_6H_6 (Fig. 2). Also the intensity of the bands due to the original vibrations should decrease.

Equation (2) deals specifically with the ground state normal modes; thus it can only be strictly considered for the hot band spectra where ground state normal coordinates act as the coupling vibration. If the modes are similar in the excited state, the hot and cold spectra should be similar. The fundamental hot band spectrum of s-$C_6H_3D_3$ is shown in Fig. 3. Two bands in this region cannot be correlated with those found in the C_6H_6 spectrum, indicating that new vibronic activity is present.

The first is a strong band at −1322 cm^{-1} from the missing origin. This band shows polarization behavior of a transition to a totally symmetric state.[13, 33] Only three fundamentals may show this polarization behavior in s-$C_6H_3D_3$: 14_1^0, 15_1^0, and 3_1^0 (the notation is that of Callomon et al.[34]); 14_1^0 and 15_1^0 are present in C_6H_6 and are found in s-$C_6H_3D_3$ at −1263 and −916 cm^{-1}, corresponding to the IR frequencies 1252 and 911 cm^{-1}, respectively.[35] The IR (infrared) frequency of ν_3 is historically 1311 cm^{-1}.[31,35] Thus, the new band at −1322 cm^{-1} can only be assigned as 3_1^0, correlating with the transformation matrix prediction. Note, however the

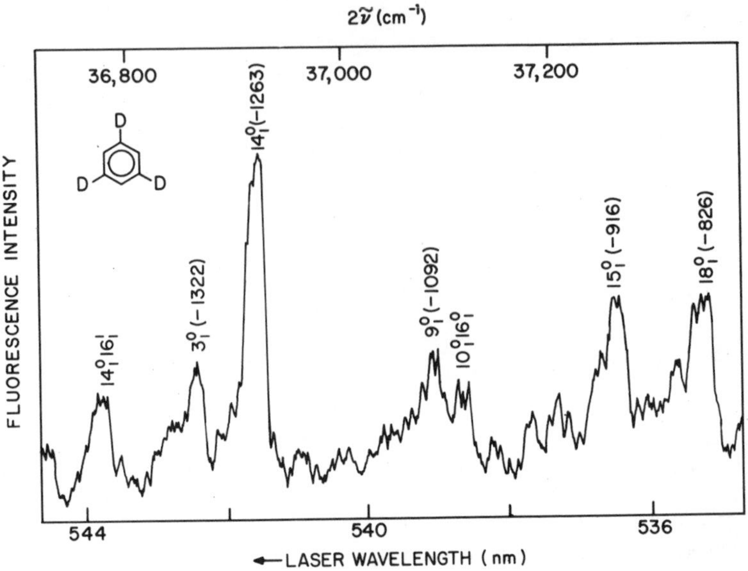

Fig. 3. Fundamental region of the two-photon hot band spectrum of s-$C_6H_3D_3$ vapor.

mode designations of ν_3 and ν_{14} would be changed from the historical IR assignments.

The second new band in s-$C_6H_3D_3$ appears at $-1092\ cm^{-1}$, and this band shows polarization behavior characteristic of a non–totally symmetric band.[33] The polarization is the same as 18^0_1, which is observed at $-826\ cm^{-1}$ (IR frequency $833\ cm^{-1}$). The IR spectrum also has a band at $1101\ cm^{-1}$, assigned to the fundamental ν_9. The correlation of frequencies and polarization behavior allows definite assignment of the new band at $-1092\ cm^{-1}$ as 9^0_1, active through harmonic mode scrambling.

The scrambling caused by isotopic substitution allows two new frequencies to become active in the hot band spectrum of s-$C_6H_3D_3$, as was predicted by the transformation matrix. We must examine the cold band spectrum, given in Fig. 4, and see if the excited state modes undergo similar scrambling effects. In this spectrum, the intensity of 14^1_0 has remained constant from its counterpart in C_6H_6 (Table II).[24, 26] In addition, no new fundamental bands can be found due to a totally symmetric transition. In fact, the entire spectrum can be correlated with the C_6H_6 spectrum except for a new strong band at $+1020\ cm^{-1}$ which shows itself to be of non–totally symmetric symmetry. Also, the intensity of 18^1_0 at $800\ cm^{-1}$ has decreased by a factor of about 2 (Table II). The polarization and hot band results require an assignment of this new band as 9^1_0, unobserved in C_6H_6.

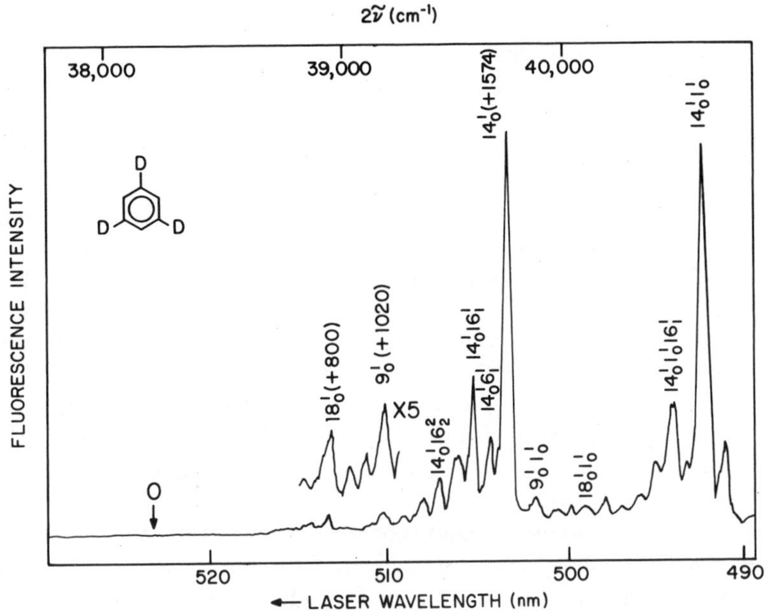

Fig. 4. Continuation of Fig. 3 into the cold band region; *O* designates the missing origin band.

TABLE II
Relative Strengths and Frequencies of Fundamental Bands in the $^1B_{2u}$ TP Spectra of Isotopic Benzenes[a, b, c]

Molecule	Symmetry	$9a^1_0$	$9b^1_0$	15^1_0	14^1_0	$18a^1_0$	$18b^1_0$
C_6H_6	D_{6h}	—[e]		c(1148)	1000(1563)	150(923)	
C_6D_6[d]	D_{6h}	—[e]		c(813)	1000(1564)	100(753)	
s-$C_6H_3D_3$	D_{3h}	60(1020)		—[f]	1000(1574)	55(800)	
p-$C_6H_4D_2$	D_{2h}	—[e]		—[f]	1000(1572)	130(921)	c(790)
p-$C_6H_2D_4$	D_{2h}	—[e]		45(970)	1000(1563)	45(764)	35(790)
C_6H_5D	C_{2v}		40(990)[g]	c(1145)	1000(1561)	40(921)	45(826)[g]
o-$C_6H_4D_2$	C_{2v}		65(947)	—[f]	1000(1570)	70(820)	—
m-$C_6H_4D_2$	C_{2v}	20(1032)	30(984)[h]		1000(1570)	25(798)	25(849)[h]
C_6HD_5	C_{2v}		40(957)	c(816)	1000(1559)	95(765)	c(751)

[a] The intensity of 14^1_0 in C_6H_6 has been designated 1000. All other 14^1_0 band intensities are measured relative to C_6H_6.

[b] Frequencies (cm^{-1}) are in parentheses.

[c] Other fundamentals lie in congested areas of the spectrum and do not allow reliable intensity measurements.

[d] The 19^1_0 band ($\omega' = 1212$ cm^{-1}) is also observed with a relative intensity of 80.

[e] Parity forbidden.

[f] Too weak for unambiguous assignment.

[g] These bands arise from strongly scrambled ν_{9b} and ν_{18b}.

[h] These bands arise from strongly scrambled ν_{9b} and ν_{18b}. Mode scrambling calculations show that the 984 cm^{-1} band alternatively can be assigned to 15^1_0.

These results indicate that the scrambling predicted by the ground state transformation is qualitatively correct for the e' excited modes only (but not quantitatively). Further consideration is still required for the a_2' excited state modes. Nevertheless the experiments demonstrate the sensitivity of the TP absorption spectrum to isotopic scrambling effects.

Lack of observation of the band 3_0^1 is not definitive evidence for lack of mode scrambling of 14^1. We now discuss an experiment[36] which proves that the mechanism for mode scrambling in the excited state involves solely the e_{1u} mode 18^1 as contrasted to mode scrambling in the ground state, which involves the b_{2u} mode 14_1 as well as 18_1. The proof is based on the polarization characteristics of the spectrum of a deuterium benzene. McClain's work demonstrates how different polarizations of incident light allow the mechanisms in Eq. (1) responsible for TP transitions to be distinguished.[33] In an appropriate example of C_{2v} symmetry ν_{14} becomes b_2, and both ν_{18} and ν_9 split into a_1 and b_2 components. The spectrum also becomes formally symmetry allowed, but no origin band is observed in any deuterium benzene, no matter what symmetry, demonstrating that the pure electronic $^1B_{2u}$ wave function sees only the D_{6h} potential of the light molecule. If a newly active b_2 mode in the spectrum derives intensity from 14^1, it will show the characteristic polarization of a totally symmetric transition (i.e., the Q branch will be strongly attenuated in circularly polarized light[11,33]). If the mode arises from mode scrambling with 18^1, it will show the polarization of a nontotally symmetric state (i.e, the entire band will be slightly enhanced in circularly polarized light).

This experiment was carried out[36] on m-$C_6H_4D_2$, where the four transitions $18a_0^1$, $18b_0^1$, $9a_0^1$, and $9b_0^1$ are all active and completely resolved (as is shown in Fig. 5a). The important point is that even though ν_{9a} (a_1) and ν_{9b} (b_2) possess different symmetries the polarization behaviors of the associated bands are expected to be completely derived from the mode scrambling mechanism, not from the formal symmetries.

Comparison of the linearly and circularly polarized light spectra for $9a_0^1$ and $9b_0^1$ in Fig. 5b, c shows no significant attenuation of either $9a_0^1$ relative to $9b_0^1$, or $18a_0^1$ relative to $18b_0^1$, or the components of 9_0^1 relative to 18_0^1. *The polarization result provides clear proof that in the excited state harmonic mode scrambling takes place with the* e_{1u} *mode* $Q_{18}^{0'}$ *and not with the* b_{2u} *mode* $Q_{14}^{0'}$.

In summary, the intensities of many of the vibronically active bands in the TP spectra corresponding to the $^1B_{2u} \leftarrow {}^1A_{1g}$ transition of deuterium benzenes are highly sensitive to mass perturbations on the hydrogen atoms (Table II). This conclusion when carried over to electrostatically perturbed benzenes requires that a portion of apparent Franck–Condon (FC) intensity (but not of the 0–0 band) will derive from the mass perturbation of the substituent alone since appearance of totally symmetric modes in the deuterium

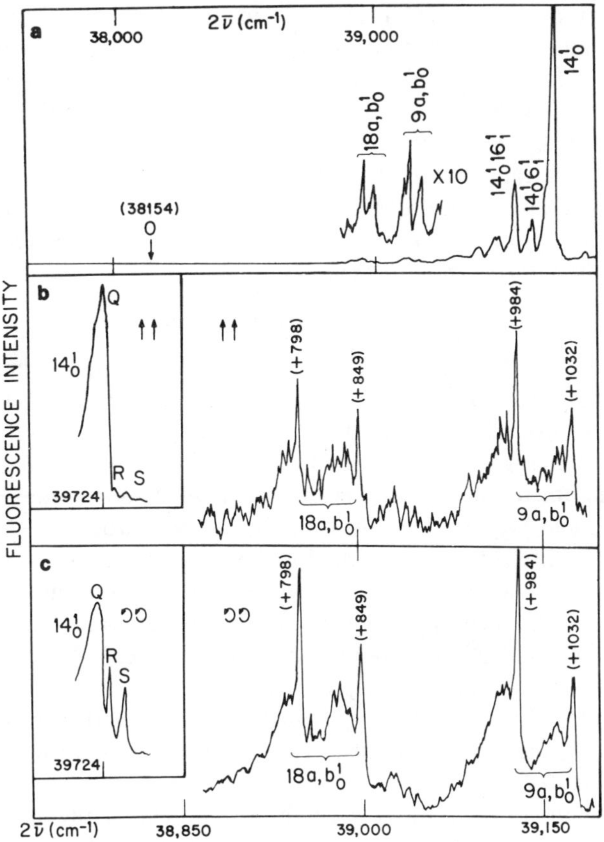

Fig. 5. Normalized two-photon fluorescence excitation spectrum of m-$C_6H_4D_2$ vapor: (a) in linearly polarized light; (b) expanded spectrum of the 9_0^1, 18_0^1, and 14_0^1 regions in linearly polarized light; and (c) in circularly polarized light.

benzenes is due ultimately to vibronic coupling (VC) effects of the Herzberg–Teller type.

The consequence of 14^1 being uninvolved in mode scrambling caused by deuterium substitution is that the intensity of 14_0^1 is invariant to external mass perturbations[24] (Table II), and thus this band may be used as an internal intensity standard for substituent effects in molecules of the C_6H_5X type (see Section IX).

Establishment of the mechanisms for the origin of new modes in isotopic benzenes allows the possibility of observation of missing excited state mode frequencies through selective labeling. One such frequency has already been observed by this method.[28] This is 12^1, identity forbidden in the C_6H_6 TP spectrum, and of incorrect symmetry in the OP spectrum. Another long-searched-for frequency is that of 8^1. Calculations[37] analogous to that given in Table I shows that this mode should be made active in the TP spectrum of $^{13}C^{12}C_5H_6$ by mode scrambling with 14^1.

Finally, the transformation matrix of Table I also shows that the only coupling mode of the OP benzene spectrum, ν_6, remains unaltered by isotope substitution. This points out the reason for the inaccessibility of these effects in the OP spectrum of benzene.

B. Duschinsky Rotation

When the geometry of a molecule and the force constants change on excitation, the normal coordinates of the excited state are in general different (i.e., rotated) from those of the ground state. This is true even if the symmetry does not change and was recognized in 1937 by Duschinsky.[38]

The experiments of Wunsch et al.[19] in 1977 demonstrated that the b_{2u} modes (ν_{14} and ν_{15}) undergo strong Duschinsky rotation (DR) upon going to the $^1B_{2u}$ excited state. The basis for this conclusion is an experiment devised by Small[39] where the relative intensity of two vibronic origins of the same symmetry, ν_X and ν_Y, are compared in the hot and cold spectra (i.e., X_1^0/Y_1^0 vs. X_0^1/Y_0^1). Wunsch showed that $14_0^1/15_0^1 \gg 14_1^0/15_1^0$ and thus these modes have become significantly altered. Robey and Schlag[40] subsequently showed that in the upper state $Q_{14}^{0'}$ is described by a mixture of $Q_{14}^{0''}$ and $Q_{15}^{0''}$ such that $Q_{14}^{0'}$ has become almost completely the carbon skeletal vibration shown in Fig. 6.

The difficulty of observing DR by this experiment is obvious. Two modes of the same symmetry class are required to be active in the hot or cold band

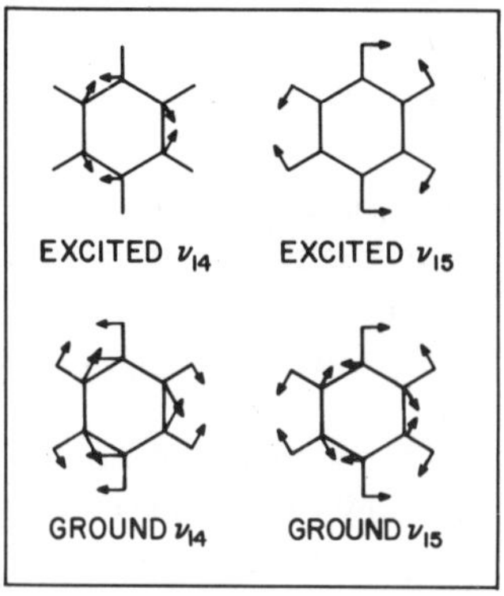

Fig. 6. Comparison of zero-point displacements (H motions are weighted by $(m_H/m_C)^{1/2}$) for the b_{2u} modes in the ground and $^1B_{2u}$ excited states of benzene. (From Robey and Schlag.[40])

spectrum, a situation which is rarely met. However, the isotopic perturbation scheme of the previous section can be used to activate new modes to probe DR. This technique widely expands the applicability of the Small experiment to symmetry classes of the parent molecule which exhibit only a uniquely active mode, or even to cases where no active vibronic origin is observed.[25]

The spectrum of s-$C_6H_3D_3$ is again used as an example. The results for the a_2' modes confirm the observation of Wunsch et al.[19] for ν_{14}. The intensity ratio of $14_0^1/3_0^1 \gg 14_1^0/3_1^0$ indicates that there is strong DR of these modes in s-$C_6H_3D_3$. Since $Q_{14}^{0'}$ is primarily skeletal,[40] $Q_3^{0'}$ has no opportunity to mix through a deuterium perturbation in the excited state. However, the hot bands show that $Q_{14}^{0''}$ has enough hydrogen character for harmonic mode scrambling to occur in the ground state and consequently 3_1^0 to strongly appear.

The procedure can be extended to the e_{1u} symmetry class of C_6H_6 which only has one active vibration, ν_{18}. In s-$C_6H_3D_3$, the e' modes ν_{18} and ν_9 are both active, with intensity ratios $(9_1^0/18_1^0) \approx 2(9_0^1/18_0^1)$. The difference in the cold and hot band ratios requires a Duschinsky effect on the e' modes in the labeled molecule, s-$C_6H_3D_3$. It is necessary to see if DR of the e' modes in s-$C_6H_3D_3$ unambiguously implies DR of the e_{1u} modes in the parent molecule, C_6H_6. The required transformations are

$$Q_k^{D'} = \sum_{l \in \Gamma(D)} J_{kl}^D Q_l^{D''} \tag{3}$$

and

$$Q_m^{0'} = \sum_{n \in \Gamma(H)} J_{mn}^H Q_n^{0''} \tag{4}$$

where $Q^{D''}$ and $Q^{D'}$ are harmonic ground and excited state normal modes, respectively, in the labeled molecule, and $Q^{0''}$ and $Q^{0'}$ are similar modes in the parent molecule. We wish to know whether there are nonzero cross-terms in (4) (i.e., DR of the parent modes) when there are nonzero cross-terms in (3) (i.e., DR of the labeled modes). The Duschinsky matrices $\mathbf{J}^D$ and $\mathbf{J}^H$ are related by the isotopic mode scrambling transformations. The ground state transformation is given by Eq. (2), the excited state transformation by

$$Q_r^{D'} = \sum_{s \in \Gamma(D)} A_{rs} Q_s^{0'} \tag{5}$$

Expressing the TP tensor for the vibronic bands through linear Herzberg–Teller theory, $S = S(Q_0) + (\partial S/\partial Q)_{Q_0} Q$ [note that $S(Q_0) = 0$ for the forbidden benzene spectrum that we are discussing]; and making use of Eqs. (1)–(5) gives

$$S(Q_r^{D'}) = \sum_{m \in \Gamma(D)} A_{rm} \left[\sum_{n \in \Gamma(H)} J_{mn}^{H} S(Q_n^{0''}) \right] \tag{6}$$

for the cold bands and

$$S(Q_i^{D''}) = \sum_{j \in \Gamma(D)} a_{ij} S(Q_j^{0''}) \tag{7}$$

for the hot ones.* If the Duschinsky cross-terms in C_6H_6 vanish ($\mathbf{J}^H = \mathbf{1}$) and mode scramblings in the ground and excited state are identical ($\mathbf{a} = \mathbf{A}$), then according to Eqs. (6) and (7) $S(Q_k^{D'}) = S(Q_k^{D''})$. For this case, we conclude that except for frequency factors, the cold and hot band spectrum will exhibit the same intensity distribution. However, there are two cases which will show a disparity between the cold and hot spectra. First, if $\mathbf{A} \neq \mathbf{a}$ then $S(Q_k^{D'}) \neq S(Q_k^{D''})$, even if $\mathbf{J}^H = \mathbf{1}$. In this case DR will be perceived in the normal modes of the labeled benzene even though none exists in the parent molecule. The converse case, $\mathbf{J}^H \neq \mathbf{1}$, automatically will generate $\mathbf{A} \neq \mathbf{a}$ and different mode scrambling perturbations will be perceived in the ground and excited states of the labeled molecule due to DR of C_6H_6 modes. This, of course, is the basis of the isotopic probing experiments that we are describing. The dependence of DR on frequency differences between coupling modes suggests that $\mathbf{a} \neq \mathbf{A}$ is sensitive to diagonal isotopic perturbations. In effect, a diagonal perturbation may induce a type of Fermi resonance between coupling modes having similar frequencies in the labeled molecule. However, since the frequency separations range from 200–600 cm^{-1} in *s*-$C_6H_3D_3$, Fermi resonance is probably not an important effect in this molecule.

Thus these experiments can be understood if the C_6H_6 mode 9^1 scrambles with 18^1 to a lesser degree than 9_1 with 18_1. Since the importance of ν_{18} in the deuterium-induced scrambling mechanism arises from the hydrogen displacements, 18^1 is thought to have reduced hydrogen wagging character compared to 18_1. Support for this conclusion is provided by the change in the ν_{18} frequency ratio upon perdeuteration, from $\omega''_{18}(H_6)/\omega''_{18}(D_6) = 1.27$ to $\omega'_{18}(H_6)/\omega'_{18}(D_6) = 1.22$, which also indicates attenuation of the hydrogen wagging character of ν_{18} in the excited state. The persistance of these results in several isotopic benzenes is also consistent with these conclusions,[25] but a force field for the excited e_{1u} modes has yet to be constructed.

*In these expressions the vibrational frequency factors, $\omega^{-1/2}$ have been omitted for simplicity.

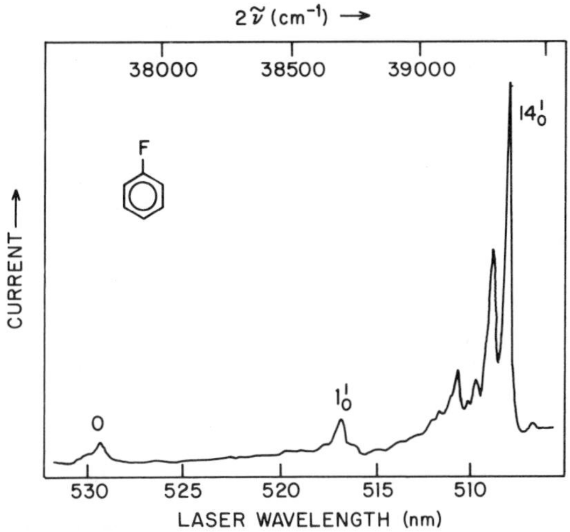

Fig. 7. Fundamental region of the normalized two-photon fluorescence excitation spectrum of the $L_b \leftarrow A$ transition in fluorobenzene vapor.

IV. TWO-PHOTON L_b SPECTRA OF TOLUENE AND FLUOROBENZENE

The fundamental region of the TP vapor fluorescence excitation spectrum for the L_b state of fluorobenzene is given in Fig. 7. The L_b TP resonant ionization spectrum[41] of toluene obtained in a supersonic jet is shown in Fig. 8. The sequence band congestion due to the low-frequency torsional rotation of the methyl group is removed by the low temperature in the jet.[42] Three important features are clearly perceived in the spectra:[43-46]

1. The spectra of both molecules contain considerable memory of the benzene spectrum, with the bulk of the intensity appearing as in benzene itself, through vibronic coupling promoted by the 1500 cm^{-1} mode, ν_{14}.
2. An origin band, forbidden in benzene, is now observed with attendant Franck–Condon progressions in ν_1 and ν_{12}.*

*Cognizance of the mixed VC, FC character of ν_{12} should be taken. Although ν_{12} is totally symmetric, it undergoes harmonic mode scrambling with ν_{18} owing to the mass perturbation of the substituent and consequently part of the strength of 12^1_0 is due to vibronic coupling. This effect, probably not important for toluene (where the origin band and thus the FC intensity is strong) may be substantial in fluorobenzene.

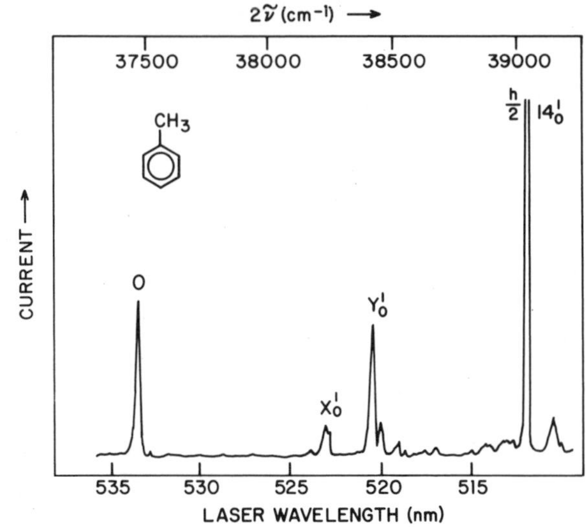

Fig. 8. Fundamental region of the normalized resonant two-photon ionization spectrum (intensity scale, $14_0^1 = h$, is twice that of Fig. 7) of toluene obtained in a supersonic jet with a rotational temperature of approximately 20 K. Here, X and Y refer to mixed modes derived from ν_1 and ν_{12}.

3. Using the intensity of 14_0^1 as an internal standard, as discussed in Section III.A, the intensity of the origin band and the progressions in ν_1 and ν_{12} are seen to be much more intense in toluene than in fluorobenzene.

The OP L_b absorption spectra of toluene and fluorobenzene contrast dramatically with the TP spectra. In OP absorption the fluorine perturbation brings in such strong allowed intensity that it completely dominates the vibronic coupling part—unlike methyl, which leaves the L_b transition only weakly allowed.

Thus fluorine, which is known from chemical evidence and valence principles to provide a strong electrostatic inductive perturbation, is ineffective in promoting intensity in the TP L_b spectrum. On the other hand, the weak hyperconjugative interaction of the methyl group is effective in inducing TP strength.

V. REGULARITIES IN THE TWO-PHOTON SPECTRA OF PERTURBED BENZENES

The labeled benzene, toluene, and fluorobenzene TP spectra (discussed in Sections III and IV) are examples of the many TP L_b spectra of substituted

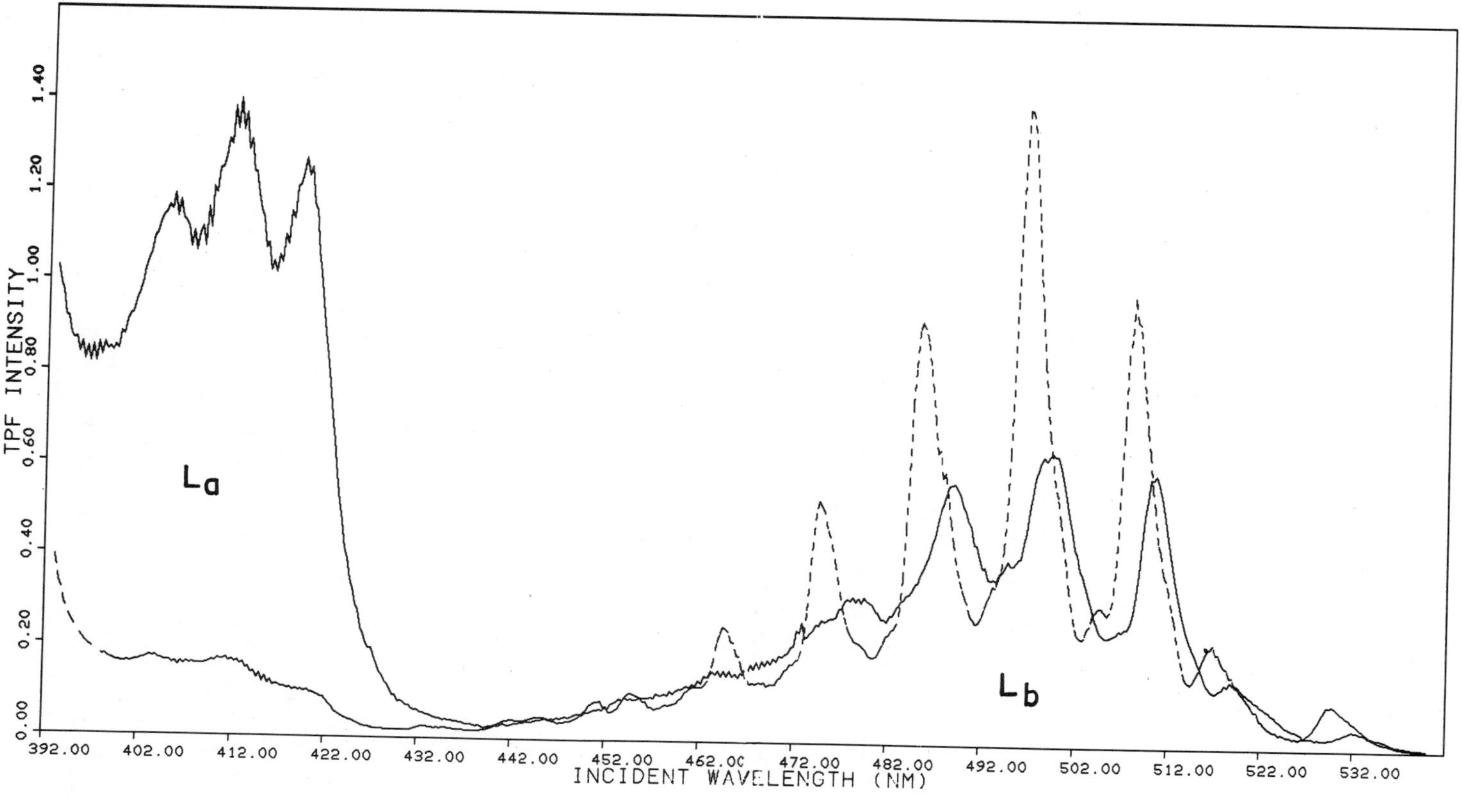

Fig. 9. Comparison of two-photon L_b and L_a fluorescence excitation spectra for benzene (dashed line) and fluorobenzene (solid line) solutions in *n*-hexane. The L_a band should be multiplied by a factor of approximately 3 to compensate for fluorescence yield variations. (Reproduced from Scott, Callis and Albrecht[53] by permission.)

benzenes that have been reported in recent years. Other substituents include Cl,[47,48] Br,[48] OH,[49] NH_2,[48,49] C≡CH,[50] CH=CH_2,[51] C≡N,[51] SiH_3,[51a] and aza nitrogens.[52] Scott et al.[53] have also recently reported the TP L_a band spectra of benzene and fluorobenzene. These spectra are shown in Fig. 9, which demonstrates that the L_a band of benzene is much weaker than the L_b band but is strongly enhanced in fluorobenzene. These derivatives provide a wide variation in the inductive and conjugative contributions to the electrostatic perturbation, and sufficient spectra have been studied to permit us to state the following rules:

1. The dominant vibrational activity in the benzene L_b spectrum involves b_{2u} modes, namely, ν_{14} (in the excited state) and ν_{15} and ν_{14} (in the ground state). This activity persists through many perturbed benzenes.
2. The forbidden L_b band of benzene is much more strongly activated by vibrational perturbations than is the L_a band.
3. The mass perturbation of the substituent alone provides a significant effect on the vibrational band intensities in the L_b spectrum.
4. The intensity of the L_b band is insensitive to inductive substituents, but the L_a band is strongly affected.
5. The intensity of the L_b band is very responsive to conjugative substituents (no data exist on the L_a band).

The last four rules are opposite to those found for the OP spectrum.

VI. THE PERTURBATION APPROACH TO PERTURBED BENZENE

The perturbation scheme which we follow parallels that described by Moffitt[2] for OP spectra of cata-condensed hydrocarbons. Though initially discussion will be limited to the benzene TP L_b spectrum, the transition to the lowest excited state, the procedures are introduced in a general form so that extension to other states and other alternate hydrocarbons can be performed easily.

The three lowest singlet π-electron states in C_6H_6 are $B_{2u}(L_b)$, $B_{1u}(L_a)$, and $E_{1u}(B_{a,b})$, in order of increasing energy, arising from promotion of an electron in the e_{1g} frontier π orbital (Φ_2, Φ_3) to the e_{2u} lowest unoccupied π orbital (Φ_4, Φ_5). This excitation results in four zeroth-order excited degenerate configurations, denoted χ_{25} (e.g., $\Phi_2 \rightarrow \Phi_5$), χ_{34}, χ_{35}, and χ_{24}. The D_{6h} benzene symmetry requires that the zeroth-order state wave functions are those given in Table III in terms of the π molecular orbitals listed in Table IV. In OP spectra, the transition to the $^1E_{1u}$ ($B_{a,b}$) state is electric dipole allowed and very intense, while for TP spectra, only transitions to E_{2g} [$C_{a,b}$ and $C'_{a,b}$ (Table III)] states are allowed but as yet have not been observed.

TABLE III
Zeroth-Order Excited State Wavefunctions[a, b]

$\Psi^0_{L_b} = 2^{-1/2}[\chi_{25} + \chi_{34}]$	$\Psi^0_{C'_a} = \chi_{15}$
$\Psi^0_{B_b} = 2^{-1/2}[\chi_{25} - \chi_{34}]$	$\Psi^0_{C'_b} = \chi_{14}$
$\Psi^0_{L_a} = 2^{-1/2}[\chi_{24} - \chi_{35}]$	$\Psi^0_{L'_a} = \chi_{16}$
$\Psi^0_{B_a} = 2^{-1/2}[\chi_{24} + \chi_{35}]$	$\Psi^0_{CT_b} = \chi_{74}$
$\Psi^0_{C_a} = \chi_{36}$	$\Psi^0_{CT_a} = \chi_{75}$
$\Psi^0_{C_b} = \chi_{26}$	$\Psi^0_{CT'_a} = \chi_{76}$

[a]Here, χ_{ij} represents the configurational wave function generated by the promotion $\Phi_i \rightarrow \Phi_j$. The molecular orbitals are given in Table IV.

[b]The charge-transfer (CT) states shown here are for a monosubstituted benzene with an electron donating substituent.

We will be concerned with vibronic and substituent induced electrostatic perturbations on the benzene ring, causing states of the same symmetry to mix to varying degrees depending on the strength of the perturbation. In addition, if the perturbation is caused by a substituent that has a low-lying π-type orbital (Φ_7 of Table IV), charge-transfer (CT) states between the substituent and the ring must be introduced into the perturbation scheme. A substituent which introduces this kind of delocalization is termed a resonance type. When the only effect of the substituent is to scramble the C_6H_6 states through the potential field effect of the substituent on the ring π electrons, the substituent is said to cause an inductive perturbation. Frequently,

TABLE IV
Molecular Orbital Numbering[a]

$$\Phi_1 = \frac{1}{\sqrt{6}}(\phi_0 + \phi_1 + \phi_2 + \phi_3 + \phi_4 + \phi_5)$$

$$\Phi_2 = \frac{1}{\sqrt{4}}(\phi_1 + \phi_2 - \phi_4 - \phi_5)$$

$$\Phi_3 = \frac{1}{\sqrt{12}}(2\phi_0 + \phi_1 - \phi_2 - 2\phi_3 - \phi_4 + \phi_5)$$

$$\Phi_4 = \frac{1}{\sqrt{4}}(\phi_1 - \phi_2 + \phi_4 - \phi_5)$$

$$\Phi_5 = \frac{1}{\sqrt{12}}(2\phi_0 - \phi_1 - \phi_2 + 2\phi_3 - \phi_4 - \phi_5)$$

$$\Phi_6 = \frac{1}{\sqrt{6}}(\phi_0 - \phi_1 + \phi_2 - \phi_3 + \phi_4 - \phi_5)$$

$$\Phi_7 = \phi_6$$

[a]Here, ϕ_r represents a $2p\pi$ atomic orbital centered on atom r. See Fig. 10 for atomic orbital numbering.

neither effect is mutually exclusive, but separation of the two effects is conceptually useful.

Following Moffitt's formalism, the matrix elements of the perturbation operator P over the molecular orbitals of Table IV may be written:

$$\mathbf{P}_{\xi\eta} = \langle \Phi_\xi | P | \Phi_\eta \rangle = \sum_r \sigma_\xi^r \sigma_\eta^r \alpha_r + \sum_r \sum_{s>r} \left(\sigma_\xi^s \sigma_\eta^r + \sigma_\xi^r \sigma_\eta^s \right) \beta_{rs} \tag{8}$$

where

$$\alpha_r = \langle \phi_r | P | \phi_r \rangle, \qquad \beta_{rs} = \langle \phi_r | P | \phi_s \rangle \tag{9}$$

are matrix elements of one-electron operators in the atomic orbital basis, and σ_j^r represents the coefficient in molecular orbital j at position r; α_r represents the difference between the substituted and unsubstituted Coulomb integral, associated with the field of the substituent at carbon atom r, and β_{rs} corresponds to changes in resonance integrals associated with either vibrational distortion or substituent conjugation with the ring.

The configurational perturbation matrix elements in terms of MO basis are given by

$$\langle \chi_{\xi\eta} | P | \chi_{\xi'\eta'} \rangle$$

The matrix elements between the states are then given by:

$$H_{i,j} = \langle \Psi_i | P | \Psi_j \rangle \tag{10}$$

As an example, consider the matrix element between L_b^0 and B_b^0 (Table V).

TABLE V
Inductive Effect Matrix Elements for Monosubstituted Benzene[a, b]

$\frac{1}{6}\alpha$	$-\frac{1}{6}\alpha$
$\langle L_b^0 \| P \| B_b^0 \rangle$	$\langle L_a^0 \| P \| C_a^0 \rangle$
$\langle L_b^0 \| P \| C_b^0 \rangle$	$\langle L_b^0 \| P \| C_b'^0 \rangle$
$\langle L_a^0 \| P \| C_a'^0 \rangle$	$\langle B_a^0 \| P \| C_a'^0 \rangle$
$\langle B_b^0 \| P \| C_b^0 \rangle$	$\sqrt{2}\langle A^0 \| P \| L_a^0 \rangle$
$\langle B_a^0 \| P \| C_a^0 \rangle$	
$\langle B_b^0 \| P \| C_b'^0 \rangle$	

[a] The high-lying $L_a'^0$ state (Table III) is ignored.

[b] Here, α is the change in the Coulomb integral at atom 0 (see Fig. 10).

Using Table III,

$$\begin{aligned} H_{L_b^0, B_b^0} &= \langle \Psi_{L_b}^0 | P | \Psi_{B_b}^0 \rangle \\ &= \tfrac{1}{2}[\langle \chi_{25} | P | \chi_{25} \rangle - \langle \chi_{34} | P | \chi_{34} \rangle] \\ &= \tfrac{1}{2}\{\mathbf{P}_{55} + \mathbf{P}_{22} - \mathbf{P}_{33} - \mathbf{P}_{44} + 2\mathbf{P}_{33} - 2\mathbf{P}_{22}\} \end{aligned} \tag{11}$$

Before moving into the tensor itself, the calculation of electric dipole matrix elements must be considered. The details again have been given by Moffitt. The π dipole matrix elements may be written

$$\begin{aligned} \mathbf{M}_{\xi\eta} &= \langle \Phi_\xi | \mu_t | \Phi_\eta \rangle \\ &= \left[\sum_m \sigma_\xi^m \sigma_\eta^m \mathbf{a}_m + \sum_m \sum_{n>m}{}' \left(\sigma_\xi^m \sigma_\eta^n + \sigma_\xi^n \sigma_\eta^m \right) \mathbf{b}_{mn} \right] \\ &\quad \times \begin{cases} \cos(m\pi/3)\mathbf{k} & \text{if } t = z \\ \sin(m\pi/3)\mathbf{j} & \text{if } t = y \end{cases} \end{aligned} \tag{12}$$

where

$$\begin{aligned} \mathbf{a}_m &= \langle \phi_m | \mu_t | \phi_m \rangle = \mathbf{r}_m \\ \mathbf{b}_{mn} &= \langle \phi_m | \mu_t | \phi_n \rangle = \tfrac{1}{2}(\mathbf{r}_m + \mathbf{r}_n) \mathbb{S}_{mn} \end{aligned} \tag{13}$$

In Eqs. (12) and (13), $m = 0$ to 5 for the atom numbering in Fig. 10; $\mathbf{r}_m, \mathbf{r}_n$ are the position vectors of the mth and nth atoms; $\mathbb{S}_{mn}$ is the overlap integral between ϕ_m and ϕ_n; and $\mathbf{k}, \mathbf{j}$ are unit vectors in the z and y directions. Parallel to the perturbation matrix elements, $\mathbf{a}_m$ occurs from substituent field effects mixing benzene states and $\mathbf{b}_{mn}$ arises from resonance-type terms. Note that the $a \leftrightarrow a$ and $b \leftrightarrow b$ excitations (e.g. $L_b \leftrightarrow C_b$) are always $t = z$ and the $a \leftrightarrow b$ excitations are $t = y$. Defining $Q_{i,j}$ analogously to $H_{i,j}$ allows generation of the dipole matrix elements between C_6H_6 states given in Table VI.

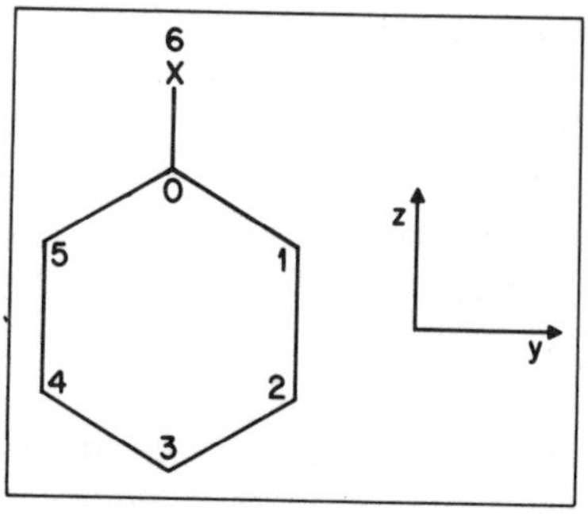

Fig. 10. Coordinate and numbering designations. The origin is at the center of the benzene ring.

TABLE VI
Electric Dipole Matrix Elements Between Benzene States[a]

$\frac{1}{2}R\mathbf{j}$	$-\frac{1}{2}R\mathbf{j}$	$\frac{1}{2}R\mathbf{k}$	$-\frac{1}{2}R\mathbf{k}$
$\langle C_a^0 \vert \mu_y \vert B_b^0 \rangle$	$\frac{1}{2}\langle A^0 \vert \mu_y \vert B_b^0 \rangle$	$\frac{1}{2}\langle A^0 \vert \mu_z \vert B_a^0 \rangle$	$\langle L_a^0 \vert \mu_z \vert C_a^0 \rangle$
	$\langle L_a^0 \vert \mu_y \vert C_b^0 \rangle$	$\langle B_a^0 \vert \mu_z \vert C_a^0 \rangle$	$\langle B_a^0 \vert \mu_z \vert C_a'^0 \rangle$
	$\langle B_a^0 \vert \mu_y \vert C_b^0 \rangle$	$\langle L_a^0 \vert \mu_z \vert C_a'^0 \rangle$	$\langle L_b^0 \vert \mu_z \vert C_b'^0 \rangle$
	$\langle L_b^0 \vert \mu_y \vert C_a^0 \rangle$	$\langle L_b^0 \vert \mu_z \vert C_b^0 \rangle$	
	$\langle L_a^0 \vert \mu_y \vert C_b'^0 \rangle$	$\langle B_b^0 \vert \mu_z \vert C_b^0 \rangle$	
	$\langle B_a^0 \vert \mu_y \vert C_b'^0 \rangle$	$\langle B_b^0 \vert \mu_z \vert C_b'^0 \rangle$	
	$\langle L_b^0 \vert \mu_y \vert C_a'^0 \rangle$		
	$\langle C_a'^0 \vert \mu_y \vert B_b^0 \rangle$		

[a] See footnote *a* of Table V. Axis designation is that shown in Fig. 10. All bond distances are assumed to be R.

As an example, consider the transition moment between an electron-donating CT_b charge-transfer state and the B_b benzene state in monosubstituted benzene where $CT_b = \chi_{74}$. The sum in Eq. (12) requires the a_m terms to be zero yielding for

$$\begin{aligned}\mathbf{Q}_{B_b^0, CT_b} &= \langle B_b^0 | \mu_z | CT_b \rangle \\ &= 2^{-1/2}\{\langle \chi_{25} | \mu_z | \chi_{74} \rangle - \langle \chi_{34} | \mu_z | \chi_{74} \rangle\} \\ &= 2^{-1/2}\{0 + \mathbf{M}_{37}\} \\ &= \left(\frac{1}{\sqrt{6}}\right)\mathbf{b}\mathfrak{S}\mathbf{k} = \left(\frac{3}{2\sqrt{6}}\right)R\mathfrak{S}\mathbf{k}\end{aligned}$$

We have taken the carbon-substituent distance to be R (the C—C bond distance of the ring), $\mathbf{b} = \frac{1}{2}|(\mathbf{r}_0 + \mathbf{r}_6)| = \frac{3}{2}R$, and $\mathfrak{S} = \mathfrak{S}_{06}$. The change in sign in M_{37} from minus to plus results from well-known rules for matrix elements involving determinantal wavefunctions,[67]

$$\langle \chi_{ik} | O | \chi_{jk} \rangle = -\langle \chi_{ki} | O | \chi_{kj} \rangle = -O_{ij} \tag{14}$$

where O represents any one-electron operator.

VII. VIBRONIC MECHANISMS IN THE BENZENE TWO-PHOTON SPECTRUM

A. The B_{2u} State

The TP $^1B_{2u}$ spectrum can only occur by mixing in *g*-parity electronic states. The analyses of the C_6H_6 TP $^1B_{2u} \leftarrow {}^1A_{1g}$ transition have shown, as was discussed in Section II, that the major intensity source is vibronic coupling to A_{1g} states through ν_{14} and to E_{2g} states through ν_{18}. INDO* distorted molecule calculations demonstrated that the primary A_{1g} state is the ground state,[24] in accord with the idea of Friedrich and McClain,[13] and predicted correctly that 18_0^1 is much weaker than 14_0^1. But neither the spectra nor the calculations give good physical insight into why ν_{14} is so much better as a coupling mode than is ν_{18}.

The perturbation approach offers such insight. In the Moffitt formalism, a nonvanishing β_{rs} in Eq. (8), can be associated with a vibrational distortion. The vibrational perturbation matrix element between the ground and excited state is $H_{A_{1g},B_{2u}}=2^{-1/2}[\mathbf{P}_{25}+\mathbf{P}_{34}]$. Since $\alpha_r=0$ in Eq. (8) because the potential field does not change in a first approximation for a vibration,

$$H_{A_{1g},B_{2u}}=\frac{2}{\sqrt{6}}[\beta_{01}-\beta_{12}+\beta_{23}-\beta_{34}+\beta_{45}-\beta_{50}] \tag{15}$$

Equation (15) shows that a b_{2u} vibration which alternately stretches and compresses ring C—C bonds will effectively couple the ground and L_b states. Robey and Schlag[40] have demonstrated that only the strongly active ν_{14} vibration possesses this type of motion in the excited state, as is shown in Fig. 6. Figure 6 also shows that both ν_{14}'' and ν_{15}'' possess alternate C—C compression and expansion character. Since only ν_{14} in the excited state fits the perturbation criteria but both the ground state forms of ν_{14} and ν_{15} fit, a particularly satisfying insight is provided into why both b_{2u} modes are active in the hot band spectrum but only ν_{14}' is active in the cold band spectrum.

The vibrational matrix element to the (C) E_{2g} state of Table III is

$$H_{E_{2g},B_{2u}}=2^{-1/2}[\mathbf{P}_{46}+\mathbf{P}_{56}] \tag{16a}$$

In Eq. (16a) the first term arises from C_a and the second from C_b. Expansion of the vibrational part of these terms yields

$$\mathbf{P}_{46}=\frac{1}{2\sqrt{6}}[\beta_{01}+2\beta_{12}+\beta_{23}-\beta_{34}-2\beta_{43}-\beta_{50}] \tag{16b}$$

and

$$\mathbf{P}_{56}=\frac{1}{2\sqrt{2}}[-\beta_{01}+\beta_{23}+\beta_{34}-\beta_{50}]. \tag{16c}$$

Therefore the perturbation requirement for an effective e_{1u} distortion is con-

*Intermediate neglect of differential overlap molecular orbital method (J. A. Pople and D. L. Beveridge, *Approximate Molecular Orbital Theory*, McGraw Hill, New York 1970).

traction of one half of the ring and expansion of the other half. However, none of the e_{1u} normal vibrations (Fig. 2) effectively approximates this motion. A pleasing consequence of the perturbation criterion given by Eqs. (16b) and (16c) is that Whiffen's normal coordinates show that the e_{1u} normal mode with the greatest contribution of these symmetry coordinates is in fact ν_{18}. The matrix element for coupling to the (C') E_{2g} state is similar to Eq. (16a) but with opposite phase and leads to perturbation requirements identical to (16b) and (16c).

Thus we have rationalized observation 1 of Section V that the major source of intensity for the B_{2u} benzene TP transition is through the b_{2u} vibration ν'_{14}, because the displacements in this mode represent nearly an optimum overlap with the perturbation requirement. The nearly complete skelatal localization of ν'_{14} (Section III.A) permits extension of this argument to benzene derivatives accounting for the high activity of ν'_{14} in these molecules.

B. The B_{1u} State

The vibrational perturbation matrix element between the ground and B_{1u} state is given by

$$H_{A_{1g}, B_{1u}} = 2^{-1/2}[\mathbf{P}_{24} - \mathbf{P}_{35}] = 0 \tag{17}$$

This matrix element is also zero if the A_{1g} state is doubly excited. We thus conclude that the vibronic coupling between the A_{1g} and B_{1u} states are much less effective than between A_{1g} and B_{2u} states, with the consequence that b_{1u} vibrational activity will be weak.

The remaining intensity mechanism for the B_{1u} state is coupling to E_{2g} states through e_{1u} vibrations. The perturbation matrix elements for this case are found to be identical to that for B_{2u}, except that the C and C' components add in phase, but as discussed in Section VII.A the normal e_{1u} vibrations are not effective distortions.

These results, which ignore energy denominator factors, qualitatively rationalize observation 2 of Section V that the TP spectrum of the benzene B_{1u} state is weak and that of the B_{2u} state strong. They are also consistent with the TP spectra of catacondensed hydrocarbons in general which usually show either very weak or unobserved TP L_a transitions.[54, 54a] Callis et al.[55] have recently drawn the same general conclusion for alternate hydrocarbons, making use of pseudoparity properties of alternate hydrocarbon wave functions.

It is interesting to consider bands involving vibronic coupling to the A_{2g} states. These involve b_{1u} vibrations in the case of the B_{2u} state, and b_{2u} vibrations in the case of the B_{1u} state, and are identity forbidden in experiments employing identical frequency photons of the same polarization. The only available A_{2g} state is provided by a doubly excited $e_{1g}^2 \rightarrow e_{2u}^2$ excitation, giving a vanishing vibronic matrix element with B_{2u}. Thus even in a two-laser experiment b_{1u} modes are expected to be weak, and indeed this is the case.[56] However, the matrix element between B_{1u} and A_{2g} does not vanish, and the perturbation distortion criterion again overlaps the motion of ν_{14}. This suggests that strong b_{2u} mode vibronic activity may arise in the $^1B_{1u}$ TP transition in a two laser experiment, employing photons of different frequencies and polarizations.

The region of the second excited singlet transition in benzene has received much less attention than $^1B_{2u}$. A weak absorption corresponding to the two-photon $^1B_{1u}$ transition has been detected in pure liquid benzene and in solutions in n-hexane by both direct absorption[56a, 56b] and fluorescence excitation.[53] When the spectrum is examined in circularly polarized light the intensity does not nearly vanish[56a] as it does for the $^1B_{2u}$ transition. This result is entirely consistent with lack of any efficient vibronic coupling routes in the $^1B_{1u}$ band involving totally symmetric states as is the case for the first singlet transition.

VIII. ELECTROSTATIC EFFECTS ON THE TWO-PHOTON SPECTRUM

A. The L_b State

The TP symmetric tensor term in Eq. (1) for the absorption $f \leftarrow g$ involving identical photons is

$$(S^s_{\rho,\sigma})_{f,g} = \sum_i \left[\frac{1}{\Delta E_i - \hbar\omega}\right]\left[\langle g|\mu_\rho|i\rangle\langle i|\mu_\sigma|f\rangle + \langle g|\mu_\sigma|i\rangle\langle i|\mu_\rho|f\rangle\right] \tag{18}$$

where i runs over virtual intermediate states; μ_σ or μ_ρ are the appropriate electric dipole components; $\rho, \sigma = x, y, z$; $\hbar\omega$ is the laser energy; and ΔE_i is the energy difference between the ground and intermediate states. For substituted benzenes of C_{2v} symmetry, $S^s_{\rho\sigma}$ involves only the in-plane S_{yz} components for a transition to the B_2 symmetry L_b state. Then

$$S_{yz} = \sum_i \left[\frac{1}{\Delta E_i - \hbar\omega}\right]\left[\langle A|\mu_y|i\rangle\langle i|\mu_z|L_b\rangle + \langle A|\mu_z|i\rangle\langle i|\mu_y|L_b\rangle\right] \tag{19}$$

By restricting the intermediate states to the basis set of Table III,* a simple physical model is provided for the TP intensities.

*Two terms involving ground and excited states as intermediate states have been omitted in this discussion. They are examined in Sections IX and X.

1. *Inductive Perturbations*

a. Weak Coupling. We first consider the weak coupling case, defined in terms of the zeroth-order basis which diagonalizes the perturbation submatrix made up of the three states B_{2u}, B_{1u}, and E_{1u}, which in turn represent splitting of the $\chi_{24}, \chi_{34}, \chi_{25}, \chi_{35}$ configurational degeneracy. Substitution of the possible eigenfunctions into Eq. (19) shows that S_{yz} vanishes for any intermediate state because all states still retain their g and u character within the weak coupling approximation and the $g \leftrightarrow g$ two-photon selection rule is unsatisfied. In the one-photon spectrum, this limit represents the Moffitt "even" perturbation case with only "intraconfigurational" interaction, that is, first-order configurational interaction. The OP forbidden B_{2u} transition becomes allowed by weak coupling because of mixing with the strongly allowed $E_{1u}(B_b)$ state.

b. Intermediate Coupling. The intermediate coupling case is considered to be the result of interconfigurational interaction, or second-order configurational interacting. One-photon transition moments are found to remain essentially unchanged from the weak coupling case, but because of the lack of g and u character the TP tensor no longer vanishes.

(i) *$B_{a,b}$ intermediate states.* We start with $B_{a,b}$ as intermediate states, derived from the very strong ${}^1E_{1u} \leftarrow {}^1A_{1g}(V \leftarrow N)$ transition in benzene, which splits into $A_1(B_a)$ and $B_2(B_b)$ components in C_{2v}. Equation (19) becomes

$$S_{yz}^{B} = \left[\frac{1}{\Delta E_B - \hbar\omega}\right]\left[\langle A|\mu_z|B_a\rangle\langle B_a|\mu_y|L_b\rangle + \langle A|\mu_y|B_b\rangle\langle B_b|\mu_z|L_b\rangle\right] \tag{20}$$

where ΔE_B is the average energy of the E_{1u} states.

The inductive effect allows TP transitions by mixing g-parity benzene analog states into L_b through substituent modification of the potential acting on the benzene π electrons. There are three g states which in principle may mix into L_b^0 and $B_{a,b}^0$, the A_{1g} ground state and the two E_{2g} states, $C_{a,b}^0$ and $C_{a,b}'^0$. (This ignores doubly excited states.) Consequently for C_{2v} and higher symmetries,

$$\begin{aligned}
\psi_{L_b} &= \psi_{L_b}^0 + \lambda(B_b)\psi_{B_b}^0 + \lambda(C_b)\psi_{C_b}^0 + \lambda(C_b')\psi_{C_b'}^0 \\
\psi_{L_a} &= \psi_{L_a}^0 + \gamma(B_a)\psi_{B_a}^0 + \gamma(C_a)\psi_{C_a}^0 + \gamma(C_a')\psi_{C_a'}^0 + \gamma(A)\psi_A^0 \\
\psi_{B_b} &= \psi_{B_b}^0 + \Lambda(L_b)\psi_{L_b}^0 + \Lambda(C_b)\psi_{C_b}^0 + \Lambda(C_b')\psi_{C_b'}^0 \\
\psi_{B_a} &= \psi_{B_a}^0 + \ell(L_a)\psi_{L_a}^0 + \ell(C_a)\psi_{C_a}^0 + \ell(C_a')\psi_{C_a'}^0 + \ell(A)\psi_A^0, \quad \text{etc.}
\end{aligned} \tag{21}$$

These mixing coefficients are, for example, $\lambda(B_b)=H_{B_b^0,L_b^0}/\Delta E_{B_b^0L_b^0}$, where $\Delta E_{B_b^0L_b^0}=E(B_b^0)-E(L_b^0)$. The relevant matrix elements, generated by use of Eqs. (8)–(11), are listed in Table V for monosubstitution.

We may now examine the expectations for the B intermediate states by substituting the wave functions of Eq. (21) into Eq. (20) and making use of the individual electric dipole transition moments for the benzene states in Table VI. The result is

$$S_{yz}^{B}=\left[\frac{1}{\Delta E_B-\hbar\omega}\right]\frac{1}{2}R^2\mathbf{jk}[-\ell(C_a)-\ell(C_a')-\Lambda(C_b)+\Lambda(C_b') \\ -2\lambda(C_b)-2\lambda(C_b')] \qquad (22)$$

where only first-order terms have been retained. The linear dependence of the mixing coefficients $\lambda(C)$, $\Lambda(C)$, and $\ell(C)$ on the perturbation matrix elements and the degeneracy of C_a^0 with C_b^0 and of $C_a'^0$ with $C_b'^0$ simplifies Eq. (22) to

$$S_{yz}^{B}=\left[\frac{-R^2}{\Delta E_B-\hbar\omega}\right]\mathbf{jk}\{[\lambda(C_b)+\lambda(C_b')]+[\Lambda(C_b)-\Lambda(C_b')]\} \qquad (23)$$

The mixing coefficients within the individual square bracketed terms in Eq. (23) are identical except for the energy denominators, but they are opposite in sign. Thus for the B intermediate states, inductive type perturbations bring in terms which to a large extent cancel out. It is therefore important to examine other possible benzene intermediate states.

(ii) *C and C' intermediate states.* When the $C_{a,b}$ states are considered as intermediate states in Eq. (19),

$$S_{yz}^{C}=\left[\frac{1}{\Delta E_C-\hbar\omega}\right][\langle A|\mu_y|C_b\rangle\langle C_b|\mu_z|L_b\rangle+\langle A|\mu_y|C_a\rangle\langle C_a|\mu_z|L_b\rangle] \qquad (24)$$

Again making use of Table VI,

$$S_{yz}^{C}=\left[\frac{R^2}{\Delta E_C-\hbar\omega}\right]\mathbf{jk}\left[\Lambda(C_b)-\frac{1}{2}\xi(B_a)\right] \qquad (25)$$

where $\xi(B_a)$ is the contamination of the ground state by B_a^0. An exactly analogous result is found for the $C'_{a,b}$ states, that is,

$$S_{yz}^{C'}=\left[\frac{-R^2}{\Delta E_{C'}-\hbar\omega}\right]\mathbf{jk}\left[\Lambda(C_b')-\frac{1}{2}\xi(B_a)\right] \qquad (26)$$

(iii) *Total inductive effect.* Summation of Eqs. (23), (25) and (26) gives the total inductive effect on the TP L_b transition tensor:

$$S_{yz}^{I} = R^2\mathbf{jk}\Bigg\{\left(\frac{-1}{\Delta E_B - \hbar\omega}\right)[\lambda(C_b') + \lambda(C_b)] + \Lambda(C_b)\left[\frac{1}{\Delta E_C - \hbar\omega} - \frac{1}{\Delta E_B - \hbar\omega}\right] - \Lambda(C_b')\left[\frac{1}{\Delta E_{C'} - \hbar\omega} - \frac{1}{\Delta E_B - \hbar\omega}\right] - \frac{1}{2}\xi(B_a)\left[\frac{1}{\Delta E_{C'} - \hbar\omega} - \frac{1}{\Delta E_C - \hbar\omega}\right]\Bigg\} \tag{27}$$

Since the mismatch of the blue-green laser light energy (used to excite the L_b transition) with each of the intermediate state terms in Eq. (27) is approximately the same, all the square-bracketed terms are small. The important conclusion is that an inductive type perturbation, "even" in the Moffitt sense, will be ineffective in imparting intensity to the L_b TP spectrum. This allows an understanding of observation 4 of Section V.

c. Strong Coupling. In the strong coupling case, perturbation theory initiated from a benzene state basis set breaks down. This may be somewhat imperfectly expressed by retention of square terms (e.g. $\ell(\mathrm{A})\lambda(\mathrm{B}_b)$) in the TP tensor. Since each of the contamination terms arising from Eq. (21) enters the tensor in a symmetric way with approximately equal magnitude, it is probable that even in this limit the inductive effect is unimportant.

2. *Resonance Perturbations*

The resonance effect is described as the ability of a substituent to extend the space over which the π electrons are delocalized. This is exemplified by admixture of CT states into benzene states. We shall limit our discussion to electron-donating substituents until Section XI.

a. Weak Coupling. This case makes no contribution to the TP tensor.

b. Intermediate Coupling. Since in the paradigm situation that we are visualizing there is no modification of the potential acting on the benzene π electrons, perturbation matrix elements over ring or substituent states alone vanish [i.e., $\alpha_r = 0$ in Eq. (8)]. The first-order perturbed wave functions become

$$\begin{aligned}\psi_{L_b} &= \psi^0_{L_b} + \lambda(\mathrm{CT}_b)\psi^0_{CT_b}\\ \psi_{B_b} &= \psi^0_{B_b} + \Lambda(\mathrm{CT}_b)\psi^0_{CT_b}\\ \psi_{B_a} &= \psi^0_{B_a} + \ell(\mathrm{CT}_a)\psi^0_{CT_a}\\ &\text{etc.}\end{aligned} \tag{28}$$

The relevant matrix elements for monosubstitution are listed in Table VII.

We again consider the various intermediate states individually, but the resonance effect leads to an additional possibility, namely, that the CT states may act as intermediate states.

(i) $B_{a,b}$ *intermediate states.* Substituting the wave functions of Eq. (28) into the basic TP tensor [Eq. (19)] with B intermediate states and again retaining only first-order terms,

$$\begin{aligned}S^B_{yz} = \left[\frac{1}{\Delta E_B - \hbar\omega}\right]&\Big[\langle A^0|\mu_z|B^0_a\rangle\{\ell(\mathrm{CT}_a)\langle \mathrm{CT}_a|\mu_y|L^0_b\rangle\\ &+\lambda(\mathrm{CT}_b)\langle B^0_a|\mu_y|\mathrm{CT}_b\rangle\}\\ &+\langle A^0|\mu_y|B^0_b\rangle\{\Lambda(\mathrm{CT}_b)\langle \mathrm{CT}_b|\mu_z|L^0_b\rangle + \lambda(\mathrm{CT}_b)\langle B^0_b|\mu_z|\mathrm{CT}_b\rangle\}\Big]\end{aligned} \tag{29}$$

In this case both the perturbation coefficients and the CT electric dipole moments depend on the number and orientation of substituents, and thus Eq. (29) is best left in its present form until a discussion of specific cases.

TABLE VII
Resonance Matrix Elements Between C_6H_6 and Charge-Transfer States for Monosubstituted Benzene[a,b]

$6^{-1/2}\beta$	$-6^{-1/2}\beta$
$\frac{1}{2}\langle A^0\|P\|\mathrm{CT}_a\rangle$	$\langle B^0_a\|P\|\mathrm{CT}_a\rangle$
$\langle L^0_a\|P\|\mathrm{CT}_a\rangle$	$\langle C'^0_a\|P\|\mathrm{CT}_a\rangle$
$\langle B^0_b\|P\|\mathrm{CT}_b\rangle$	$\langle L^0_b\|P\|\mathrm{CT}_b\rangle$
	$\langle C'^0_b\|P\|\mathrm{CT}_b\rangle$

[a]The (high-energy) CT'_a state (Table III) is ignored.
[b]Here, β is the resonance integral between π-type orbitals on atoms 0 and 6 (see Fig. 10).

Note, though, that the sign relationships in Table VII for the perturbation matrix elements, as well as those for CT-type dipole moments in Table VIII, require that for monosubstitution all the terms add in phase. The important conclusion is that resonance type substituents can impart large intensity to the L_b TP spectrum.

(ii) *C intermediate states.* The role of the $C_{a,b}$ and $C'_{a,b}$ states as intermediate states are now considered. Substituting the wave functions (28) into the TP tensor with $C_{a,b}$ intermediate states shows that

$$S_{yz}^{C} = \left[\frac{1}{\Delta E_C - \hbar\omega}\right]\left[\langle C_b^0|\mu_z|L_b^0\rangle\xi(\mathrm{CT}_a)\langle \mathrm{CT}_a|\mu_y|C_b^0\rangle + \langle C_a^0|\mu_y|L_b^0\rangle\xi(\mathrm{CT}_a)\langle \mathrm{CT}_a|\mu_z|C_a^0\rangle\right] \tag{30}$$

Similarly, with $C'_{a,b}$ intermediate states

$$S_{yz}^{C'} = \left[\frac{1}{\Delta E_{C'} - \hbar\omega}\right]\left[\langle C_b'^0|\mu_z|L_b^0\rangle\{\langle A^0|\mu_y|\mathrm{CT}_b\rangle\Omega'(\mathrm{CT}_b) + \xi(\mathrm{CT}_a)\langle \mathrm{CT}_a|\mu_y|C_b'^0\rangle\} + \langle C_a'^0|\mu_y|L_b^0\rangle\{\langle A^0|\mu_z|\mathrm{CT}_a\rangle\Omega'(\mathrm{CT}_a) + \xi(\mathrm{CT}_a)\langle \mathrm{CT}_a|\mu_z|C_a'^0\rangle\}\right] \tag{31}$$

where Ω' is the perturbation parameter measuring the amount of CT character in the C' wavefunction.

(iii) *CT intermediate states.* When we are considering resonance-type perturbations, an additional type of intermediate state must be allowed into

TABLE VIII
Electric Dipole Matrix Elements Between Benzene and Charge-Transfer States in Monosubstituted Benzene[a,b]

$3/2(6^{-1/2})R\mathbb{S}\mathbf{k}$	$-3/2(6^{-1/2})R\mathbb{S}\mathbf{k}$
$\frac{1}{2}\langle A^0|\mu_z|\mathrm{CT}_a\rangle$	$\langle B_a^0|\mu_z|\mathrm{CT}_a\rangle$
$\langle L_a^0|\mu_z|\mathrm{CT}_a\rangle$	$\langle C_a'^0|\mu_z|\mathrm{CT}_a\rangle$
$\langle B_b^0|\mu_z|\mathrm{CT}_b\rangle$	$\langle L_b^0|\mu_z|\mathrm{CT}_b\rangle$
	$\langle C_b'^0|\mu_z|\mathrm{CT}_b\rangle$

[a]See footnotes *a* and *b* of Table VI and *a* of Table VII.

[b]Here, $\mathbb{S}$ represents the overlap integral between π-type orbitals on atoms 0 and 6 (see Fig. 10).

the sum of Eq. (19)—the CT states. These intermediate states lead to two types of terms:

First, there are zeroth-order terms, $S^{\mathrm{CT}^{(0)}}$, which do not require any perturbation matrix elements but contain a product of two CT transition moments in the tensor. These may be written

$$S_{yz}^{\mathrm{CT}^{(0)}} = \left[\frac{1}{\Delta E_{\mathrm{CT}} - \hbar\omega}\right]\left[\langle A^0|\mu_y|\mathrm{CT}_b\rangle\langle \mathrm{CT}_b|\mu_z|L_b^0\rangle + \langle A^0|\mu_z|\mathrm{CT}_a\rangle\langle \mathrm{CT}_a|\mu_y|L_b^0\rangle\right] \tag{32}$$

We shall show in subsequent sections that these terms are generally zero for most substitutions and can be ignored for others.

The second type of term contain the first-order corrections to the CT wave functions:

$$\begin{aligned} \psi(\mathrm{CT}_b) &= \psi^0_{\mathrm{CT}_b} - \lambda(\mathrm{CT}_b)\psi^0_{L_b} - \Lambda(\mathrm{CT}_b)\psi^0_{B_b} - \Omega'(\mathrm{CT}_b)\psi^0_{C'_b} \\ \psi(\mathrm{CT}_a) &= \psi^0_{\mathrm{CT}_a} - \gamma(\mathrm{CT}_a)\psi^0_{L_a} - \ell(\mathrm{CT}_a)\psi^0_{B_a} - \Omega'(\mathrm{CT}_a)\psi^0_{C'_a} \end{aligned} \tag{33}$$

First-order perturbation theory requires that the same coefficients enter Eq. (33) as in (28) but with reversed sign. Substituting the wave functions of Eqs. (28) and (33) into the TP tensor [eq. (19)] with CT intermediate states gives

$$\begin{aligned} S_{yz}^{\mathrm{CT}^{(1)}} = \left[\frac{1}{\Delta E_{\mathrm{CT}} - \hbar\omega}\right]\Big[&\langle A^0|\mu_y|\mathrm{CT}_b\rangle\{-\Omega'(\mathrm{CT}_b)\langle C_b'^0|\mu_z|L_b^0\rangle \\ &+ \lambda(\mathrm{CT}_b)\langle \mathrm{CT}_b|\mu_z|\mathrm{CT}_b\rangle\} \\ &+ \langle \mathrm{CT}_b|\mu_z|L_b^0\rangle\{-\Lambda(\mathrm{CT}_b)\langle A^0|\mu_y|B_b^0\rangle\} \\ &+ \langle A^0|\mu_z|\mathrm{CT}_a\rangle\{-\Omega'(\mathrm{CT}_a)\langle C_a'^0|\mu_y|L_b^0\rangle\} \\ &+ \langle \mathrm{CT}_a|\mu_y|L_b^0\rangle\{-\ell(\mathrm{CT}_a)\langle A^0|\mu_z|B_a^0\rangle \\ &+ \xi(\mathrm{CT}_a)\langle \mathrm{CT}_a|\mu_z|\mathrm{CT}_a\rangle\}\Big] \end{aligned} \tag{34}$$

We see two types of terms in Eq. (34). The first contains contributions which add out of phase with the benzene intermediate state terms, that is, $-\Lambda(\mathrm{CT}_b)$, $-\Omega'(\mathrm{CT}_a)$, $-\ell(\mathrm{CT}_a)$, $-\Omega'(\mathrm{CT}_b)$ terms. The second contains the dipole moments of the CT states, e.g., $\mu_z(\mathrm{CT}_a) = \langle \mathrm{CT}_a|\mu_z|\mathrm{CT}_a\rangle$. These contributions arise from both CT contamination of the ground and excited state wave functions and may yield an important intensity mechanism for certain resonance-type substituents.

(iv) *Total resonance effect.* We tend to favor the expression due to the $B_{a,b}$ intermediate states as being the major source of intensity [Eq. (29)] because the other states are generally considered Rydberg-like,[56c] with small transition moments and perturbation matrix elements. However, it is instructive to see the total result for the resonance effect by summing Eqs. (29), (30), (31), and (34). The result is

$$
\begin{aligned}
S_{yz} = {} & \left[\frac{R}{\Delta E_B - \hbar\omega}\right]\left[\lambda(\mathrm{CT}_b)\left(\langle B_a^0|\mu_y|\mathrm{CT}_b\rangle\mathbf{k} - \langle B_b^0|\mu_z|\mathrm{CT}_b\rangle\mathbf{j}\right)\right] + \left[\frac{1}{\Delta E_{\mathrm{CT}} - \hbar\omega}\right] \\
& \times\left[\lambda(\mathrm{CT}_b)\mu_z(\mathrm{CT}_b)\langle A^0|\mu_y|\mathrm{CT}_b\rangle + \xi(\mathrm{CT}_a)\mu_z(\mathrm{CT}_a)\langle \mathrm{CT}_a|\mu_y|L_b^0\rangle\right] \\
& + R\Lambda(\mathrm{CT}_b)\langle \mathrm{CT}_b|\mu_z|L_b^0\rangle\left\{\frac{1}{\Delta E_{\mathrm{CT}} - \hbar\omega} - \frac{1}{\Delta E_B - \hbar\omega}\right\}\mathbf{j} \\
& + R\ell(\mathrm{CT}_a)\langle \mathrm{CT}_a|\mu_y|L_b^0\rangle\left\{\frac{1}{\Delta E_B - \hbar\omega} - \frac{1}{\Delta E_{\mathrm{CT}} - \hbar\omega}\right\}\mathbf{k} \\
& - 2^{-1}R\Omega'(\mathrm{CT}_b)\langle A^0|\mu_y|\mathrm{CT}_b\rangle\left\{\frac{1}{\Delta E_C' - \hbar\omega} - \frac{1}{\Delta E_{\mathrm{CT}} - \hbar\omega}\right\}\mathbf{k} \\
& - 2^{-1}R\Omega'(\mathrm{CT}_a)\langle A^0|\mu_z|\mathrm{CT}_a\rangle\left\{\frac{1}{\Delta E_C' - \hbar\omega} - \frac{1}{\Delta E_{\mathrm{CT}} - \hbar\omega}\right\}\mathbf{j} \\
& + 2^{-1}R\xi(\mathrm{CT}_a)\left[\left[\frac{1}{\Delta E_C - \hbar\omega}\right]\left(\langle \mathrm{CT}_a|\mu_y|C_b^0\rangle\mathbf{k} - \langle \mathrm{CT}_a|\mu_z|C_a^0\rangle\mathbf{j}\right)\right. \\
& \left. - \left[\frac{1}{\Delta E_C' - \hbar\omega}\right]\left(\langle \mathrm{CT}_a|\mu_y|C_b'^0\rangle\mathbf{k} + \langle \mathrm{CT}_a|\mu_z|C_a'^0\rangle\mathbf{j}\right)\right]
\end{aligned}
\tag{35}
$$

where the benzene transition moments of Table VI have been used and the zeroth-order CT intermediate state terms left out.

The terms in braces in Eq. (35) are small since the energy denominators are similar. This enormously simplifies the final result. The first term in square brackets in (35) represents the introduction of CT character into the L_b state, and the opposite signs of the transition moments between CT_b and B_a^0, B_b^0 will make this term large. The second term in (35) contains the contribution of the permanent dipole moments of the CT states, CT_a and CT_b. The last term has as additional contributions the CT character in the ground state. For substituents with relatively low-lying CT states, the $\xi(\mathrm{CT}_a)$ term will be unimportant compared to excited state contaminations and Eq. (35) reduces to the physically useful form

$$
\begin{aligned}
S_{yz} = {} & \left[\frac{R}{\Delta E_B - \hbar\omega}\right]\left[\lambda(\mathrm{CT}_b)\{\langle B_a^0|\mu_y|\mathrm{CT}_b\rangle\mathbf{k} - \langle B_b^0|\mu_z|\mathrm{CT}_b\rangle\mathbf{j}\}\right] \\
& + \left[\frac{1}{\Delta E_{\mathrm{CT}} - \hbar\omega}\right]\left[\lambda(\mathrm{CT}_b)\mu_z(\mathrm{CT}_b)\langle A^0|\mu_y|\mathrm{CT}_b\rangle\right]
\end{aligned}
\tag{36}
$$

An important conclusion can be drawn from Eq. (36): there is large intensity induced into the TP L_b transition through CT contamination of the L_b state. This is the same result that was deduced from Eq. (29) and rationalizes observation 5 of Section V.

c. **Strong Coupling.** In this limit, energetically low-lying CT configurations cause the wave functions to take on a strong C_{2v} complexion, and the TP transition is expected to reflect the complete lack of parity due to the large CT character. Consequently, the TP spectrum is predicted to become strong and to a high degree resemble the OP spectrum, which in the strong coupling case contains substituent modes. We can express this case to a considerable extent by retaining the square terms [e.g. $\lambda(\mathrm{CT}_b)\Lambda(\mathrm{CT}_a)$] neglected in the intermediate coupling case (some of the neglected linear terms also may become important). There are many such terms, but in the strong coupling limit the CT_b state is coming into resonance with a benzene state. The leading term is then that which arises from simultaneous CT_b contamination of the intermediate and final states, e.g. for B_b:

$$\left[\frac{R\mathbf{j}}{\Delta E_B - \hbar\omega}\right]\Lambda(\mathrm{CT}_b)\lambda(\mathrm{CT}_b)\mu_z(\mathrm{CT}_b) \tag{37}$$

Here, $\mu_z(\mathrm{CT}_b) \simeq \mu_g - 2e\mathbf{R}$ to a point charge approximation and is especially large when the permanent dipole moment is small. The correction term [Eq. (37)] can become important if $\Lambda\lambda$ is significant, and this may be used as a diagnostic test for the validity of the first-order theory of Section VIII.A.2.*b*.

B. The L_a State

Though most TP studies on perturbed benzenes have concentrated on the lowest $\pi\pi^*$ singlet L_b state, the first reports of TP spectra for the second singlet L_a state have been made.[53] We will show through the Moffitt formalism that the high influence on the L_a TP spectrum by inductive-type substituents is largely brought about by the ground state.

1. *Inductive Perturbations*

a. **Weak Coupling.** The TP intensity vanishes in this approximation because of parity retention. The OP spectrum is also unaffected by the "even"-type perturbation, which may be associated with an inductive effect [i.e., $H_{L_a^0, B_a^0} = 0$ from Eqs. (8)–(10) and (14)].

b. **Intermediate Coupling.** We proceed as we did for the L_b state, restricting the possible intermediate states to the basis set of Table III. The S_{yy} and S_{zz} components are now relevant instead of S_{yz}, the sole component for the L_b state. Starting with the $B_{a,b}$ intermediate states, we find

$$S_{yy} = \left[\frac{2^{-1}R^2}{\Delta E_B - \hbar\omega}\right]\{2\gamma(A) - [\gamma(C_a) - \gamma(C_a')] + [\ell(C_a) - \ell(C_a')]\}\mathbf{jj}$$

and (38)

$$S_{zz} = \left[\frac{2^{-1}R^2}{\Delta E_B - \hbar\omega}\right]\{2\gamma(A) + [\gamma(C_a) - \gamma(C_a')] - [\ell(C_a) - \ell(C_a')]\}\mathbf{kk}$$

Remember that γ, ℓ indicate mixing into the L_a and B_a states, respectively. Equations (38) show the possible strong role of mixing of the ground state with L_a through the term $\gamma(A)$ in inducing TP intensity by an inductive perturbation. The total inductive effect expression becomes

$$\begin{aligned} S_{yy} = 2^{-2}R^2\mathbf{jj}\Big\{ &\gamma(A)\left[\frac{4}{\Delta E_B - \hbar\omega} - \frac{1}{\Delta E_C - \hbar\omega} - \frac{1}{\Delta E_{C'} - \hbar\omega}\right] \\ &-2\left[\frac{1}{\Delta E_B - \hbar\omega}\right][\gamma(C_a) - \gamma(C_a')] \\ &+2\ell(C_a)\left[\frac{1}{\Delta E_B - \hbar\omega} - \frac{1}{\Delta E_C - \hbar\omega}\right] \\ &-2\ell(C_a')\left[\frac{1}{\Delta E_B - \hbar\omega} - \frac{1}{\Delta E_{C'} - \hbar\omega}\right] \\ &-\ell(A)\left[\frac{1}{\Delta E_C - \hbar\omega} + \frac{1}{\Delta E_{C'} - \hbar\omega}\right]\Big\} \end{aligned} \tag{39}$$

$$\begin{aligned} S_{zz} = 2^{-2}R^2\mathbf{kk}\Big\{ &\gamma(A)\left[\frac{4}{\Delta E_B - \hbar\omega} - \frac{1}{\Delta E_C - \hbar\omega} - \frac{1}{\Delta E_{C'} - \hbar\omega}\right] \\ &+2\left[\frac{1}{\Delta E_B - \hbar\omega}\right][\gamma(C_a) - \gamma(C_a')] \\ &+2\ell(C_a)\left[\frac{1}{\Delta E_C - \hbar\omega} - \frac{1}{\Delta E_B - \hbar\omega}\right] \\ &-2\ell(C_a')\left[\frac{1}{\Delta E_{C'} - \hbar\omega} - \frac{1}{\Delta E_B - \hbar\omega}\right] \\ &+\ell(A)\left[\frac{1}{\Delta E_C - \hbar\omega} + \frac{1}{\Delta E_{C'} - \hbar\omega}\right]\Big\} \end{aligned} \tag{40}$$

These tedious expressions indicate that the most important terms are those due to the ground state mixing into the L_a and B_a states, through $\gamma(A)$ and $\ell(A)$ respectively, and the C states mixing into the L_a state, through $\gamma(C_a) - \gamma(C_a')$. It is important to recall that $\gamma(A), \gamma(C_a'), \ell(C_a)$ are negative and $\ell(A)$, $\gamma(C_a)$, and $\ell(C_a')$ are positive. Thus all terms for S_{yy} in Eq. (39) have the same $(-)$ sign, while the last four terms in the expression for S_{zz} [Eq. (40)] are positive and the first term is negative. These interferences between the $\gamma(A)$ and $\ell(A)$ terms (and the $\gamma(C_A)$ and $\gamma(C_a')$ terms), constructive in Eq. (39) and destructive in Eq. (40) insure that S_{yy} is predominant over S_{zz}. (This result can be seen most simply by retaining only the leading $\gamma(A)$, $\ell(A)$, $\gamma(C_A)$, and $\gamma(C_A')$ terms.) The experimental result of Scott et al.[53] that a fluorine perturbation does induce major intensity into the L_a transition (observation 4 of Section V), is consistent with the theoretical conclusion that an inductive perturbation induces large intensity into the L_a TP spectrum.

2. *Resonance Perturbations*

a. Weak Coupling. This case makes no contribution to the TP tensor.

b. Intermediate Coupling. The resonance terms yield a result completely analogous to Eq. (18) for the L_b state for each tensor component S_{yy} and S_{zz} of the L_a state. This suggests that resonance type substituents, such as methyl, may also impart important intensity to the L_a TP spectrum.

For the L_a TP spectrum, it appears that resonance and inductive effects on the cross section will be heavily mixed, in contrast to the L_b TP spectrum where resonance contributions are dominant. Confirmation of the effects of resonance substituents on the L_a TP spectrum awaits further experimental work.

IX. TWO-PHOTON L_b SPECTRA OF MONO-SUBSTITUTED BENZENES

Making use of the conclusion of Section VIII, the resonance effect represents the dominating mechanism for inducing intensity into the L_b spectrum. For a monosubstituted benzene, short axis (y) polarized transition moments involving CT configurations are negligible compared to long-axis ones, and Eq. (36) reduces to the particularly simple form

$$S_{yz} = \left[\frac{R\mathbf{j}}{\Delta E_B - \hbar\omega}\right]\left[\lambda(\mathrm{CT}_b)\langle B_b|\mu_z|\mathrm{CT}_b\rangle\right]$$

$$\approx \frac{R^2}{4}\mathbf{jk}\beta\mathbb{S}\left[\frac{1}{\Delta E_B - \hbar\omega}\right]\left[\frac{1}{\Delta E_{\mathrm{CT}_b, L_b^0}}\right] \qquad (41)$$

The diagnostic term for strong coupling [Eq. (37)], and for the breakdown of the linear theory inherent in Eq. (36) becomes

$$\frac{\beta^2 R\mathbf{j}}{6}\left[\boldsymbol{\mu}_g - 2e\mathbf{R}\right]\left[\frac{1}{\Delta E_B - \hbar\omega}\right]\left[\frac{1}{\Delta E_{\mathrm{CT}_b, B_b^0}}\right]\left[\frac{1}{\Delta E_{\mathrm{CT}_b, L_b^0}}\right]. \tag{42}$$

Consider, as an example the spectrum of phenol, shown in Fig. 11*a*. The Franck–Condon intensity is somewhat less than the VC intensity. Using reasonable values for $\beta \approx 1.2$ eV, $E_{\mathrm{CT}_b} \cong 8$ eV, and the observed ground state dipole moment ≈ 4 D (≈ 1 Å), the strong coupling correction, Eq. (42) is $0.05R^2[1/(\Delta E_B - \hbar\omega)]$. This actually exceeds the magnitude of the first-order S_{yz} term [Eq. (41)] estimated as $0.02R^2[1/(\Delta E_B - \hbar\omega)]$ giving $\delta = 2\times 10^{-50}$ cm^4 sec molecule^{-1} photon^{-1}. Although the dipole moment term is difficult to calculate accurately and is probably overestimated, we see even in the case of phenol, where the magnitude of the FC intensity still has not reached the magnitude of the parent benzene VC intensity, that the dipole moment term is important. The most recent measurement for the intensity of the VC band 14_0^1 in benzene is in the crystal at 4 K and is $\delta = 8.7\times 10^{-50}$ cm^4 sec molecule^{-1} photon^{-1}, equivalent to $S = 0.05R^2[1/(\Delta E_B - \hbar\omega)]$.[23] These perturbation terms thus correctly estimate the qualitative magnitude of the intensities expected. The beginning of activity in the phenol spectrum of the substituent sensitive mode, ν_{6a}, is consistent with the onset of strong coupling interactions.

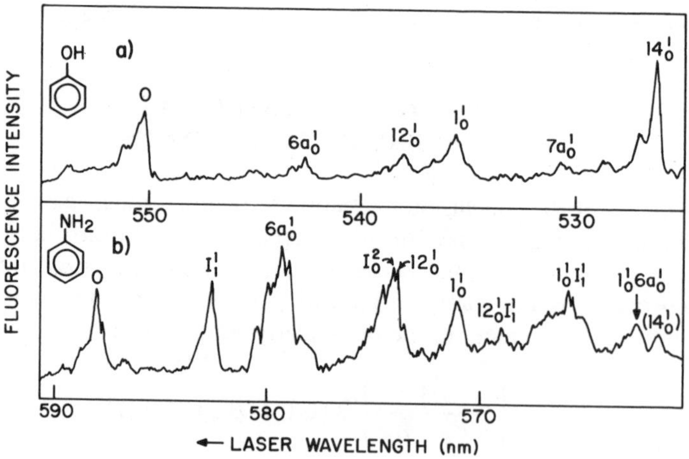

Fig. 11. Fundamental region of normalized two-photon L_b vapor spectra for: (*a*) the intermediate coupling molecule, phenol; and (*b*) the strong coupling molecule, aniline. The vibronic coupling band, 14_0^1 has approximately the same intensity in both spectra, the remaining bands representing the Franck–Condon (allowed) intensity. I is the NH_2 inversion mode in aniline.

A particularly lucid illustration of the different electrostatic perturbation effects on the TP and OP intensities is found in the L_b spectra of the halobenzenes. A comparison of the TP spectra of fluoro-, chloro-, and bromobenzene is shown in Fig. 12 with the intensities given in Table IX. (The spectrum of iodobenzene has not been reported because of problems with the photochemistry).

The 14_0^1 VC band remains the dominant band in the TP spectrum throughout the halobenzene series, indicating that strong coupling interactions are not preponderant. The allowed strength is observed through FC vibrations ν_1 (ring breathing), ν_{12} (Kekulé distortion), and the origin band.

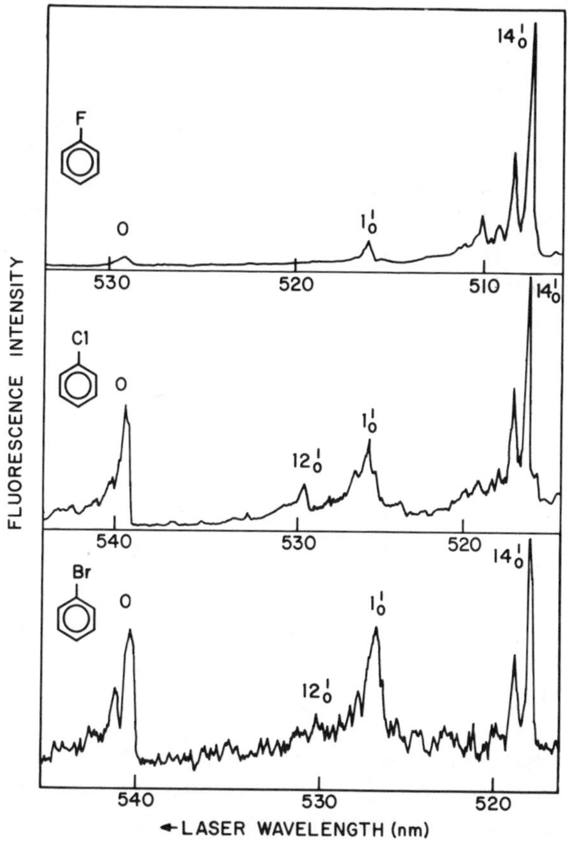

Fig. 12. Fundamental region of normalized two-photon L_b vapor spectra for the halobenzenes (the benzene-like vibronic coupling band 14_0^1 is approximately constant intensity for all of the spectra). The spectra shows the growth of Franck–Condon intensity from fluorobenzene to bromobenzene demonstrating the insensitivity of the spectra to the inductive effect.

This increases in the order Br > Cl > F, that is, the halobenzene with the largest inductive effect has the smallest TP intensity. There is, however, a direct correlation between the FC intensity and the energy of the CT configurations, that is, the halobenzene with the lowest energy CT states has the largest TP intensity.

The OP FC intensity ordering on the other hand is just opposite to that of TP, as is shown in Table IX: it correlates with the strength of the inductive perturbation. The complementary pattern of the two spectra provides strong support for the conclusions expressed in Section VIII.

Finally, in the case of aniline, the strong coupling correction becomes so important $[\lambda(CT_b)\,\Lambda(CT_b) > 0.1]$ that it probably dominates the first order term, with the consequence that Eq. (41) is no longer valid. The observed TP spectrum, shown in Fig. 11*b*, shows strong allowedness, with FC bands having much greater strength than the VC band 14_0^1. The spectrum is virtually identical to the OP spectrum, showing little relationship to the parent benzene spectrum.

The large magnitude of the strong coupling correction in aniline suggests that further insight into the aniline spectrum may be obtained by utilizing benzyl anion wavefunctions which exhibit complete delocalization as the starting point. In the simple l.c.a.o. treatment the two lowest energy configurations of benzyl anion are:[4]

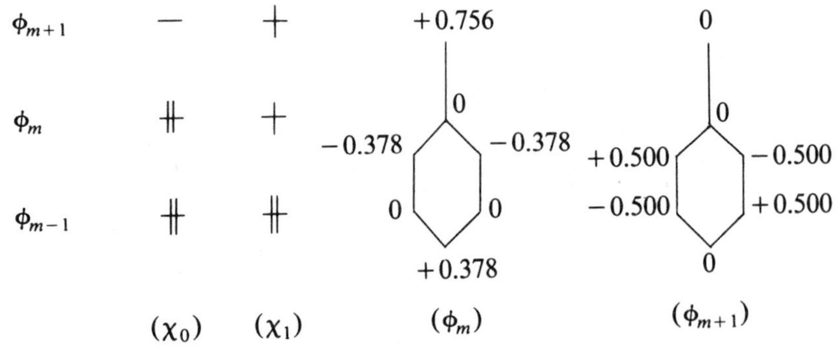

χ_0 represents the ground state configuration; χ_1 is the excited configuration $\phi_{m-1}^2\phi_m\phi_{m+1}$, where ϕ_m and ϕ_{m+1} are the frontier and lowest vacant orbitals respectively of the anion. The transition of interest is the charge transfer transition $\chi_1 \leftarrow \chi_0$, with which a large dipole moment change can be associated.

Large dipole moment changes between the ground and final state raise the question of participation of these states as intermediate states in the two-

TABLE IX
Two-Photon Absorptivities[a] and One-Photon Oscillator Strengths[b] for Monosubstituted Benzene ^1L_b Bands

Molecule	$\delta(0\text{–}0)$	$\delta(1_0^1)$	$f(0\text{–}0)$
Benzene	0	0	0
Fluorobenzene	5×10^{-51}	7×10^{-51}	0.0072
Phenol	7×10^{-50}	4×10^{-50}	0.0154
Aniline	5×10^{-49}	3×10^{-49}	0.0224
Chlorobenzene	5×10^{-50}	3×10^{-50}	0.0012
Bromobenzene	7×10^{-50}	6×10^{-50}	0.0010

[a]The TP vapor phase absorptivities (δ = cm^4 sec molecule^{-1} photon^{-1}) have been determined by using 14_0^1 ($\delta=1.0\times10^{-49}$) as a standard, as discussed in Section III.A.

[b]From Ref. 7.

photon process.[57] Mortensen and Svendsen have shown that these states should be included in any second-order perturbation treatment of the interaction of light and matter, but also point out that these terms are small in Raman scattering.[58] However, Dick and Hohlneicher have recently suggested that the ground and final states may serve as important intermediate states in the TP spectroscopy of polar molecules.[57]

The contribution of these terms to the symmetric tensor is[59]

$$S_{yz}=\frac{1}{\hbar\omega}[\mu_f-\mu_g]\langle A|\mu_y|f\rangle \tag{43}$$

where μ_f and μ_g are the permanent dipole moments of the final and ground state, respectively. Since $\mu_f-\mu_g\approx R\mathbf{k}$ and $\langle\chi_0|\mu_y|\chi_1\rangle\approx\frac{1}{2}R\mathbf{j}$ (these are calculated to be ~4 and ~2 Debyes respectively using Huckel orbitals), both factors in Eq. (43) are large for benzyl anion, on the order of one-photon $V\leftarrow N$ transition moments. Consequently the permanent dipole contribution to the $\chi_1\leftarrow\chi_0$ two-photon transition is expected to be large. Numerical calculation using Huckel orbitals suggest that this mechanism accounts for the largest single contribution, that is, 10 gm*, to the two-photon cross section of the benzyl anion.

The analagous term in aniline is only 0.05 gm, based on the observed dipole moment change[60] and the oscillator strength. The large reduction from benzyl anion arises largely because of lowered charge transfer character and

*1 Goeppert-Mayer (gm) = 1×10^{-50} cm^4 sec molecule^{-1} photon^{-1}.

higher transition energy. Virtual intermediate state terms are therefore calculated to be much more important. The common feature of the two approaches is that the great strength of the aniline transition ($\sim 10^2$ gm; Table IX) can be interpreted as deriving from the electron-transfer character in the excited state. Thus, even though the simple mathematics of the intermediate coupling model is lost for a strong substituent, such as amino, the underlying source of TP intensity remains charge-transfer in character.

X. POLYSUBSTITUTION

Just as the mechanism responsible for inducing intensity in the OP spectra leads to simple orientation rules,[5-7] for identical substituents the TP mechanisms discussed in Section VIII lead to an analogous result. We make use of our conclusion of Section VIII.A2 that intensity in the TP L_b spectrum is preponderantly induced by the resonance effect rather than by the inductive effect. By assuming (1) each carbon-substituent bond is equivalent regardless of the number or orientation of substituents, (2) the carbon ring is undistorted, and (3) neglect of interaction between substituents, we may take the perturbation operator for polysubstitution as $P=\Sigma_r P_r$, where P_r, is the equivalent operator for monosubstitution at position r. The additive nature of P allows the effect of multiple substitution to be additive:

$$S=\sum_r S_r \tag{44}$$

For the L_b state of the polysubstituted benzenes only the symmetric tensor components [S^s of Eq. (1)] contribute to Eq. (44) for 1, 1,3, 1,4, 1,2,3, and 1,3,5 orientations because the L_b excited state is nontotally symmetric. For 1,2 and 1,2,4 orientations, which have totally symmetric excited states, there is the possible contribution of the scalar component (S^0) in addition to S^s. In Eq. (44) S_r^s is S_{yz} from Eq. (35). Rather than go through the rather laborious algebra of summing the S_{yz} terms, we extend a symmetry argument developed by Petruska[7] for OP spectra which leads to the same result. Petruska showed that the intensity induced in a symmetry forbidden benzene analog transition exhibits a characteristic polysubstitution pattern determined by the number of radial nodal planes, M, normal to the ring in the perturbing states. The nodal planes in the CT states responsible for the TP L_b intensity transform like E_{2g} ($M=2$) and consequently the cross section, δ, relative to monosubstitution, $\delta(1)$ is

$$\frac{\delta}{\delta(1)}=\left|\sum_r \exp\left(\frac{ir\pi}{3}\right)\right|^2 \tag{45}$$

TABLE X
Effect of Substitution on Two-Photon Intensities of the L_b Benzene Bands[a, b]

Disubstitution	Enhancement[b]	Trisubstitution	Enhancement
ortho	3	1,2,3	4
meta	1	1,3,5	0
para	0	1,2,4	1

[a]Symmetry and the perturbation theory approach require holes to show the same intensity as substituents (e.g., penta is equivalent to monosubstitution).
[b]The enhancement factor is taken relative to the effect of monosubstitution.

For example, for *meta* disubstitution $r = 0, 2$, $\delta/\delta(1) = |1+\exp(i2\pi/3)|^2 = 1$.

Table X, as derived from either of the aforementioned methods, indicates that the TP L_b intensities of homosubstituted benzenes should follow the rule *1,2,3 > ortho > mono = meta = 1,2,4 > para = 1,3,5* (forbidden), with relative intensities 4:3:1:1:1:0:0. Comparison of the one- and two-photon rules shows that there is no resemblance. The spectra for fluoro- and methyl-polysubstituted benzenes[59] shown in Figs. 13 and 14 and the intensities given in Table XI are in essential agreement with the rules and demonstrate that there is a striking orientation effect on the intensities. Several points may be made: 1,2,3-substitution is predicted to induce the largest FC intensity in TP absorption to the L_b state—just opposite to the effect on the OP spectrum, where this orientation causes the transition to vanish. The analog of the effect of 1,2,3-substitution on OP spectra is the symmetry allowed $A_2'(L_b) \leftarrow A_1'$ TP transition in the 1,3,5-trisubstituted molecule. Despite the fact that 1,3,5 represents a symmetry-allowed TP transition, it is both perturbation and identity forbidden. Another striking difference between the OP and TP orientation rules, confirmed by the spectra, is the effect of *ortho* substitution:* in TP spectroscopy, *ortho* is three times more intense than *mono*; in OP spectroscopy, it is equal.

Symmetry dictates that in the asymmetrically substituted $C_s(1,2,4)$ molecule the scalar tensor term S^0 of Eq. (1) must be considered as a possible intensity mechanism. Calculation within the framework of Eq. (44), however, shows that S^0 vanishes and Eq. (45) applies.[59] A striking polarization confirmation of this prediction has been carried out, based on the strong attenuation of the VC 14_0^1 band in circularly polarized light in benzene due to its vibronic coupling to the totally symmetric ground state. Despite identical

*Taking into account the π-electron assumption the S^0 term for the ortho derivatives reduces to $S_{yy} + S_{zz}$. Assuming the same virtual intermediate states as in Section VIII leads to $S_{yy} = -S_{zz}$. Thus the scaler term vanishes and the relative intensity is given by Eq. (45).

symmetry of the 14^1_0, origin, and 1^1_0 FC transitions in *ortho* difluoro and in 1,2,4-trifluorobenzene, the origin and 1^1_0 bands have opposite polarization behavior to 14^1_0.[59] The lack of attenuation of the FC bands in circularly polarized light demonstrate that they obtain intensity through the off-diagonal tensor component, S^s. The strong attenuation of 14^1_0 shows that it continues to obtain intensity through S^0. This experiment provides strong confirmation of the validity of the basic ideas inherent in Eqs. (44) and (45).

The polysubstitution experiments eliminate certain of the TP tensor terms (considered in Section VIII), as being unimportant. The first are the zeroth-order terms that arise from CT intermediate states. These terms predict very small intensity for *mono-*, *meta-*, and 1,2,3-substitution, while adding an important intensity contribution to the *ortho* case. The experimental result for methyl and fluorobenzene spectra that *ortho* is less intense than 1,2,3 (Table XI) indicates that the zeroth-order CT term is unimportant.

A second type of term for which polysubstitution spectra provides a selective discrimination involves the ground and final states as intermediates states. Some resolution of this problem is provided by the TP $L_b \leftarrow A$ spec-

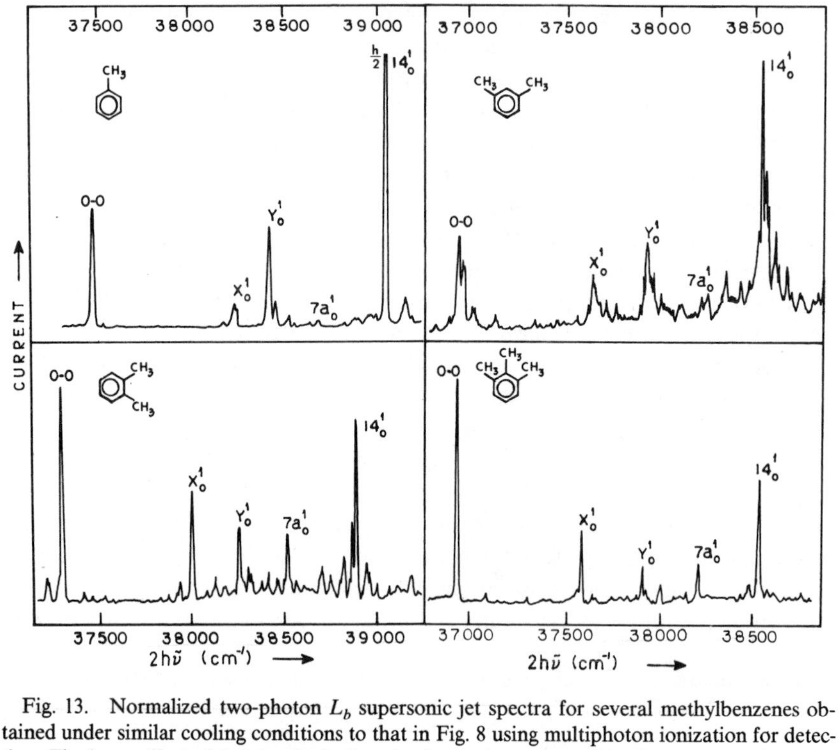

Fig. 13. Normalized two-photon L_b supersonic jet spectra for several methylbenzenes obtained under similar cooling conditions to that in Fig. 8 using multiphoton ionization for detection. The large effect of 1,2,3 substitutions is clearly demonstrated in these spectra (intensity scale for toluene is twice that of the others).

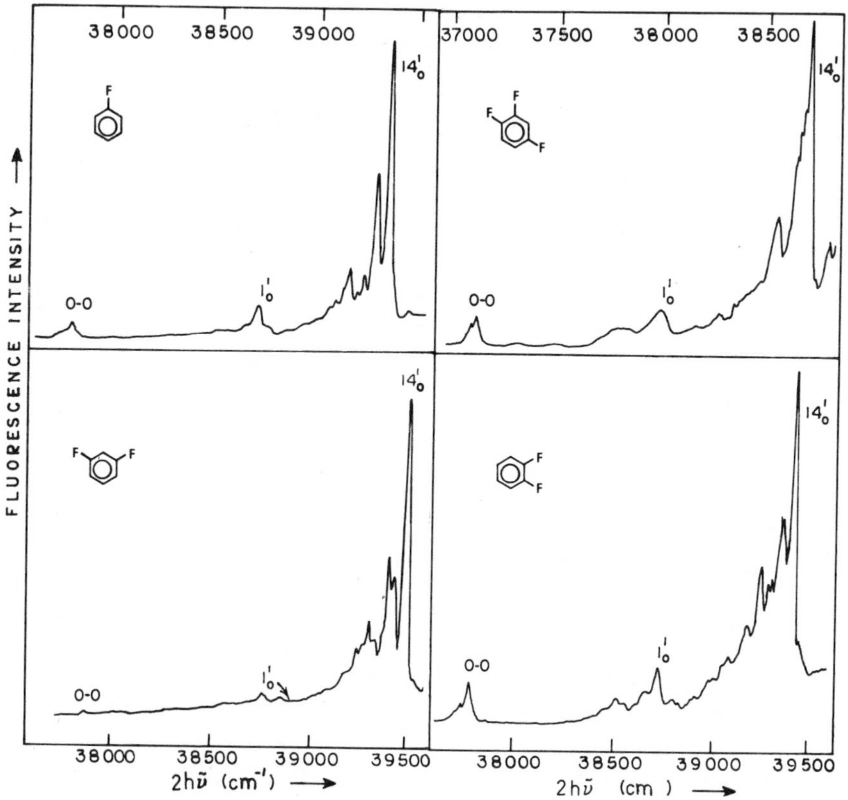

Fig. 14. Normalized two-photon L_b vapor spectra for several fluorobenzenes demonstrating the orientation effect of substituents on the L_b spectra and the lack of special intensity effects by asymmetric 1,2,4 substitution.

tra of the polar polysubstituted fluorobenzenes. The orientation rules for this term [given by Eq. (43)] depend on those due to the OP moment, $\langle A|\mu_y|L_b\rangle$ and the difference between the final and ground state dipole moments (i.e., terms transforming like $M=1$ and $M=0$, respectively, in the Petruska symmetry discussion). These terms do not contribute to the intensity of 1,2,3-substituted molecules since $\langle A|\mu_y|L_b\rangle = 0$, but an intensity contribution is possible for both *ortho* and 1,2,4 orientations. Since the OP transition dipole for fluorobenzene is substantial, because of the large inductive effect, and because of the sizeable difference between the final and ground state permanent dipole moments (0.3 D),[60] the fluorobenzenes provide a test of the importance of the ground and final states as intermediate states* to benzene analog transitions. The 1,2,3 spectrum has not been reported, but the low intensity of 1,2,4 trifluorobenzene compared to that of *ortho* (Table XI),

TABLE XI
Relative Two-Photon Franck–Condon Intensities for Fluoro- and Methylbenzene L_b Bands[a]

Orientation	Fluoro	Methyl
mono	(1)[b]	(1)[b]
ortho	3.7	2.7
meta	0.4	0.7
para	0.0	0.0
1,2,3	—	2.9
1,2,4	1.2	—
1,3,5	0.0	0.0

[a]Intensities represent the total Franck–Condon band areas measured for the vapor spectra relative to the vibronic coupling band 14_0^1.
[b]The intensity in the monosubstituted molecule is taken as 1.

and the polarization behavior in both these derivatives provide evidence that the permanent dipole terms are not significant in the TP absorption to the L_b state of substituted benzenes. The small observed influence of the permanent dipole term is not surprising since Eq. (43) shows that for this term to become important f needs to approach 1 and $\Delta\mu$ to exceed 1 Debye.

XI. PHENYLACETYLENE, BENZONITRILE, AND STYRENE

We now turn to the TP spectrum of the near-alternant hydrocarbon, phenylacetylene (near alternant because of nonequivalence of acetylenic and benzene carbons).[50] We proceed as we have in Section VIII; but in phenylacetylene, ring → acetylene (R → A*) CT states promotions are introduced into the benzene wave functions as well as the acetylene → ring (A → R*) promotions. Restricting ourselves to the $B_{a,b}$ intermediate states, the TP tensor (19) for L_b absorption becomes

$$S_{yz} = \left[\frac{2^{-1/2}}{\Delta E_B - \hbar\omega}\right]\Big[\langle A^0|\mu_y|B_b^0\rangle\{(\lambda(\mathrm{R}\rightarrow \mathrm{A}^*)+\Lambda(\mathrm{R}\rightarrow \mathrm{A}^*))\langle \Phi_5|\mu_z|\Phi_8\rangle$$
$$-(\Lambda(\mathrm{A}\rightarrow \mathrm{R}^*)-\lambda(\mathrm{A}\rightarrow \mathrm{R}^*))\langle \Phi_7|\mu_z|\Phi_3\rangle\}\Big] \tag{46}$$

*The calculated contribution of the permanent dipole term to the TP absorbtivity for fluorobenzene ($f = 0.0072$ (Table IX), $\Delta\mu = 0.3$D, and $\Delta E \approx 5$eV) from Eq. (43) is 2×10^{-4} gm, negligible compared to the observed total cross-section (Table IX). However, this is the largest calculated *fractional* cross-section arising from permanent dipole terms for any second-row substituted benzene.

where Φ_7 of Table IV is replaced by the acetylene HOMO $\Phi_7 = (2^{-1/2})(\phi_7 + \phi_8)$ (ϕ_7, ϕ_8 are the acetylene carbon atom p_x orbitals), and Φ_8 represents the acetylene π-LUMO, $\Phi_8 = (2^{-1/2})(\phi_7 - \phi_8)$. This term predicts contributions from both types of CT states so that there is a push–pull mechanism based on migration of charge into and out of the ring. It might appear that the V ← N acetylene transition χ_{78} can also serve as an intermediate state; however, terms in the L_b tensor involving this state vanish because of two-electron orthogonality. Equation (46) may be simplified to the analog of Eq. (36) by making use of the approximate equality $\langle \Phi_5 | \mu_z | \Phi_8 \rangle \approx \langle \Phi_7 | \mu_z | \Phi_3 \rangle$)

$$S_{yz} = \left[\frac{-6^{1/2}}{\Delta E_B - \hbar\omega} \right] \frac{R^2}{4} \mathfrak{S}\,\mathbf{jk}[\{\lambda(\mathrm{R} \to \mathrm{A}^*) + \lambda(\mathrm{A} \to \mathrm{R}^*)\}$$
$$+ \{\Lambda(\mathrm{R} \to \mathrm{A}^*) - \Lambda(\mathrm{A} \to \mathrm{R}^*)\}] \tag{47}$$

The alternant hydrocarbon condition sets

$$\left. \begin{aligned} -\lambda(\mathrm{A} \to \mathrm{R}^*) = \lambda(\mathrm{R} \to \mathrm{A}^*) = \lambda \\ \Lambda(\mathrm{A} \to \mathrm{R}^*) = \Lambda(\mathrm{R} \to \mathrm{A}^*) = \Lambda \end{aligned} \right\} \quad \lambda, \Lambda > 0 \tag{48}$$

and S_{yz} vanishes.

It becomes necessary then to examine the strong coupling correction (37) to the TP tensor. The extension of Eq. (37) to phenylacetylene is $\lambda(\mathrm{R} \to \mathrm{A}^*)\Lambda(\mathrm{R} \to \mathrm{A}^*)\mu(\mathrm{R} \to \mathrm{A}^*) + \lambda(\mathrm{A} \to \mathrm{R}^*)\Lambda(\mathrm{A} \to \mathrm{R}^*)\mu(\mathrm{A} \to \mathrm{R}^*)$. Since μ_g for an alternant hydrocarbon is small, $|\mu(\mathrm{R} \to \mathrm{A}^*)| \approx -|\mu(\mathrm{A} \to \mathrm{R}^*)|$, and the leading term in the strong coupling correction becomes $2\lambda\Lambda|\mu|$, that is, predominant. Note, however, that just the opposite is true for the second-order components of $\lambda(B_b)$ owing to the A → R* and R → A* charge migrations: these just cancel, with the consequence that the OP transition is predicted to be forbidden.

The TP and OP L_b spectra are compared in Fig. 15, and the comparison reveals the different kinds and strengths of interactions influencing TP and OP spectra. The OP spectrum has a weak origin band with weak FC bands, in contrast to the strongly allowed TP spectrum (Table XII). The allowed intensity of the OP spectrum approximates toluene (Table XII), so that the OP spectrum is actually nearly forbidden. In dramatic contrast, the TP intensity approaches that of aniline—an example of strong coupling. These differences are in accord with the foregoing discussion.

Although the weakness of the FC portion of the OP spectrum resembles toluene,[61] the VC bands have markedly changed from those observed in toluene, which mainly involves ν_{6b}, the active mode in benzene. In phenylacetylene, VC arises mainly from ν_{35}, an in-plane C–C≡C bending mode (ν_A

in Fig. 15), and the VC intensity, as is shown in Table XII, has doubled compared to that in toluene. The implication is that the interactions controlling the VC bands in phenylacetylene are very different from the ones in the weak coupling molecule toluene, even though the FC spectrum resembles that of toluene.

The conclusion that may be drawn from this discussion is that the high phenylacetylene TP strength arises from complete delocalization of the π electrons over the benzene and acetylene parts of the molecule. Since this is

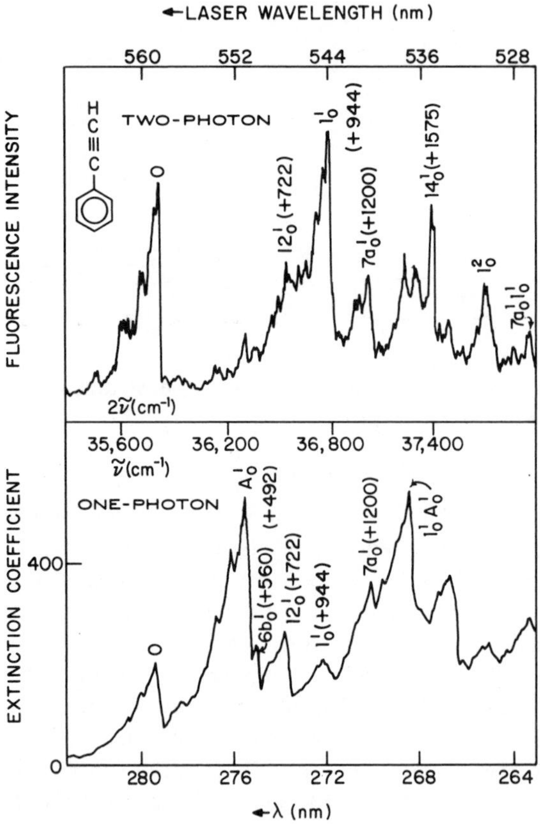

Fig. 15. Comparison of phenylacetylene normalized two-photon (upper) and one-photon (lower) vapor phase spectra in the green and near-ultraviolet regions, respectively. In the two-photon spectrum 14_0^1 is the vibronic coupling band carried over from benzene and the remaining bands are Franck–Condon modes. In the one-photon spectrum $6b_0^1$ and A_0^1 are vibronic coupling bands, with $6b_0^1$ being carried over from benzene and A_0^1 arising from an acetylenic distortion.

TABLE XII
Comparison of One- and Two-Photon Intensities for the $^1L_b \leftarrow {}^1A$ Transition in Toluene, Phenylacetylene, Benzonitrile and Styrene[a]

One photon	$\varepsilon(0\text{–}0)$	f(FC)	f(VC)	f(total)
Toluene[b]	225	0.0001	0.0002	0.0003
Phenylacetylene	220	0.0001	0.0004	0.0005
Benzonitrile[c]	2300	0.01		0.01
Styrene[c]	750	0.005		0.005
Two photon	$\delta(0\text{–}0)$	δ(FC)	δ(total)	
Toluene	6×10^{-50}	9×10^{-50}	2×10^{-49}	
Phenylacetylene	2×10^{-49}	6×10^{-49}	12×10^{-49}	
Benzonitrile	2×10^{-49}	7×10^{-49}	13×10^{-49}	
Styrene	2×10^{-49}	6×10^{-49}	12×10^{-49}	

[a]See footnote *a* and units of Table IX.
[b]One-photon spectral data for toluene from Ref. 7.
[c]From Ref. 51.

clearly an example of strong coupling, it is probably not profitable to focus on perturbing the benzene wave functions.

Callis et al.[55] have predicted from properties of alternant hydrocarbon wave functions[60a] that the L_b TP transition in alternant hydrocarbons will in general be insensitive to inductive perturbations. We have shown how this insensitivity in benzene can be derived from the special symmetry properties of the benzene wave functions. A test of the general perturbation rule can be made by comparing the spectra of phenylacetylene and benzonitrile since there are no special factors of symmetry left in phenylacetylene. The nitrile inductive effect is the normal one for an L_b transition enhancing the phenylacetylene OP intensity ~ 50-fold (Table XII). The similarity of the TP spectra (the same FC and VC couplings modes, origin band strengths, and nearly the same total absorbtivities)[51] indicates that there is a general insensitivity of TP L_b spectra to inductive perturbations.

Another kind of perturbation is provided by the vinyl group of styrene. In this C_s symmetry alternant hydrocarbon, the topological perturbation of the skewed vinyl group has switched the short y-axis L_b transition moment of phenylacetylene to long axis (z) in styrene.[66] An intriguing question is then raised: Will the TP spectra of phenylacetylene and styrene strongly differ? The TP L_b spectra are again found to be very similar,[51] and in particular the similar polarization behavior shows that the TP transition in styrene obtains most of its intensity from the symmetric tensor term in Eq. (1) despite the totally symmetric symmetry of the L_b state. This invariance to topological perturbations suggests that orbitally allowed L_b TP spectra will in general be very strong in alternate hydrocarbons.

XII. CONCLUSIONS

The thrust of this chapter has been to establish and explain the complementary behavior of the one- and two-photon spectra of perturbed benzenes. We have discussed the contrasting behaviors of the two spectra to isotopic, vibronic, inductive, conjugative, and polysubstitution perturbations.

Our first conclusion, regarding isotope effects, is that the vibronic structure of the benzene L_b TP spectrum is, unlike the OP spectrum, strongly affected by the mode-scrambling perturbations caused by deuterium labeling. Since totally symmetric modes are involved, a consequence is that part of the apparent Franck–Condon intensity in substituted benzenes is derived from the mass perturbation alone. This effect has no counterpart in the OP spectra of substituted benzenes and is most significant for a substituent like fluorine, where the electrostatically induced TP intensity is small.

The "new" active modes present in the spectra of isotopically labeled benzenes, in addition to those observed in the parent C_6H_6 molecule, allow isotopic probing experiments for Duschinsky effects on benzene modes that cannot be carried out on benzene itself. These experiments demonstrate important DR of both e_{1u} and b_{2u} modes. As a consequence of the massive DR established for the b_{2u} modes and the ascendency of the ground state role in the intensity mechanism for the activity of ν_{14}, there is an insensitivity of the associated vibronic band 14_0^1 to mass and electrostatic perturbations. This insensitivity allows its strength, $\delta = 8.7 \times 10^{-50}$ cm^4 sec molecule^{-1} photon^{-1} to be used as an internal intensity standard for many benzene derivatives.

The second conclusion concerns vibronic activity in the TP spectra. The great vibronic strength of the B_{2u} benzene spectrum arises from the close overlap of alternating C—C compression–expansion displacements in the b_{2u} mode, ν_{14}, with the optimum perturbation displacements required for interaction between the ground and B_{2u} final states. But there is no mode which overlaps with the perturbation displacements required for the B_{1u} state.

Our third conclusion is that the effect of inductive perturbations, weak on L_b TP spectra and strong on L_a, are reversed from the OP case. The explanation of these different TP behaviors involves inducement of intensity by out-of-phase contamination of the intermediate and final states by the E_{2g} benzene states for L_b. However, the L_a spectrum involves in-phase contamination by E_{2g} states as well as by the ground state.

The resonance effect induces strong intensity into the TP L_b bands and is also predicted to be effective for L_a. The mechanism for the strong enhancement is attributed to contamination of the excited and intermediate state wave functions by charge-transfer states involving substituent and ring orbitals. A significant contribution to the TP tensor is made by CT state dipole moments.

Finally, polysubstitution rules for TP spectra are shown to be strikingly simple. Two particularly intriguing results are that the 1,2 and 1,2,3 orientations induce more intensity into L_b spectra than do other orientations and that 1,3,5 reverts to a forbidden transition despite symmetry allowedness.

At the time of writing many spectra have been obtained for the L_b state and there is extensive interest in TP spectra of higher energy states. Scott et al's[53] work on the L_a state is the first of what will undoubtedly be many experimental studies on the L_a spectra. However, the L_a state does not appear to give rise to the complementary TP and OP spectra that have been demonstrated to allow probing of separate chemical bonding effects in the case of the L_b transition.[49]

Extension of these ideas to naphthalene has proven fruitful. The carryover of the regularities mentioned here to the TP spectra of perturbed naphthalenes will be discussed in subsequent articles.[62]

Future work will make clear whether other effects play an important role in TP spectra. Metz[63] implied that doubly excited configurations are significant, for example, and INDO and CNDO/S calculations seem to show that they do have an important effect.[24,64,65] It is not clear, however, at this time, whether their importance results from the parameterization in the CNDO and INDO procedures or whether they are fundamentally necessary. Also needed are critical experiments to assess the controversial permanent dipole mechanism[57] for two-photon processes in very polar molecules.

Ignored in this chapter are perturbation effects on TP rovibronic spectra. These studies are only beginning to be made[43,45,68] and systematics are still around the corner.

XIII. FREQUENTLY USED ABBREVIATIONS AND SYMBOLS

A. Abbreviations

CNDO, CNDO/S, INDO.	molecular orbital approxmations (see J. A. Pople and D. L. Beveridge, *Approximate Molecular Orbital Theory*, McGraw-Hill, New York, 1970).
CT	charge-transfer
DR	Duschinsky rotation
FC	Franck-Condon

gm	one Goeppert-Mayer $= 1 \times 10^{-50}$ cm^4 sec molecule^{-1} photon^{-1}, a unit of two-photon absorption strength, δ.
INDO	see CNDO
OP	one-photon
TP	two-photon
VC	vibronic coupling

B. Symbols

a_n^m	a transition from n quanta of Q_a'' in the electronic ground state to m quanta of Q_a' in the electronic excited state.
α_r	$= \langle \phi_r \| P \| \phi_r \rangle$
$B_{a,b}$	π-electron states derived from the single noded dipole-allowed degenerate E_{1u} state of benzene (a is nodal through bonds and b through atoms).
β_{rs}	$= \langle \phi_r \| P \| \phi_s \rangle$, $r \neq s$
$C_{a,b}, C'_{a,b}$	π-electron states derived from the doubly noded dipole-forbidden degenerate E_{2g} states of benzene.
ΔE_i	= energy difference between the ground and virtual intermediate state in the TP tensor.
$\Delta E_{i^0, j^0}$	$= E(i^0) - E(j^0)$, energy difference between benzene electronic states i^0 and j^0.
ψ_i	substituted benzene electronic state wavefunction
ψ_i^0	benzene electronic state wavefunction
ψ_i	electronic state wavefunction
δ	(two-photon absorptivity) $= \left(\frac{e}{m}\right)^4 \frac{4\pi^2 g(\omega_1+\omega_2)}{C^2 \omega_1 \omega_2} \|S\|^2$ in a linearly polarized laser beam, where $g(\omega_1+\omega_2)$ is the line shape function.
$\gamma(k)$	$= \langle \psi_{L_a} \| \psi_k^0 \rangle$
H_{ij}	$= \langle \psi_i^0 \| P \| \psi_j^0 \rangle$
$L_{a,b}$	π-electron states derived from the triply noded dipole forbidden B_{1u} (nodes through bonds) and B_{2u} (nodes through atoms) states of benzene, respectively.
$l(k)$	$= \langle \psi_{B_a} \| \psi_k^0 \rangle$
$\lambda(k)$	$= \langle \psi_{L_b} \| \psi_k^0 \rangle$

$\Omega(k), \Omega'(k)$	$= \langle \psi_{Ca,b} \vert \psi_k^0 \rangle$ and $\langle \psi_{C'a,b} \vert \psi_k^0 \rangle$ respectively.
ω	laser frequency ($= \omega_1, \omega_2$ if two lasers are involved).
ω_a', ω_a''	harmonic frequencies for modes Q_a' and Q_a'', respectively.
ϕ_r	atomic orbital at position r
Φ_j	molecular orbital
P	perturbation operator
P_{ij}	$\langle \Phi_i \vert P \vert \Phi_j \rangle$
ρ, σ	$= x, y, z$
$Q_i^{0\prime\prime}, Q_i^{0\prime}$	normal coordinates for benzene vibrational modes in the ground and excited electronic states, respectively.
$Q_i^{D\prime\prime}, Q_i^{D\prime}$	normal coordinates for labeled benzene vibrational modes in the ground and excited electronic states, respectively.
R	C—C bond distance in benzene
S	total tensor in the two-photon transition amplitude
S°	$= \frac{1}{3} \sum_i \langle f \vert \mu_\sigma \vert i \rangle \langle i \vert \mu_\sigma \vert g \rangle [2\,\Delta E_i - \hbar(\omega_1 + \omega_2)]/(\Delta E - \hbar\omega_1)(\Delta E_i - \hbar\omega_2)$ where g, i, and f are ground, virtual intermediate, and final states, respectively.
S^s	$= \frac{1}{2} \sum_i [\langle f \vert \mu_\sigma \vert i \rangle \langle i \vert \mu_\rho \vert g \rangle + \langle f \vert \mu_\rho \vert i \rangle \langle i \vert \mu_\sigma \vert g \rangle] \left[\frac{2\,\Delta E_i - \hbar(\omega_1 + \omega_2)}{(\Delta E_i - \hbar\omega_1)(\Delta E_i - \hbar\omega_2)} \right] - \frac{1}{3} S^\circ$
S^a	$= \frac{1}{2} \sum_i [\langle f \vert \mu_\sigma \vert i \rangle \langle i \vert \mu_\rho \vert g \rangle - \langle f \vert \mu_\rho \vert i \rangle \langle i \vert \mu_\sigma \vert g \rangle] \hbar(\omega_1 - \omega_2)/(\Delta E_i - \hbar\omega_1)(\Delta E_i - \hbar\omega_2)$
$\mathfrak{S}$ mn	$= \langle \phi_{\mathrm{m}} \vert \phi_{\mathrm{n}} \rangle$
σ	see ρ
$\xi(k)$	$= \langle \psi_g \vert \psi_k^0 \rangle$
μ	dipole moment operator
ν_a'', ν_a'	vibrational mode frequencies in the ground and excited electronic states, respectively.

Acknowledgments

Financial Support from the National Science Foundation and The Rutgers Research Council is gratefully acknowledged. The authors thank Professor A. C. Albrecht for communicating the spectra shown in Fig. 9 in advance of publication, and Professors Patrick Callis and George Hohlneicher for critically reading the manuscript.

References

1. J. R. Platt, *J. Chem. Phys.* **17**, 480 (1949); **19**, 263 (1951).
2. W. Moffitt, *J. Chem. Phys.* **22**, 320 (1954).
3. E. B. Wilson, Jr., *Phys. Rev.* **45**, 706 (1934).
4. J. N. Murrell and H. C. Longuet-Higgins, *Proc. Phys. Soc.* (London) **A68**, 329, 601, 969 (1955).
5. A. L. Sklar, *J. Chem. Phys.* **10**, 135 (1942).
6. T. Förster, *Z. Naturforsch.* **2a**, 149 (1947).
7. J. Petruska, *J. Chem. Phys.* **34**, 1111; 1120 (1961).
8. D. M. Friedrich and W. M. McClain, *Ann. Rev. Phys. Chem.* **31**, 559–577 (1980).
9. W. M. McClain, *Acct. Chem. Res.* **7**, 129 (1974).
10. H. Mahr, in *Quantum Electronics: A Treatise*, H. Rabin and C. L. Tang, eds., Vol. 1, Pt. A, Academic Press, New York, 1975, pp. 285–361.
11. W. M. McClain and R. Harris, in *Excited States*, E. C. Lim, ed., Vol. 3, Academic Press, New York, 1977, pp. 2–56.
12. A. C. Albrecht, *J. Chem. Phys* **33**, 156, 169 (1960).
13. D. M. Friedrich and W. M. McClain, *Chem. Phys. Lett.* **32**, 541 (1975).
14. R. M. Hochstrasser, H. N. Sung, and J. E. Wessel, *J. Am. Chem. Soc.* **95**, 8179 (1973).
15. L. Wunsch, H. J. Neusser, and E. W. Schlag, *Chem. Phys. Lett.* **31**, 433 (1975).
16. J. R. Lombardi, D. M. Friedrich, and W. M. McClain, *Chem. Phys. Lett.* **38**, 213 (1975).
17. J. R. Lombardi, R. Wallenstein, T. W. Hansch, and D. M. Friedrich, *J. Chem. Phys.* **65**, 2537 (1976).
18. L. Wunsch, H. J. Neusser, and E. W. Schlag, *Chem. Phys. Lett.* **38**, 216 (1976).
19. L. Wunsch, F. Metz, H. J. Neusser, and E. W. Schlag, *J. Chem. Phys.* **66**, 386 (1977).
20. R. M. Hochstrasser, J. E. Wessel, and H. N. Sung, *J. Chem. Phys.* **60**, 317 (1974).
21. R. M. Hochstrasser, C. M. Klimak, and G. R. Meredith, *J. Chem. Phys.* **70**, 870 (1979).
22. R. M. Hochstrasser, C. M. Klimak, and G. R. Meredith, *J. Chem. Phys.* **72**, 3440 (1980).
23. R. M. Hochstrasser, G. R. Meredith, and H. P. Trommsdorf, *J. Chem. Phys.* **73**, 1009 (1980).
24. R. P. Rava, L. Goodman, and K. Krogh-Jespersen, *J. Chem. Phys.* **74**, 273 (1981).
25. R. P. Rava and L. Goodman, *J. Phys. Chem.* **86**, 480 (1982).
26. L. Goodman and R. P. Rava, *Advances in Laser Spectroscopy*, B. A. Garetz and J. R. Lombardi, eds., Vol. 1, Heyden, Philadelphia, 1982, pp. 21–53.
27. R. P. Rava and L. Goodman, *Chem. Phys. Lett.* **76**, 234 (1980).
28. R. P. Rava, L. Goodman, and K. Krogh-Jespersen, *Chem. Phys. Lett.* **68**, 337 (1979).

28a. A. Sur, J. Knee, and P. Johnson, *J. Chem. Phys.* **77**, 654 (1982).

29. E. B. Wilson, Jr., J. C. Decius, and P. C. Cross, *Molecular Vibrations*, McGraw-Hill, New York, 1955, p. 182.
30. See G. Herzberg, *Spectra of Diatomic Molecules*, 2d. ed., Van Nostrand Reinhold, New York, 1950, p. 162.

31. A. C. Albrecht, *J. Mol. Spectrosc.* **5**, 236 (1960).
32. D. H. Whiffen, *Phil. Trans. Roy. Soc.* (London) **A248**, 131 (1955).
33. W. M. McClain, *J. Chem. Phys.* **55**, 2789 (1971).
34. The sub- and superscripts indicate the number of quanta excited in the lower state and upper state, respectively. J. H. Callomon, T. M. Dunn, and I. M. Mills, *Phil. Trans. Roy. Soc.* (London) **A259**, 499 (1966).
35. C. R. Bailey, J. B. Hale, and N. Herzfeld, C. K. Ingold, A. H. Leckie, and H. G. Poole, *J. Chem. Soc.*. 255 (1946).
36. R. P. Rava and L. Goodman, *J. Chem. Phys.* **75**, 4734 (1981).
37. K. Krogh-Jespersen, R. P. Rava, and L. Goodman, unpublished data.
38. F. Duschinsky, *Acta Physicochim.* **7**, 551 (1937).
39. G. J. Small, *J. Chem. Phys.* **54**, 3300 (1971).
40. M. J. Robey and E. W. Schlag, *J. Chem. Phys.* **67**, 2775 (1977).
41. P. M. Johnson, *Acct. Chem. Res.* **13**, 20 (1980).
42. M. A. Leugers, and C. J. Seliskar, *J. Mol. Spectry.* **91**, 150 (1982).
43. R. Vasudev and J. C. D. Brand, *J. Mol. Spectrosc.* **75**, 288 (1979).
44. K. Krogh-Jespersen, R. P. Rava, and L. Goodman, *Chem. Phys.* **47**, 321 (1980).
45. R. Vasudev and J. C. D. Brand, *Chem. Phys.* **37**, 211 (1979).
46. K. Krogh-Jespersen, R. P. Rava, and L. Goodman, *Chem. Phys.* **44**, 295 (1979).
47. J. Murakami, K. Kaya, and M. Ito, *J. Chem. Phys.* **72**, 3263 (1980).
48. L. Goodman and R. P. Rava, *J. Chem. Phys.* **74**, 4826 (1981).
49. R. P. Rava and L. Goodman, *J. Am. Chem. Soc.* **104**, 3815 (1982).
50. L. Chia and L. Goodman, *J. Chem. Phys.* **76**, 4745; **77**, 3292 (1982).
51. L. Chia, L. Goodman, and J. G. Philis, *J. Chem. Phys.* (submitted).
51a. L. Chia, J. G. Philis, and L. Goodman, *J. Phys. Chem.* (submitted).
52. P. R. Callis, T. W. Scott, and A. C. Albrecht, *J. Chem. Phys.* **75**, 5640 (1981).
53. T. W. Scott, P. R. Callis, and A. C. Albrecht, *Chem. Phys. Lett.* **93**, 111 (1982).
54. B. Dick and G. Hohlneicher, *Chem. Phys. Lett.* **84**, 471 (1981).
54a. B. Dick, H. Gonska, and G. Hohlneicher, *Ber. Bunsenges Phys. Chem.* **85**, 746 (1981).
55. P. R. Callis, T. W. Scott, and A. C. Albrecht, *J. Chem. Phys.* **78**, 16 (1983).
56. W. Hampf, H. J. Neusser, and E. W. Schlag, *Chem. Phys. Lett.* **46**, 406 (1977).
56a. A. S. Twarowski and D. S. Kliger, *Chem. Phys.* **20**, 259 (1977).
56b. L. D. Ziegler and B. S. Hudson, *Chem. Phys. Lett.* **71** 113 (1980).
56c. P. J. Hay and L. Shavitt, *J. Chem. Phys.* **60**, 2865 (1974).
57. G. Hohlneicher and B. Dick, *J. Chem. Phys.* **76**, 5755 (1982).
58. O. S. Mortensen and E. N. Svendsen, *J. Chem. Phys.* **74**, 3185 (1981).
59. R. P. Rava, L. Goodman, and J. G. Philis, *J. Chem. Phys.* **77**, 4912 (1982).
60. K. T. Huang and J. R. Lombardi, *J. Chem. Phys.* **52**, 5613 (1970).
60a. R. Pariser, *J. Chem. Phys.* **24**, 250 (1956).
61. G. W. King and S. P. So, *J. Mol. Spectrosc.* **36**, 468 (1970); **37**, 535 (1971); **37**, 543 (1971).
62. R. P. Rava and L. Goodman, to be published.
63. F. Metz, *Chem. Phys. Lett.* **34**, 109 (1975).

64. G. Hohlneicher and B. Dick, *J. Chem. Phys.* **70**, 5427 (1979).
65. B. Dick and G. Hohlneicher, *Theoret. Chim. Acta* (Berlin) **53**, 221 (1979).
66. A. Hartford, Jr. and J. R. Lombardi, *J. Mol. Spec.* **35**, 413 (1970).
67. E. U. Condon and G. H. Shortley, *The Theory of Atomic Spectra*, Cambridge Univ. Press, New York, 1951, pp. 169–174.
68. S. N. Thakur and L. Goodman, *J. Chem. Phys.* **78**, 4356 (1983).

INHOMOGENEOUS RELATIVISTIC ELECTRON SYSTEMS: A DENSITY-FUNCTIONAL FORMALISM*

M. V. RAMANA AND A. K. RAJAGOPAL

Department of Physics and Astronomy
Louisiana State University
Baton Rouge, Louisiana

CONTENTS

*Based on the Doctoral Dissertation of M.V.R., submitted to the Department of Physics and Astronomy, Louisiana State University, Baton Rouge, La. (1981). Present address: Department of Physics, University of West Virginia, Morgantown, W. Va.

I. INTRODUCTION

A. Notations, Orders of Magnitude, and the Need for Relativistic Treatment

The most easily observed relativistic effect even in light atoms is the fine structure of spectral lines due to spin-orbit splitting. This was first explained by Sommerfeld[1] using the old quantum theory with appropriate relativistic generalization. These relativistic effects are now commonplace in spectroscopic studies of more complex systems where absorption lines of third-row transition metal complexes designated according to single-group symmetry are split by spin-orbit interactions. A variety of other effects such as differences in magnetic properties between the first-row transition metal complexes and the third-row transition metal complexes are all attributed to relativistic effects. Thus the relativistic aspects of many-electron systems—atoms, molecules, and solids—have been studied with increased interest over the past decade. It appears that relativity plays a significant role in atomic properties, in chemical bonding, and in the properties of solids containing heavy elements. For a recent review of these effects with extensive references to the literature, see Pyykkö.[2] We will here give a brief survey of what may be expected on simple physical grounds. In order to bring out the salient feature of relativity, we use relativistic units, $\hbar = 1 = c$, with notations as in Björken and Drell.[3] Since our interest lies in atomic and solid state systems, we choose our unit of length as the Bohr radius of the hydrogen atom: $a_0 = \hbar^2/me^2 = 1$. In these units the electromagnetic coupling constant $e^2 = \alpha$(equals $1/137.037$). Then the energy of the first level of the hydrogen atom, $e^4 m/2\hbar^2$, the Rydberg (constant or unit), equals $\frac{1}{2}e^2$ relativistic units. In the solid state context the velocity of the electron at the Fermi surface is a typical electronic velocity in the system and so our dimensionless relativistic parameter is chosen to be $\beta = k_F/m$, with k_F being the Fermi momentum. This is related to a more conventional dimensionless density parameter r_s, viz.

$$\left(\tfrac{4}{3}\pi r_s^3 a_0^3\right)^{-1} = \frac{k_F^3}{3\pi^2}$$

by

$$\beta = \frac{(9\pi/4)^{1/3}}{\alpha r_s} \cong \frac{1}{71.4 r_s}$$

A density corresponding to $\beta = 1$, where relativistic effects may be expected to be significant, then corresponds to $r_s \cong 0.014$. This may be compared with the region of "high density" in the nonrelativistic electron gas theory for which $r_s < 1$. The corresponding electron densities are $n_0 \cong 6.15 \times 10^{29}$ cm^{-3} ($r_s \cong 0.014$) and $n_0 \cong 1.69 \times 10^{24}$ cm^{-3} ($r_s \cong 1$).

To get some physical feeling for when the relativistic effects may be significant in an actual solid state context, we take the charge density at the nucleus provided by Moruzzi et al.[4] for indium ($Z = 49$) as $8 \times 10^4/(\text{Bohr radius})^3$ for which we deduce a β value of approximately 1. Another estimate could be arrived at by calculating the condition of $\beta = 1$ for a system of electrons where a pair of electrons lie inside a sphere of volume a_0/Z. This gives $Z \cong 57$. Thus, one may expect significant relativistic effects to appear for systems containing atoms with $Z \gtrsim 40$.

Apart from the spin effects, other relativistic effects appear as we move up in Z. The velocity of the core electron becomes comparable to that of light and hence its mass increases, and, since the mass appears in the denominator of the Bohr radius, its orbit shrinks. All other s orbitals which must remain orthogonal to the core shrink as much. The inner p orbitals are similarly affected. Pyykkö[2] estimates this to be about 20% in mercury ($Z = 80$). These two effects, the spin-orbit splitting and the relativistic contraction of the core, are called direct effects. Because of the contraction of s, p levels the nucleus is better shielded from the outer electrons, which in turn leads to an expansion of the outer valence shells. This is known as the indirect effect. The order of magnitude of this expansion is expected to be as large as the contraction. For the inner shells of the large Z atoms, the quantum electrodynamic effects become important too. Vacuum polarization effects due to the nuclear charge, including higher order corrections, were calculated by Huang et al.,[5] and the 1s shell of mercury acquires a contribution of about 1.5 a.u. Electron self-energy terms due to zero-point fluctuations of the electromagnetic field contribute about 7 a.u. to the 1s shell.[5]

A brief survey of some of the earlier calculations which exhibited the above expectations will now be given. More detailed accounts may be found in several review articles and summaries by Pyykkö,[2] Pyykkö and Desclaux,[6] Pitzer,[7] Desclaux,[8] Andersen et al.,[9] and an early one by Grant.[10]

To give a more detailed picture of the role of relativity, let us consider gold. Here the relativistic effects dominate other competing effects such as shell structure expansion and Lanthanoid contraction.[2] (For example, in cesium these effects cancel each other, making it the largest atom in nature.) Its properties are quite different from what can be expected from its lighter analogs, Ag and Cu. For example, the first ionization energy for Au is 9.22 eV, higher than that of Ag (7.574) or Cu (7.724). This is attributed to the relativistic contraction of the ns shell in Au. This contraction implies further solid state and molecular effects. The electron affinity of gold is much higher and allows it to form the Au^- ion in CsAu and RbAu, which are semiconductors, unlike Ag and Cu; the dissociation energy in Au_2 is higher, and the bond lengths in the hydrides are shorter. The self-consistent field expansion of the outer *d* orbitals may serve to explain why the second ionization energy of Au is smaller than in Ag or Cu, as well as its trivalency and pentavalency and its yellow color. The expansion of the outer *d* orbital places the transitions between 5*d* to the Fermi level at about 2.3 eV; thus, gold absorbs strongly in the blue and violet and reflects in the red and yellow wavelengths. Note that nonrelativistic calculations predict similar properties for both silver and gold, whereas the relativistic calculations give qualitative agreement with experimental data. (See Pyykkö[2] and references therein for more details.) While gold is the most spectacular in exhibiting the relativistic effects, the difference in behavior between the fifth-row elements and the corresponding sixth-row elements may be largely understood on the basis of these effects.

Relativistic calculations of electronic structure of heavy atoms have been attempted since 1940. They have become increasingly more sophisticated and complex, culminating in the multiconfiguration Dirac–Fock (MCDF) scheme. These calculations are computationally enormous even for atomic systems, and in the case of molecules and solids the size and complexity of the calculations become prohibitive. For treating the nonrelativistic many-electron systems, a reliable method with reasonable computational effort is available in the density-functional scheme. This theory was developed by Hohenberg and Kohn,[11] Kohn and Sham,[12] and Sham and Kohn.[13] Within a local approximation scheme, this method has been very successful in dealing with a wide variety of nonrelativistic inhomogeneous systems. For a review of this work see Rajagopal[14] and Kohn and Vashishta.[15] The accuracy of the results are no better than 1%. These methods have not been tried very widely in the relativistic cases. This, we believe, was due to the lack of proper treatment of exchange and correlation effects in a relativistic density-functional theory. The aim of the present review is to provide such a treatment. Only a brief account of other methods will be given in the next section; the reader is referred to reviews cited herein for details. The present article may be considered as a relativistic sequel to that by Rajagopal[14] in this series.

B. Survey of Earlier Work for Relativistic Many-Electron Systems

When two electrons are moving slowly with nonrelativistic speeds, their mutual interaction is just the Coulomb repulsion between them. A theory of a collection of such electrons with neutralizing positive charge background, the well-known electron gas model, has been the basis for many calculational methods for studying the electronic structure of inhomogeneous many-electron systems. Breit[16] discovered that, when two electrons move at speeds comparable to that of light, a relativistic two-electron interaction (now known as the Breit interaction) occurs, correct to order $(v/c)^2$, where v is the velocity of the electron, besides the usual Coulomb repulsion that must be taken into account. Thus, a relativistic treatment of a many-electron system requires not only the use of Dirac formalism but also a proper theory of mutual interactions. In much of the early work, the Coulomb interactions were taken into account appropriately but the Breit interaction was treated in a first-order perturbation approximation.

As regards the nonrelativistic many-electron theory, for years one of the most frequently used means of studying electronic properties was the Hartree–Fock (HF) self-consistent field method. In this method one reduces the many-electron problem to an effective one-electron Schrödinger problem wherein an electron moves in a self-consistent potential.[14] This was generalized to the relativistic system by Swirles,[17] who derived the Dirac–Hartree–Fock self-consistent equations, which are analogous to the nonrelativistic HF equations. This theory reduces the relativistic many-electron problem to a one-electron scheme where it is described by a Dirac-like equation with a self-consistent relativistic potential. Williams[18] was the first to follow Swirles's proposal to study the Cu^+ ion. He omitted the exchange potential from his considerations and found contractions of the inner shell charge densities but none in energy. Copper was too light an atom to show significant relativistic effect. Much later, Mayers[19] studied Hg, and Cohen[20] investigated Hg, Hg^{++}, W, Pt, and U; they reported results of relativistic calculations, again omitting exchange, but showed substantial improvements over the nonrelativistic calculations. In all the preceding early work, the Breit interaction was not considered.

Since the early 1960s there has been a rapid burgeoning of relativistic calculations for atoms, molecules, and solids. The Dirac–Fock (DF) method was improved upon by Roothaan[21] and was used most recently by Kagawa[22] to develop a multiconfiguration relativistic Hartree–Fock–Roothaan (HFR) theory for atomic systems. The method was used by Grant[23] and Mann;[24] a multiconfiguration Dirac-Fock (MCDF) method was used by Descalux.[25] Mann and Johnson[26] have made an extensive evaluation of the role of the Breit interaction and various forms of it (see, for example, Bethe and Salpeter[27]), including the full transverse interaction in the DF scheme. Huang et al.[5] have done almost complete calculations of atomic binding energies for atoms with $2 \leqslant Z \leqslant 106$, though they use Breit interaction and omit correlation effects. Their calculations compared with experimental results of Deslattes et al.,[28] and earlier nonrelativistic estimates of Cowan[29] suggest that correlation corrections to binding energies may be of the order of a few eV. To our knowledge, no systematic calculation of the correlation contribution in atoms in the relativistic case has been reported. (The correlation effects may be defined as those which when added to the DF calculations make the result exact.)

These calculations involve large numerical codes. In the nonrelativistic theory, the exchange effects were simplified by Slater to a local form, making the calculations relatively less extensive. When the nonrelativistic exchange scheme is used in the relativistic context, the method is known as Dirac–Fock–Slater (DFS) scheme. This method was first advocated by Liberman et al.[30] It has recently been used by Ziegler et al.[31] for complexes of heavy elements ($Z > 40$). These authors show that the HFS method is a viable alternative to more involved treatments. Calculations on periodic trends due to relativistic effects have been reported by these authors. Relativistic corrections are shown to be essential for compounds containing heavy elements and result in a contraction of bond distances and substantial changes in band energies. The density-functional scheme reviewed here, being more complete but in the same spirit as the HFS method, is expected to be equally viable as an alternative to the involved MCDF methods. Moreover the correlation effects can be included quite easily in this scheme, as we shall show later.

The most recent relativistic molecular calculations were reported for Au complexes by Guenzenberger and Ellis,[32] using a statistical exchange model. No full MCDF calculations, we believe, have been carried out for molecules. Almost all nonrelativistic band structure schemes have been generalized to the relativistic case. The earliest one was by Callaway et al.[33] Recent calculations are by Skriver et al.,[34,35] Glötzel,[36] Schwarz and Herzig,[37] Andersen et al.,[9] and MacDonald et al.[38] For reviews, see an excellent overview by Pyykkö[6] and references therein, articles by Grant[10] and Desclaux,[8] and the recent *Proceedings of the International Conference on the Physics of Actionides and Related 4f Materials.*[39]

In their recent work on exotic atoms, Pyper and Grant[40] have examined the atomic structure of the superheavy elements by MCDF theory. Nuclei in the region of $Z = 114$ are expected to be relatively stable, and these authors have calculated the ionization potentials, excitation energies, and orbital radii for the $7p$ series of superheavy elements with $Z = 113$ to 118. Such calculations would be of interest when experimental investigations of these systems are made. The effects of correlations and the like will be of importance in these cases, and our density-functional scheme is expected to be useful in preliminary investigations of these systems.

The energy spectrum of the relativistic many-electron system described by a Hamiltonian containing a sum of Dirac one-particle Hamiltonians representing the kinetic energy and attraction to nuclei plus a two-electron interaction term involving Coulomb repulsion (plus Breit terms) is not bounded from below. This is because the Coulomb interaction between electrons mixes the positive and negative energy states of the one-electron orbitals of the Dirac one-particle equation. Formally, therefore, a difficulty exists in applying the usual variational methods of computing the lowest energy state of this system as has been done in HFR and MCDF schemes described thus far. If one works within a fully field theoretic framework this difficulty does not arise, as was first pointed out by Brown and Ravenhall.[41] The question then arises as to how exactly should one go about this problem in a practical calculation. It turns out that by a suitable reformulation of the problem the numerical results obtained thus far can be fully justified formally. Mittleman[42] and Sucher[43] have independently put forward formal schemes which place the existing calculations on a blemishless foundation. Since there appears to be no real change in the structure of the calculations warranted by these important formal works, we shall not elaborate them any further here.

There are other practical difficulties associated with variational calculations for relativistic systems, requiring some subtle care in actual computations. For a discussion of these, see the recent papers of Drake and Goldman[44] and Bagayoko.[45]

The next section contains a brief summary of the density-functional formalism in relativistic systems. Since the density-functional formalism to be presented in this review is based on a field theoretic foundation, the formal objections stated above are not applicable.

C. The Density-Functional Formalism in Relativistic Systems

The density-functional formalism of Hohenberg and Kohn[11] and the local-density version of Kohn and Sham,[12] Sham and Kohn,[13] and their generalizations (Mermin[46] and recently Gupta and Rajagopal[47] for finite temperature; von Barth and Hedin[48] for spin-polarized cases; Pant and Rajagopal[49] for spin-polarized situations; Rajagopal and Callaway[50] for spins in a relativistic theory; and Gunnarsson and Lundqvist[51] for spin-polarized cases) have met with great success in providing a theoretical framework and basis for practical applications in dealing with systems of many interacting electrons. A recent review by Rajagopal[14] in this series succinctly summarizes the current status of the theory. A fundamental limitation of the density-functional formalism, namely, its inability to deal adequately with relativistic systems, was removed by Rajagopal and Callaway[50] and Rajagopal[52] and independently by MacDonald and Vosko.[53] In a local scheme, using the exchange energy of a uniform relativistic electron gas, the exchange part of the effective potential was deduced. Using this exchange-only approximation for uranium and some atoms of the lithium isoelectronic sequence, an effective one-particle Dirac equation was solved self-consistently by Das et al.[54] and by Das[55] for fermium. The contribution to the exchange term due to the transverse photon interaction, left out by Ellis[56] in his treatment, was found to be significant. It must be mentioned that the nonrelativistic Kohn–Sham exchange was used previously in a one-particle Dirac equation (Liberman et al.[30]). The Breit interaction and its nonlocal generalization of Mann and Johnson[26] (which corresponds to our transverse interaction) were evaluated perturbatively by Mann and Johnson in the Dirac–Fock scheme. They found that the Breit interaction as an approximation to the full transverse interaction was larger by about a Rydberg in mercury. The Breit interaction is obtained from the full transverse interaction by dropping a term in the denominator that represents the energy transfer by the transverse photon. We perform a similar evaluation in the case of the local approximation in density-functional theory.

In the following we further develop the relativistic density-functional scheme in two directions. We calculate the correlation part of the exchange correlation potential, based on a calculation of the ground state energy of the homogeneous electron gas, within a ring-diagram-sum approximation which is a generalization of the nonrelativistic result of Gell-Mann and Brueckner[57] (GB) and apply this to an atomic case. We extend the nonrelativistic spin-density-functional (SDF) formalism of von Barth and Hedin[48] to the relativistic case, based on the theorems proved by Rajagopal and Callaway.[50] We deal here with the problem of spin not being a good quantum number. We present the exchange potentials for the inhomogeneous polarized electron gas derived from the calculation of the exchange energy for the uniform polarized gas.

In the nonrelativistic formalism, it is known that the correlation contribution to the exchange correlation potential is important and that it profoundly affects the behavior of systems with uncompensated spin (Gunnarsson and Lundqvist[51]). In relativistic systems correlations were not studied in any serious way. Some correlation was included in some MCDF calculations. Freedman et al.[58] and Fricke et al.,[59] using an extrapolation of the nonrelativistic GB result, estimate the correlation effects in fermium to be less than 1 eV. We show from our calculation that this contribution, by omitting relativistic effects, is underestimated and in fact may be of the order of 10 eV.

D. Spin-Density-Functional Formalism in Relativistic Systems

In nonrelativistic systems with uncompensated spin it is known that the application of the density functional scheme, with the results of the unpolarized gas as input, give results seriously in error. The SDF scheme greatly improves this (Gunnarsson and Lundqvist[51]). Similarly, in relativistic systems with uncompensated spin we expect the need for an SDF scheme.

Whereas the required theorems have been proved for a relativistic density-functional (RSDF) by Rajagopal and Callaway,[50] and the effective one-particle equations written down by MacDonald and Vosko,[53] to our knowledge no calculation of the effective one-particle potentials has been reported. We therefore fill this gap. Using the results of the relativistic homogeneous spin-polarized gas as input in an exchange-only approximation, we calculate the effective one-particle potentials in a local-density scheme. While dealing with the relativistic spin-polarized gas, we encounter some novel and interesting conceptual problems.

In Section II we develop the theory for calculating the correlation contribution to the relativistic case. We outline a many-electron field theoretic formalism so that the proper interactions and other approximations can be clearly brought out. We point out the approximations and renormalizations involved. We present explicitly the longitudinal and transverse parts of the correlation energy and potential. In Section III we present and discuss the numerical results and compare them with the approximations of Jancovici,[60] and we report the results of an actual self-consistent calculation for various atoms with correlation included. Since a calculation with the full exchange has already been published, a comparison would enable us to assess the correlation contribution. In passing, we also evaluate the Breit interaction vis-à-vis the full transverse interaction in the local approximation. We find, upon comparison with the results of Mann and Johnson,[26] that nonlocal effects may be important.

In Section IV, the second part of the review, we discuss the spin-polarized homogeneous relativistic electron gas and develop its Feynmann propagator. We use this propagator to calculate some interesting physical quantities of the homogeneous gas and discuss their significance. We follow this with the effective one-particle equation for the inhomogeneous system and calculate the effective potentials. In Section V we summarize our results.

E. A Note on Renormalizations

In relativistic many-body theory there are two kinds of renormalizations that we have to deal with. The first is the usual many-body renormalizations due to interacting particles in the Fermi sea. The second is the quantum electrodynamic (QED) renormalizations due to the infinite Dirac sea. We are interested in the former; the latter has been dealt with adequately in the literature, even in the context of relativistic many-particle systems (see, for example, the references in the beginning of Section II.B). Essentially, in our formal calculations we encounter terms involving electron density (Fermi function) and terms that are divergent and independent of the density, where we assume that the infinite negative energy sea is filled. The latter terms give rise to QED renormalizations and may be dropped if we use the physical charge and mass. For further discussion see MacDonald[61] and a more recent paper by Sapirstein.[62]

II. FORMAL THEORY

A. Density-Functional Theory for the Inhomogeneous Relativistic Electron System

The Hamiltonian for our system is

$$
\begin{aligned}
H = & \int d\mathbf{r}\, \bar{\psi}(x)[\boldsymbol{\gamma}\cdot\mathbf{p} + m(1-\gamma_0)]\psi(x) + H_{\text{rad}}^{\text{tr}} \\
& + \frac{1}{2}\iint d\mathbf{r}\, d\mathbf{r}' \frac{e^2}{|\mathbf{r}-\mathbf{r}'|} \quad :\bar{\psi}(x')\gamma_0\psi(x): \quad :\bar{\psi}(x')\gamma_0\psi(x'): \\
& - e\int d\mathbf{r} \quad :\bar{\psi}(x)\gamma_i\psi(x): \quad :A_{\text{tr}}^i(x): \\
& - e\int d\mathbf{r} \quad :\bar{\psi}(x)\gamma_\mu\psi(x): \quad A_{\text{ext}}^\mu(r)
\end{aligned}
\tag{2.1}
$$

The notations are as in Björken and Drell[3] and Rajagopal and Callaway.[50] In (2.1) the ψ's are the four-component second-quantized Dirac spinor operators, the γ's are the 4×4 Dirac matrices, and $H_{\text{rad}}^{\text{tr}}$ is the Hamiltonian for the transverse radiation field. In the first term the rest mass m of the electron has been subtracted out so that the energies are measured relative to the rest mass. The usual Coulomb interaction between electrons appears explicitly in the third term because we use the radiation gauge, $\nabla\cdot\mathbf{A} = 0$. The colons represent normal ordering where all creation operators appear to the left of all destruction operators; this eliminates the infinite mass and charge of the vacuum. The fourth term represents the interaction of the current with the transverse radiation field, a contribution which is negligible in the nonrelativistic theory. It is this which leads to additional electron interactions due to relativity. As a general convention, any repeated indices are summed over with Roman indices running from 1 through 3 and Greek indices running from 0 through 3. In the last term we represent the interaction of the system with our external probe A_{ext}^μ, which is a classical entity independent of time. Here A_{ext}^μ may represent the field generated by the background positive charge, for example.

The generalization of the Hohenberg–Kohn theorem as proved in Rajagopal and Callaway may now be stated.

Relativistic Hohenberg–Kohn theorem. The nondegenerate ground state energy per unit volume of the system described by the Hamiltonian (2.1) is a functional of the ground state expectation value of the four-current density, $J_\mu(\mathbf{r}) = \langle :\bar{\psi}(x)\gamma_\mu\psi(x): \rangle$

$$\frac{\langle H \rangle}{\Omega} = F[J_\mu] + \frac{e}{\Omega}\int d\mathbf{r}\, J_\mu(\mathbf{r}) A^\mu_{\text{ext}}(\mathbf{r}) \equiv E_{A^\mu_{\text{ext}}}[J_\mu] \tag{2.2}$$

where $F[J_\mu]$ is a universal functional independent of A^μ_{ext}. The correct J_μ minimizes $E_{A^\mu_{\text{ext}}}[J_\mu]$ subject to the condition

$$\partial^\mu J_\mu = 0 \tag{2.3}$$

and the total number of electrons is a given constant,

$$\int d\mathbf{r}\, J_0(\mathbf{r}) = eN \tag{2.4}$$

The continuity Eq. (2.3) is a reflection of the gauge invariance of the Hamiltonian (2.1). The expectation value in the ground state of any operator θ is $\langle \theta \rangle$. Here, Ω is the volume of the system. The proof of this theorem[50] is a straightforward generalization of the proof of Hohenberg and Kohn.[11] The theorem of Kohn and Sham[12] can also be generalized, and we can write down the effective one-particle Dirac equation of the system which must be solved self-consistently to determine the ground state of the system. As in the nonrelativistic case, we introduce a functional

$$F[J_\mu] = T[J_\mu] + \frac{1}{2\Omega}\iint d^3r\, d^3r' \frac{e^2}{|\mathbf{r}-\mathbf{r}'|} n(\mathbf{r}) n(\mathbf{r}')$$
$$- e\int \langle \mathbf{J}(\mathbf{r}) \rangle \cdot \langle :\mathbf{A}_{\text{tr}}(\mathbf{r}): \rangle\, d^3r + E_{xc}[J_\mu] \tag{2.5}$$

where the kinetic energy functional

$$T[J_\mu] = \frac{1}{\Omega}\int d^2r \langle \bar{\psi}(x)[\boldsymbol{\gamma}\cdot\mathbf{p} + m(1-\gamma_0)]\psi(x) \rangle \tag{2.5a}$$

and $E_{xc}[J_\mu]$ is the exchange correlation contribution from the Coulomb interaction and the transverse photon interaction of the electrons and the energy of the transverse photons. Thus

$$
\begin{aligned}
E_{xc}[J_\mu] = \frac{1}{2\Omega} \iint d^3r\, d^3r' \frac{e^2}{|\mathbf{r}-\mathbf{r}'|} \\
&\times \langle : \bar{\psi}(x)\gamma_0\psi(x): \;\; :\bar{\psi}(x')\gamma_0\psi(x'): \\
&- \langle :\bar{\psi}(x)\gamma_0\psi(x):\rangle \langle :\bar{\psi}(x')\gamma_0\psi(x')\; :\rangle\rangle \\
&- \frac{e}{\Omega} \iint d^3r\, d^3r' \langle :\bar{\psi}(x)\gamma_i\psi(x): \;\; :A^i_{\text{tr}}(x): \\
&- \langle :\bar{\psi}(x)\gamma_i\psi(x):\rangle \langle :A^i_{\text{tr}}(x):\rangle\rangle \\
&+ \frac{1}{\Omega} \langle H^{\text{tr}}_{\text{rad}} \rangle
\end{aligned}
\tag{2.6}
$$

Relativistic Kohn–Sham theorem. The second theorem is then: A Dirac equation for a set of spinors $\{\phi_i(\mathbf{r})\}$ can be defined such that

$$
\left(-i\boldsymbol{\alpha}\cdot\nabla - m(1-\beta) + V_{\text{eff}}[\mathbf{r}; J_\mu]\right)\phi_i(\mathbf{r}) = \varepsilon_i \phi_i(\mathbf{r}) \tag{2.7}
$$

where

$$
\begin{aligned}
V_{\text{eff}}[\mathbf{r}; J_\mu] = -\left(v_{\text{ext}}(\mathbf{r}) + e^2 \int \frac{n(\mathbf{r}')}{|\mathbf{r}-\mathbf{r}'|} d^3r' + \frac{\delta E_{xc}}{\delta n(\mathbf{r})} \right) \\
- e\boldsymbol{\alpha}\cdot\left(\mathbf{A}_{\text{ext}}(\mathbf{r}) + e\int \frac{\mathbf{J}(\mathbf{r}')\, d^3r'}{|\mathbf{r}-\mathbf{r}'|} + \frac{\delta E_{xc}}{\delta \mathbf{J}(\mathbf{r})} \right)
\end{aligned}
\tag{2.8}
$$

The ground state particle and current densities are given by

$$
n(\mathbf{r}) = \sum_{i(\text{occ})} \text{Tr}(\phi_i^*(\mathbf{r})\phi_i(\mathbf{r})) \tag{2.9}
$$

and

$$
J_k(\mathbf{r}) = \sum_{i(\text{occ})} \text{Tr}(\phi_i^*(\mathbf{r})\alpha_k\phi_i(\mathbf{r})) \tag{2.10}
$$

Here, Tr denotes a trace over the spinors. The construction preserves the continuity equation $\nabla\cdot\mathbf{J} = 0$. Equations (2.7)–(2.10) have to be solved self-consistently.

The problem now is to find an expression for E_{xc} so that V_{eff} may be used in an actual calculation. There have been several such practical schemes, as discussed in detail by Rajagopal.[14] One of the more successful schemes is the local scheme in the nonrelativistic theory, where E_{xc} is calculated for a homogeneous electron gas and used as input for a real system with a "slowly" varying density. It may be assumed in such systems that the density is locally homogeneous. In practice, the scheme has been extremely successful, along with its spin-generalized version, even for atoms and molecules. We now discuss the local scheme for the relativistic case.

B. The Relativistic Gas

The problem of the uniform relativistic electron system was studied from various viewpoints before because this model is of great interest in many fields—nuclear physics (Chin[63]), astrophysics (Akhieser and Peletminskii[64]), many-electron physics (Fradkin;[65] Bowers et al.[66]), and plasma physics (Tsytovich;[67] Bezzerides and Dubois[68]).

In calculations in a local scheme prior to 1980, for relativistic systems of interest an ad hoc use of the results of nonrelativistic gas for E_{xc} was employed (Connolly[69]). The results of Rajagopal[52] and MacDonald and Vosko[53] for the exchange contribution of E_{xc} yielded reasonable results for heavy atomic systems (Das et al.[54]). We shall now outline the derivation of the result of the exchange-only approximation and then follow it with a calculation of the correlation contribution, the first of the two main results of this review.

To evaluate Eq. (2.6) we have to resort to approximation schemes as in the nonrelativistic theory. In the nonrelativistic case where the velocity of light is infinite, the only expansion parameter is $1/a_0 k_F = me^2/k_F$ and the theory is good at high densities. In the relativistic case where the densities are much higher, the electromagnetic coupling constant $e^2 \cong 1/137$ is still small and is a natural expansion parameter as in quantum electrodynamics. The theory is therefore valid for all densities such that $me^2/k_F \ll 1$ (Jancovici[60]). We may state that for high densities $\gtrsim 6.2 \times 10^{29}\ \text{cm}^{-3}$, a relativistic theory is needed, as indicated in Section I.A.

In the ground state of the system there are no free photons present, so that the last term in (2.6) is zero. The exchange-only approximation to the lowest order is easily accomplished by using the plane wave expressions for the Dirac spinors $\bar{\psi}(x)$ and $\psi(x)$ and calculating the ground state expectation values. The contribution of the Coulomb term is calculated exactly as in the nonrelativistic theory, the only additional complication being a trace to be taken over the Dirac γ matrices. The contribution of the transverse photon interaction is similarly evaluated; A^i_{tr} may be expressed, using the wave equation with a source term, as an integral over the transverse photon propagator for the free field (to be defined later) and a Dirac current. We have only outlined the derivation since the details have appeared elsewhere (Rajagopal,[52] MacDonald and Vosko[53]). Since the currents are of order $1/c$, at low densities, that is, the nonrelativistic limit, the transverse contribution is negligible and of order $1/c^2$. It may be pointed out here that these same terms lead to relativistic interactions of the Breit type as we will show presently. To this order, we thus arrive at the relativistic Hartree–Fock (exchange contributions) approximation.

If now, in an effort to calculate the correlation energy which is defined as

$$E_c = E_{\text{exact}} - E_{\text{Hartree-Fock}}$$

by taking the next term in the perturbation expansion, we are faced with a bad divergence due to the long range of the Coulomb interactions. The same situation occurs in the nonrelativistic theory and was circumvented by Gell-Mann and Brueckner (GB),[57] who found that a class of divergent diagrams may be resummed to give a finite value. This was the famous sum of ring diagrams, also known as the random phase approximation. We may try such a summation in the relativistic case too. At this point we may define a model Hamiltonian such as Bohm–Pines collective electron theory that will generate the set of ring diagrams and, by using the Lindhard dielectric function formalism, calculate the correlation energy as was done by Jancovici[60] for the relativistic electron gas. We, however, follow the Green function method and use the Schwinger–Dyson approach to calculate the correlation energy. This approach was originally used to derive the well-known renormalized electrodynamic theory of the electron to which our theory reduces in the limit of zero electron density. Thus the Schwinger method applied to our problem will give the renormalizations due to both electrodynamical and density-dependent effects. To this end we define several quantities that appear in our calculations.

The free Dirac operators $\psi(x)$ and $\psi^+(x)$ have the momentum space expansions (see Björken and Drell[3])

$$\psi(\mathbf{x}, t) = \sum_{\pm s} \int \frac{d^3p}{(2\pi)^{3/2}} \sqrt{\frac{m}{E_p}} \left[b(p,s)u(p,s)e^{-ip\cdot x} + d^+(p,s)v(p,s)e^{ip\cdot x}\right]$$

$$\psi^+(\mathbf{x}, t) = \sum_{\pm s} \int \frac{d^3p}{(2\pi)^{3/2}} \sqrt{\frac{m}{E_p}} \left[b^+(p,s)\bar{u}(p,s)\gamma_0 e^{ip\cdot x} + d(p,s)\bar{v}(p,s)\gamma_0 e^{-ip\cdot x}\right] \tag{2.11}$$

Here

$$E_p = p_0 = +\sqrt{|\mathbf{p}|^2 + m^2}$$

$$p \cdot x = p_\mu x^\mu \equiv p^\mu x_\mu = E_p t - \mathbf{p} \cdot \mathbf{x}$$

$$\bar{u} \equiv u^+ \gamma_0, \qquad \bar{v} = v^+ \gamma_0$$

The spin vector s has components in the rest frame $s = (0, \hat{n})$, where $\hat{n}$ is a unit vector in the direction of spin. The positive and negative energy spinors u, $\bar{u}$ and v, $\bar{v}$ satisfy the Dirac equations

$$\begin{aligned} (\not{p} - m)u(p,s) &= 0 \qquad \bar{u}(p,s)(\not{p} - m) = 0 \\ (\not{p} + m)v(p,s) &= 0 \qquad \bar{v}(p,s)(\not{p} + m) = 0 \end{aligned} \tag{2.12}$$

For any four vector a, $\not{a} = \gamma \cdot a$. They have the orthogonality relations

$$\begin{aligned} \bar{u}(p,s)u(p,s') &= \delta_{ss'} = -\bar{v}(p,s)v(p,s') \\ u^+(p,s)u(p,s') &= \frac{E_p}{m}\delta_{ss'} = v^+(p,s)v(p,s') \\ \bar{v}(p,s)u(p,s') &= 0 = v^+(p,s)u(-p,s') \end{aligned} \tag{2.13}$$

The notation $u(-p, s)$ means $u(\sqrt{p^2 + m^2}, -\mathbf{p}, s)$. They satisfy the completeness relations

$$\begin{aligned} &\sum_{\pm s} [u_\alpha(p,s)\bar{u}_\beta(p,s) - v_\alpha(p,s)\bar{v}_\beta(p,s)] = \delta_{\alpha\beta} \\ &\sum_{\pm s} u_\alpha(p,s)\bar{u}_\beta(p,s) = \left(\frac{\not{p} + m}{2m}\right)_{\alpha\beta} \equiv (\Lambda_+(p))_{\alpha\beta} \\ &-\sum_{\pm s} v_\alpha(p,s)\bar{v}_\beta(p,s) = \left(\frac{m - \not{p}}{2m}\right)_{\alpha\beta} \equiv (\Lambda_-(p))_{\alpha\beta} \end{aligned} \tag{2.14}$$

where α, β are spinor indices. The Fock space operators $b(p, s)$, $(b^+(p, s))$ annihilate (or create) electrons of a momentum $\mathbf{p}$ and spin s and $d(p, s)$, $(d^+(p, s))$ annihilate (or create) positrons of momentum $\mathbf{p}$ and spin s. They satisfy the anticommutation relations

$$
\begin{aligned}
\{b(p,s), b^+(p',s')\} &= \delta_{ss'}\delta^3(\mathbf{p}-\mathbf{p}') \\
\{d(p,s), d^+(p',s')\} &= \delta_{ss'}\delta^3(\mathbf{p}-\mathbf{p}') \\
\{b(p,s), b(p',s')\} &= 0 = \{d(p,s), d(p',s')\} \\
\{b^+(p,s), b^+(p',s')\} &= 0 = \{d^+(p,s), d^+(p',s')\} \\
\{b(p,s), d(p',s')\} &= 0 = \{b(p,s), d^+(p',s')\} \\
\{d(p,s), b(p',s')\} &= 0 = \{d(p,s), b^+(p',s')\}
\end{aligned}
\tag{2.15}
$$

The anticommutation relations for the fields follow:

$$
\begin{aligned}
\{\psi_\alpha(x,t), \psi_\beta^+(x',t)\} &= \delta^3(\mathbf{x}-\mathbf{x}')\delta_{\alpha\beta} \\
\{\psi(x,t), \psi(x',t)\} &= 0 = \{\psi^+(x,t), \psi^+(x',t)\}
\end{aligned}
\tag{2.16}
$$

Notice that these relations are defined only at equal times.

The interacting field satisfies the equation

$$\{(i\hbar\partial\!\!\!/ - A\!\!\!/) - mc\}\psi(x) = 0 \tag{2.17}$$

where the electromagnetic field A appears through the minimum coupling. The electromagnetic field itself satisfies the equation

$$
\begin{aligned}
(\Box A^\mu - \partial^\mu \partial_\nu A^\nu) &= \frac{4\pi}{c}\{j^\mu(x) + j^\mu_{\text{ext}}(x)\} \\
\Box &= \nabla^2 - \frac{1}{c^2}\frac{\partial^2}{\partial t^2}
\end{aligned}
\tag{2.18}
$$

$j^\mu_{\text{ext}}(x)$ is a c-number external current designed to probe the system and is set equal to zero at the end of the calculation. We define the Green function for the electron system (Fradkin;[65] Bowers et al.;[66] Chin[63]):

$$S_F(1,2) = -\frac{i}{\hbar}\langle T(\psi(1)\bar{\psi}(2))\rangle \tag{2.19}$$

where for brevity we have indicated the space–time point $x_1 \equiv 1$; T indicates the usual time ordering where for fermions

$$T(\psi(1)\bar{\psi}(2)) = \psi(1)\bar{\psi}(2)\theta(t_1 - t_2) - \bar{\psi}(2)\psi(1)\theta(t_2 - t_1) \tag{2.20}$$

the minus sign appears because of the anticommutativity of the fermion fields; θ is the usual step function

$$\theta(t)=\begin{cases}1 & t>0\\ 0 & t<0\end{cases} \tag{2.21}$$

We define two other functions

$$S_F^{>}(1,2)=\langle\psi(1)\psi^{+}(2)\rangle$$

and

$$S_F^{<}(1,2)=-\langle\psi^{+}(2)\psi(1)\rangle \tag{2.22}$$

which are related to the Green function in an obvious way. The induced current is then

$$j^{\mu}(1)=-iec\hbar\,\mathrm{tr}\{\gamma^{\mu}S_F(1,1^{+})\} \tag{2.23}$$

where t_1^{+} is an infinitesimally later time than t_1. We also define the photon field correlation functions:

$$\frac{4\pi}{c}D^{\mu\nu}(12)=\frac{1}{i\hbar c}[\langle T(A^{\mu}(1)A^{\nu}(2))\rangle-\langle A^{\mu}(1)\rangle\langle A^{\nu}(2)\rangle] \tag{2.24}$$

We make use of the Schwinger identity, which states that for any operator O in the presence of an external probe:

$$\frac{\delta\langle O(1)\rangle_{\mathrm{ext}}}{\delta j_{\nu}^{\mathrm{ext}}(2)}=\frac{1}{ic\hbar}[\langle T(O(1)A^{\nu}(1))\rangle-\langle O(1)\rangle\langle A^{\nu}(1)\rangle]. \tag{2.25}$$

Applying this to the field A^{μ}, we find

$$\frac{\delta\langle A^{\mu}(1)\rangle}{\delta j_{\nu}^{\mathrm{ext}}(2)}=\frac{4\pi}{c}D^{\mu\nu}(12) \tag{2.26}$$

The electron propagator $S_F(12)$ satisfies the equation

$$[i\hbar\partial\!\!\!/_1-mc]S_F(12)-\frac{e}{ic\hbar}\langle T(A\!\!\!/(1)\psi(1)\bar{\psi}(2))\rangle=\delta(12) \tag{2.27}$$

as can be verified by using (2.17) and (2.19). The last term in (2.27) can be expressed in terms of S_F and variational derivatives

$$\frac{\delta}{\delta j^{\nu}_{\text{ext}}(1)}\langle T(\psi(1)\bar{\psi}(2))\rangle = \frac{1}{ic\hbar}\langle T(A_\nu(1)\psi(1)\bar{\psi}(2))\rangle - \langle A_\nu(1)\rangle\langle T(\psi(1)\bar{\psi}(2))\rangle$$

so that

$$\frac{1}{i\hbar}\langle T(A_\nu(1)\psi(1)\bar{\psi}(2))\rangle = ic\hbar\left\{\frac{\delta}{\delta j^{\nu}_{\text{ext}}(1)} + \langle A_\nu(1)\rangle\right\}S_F(12)$$

Using this in (2.27)

$$\left[\left(i\hbar\partial\!\!\!/ - \frac{e}{c}\langle A\!\!\!/(1)\rangle\right) - mc\right]S_F(12) - \gamma^{\mu}\frac{ie\hbar}{c}\frac{\delta}{\delta j^{\mu}_{\text{ext}}(1)}S_F(12) = \delta(12) \tag{2.28}$$

The response of the induced currents j^{μ} to the fields A^{λ} is described by the polarizability tensor

$$\frac{4\pi}{c}\frac{\delta\langle j^{\mu}(1)\rangle}{\delta\langle A^{\lambda}(2)\rangle} \equiv Q^{\mu}_{\lambda}(12) \tag{2.29}$$

using (2.23)

$$Q^{\mu}_{\lambda}(1,2) = -\frac{4\pi e}{c}\,\text{tr}\left\{\gamma^{\mu}ic\hbar\frac{\delta S_F(11^+)}{\delta\langle A^{\lambda}(2)\rangle}\right\} \tag{2.30}$$

Now,

$$\frac{\delta S_F(12)}{\delta j^{\text{ext}}_{\mu}(1)} = \int\frac{\delta S_F(12)}{\delta\langle A^{\lambda}(3)\rangle}\frac{\delta\langle A^{\lambda}(3)\rangle}{\delta j^{\text{ext}}_{\mu}(1)}d^43$$
$$= \frac{4\pi}{c}\int D^{\lambda\mu}(31)\,d^43\,\frac{\delta S_F(12)}{\delta\langle A^{\lambda}(3)\rangle} \tag{2.31}$$

We may now make use of the definition of the inverse Green function, S_F^{-1},

$$\int S_F(12)S_F^{-1}(22')\,d^42 = \int S_F^{-1}(12)S_F(22')\,d^42 = \delta(12') \tag{2.32}$$

to get

$$\int \left[\frac{\delta S_F(1\bar{2})}{\delta \langle A^\lambda(3) \rangle} S_F^{-1}(\bar{2}2') + S_F(1\bar{2}) \frac{\delta S_F^{-1}(\bar{2}2')}{\delta \langle A^\lambda(3) \rangle} \right] d\bar{2} = 0$$

and

$$\frac{\delta S_F(12)}{\delta \langle A^\lambda(3) \rangle} = -\int S_F(1\bar{1}) \frac{\delta S_F^{-1}(\bar{1}\bar{2})}{\delta \langle A^\lambda(3) \rangle} S_F(\bar{2}2)\, d^4\bar{1}\, d^4\bar{2}$$

This equation along with (2.31) may be used in (2.28)

$$\begin{aligned} &\left\{ \left(i\hbar \not{\partial}_\mu - \frac{e}{c} \langle A_\mu(1) \rangle \right) - mc \right\} S_F(12') \\ &\quad + ie\hbar \frac{4\pi}{c} \gamma_\mu \int D^{\lambda\mu}(\bar{3}1) S_F(1\bar{1}) \frac{\delta S_F^{-1}(\bar{1}\bar{2})}{\delta \langle A^\lambda(\bar{3}) \rangle} S_F(\bar{2}2')\, d^4\bar{1}\, d^4\bar{2}\, d^4\bar{3} \\ &= \delta(12') \end{aligned} \tag{2.33}$$

Multiplying both sides by $i\int S_F^{-1}(2'2) d^4 2'$ gives an expression for S_F^{-1}

$$\begin{aligned} S_F^{-1}(12) = &\left[i\hbar\, \not{\partial}_1 - \frac{e}{c} \langle \not{A} \rangle - mc \right] \delta(12) + \frac{4\pi e\hbar}{c} i\gamma_\mu \\ &\cdot \int S_F(1\bar{1}) \frac{\delta S_F^{-1}(\bar{1}2)}{\delta \langle A^\lambda(\bar{2}) \rangle} D^{\lambda\mu}(\bar{2}1)\, d^4\bar{1}\, d^4\bar{2} \end{aligned} \tag{2.34}$$

In the lowest order

$$\frac{\delta S_F^{-1}(\bar{1}2)}{\delta \langle A^\lambda(\bar{2}) \rangle} = -\frac{e}{c} \gamma_\lambda \delta(\bar{1}\bar{2}) \delta(\bar{1}2)$$

$$\begin{aligned} S_F^{-1}(12) \simeq &\left[i\hbar\, \not{\partial} - \frac{e}{c} \langle \not{A} \rangle - mc \right] \delta(12) \\ &- \frac{4\pi e^2 \hbar}{c^2} i\gamma_\mu S_F(12) \gamma_\lambda D^{\lambda\mu}(21) \end{aligned} \tag{2.35}$$

We now turn our attention to the electromagnetic propagator. Let $D_0^{\lambda\mu}(12)$ represent the noninteracting electromagnetic propagator. We may define its inverse as in (2.32). Then from (2.18)

$$\int d2\, D_0^{\lambda\mu}(12)^{-1} \langle A_\mu(2) \rangle = \frac{4\pi}{c} \{ \langle j^\lambda(1) \rangle + j^\lambda_{\text{ext}}(1) \} \tag{2.36}$$

or

$$\int d2\, D_0^{\lambda\mu}(12)^{-1} D_{\mu\nu}(23) = \frac{\delta\langle j^\lambda(1)\rangle}{\delta j_{\text{ext}}^\nu(3)} + \delta(13)\delta_\nu^\lambda \tag{2.37}$$

upon taking the functional derivative of (2.36) w.r.t. $j_\nu^{(\text{ext})}(3)$. Now

$$\frac{\delta\langle j^\lambda(1)\rangle}{\delta j_{\text{ext}}^\nu(3)} = \int d^4\bar{2} \frac{\delta\langle j^\lambda(1)\rangle}{\delta\langle A^k(\bar{2})\rangle} \frac{\delta\langle A^k(\bar{2})\rangle}{\delta j_{\text{ext}}^\nu(3)}$$

$$= \int Q_\kappa^\lambda(1\bar{2})\, D_\nu^\kappa(\bar{2}3)\, d\bar{2}$$

$$\int d2 D_0^{\lambda\mu}(12)^{-1} D_{\mu\nu}(23) = \int Q_\kappa^\lambda(1\bar{2})\, D_\nu^\kappa(\bar{2}3)\, d\bar{2} + \delta(13)\delta_\nu^\lambda \tag{2.38}$$

which gives us the Dyson equation for the photon propagator in the position space:

$$D^{\lambda\mu}(12) = D_0^{\lambda\mu}(12) + \int D_0^{\lambda\kappa}(1\bar{1}) Q_{\kappa\nu}(\bar{1}\bar{2}) D^{\nu\mu}(\bar{2}2)\, d\bar{1}\, d\bar{2} \tag{2.39}$$

We may express both (2.35) and (2.39) formally in momentum space

$$S_F^{-1}(p) = (\hbar \not{p} - mc) - \frac{4\pi e^2 \hbar}{c^2} i \int \frac{d^4k}{(2\pi\hbar)^4} \gamma_\mu S_F^{(0)}(k+p)\gamma_\lambda D^{\lambda\mu}(k) \tag{2.40}$$

and

$$D^{\lambda\mu}(q) = D_0^{\lambda\mu}(q) + D_0^{\lambda\kappa}(q) Q_{\kappa\nu}(q) D^{\nu\mu}(q) \tag{2.41}$$

The third term in (2.40) and the second term in (2.41) represent the self-energies due to interactions of the electron and photon, respectively. This leads us to the problem of evaluating the polarizability tensor $Q_{\kappa\nu}$. To the lowest order

$$Q_{\kappa\nu}^{(0)}(q) = -\frac{4\pi e^2 \hbar}{c} i \int \frac{d^4p}{(2\pi\hbar)^4} \operatorname{tr}\left(\gamma_\kappa S_F^{(0)}(p+q)\gamma_\nu S_F^{(0)}(p)\right) \tag{2.42}$$

where $S_F^{(0)}$ is the noninteracting fermion propagator; $S_F^{(0)}$ may be easily evaluated using the plane wave expansions for the spinor fields or by dropping from (2.40) the last term representing interactions and doing some algebra

$$S_F^{(0)}(p) = \frac{(\not{p}_+ + m)}{2E_\mathbf{p}}\left(\frac{n_F(|\mathbf{p}|)}{p_0 - E_\mathbf{p} - i\eta} + \frac{1 - n_F(|\mathbf{p}|)}{p_0 - E_\mathbf{p} + i\eta}\right)$$
$$- \frac{(\not{p}_- + m)}{2E_\mathbf{p}}\left(\frac{1 - \bar{n}_F(|\mathbf{p}|)}{p_0 + E_\mathbf{p} - i\eta} + \frac{\bar{n}_F(|\mathbf{p}|)}{p_0 + E_\mathbf{p} + i\eta}\right)$$
$$\not{p}_+ = \{+E_\mathbf{p}, \mathbf{p}\}, \qquad \not{p}_- = \{-E_\mathbf{p}, \mathbf{p}\} \tag{2.43}$$

where $n_F(|\mathbf{p}|)$ is the Fermi function for positive energy particles and $\bar{n}_F(|\mathbf{p}|)$ for negative energy particles

$$\bar{n}_F(|\mathbf{p}|) = 1 - n_F(|\mathbf{p}|) \tag{2.44}$$

In our model the negative energy Dirac sea is completely filled so that $\bar{n}_F(|\mathbf{p}|) = 0$ and we have ($\hbar = c = 1$)

$$S_F^0(p) = \frac{(\not{p}_+ + m)}{2E_\mathbf{p}}\left\{\frac{n_F(|\mathbf{p}|)}{p_0 - E_\mathbf{p} - i\eta} + \frac{1 - n_F(|\mathbf{p}|)}{p_0 - E_\mathbf{p} + i\eta}\right\}$$
$$- \left\{\frac{1}{p_0 + E_\mathbf{p} - i\eta}\right\}\frac{(\not{p}_- + m)}{2E_\mathbf{p}} \tag{2.45}$$

The first term in this expression for the propagator describes positive energy particles; the second term describes positive energy holes, or "Fermi holes," which should be distinguished from holes in the Dirac sea (i.e., the positrons). The last term without a Fermi function leads to divergent integrals of QED and is eliminated by renormalizing the charge and mass. We accomplish this by using the physical charge and physical mass for the electron. We still have our familiar many-body renormalizations. Using (2.45) in (2.42) we have

$$Q_{\lambda\mu}^0(q) = -4\pi e^2 i \int \frac{d^4p}{(2\pi)^4} \left.\frac{\mathrm{tr}\left[\gamma_\lambda(\not{p} + \not{q} + m)\gamma_\mu(\not{p} + m)\right]}{4E_\mathbf{p}E_{\mathbf{p}+\mathbf{q}}}\right|_{\substack{p_0 = E_\mathbf{p} \\ p_0 + q_0 = E_{\mathbf{p}+\mathbf{q}}}}$$
$$\cdot\left\{\frac{1 - n_F(|\mathbf{p}+\mathbf{q}|)}{p_0 + q_0 - E_{\mathbf{p}+\mathbf{q}} + i\eta} + \frac{n_F(|\mathbf{p}+\mathbf{q}|)}{p_0 + q_0 - E_{\mathbf{p}+\mathbf{q}} - i\eta}\right\}$$
$$\cdot\left\{\frac{1 - n_F(|\mathbf{p}|)}{p_0 - E_\mathbf{p} + i\eta} + \frac{n_F(|\mathbf{p}|)}{p_0 - E_\mathbf{p} - i\eta}\right\}$$
$$+ 4\pi e^2 i \int \frac{d^4p}{(2\pi)^4} \left.\frac{\mathrm{tr}\left[\gamma_\lambda(\not{p} + \not{q} + m)\gamma_\mu(\not{p} + m)\right]}{4E_\mathbf{p}E_{\mathbf{p}+\mathbf{q}}}\right|_{\substack{p_0 = E_\mathbf{p} \\ p_0 + q_0 = -E_{\mathbf{p}+\mathbf{q}}}}$$

$$\cdot\left\{\frac{1-n_F(|\mathbf{p}|)}{p_0-E_{\mathbf{p}}+i\eta}+\frac{n_F(|\mathbf{p}|)}{p_0-E_{\mathbf{p}}-i\eta}\right\}\frac{1}{\left[p_0+q_0+E_{\mathbf{p+q}}-i\eta\right]}$$

$$+4\pi e^2 i\int\frac{d^4p}{(2\pi)^4}\left.\frac{\mathrm{tr}\left[\gamma_\lambda(\not{p}+\not{q}+m)\gamma_\mu(\not{p}+m)\right]}{4E_{\mathbf{p}}E_{\mathbf{p+q}}}\right|_{\substack{p_0=-E_{\mathbf{p}}\\ p_0+q_0=E_{\mathbf{p+q}}}}$$

$$\cdot\left\{\frac{1-n_F(|\mathbf{p+q}|)}{p_0+q_0-E_{\mathbf{p+q}}+i\eta}+\frac{n_F(|\mathbf{p+q}|)}{p_0+q_0-E_{\mathbf{p+q}}-i\eta}\right\}\frac{1}{(p_0+E_{\mathbf{p}}-i\eta)}$$

$$-4\pi e^2 i\int\frac{d^4p}{(2\pi)^4}\left.\frac{\mathrm{tr}\left[\gamma_\lambda(\not{p}+\not{q}+m)\gamma_\mu(\not{p}+m)\right]}{4E_{\mathbf{p}}E_{\mathbf{p+q}}}\right|_{\substack{p_0=-E_{\mathbf{p}}\\ p_0+q_0=-E_{\mathbf{p+q}}}}$$

$$\cdot\frac{1}{(p_0+q_0+E_{\mathbf{p+q}}-i\eta)}\frac{1}{(p_0+E_{\mathbf{p}}-i\eta)} \tag{2.46}$$

In evaluating the contour integrals in the above expression, the terms for which the two simple poles occur on the same side of the real axis are zero. This is because the contour can be closed in the other half plane.

$$Q^{(0)}_{\lambda\mu}(q)=4\pi e^2\int\frac{d^3p}{(2\pi)^3}\left.\frac{\mathrm{tr}\left[\gamma_\lambda(\not{p}+\not{q}+m)\gamma_\mu(\not{p}+m)\right.}{4E_{\mathbf{p}}E_{\mathbf{p+q}}}\right|_{\substack{p_0=E_{\mathbf{p}}\\ p_0+q_0=E_{\mathbf{p+q}}}}$$

$$\cdot\left\{\frac{(1-n_F(|\mathbf{p+q}|))n_F(|\mathbf{p}|)}{\omega+E_{\mathbf{p}}-E_{\mathbf{p+q}}+i\eta}-\frac{n_F(|\mathbf{p+q}|)(1-n_F(|\mathbf{p}|)}{\omega+E_{\mathbf{p}}-E_{\mathbf{p+q}}-i\eta}\right\}$$

$$+4\pi e^2\int\frac{d^3p}{(2\pi)^3}\left.\frac{\mathrm{tr}\left[\gamma_\lambda(\not{p}+\not{q}+m)\gamma_\mu(\not{p}+m)\right]}{4E_{\mathbf{p}}E_{\mathbf{p+q}}}\right|_{\substack{p_0=E_{\mathbf{p}}\\ p_0+q_0=-E_{\mathbf{p+q}}}}$$

$$\cdot\frac{(1-n_F(|\mathbf{p}|))}{(\omega+E_{\mathbf{p}}+E_{\mathbf{p+q}}-i\eta)}$$

$$-4\pi e^2\int\frac{d^3p}{(2\pi)^3}\left.\frac{\mathrm{tr}\left[\gamma_\lambda(\not{p}+\not{q}+m)\gamma_\mu(\not{p}+m)\right]}{4E_{\mathbf{p}}E_{\mathbf{p+q}}}\right|_{\substack{p_0=-E_{\mathbf{p}}\\ p_0+q_0=-E_{\mathbf{p+q}}}}$$

$$\cdot\frac{(1-n_F(|\mathbf{p+q}|))}{\omega-E_{\mathbf{p}}-E_{\mathbf{p+q}}+i\eta} \tag{2.47}$$

Upon evaluating the traces, we have

$$\mathrm{Re}\big(Q^{(0)}_{\lambda\mu}(q)\big) = 4\pi e^2 \int \frac{d^3p}{(2\pi)^3} \left.\frac{\big[(p+q)_\lambda p_\lambda + (p+q)_\mu p_\lambda - g_{\lambda\mu} p\cdot q\big]}{E_{\mathbf{p}} E_{\mathbf{p}+\mathbf{q}}}\right|_{\substack{p_0 = E_{\mathbf{p}} \\ q_0 = E_{\mathbf{p}+\mathbf{q}} - E_{\mathbf{p}}}}$$

$$\cdot \frac{n_F(|\mathbf{p}|) - n_F(|\mathbf{p}+\mathbf{q}|)}{\omega + E_{\mathbf{p}} - E_{\mathbf{p}+\mathbf{q}}}$$

$$+4\pi e^2 \int \frac{d^3p}{(2\pi)^3} \left.\frac{\big[(p+q)_\lambda p_\mu + (p+q)_\mu p_\lambda - g_{\lambda\mu} p\cdot q\big]}{E_{\mathbf{p}} E_{\mathbf{p}+\mathbf{q}}}\right|_{\substack{p_0 = E_{\mathbf{p}} \\ q_0 = -E_{\mathbf{p}} - E_{\mathbf{p}+\mathbf{q}}}}$$

$$\cdot \frac{1 - n_F(|\mathbf{p}|)}{\omega + E_{\mathbf{p}} + E_{\mathbf{p}+\mathbf{q}}}$$

$$-4\pi e^2 \int \frac{d^3p}{(2\pi)^3} \left.\frac{\big[(p+q)_\lambda p_\mu + (p+q)_\mu p_\lambda - g_{\lambda\mu} p\cdot q\big]}{E_{\mathbf{p}} E_{\mathbf{p}+\mathbf{q}}}\right|_{\substack{p_0 = -E_{\mathbf{p}} \\ q_0 = E_{\mathbf{p}} + E_{\mathbf{p}+\mathbf{q}}}}$$

$$\cdot \frac{1 - n_F(|\mathbf{p}+\mathbf{q}|)}{\omega - E_{\mathbf{p}} - E_{\mathbf{p}+\mathbf{q}}} \tag{2.48}$$

We now compute $Q^{(0)}_{00}(q)$ and $Q^{(0)}_{ij}(q)$ from the above expression. Since we are ultimately interested in $Q_{\mu\nu}D^{\nu\mu}$ and $D^{i0} = 0 = D^{0i}$ in the Coulomb gauge, we need not calculate Q_{0i}. We can write $Q^{(0)}_{ij}$ in terms of longitudinal and transverse components. The only symmetric tensors that are available are $q_i q_j$ and δ_{ij}. Therefore

$$Q^{(0)}_{ij}(q) = \frac{q_i q_j}{|\mathbf{q}|^2} Q^{(0)}_L(q) + \left(\delta_{ij} - \frac{q_i q_j}{|\mathbf{q}|^2}\right) Q^{(0)}_T(q) \tag{2.49}$$

where

$$Q^{(0)}_L(q) = \frac{q_i q_j}{|\mathbf{q}|^2} Q^{(0)}_{ij}(q) \tag{2.50a}$$

and

$$Q^{(0)}_T(q) = \frac{1}{2}\left(\delta_{ij} - \frac{q_i q_j}{|\mathbf{q}|^2}\right) Q^{(0)}_{ij}(q) \tag{2.50b}$$

Now consider the Dyson equation for $D_{\mu\nu}$. In the Coulomb gauge,

$$D^{(0)}_{i0} = 0 = D^{(0)}_{0i}$$

and

$$D_{00}^{(0)}(k) = \frac{\hbar^2}{|\mathbf{k}|^2}$$

$$D_{ij}^{(0)}(k) = -\frac{\hbar^2}{k_0^2 - |\mathbf{k}|^2}\left(\delta_{ij} - \frac{k_i k_j}{|\mathbf{k}|^2}\right) \tag{2.51}$$

Then

$$\begin{aligned} D_{00}(q) &= D_{00}^{(0)}(q) + D_{0\lambda}^{(0)} Q^{(0)\lambda\kappa}(q) D_{\kappa 0}(q) \\ &= D_{00}^{(0)}(q) + D_{00}^{(0)} Q^{(0)0\kappa}(q) D_{\kappa 0}(q) \end{aligned}$$

and

$$D_{00}(q) = D_0^{(0)}(q) + D_{00}^{(0)}(q) Q_{00}^{(0)}(q) D_{00}(q) \tag{2.52}$$

where in going to the last line we have used the condition $D_{i0} = 0$:

$$D_{00}(q) = \frac{\hbar^2}{|\mathbf{q}|^2} \cdot \frac{1}{\left[1 - (\hbar^2/|q|^2) Q_{00}^{(0)}\right]} \tag{2.53}$$

This is just the Coulomb interaction, appropriately screened by the longitudinal dielectric function of the medium.

$$\begin{aligned} D_{ij}(q) &= D_{ij}^{(0)}(q) + D_{i\lambda}^{(0)}(q) Q^{(0)\lambda\kappa}(q) D_{\kappa j}(q) \\ &= D_{ij}^{(0)}(q) + D_{il}^{(0)}(q) Q^{(0)l\kappa}(q) D_{\kappa j}(q) \end{aligned}$$

and again using $D_{il}^{(0)} q^l = 0$

$$D_{ij}(q) = D_{ij}^{(0)}(q) + D_{il}^{(0)}(q) Q^{(0)lm}(q) D_{mj}(q) \tag{2.54}$$

Using $D_L(q) = 0$, we may write this in terms of transverse contributions only.

$$D_T(q) = D_T^{(0)}(q) + D_T^{(0)}(q) Q_T^{(0)}(q) D_T(q)$$

or

$$D_T(q) = \frac{D_T^{(0)}}{1 - D_T^{(0)} Q_T^{(0)}(q)} \tag{2.55}$$

This is the additional effective two-electron interaction arising from the electron–transverse photon interaction, which is appropriately screened by the system of electrons. The various polarizabilities appearing here have been defined in the literature, in slightly different ways and the final results appear different. But these expressions are all equivalent, as they must be, as is demonstrated in Appendix A.

C. The Calculation of Exchange-Correlation Energy

We now calculate the interaction energy of the system using the Pauli trick appropriately generalized to the relativistic case:

$$\langle H \rangle = \langle H_0 \rangle + \int_0^e \frac{de'}{e'} \langle H_{\text{int}} \rangle_{e'} \tag{2.56}$$

where $\langle H_0 \rangle$ is the kinetic energy in the noninteracting ground state and $\langle H_{\text{int}} \rangle_{e'}$ is the interaction energy for an arbitrary value of the coupling constant e'. Thus

$$E_{xc} = \int_0^e \frac{de'}{e'} \langle H_{\text{int}} \rangle_{e'} - E_{\text{Hartree}} \tag{2.57}$$

$$\langle H_{\text{int}} \rangle_{e'} = e' \int d^3 r_1 \langle \bar{\psi}(1) \gamma_\mu \psi(1) A^\mu(1) \rangle_{e'} \tag{2.58}$$

Using (2.23) and (2.25), this may be written as

$$\begin{aligned} \frac{\langle H_{\text{int}} \rangle_{e'}}{e'} &= i \int Q_{\nu\mu}(12) D^{\mu\nu}(21^+) d^4 2\, d^3 r_1 \\ &\quad + \frac{1}{c} \int d^3 r_1 \langle j_\nu(1) \rangle \langle A^\nu(1) \rangle \end{aligned} \tag{2.59}$$

$$\begin{aligned} E_{xc} &= \frac{1}{\Omega} \langle H_{\text{int}} \rangle - E_{\text{Hartree}} \\ &= \frac{1}{2} \int_0^{e^2} \frac{de'^2}{e'^2} Q^{(0)}_{\nu\mu}(q) D^{\mu\nu}(q) \frac{d^4 q}{(2\pi)^4} \end{aligned}$$

In the Coulomb gauge

$$\begin{aligned} E_{xc} &= \frac{i}{2} \int_0^{e^2} \frac{de'^2}{e'^2} \int Q_{00}(q) D^{00}(q) \frac{d^4 q}{(2\pi)^4} \\ &\quad + i \int_0^{e^2} \frac{de'^2}{e'^2} Q_T^{(0)} D_T \frac{d^4 q}{(2\pi)^4} \end{aligned} \tag{2.60}$$

Note that the second term, the transverse photon–electron interaction energy, carries a factor of 1 in contrast to $\frac{1}{2}$ going with the Coulomb interaction energy. By using (2.55) and (2.53) and doing the e^2 integration, we have

$$E_{xc} = \frac{1}{2}\int \frac{d^3q}{(2\pi)^3}\int_{-\infty}^{\infty}\frac{d\omega}{2\pi}\ln\left[1+\frac{4\pi e^2}{q^2}Q_{00}(q,\omega)\right] + \int \frac{d^3q}{(2\pi)^3}\int_{-\infty}^{\infty}\frac{d\omega}{2\pi}\ln\left[1-\frac{8\pi e^2}{q^2+\omega^2}Q_T(q,\omega)\right] \tag{2.61}$$

where in going from (2.60) to (2.61), ω has been replaced by $i\omega$, which follows on shifting the real ω-line integral in (2.60) into the imaginary ω-line integral; Q_{00} and Q_T may be evaluated using (2.48) and (2.50)

$$Q_{00}(q,\omega) = \int_{k<k_F}\frac{d^3k}{(2\pi)^3}\left[\frac{(E_{\mathbf{k}+\mathbf{q}}-E_k)\left[(E_{\mathbf{k}+\mathbf{q}}+E_k)^2-q^2\right]}{E_kE_{\mathbf{k}+\mathbf{q}}\left[(E_{\mathbf{k}+\mathbf{q}}-E_k)^2+\omega^2\right]} + \frac{(E_{\mathbf{k}+\mathbf{q}}+E_k)\left[(E_{\mathbf{k}+\mathbf{q}}-E_k)^2-q^2\right]}{E_kE_{\mathbf{k}+\mathbf{q}}\left[(E_{\mathbf{k}+\mathbf{q}}+E_k)^2+\omega^2\right]}\right] \tag{2.62a}$$

$$Q_T(q,\omega) = \int_{k<k_F}\frac{d^3k}{(2\pi)^3}\left[\frac{(E_{\mathbf{k}+\mathbf{q}}-E_k)\left[E_{\mathbf{k}+\mathbf{q}}E_k-\frac{(\mathbf{k}\cdot\mathbf{q})^2}{q^2}-(\mathbf{k}\cdot\mathbf{q})-m^2\right]}{E_kE_{\mathbf{k}+\mathbf{q}}\left[(E_{\mathbf{k}+\mathbf{q}}-E_k)^2+\omega^2\right]} - \frac{(E_{\mathbf{k}+\mathbf{q}}+E_k)\left[E_{\mathbf{k}+\mathbf{q}}E_k+\frac{(\mathbf{k}\cdot\mathbf{q})^2}{q^2}+(\mathbf{k}\cdot\mathbf{q})+m^2\right]}{E_kE_{\mathbf{k}+\mathbf{q}}\left[(E_{\mathbf{k}+\mathbf{q}}+E_k)^2+\omega^2\right]}\right] \tag{2.62b}$$

Finally, scaling

$$\mathbf{k} = k_F\mathbf{x}; \qquad \mathbf{q} = k_F\mathbf{y}; \qquad \omega = k_F y u; \qquad E_k = mE_x \equiv m\sqrt{1+\beta^2x^2}$$

$$E_{xc} = \frac{3Nk_F}{8\pi}\int_0^{\infty}y^3\,dy\int_{-\infty}^{\infty}du\ln\left\{1+\frac{e^2\beta}{\pi y^2}\tilde{Q}_{00}(y,u)\right\} + \frac{3Nk_F}{4\pi}\int_0^{\infty}y^3\,dy\int_{-\infty}^{\infty}du\ln\left\{1-\frac{2e^2\beta}{\pi y^2(1+u^2)}\tilde{Q}_T(y,u)\right\} \tag{2.63}$$

where

$$Q(k_F y, k_F yu) = \frac{k_F^3}{4\pi^2 m} \tilde{Q}(y, u)$$

In the limit $\beta \to 0$ the expression (2.62a) goes over to the $Q_q(u)$ of GB and (2.62b) is of order $0(\beta^2)$. The statement that the GB result is accurate in the high-density limit is true only in the nonrelativistic theory.

The exchange-correlation potential is then given by

$$V_{xc} = \frac{\partial E_{xc}^{(\text{Total})}}{\partial n} = \frac{\pi^2}{k_F^2} \frac{\partial E_{xc}}{\partial k_F} \tag{2.64}$$

Separating out the exchange part, which is linear in e^2, we may express the correlation energy and potential as

$$E_c^{(\text{Rings})} = E_c^{(c)(\text{Rings})} + E_c^{(\text{tr})(\text{Rings})} \tag{2.65a}$$

and

$$V_c^{(\text{Rings})} = V_c^{(c)(\text{Rings})} + V_c^{(\text{tr})(\text{Rings})} \tag{2.65b}$$

where

$$\frac{E_c^{(c)(\text{Rings})}}{N} = \frac{3k_F}{8\pi} \int_0^\infty y^3\,dy \int_{-\infty}^\infty du \left\{ \ln\left[1 + \frac{e^2\beta}{\pi y^2} \tilde{Q}_{00}(y, u)\right] - \frac{e^2\beta}{\pi y^2} \tilde{Q}_{00}(y, u) \right\} \tag{2.66a}$$

$$\frac{E_c^{(\text{tr})(\text{Rings})}}{N} = \frac{3k_F}{4\pi} \int_0^\infty y^3\,dy \int_{-\infty}^\infty du \cdot \left\{ \ln\left[1 - \frac{2e^2\beta}{\pi y^2(1+u^2)} \tilde{Q}_T(y, u)\right] + \frac{2e^2\beta}{\pi y^2(1+u^2)} \tilde{Q}_T(y, u) \right\} \tag{2.66b}$$

and

$$V_c^{(c)(\text{Rings})} = \frac{e^2\beta k_F}{8\pi^2}\int_0^\infty y\,dy\int_{-\infty}^\infty du\,\tilde{Q}_{00}^{(1)}(y,u) \cdot\left\{\frac{1}{1+(e^2\beta/\pi y^2)\tilde{Q}_{00}(y,u)}-1\right\} \tag{2.67a}$$

$$V_c^{(\text{tr})(\text{Rings})} = \frac{e^2\beta k_F}{2\pi^2}\int_0^\infty y\,dy\int_{-\infty}^\infty \frac{du}{1+u^2}Q_T^{(1)}(y,u) \cdot\left\{\frac{1}{1-\left[2e^2\beta/\pi y^2(1+u^2)\right]\tilde{Q}_T(y,u)}-1\right\} \tag{2.67b}$$

where

$$\frac{\partial Q_{00}(q,\omega)}{\partial k_F} = \frac{k_F^2}{4\pi^2 m}\tilde{Q}_{00}^{(1)}(y,u)$$

$$\frac{\partial Q_T(q,\omega)}{\partial k_F} = \frac{k_F^2}{4\pi^2 m}\tilde{Q}_T^{(1)}(y,u) \tag{2.68}$$

Here $\tilde{Q}^{(1)}(y,u)$ is the same as $\tilde{Q}(y,u)$ except that the integrand is evaluated at $x=1$ and the integral over the angle remains.

Expressions for $Q_{\lambda\mu}(q;\omega)$ were given in different forms by the authors quoted earlier in this chapter, and their mutual equivalence is demonstrated in Appendix A. The forms (2.66*a*) and (2.66*b*) were first given by Jancovici.[60]

III. SOME NUMERICAL CALCULATIONS

A. Numerical Evaluation of the Correlation Energies

In the previous section we derived the expressions for the correlation contribution to the energy and potential. In this section we carry out these integrations numerically and compare the results with nonrelativistic ones and some approximations of Jancovici.

As in the nonrelativistic theory, the major contributions to the correlation energy are from regions of small energy and momentum transfers ($q < k_F$, $\omega < E_F$). If such an approximation is made we find (see Appendix B) the following expression for ε_c:

$$\varepsilon_c^{(\text{Rings})} \equiv \frac{E_c^{(\text{Rings})}}{N} \simeq \frac{3(k_F^2/m)}{32\pi\sqrt{1+\beta^2}} \int_{-\infty}^{\infty} dv \{\ln(1+R_L(v))$$

$$- R_L(v) - (R_L(v))^2(\ln(1+R_L(v)) - \ln(R_L(v)))\}$$

$$+ \frac{3k_F}{16\pi}\int_{-\infty}^{\infty} dv \{\ln(1+R_{\text{tr}}(v)) - R_{\text{tr}}(v)$$

$$-(R_{\text{tr}}(v))^2(\ln(1+R_{\text{tr}}(v)) - \ln R_{\text{tr}}(v))\} \tag{3.1}$$

where

$$R_L(v) = \frac{4e^2\sqrt{1+\beta^2}}{\pi\beta}\left(1 - v\tan^{-1}\frac{1}{v}\right) \tag{3.2a}$$

$$R_{\text{tr}}(v) = \frac{2e^2\beta}{\pi\sqrt{1+\beta^2}\,(1+v^2)}\left[-v^2\frac{\beta^2+1}{\beta^2} + v\frac{\sqrt{1+\beta^2}}{\beta}\right.$$

$$\left.\cdot\left(1+\frac{v^2(\beta^2+1)}{\beta^2}\right)\tan^{-1}\left(\frac{\beta}{v\sqrt{\beta^2+1}}\right)\right] \tag{3.2b}$$

and

$$v_c^{(\text{Rings})} = \varepsilon_c^{(\text{Rings})} + \frac{\beta}{3}\frac{\partial}{\partial\beta}\varepsilon_c^{(\text{Rings})} \tag{3.3}$$

These integrals are evaluated numerically. For $\beta \ll 1$, they reduce to the correct nonrelativistic expressions and $\varepsilon_c^{(\text{Rings})} \simeq \varepsilon_c^{(\text{GB})}$. Jancovici makes a further approximation keeping only terms to the order $e^4 \ln e^2$ and gets

$$\varepsilon_c^{(c)} = \frac{(1-\ln 2)}{\pi^2}\sqrt{1+\beta^2}\, me^4 \ln e^2 \tag{3.4a}$$

and

$$\varepsilon_c^{(\mathrm{tr})} = \frac{3m}{4\pi^3} \frac{\beta^3 e^4 \ln e^2}{(1+\beta^2)} \int_{-\infty}^{\infty} \frac{du}{(1+u^2)^2} \cdot \left\{ -\frac{u^2(1+\beta^2)}{\beta^2} + \frac{u\sqrt{1+\beta^2}}{\beta} \left[1 + \frac{u^2}{\beta^2}(1+\beta^2) \right] \cdot \tan^{-1}\left(\frac{\beta}{u\sqrt{1+\beta^2}} \right) \right\}^2 \tag{3.4b}$$

We have plotted the correlation energy contributions per particle as a function of β in Fig. 1. The full line curves are our integrals computed

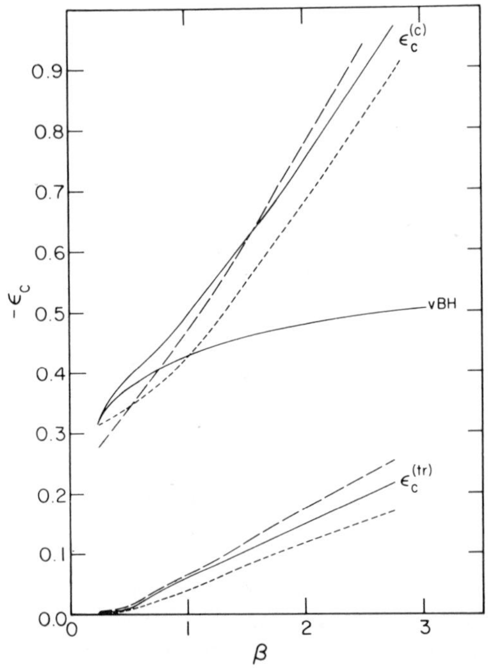

Fig. 1. The correlation energy expressed in Rydbergs versus β ($= \hbar k_F/mc$). The term $\varepsilon_c^{(c)}$ stands for Coulomb, $\varepsilon_c^{(\mathrm{tr})}$ for the transverse photon contributions. The full line curves are our results for $\varepsilon_c^{(c)}$ and $\varepsilon_c^{(\mathrm{tr})}$; the long dashed curves are the result of eq. (3.1); and the short dashed curves are the $e^4 \ln e^2$ approximation of Jancovici. The nonrelativistic answer is labeled vBH. (From Ref. 71.)

numerically and the nonrelativistic result of von Barth and Hedin (vBH).[48] The short dashed curves are the $e^4 \ln e^2$ approximation of Jancovici and the long dashed curves are without such an approximation. The Jancovici (JGB) approximation is in the same spirit as the Gell – Mann and Brueckner (GB) one, involving a small q approximation for the Q_0, Q_T but retaining the ω dependence.

It can be seen from the figure that our numerically integrated result merges smoothly with the vBH curve for $\beta < 0.25$. The JGB approximation intersects our curve at about $\beta = 1.6$, lying above it for larger β and below it for smaller values. The disagreement ranges from 15% at $\beta = 0.25$ to 8% at $\beta = 2.75$. The q dependence is not as important as the nonrelativistic case where the GB results differ by 20–30% from vBH answers at low densities. However, for the transverse part of the correlation energy the JGB lies above our result for the entire β range. Thus for the total correlation energy the q dependence is not as important as in the nonrelativistic case. The $e^4 \ln e^2$ approximation, on the other hand, lies below our curve for both $\varepsilon_c^{(c)}$ (8% for $\beta = 2.75$ and 3% for $\beta = 0.75$) and $\varepsilon_c^{(\mathrm{tr})}$ (21% for $\beta = 2.75$) in the entire β range. At low densities the $e^4 \ln e^2$ scheme amd (JGB) do not go to the vBH answer. However, the JGB goes into the GB as it should. The potential behaves like the energy and similar remarks apply to it. We show our results for the potential in Fig. 3 (see below).

The relativistic result begins to deviate from the nonrelativistic one for $\beta > 0.25$. Whereas the vBH answer rises logarithmically, the relativistic answer is linear for large β. In Fig. 2 we have plotted the total correlation energy per particle versus $\log_{10} n$, where n is the electron density in atomic units (density $\times\, a_0^3$). The correlation potential as a function of β is given in Fig. 3, where the individual Coulomb and transverse contributions are plotted separately, as well as the vBH answer for comparison. In Fig. 4 the total correlation potential as a function of $\log_{10} n$ is given. The departure from the nonrelativistic result begins at $\beta > 0.25$ as in the energy case. It should be pointed out that for both energy and potential the transverse photon contribution, though small, is found to be significant.

In Table I, we have given the numerical values for energy per particle and the potential for some values of β along with the results of von Barth and Hedin. As a useful numerical check, we have calculated the correlation energy and potential for very small values of $\beta (\beta = 0.007, 0.014 \Rightarrow r_s = 2, 1$, respectively) using our full relativistic expressions and find that they compare very well with the nonrelativistic results of vBH. We also give the separate Coulomb and transverse contributions in the table. Unlike the case of exchange[52] where the transverse contribution was opposite in sign to the Coulomb and large enough to cancel it, the transverse correlation contribution is of the same sign as the Coulomb part. The correlation contribution

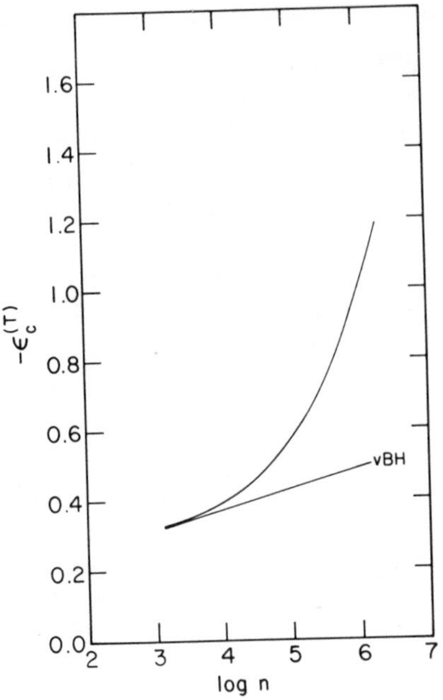

Fig. 2. The total correlation energy in Rydbergs as a function of $\log n$ where $n =$ density $\times a_0^3$, a_0 being the Bohr radius. The corresponding nonrelativistic result is labeled vBH. (From Ref. 71).

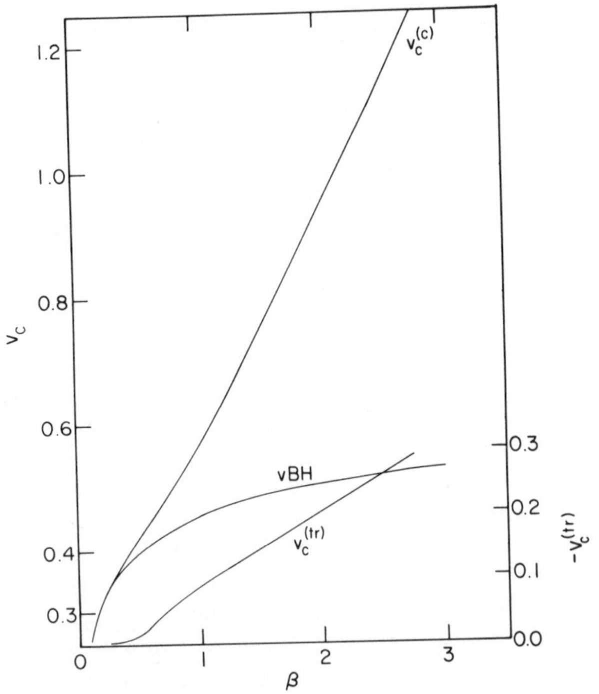

Fig. 3. The correlation potential in Rydbergs versus β. The terms $V_c^{(c)}$, $V_c^{(tr)}$, and vBH stand for Coulomb, transverse photon, and von Barth–Hedin results. The scale on the right should be used for $V_c^{(tr)}$. (From Ref. 71.)

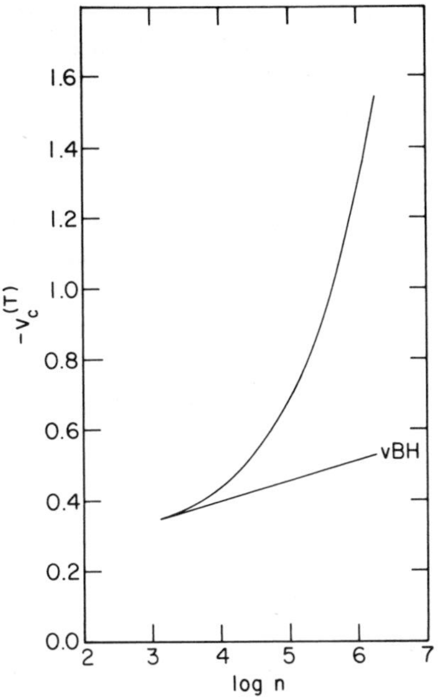

Fig. 4. The total correlation potential versus ln n'. The corresponding nonrelativistic approximation is labeled vBH. (From Ref. 71.)

TABLE I
The Values of Correlation Energy and Potential in Rydbergs Are Given for a Set of β Values of Interest to Heavy Atomic Systems:[a]
$\varepsilon_c^{(T)} = \varepsilon_c^{(c)} + \varepsilon_c^{(tr)}$ and $V_c^{(T)} = V_c^{(c)} + V_c^{(tr)}$

β	$-\varepsilon_c^{(vBH)}$	$-\varepsilon_c^{(c)}$	$-\varepsilon_c^{(tr)}$	$-\varepsilon_c^{(T)}$	$-v_c^{(vBH)}$	$-v_c^{(c)}$	$-v_c^{(tr)}$	$-v_c^{(T)}$
0.007	0.1234	0.1235	—	0.1235	0.140	0.1389	—	0.1389
0.014	0.1573	0.1574	—	0.1574	0.173	0.174	—	0.174
0.25	0.3252	0.3262	0.0036	0.3298	0.347	0.3453	0.0064	0.3517
0.50	0.3739	0.3924	0.0075	0.3999	0.399	0.4209	0.0131	0.4340
0.75	0.4046	0.4370	0.0369	0.4739	0.430	0.4918	0.0586	0.5504
1.0	0.4269	0.4997	0.0595	0.5592	0.453	0.5721	0.0967	0.6688
1.25	0.4442	0.5546	0.0828	0.6374	0.470	0.6581	0.1218	0.7799
1.5	0.4581	0.6188	0.1062	0.7250	0.483	0.7489	0.1530	0.9019
1.75	0.4697	0.6857	0.1292	0.8149	0.494	0.8336	0.1825	1.0161
2.0	0.4794	0.7545	0.1519	0.9164	0.503	0.9373	0.2119	1.1492
2.25	0.4878	0.8249	0.1740	0.9989	0.511	1.0356	0.2409	1.2765
2.5	0.4951	0.8965	0.1964	1.0929	0.518	1.1339	0.2695	1.4034
2.75	0.5015	0.9690	0.2182	1.1872	0.523	1.2329	0.2979	1.5308

[a]For comparison we have also given the von Barth–Hedin results. (From Ref. 72.)

to the potential for the electron gas is about 5% of the total exchange at $\beta = 2.5$ and drops to half this value at $\beta = 1$. In the next subsection we incorporate our potentials in a self-consistent relativistic atomic calculation and compare our results with some previous estimates.

Before doing this, however, we report on the results of an atomic calculation to assess the contribution of the Breit interaction versus the full transverse interaction. This has been done in the case of Dirac–Fock calculations by Mann and Johnson[26] but not for the local scheme to our knowledge.

If in the calculation of the transverse energy the energy transfer by the photon is assumed negligible, then we obtain the Breit energy. That the transverse photon interaction reduces to the Breit interaction was shown by MacDonald.[61] The integrals involved may be done analytically (see Appendix C), with the results

$$\varepsilon_x^{(\mathrm{Breit})} = \frac{3}{4\pi}\beta\left\{1 + \frac{2}{\beta^4}[\ln(\beta + E_F)]^2 - \frac{4E_F}{\beta^4}[\beta\ln(E_F + \beta) - E_F \ln\beta]\right\} \tag{3.5}$$

$$v_x^{(\mathrm{Breit})} = \frac{\beta}{\pi}\left\{1 - \frac{2}{\beta E_F}\ln(\beta + E_F) + \frac{2}{\beta^2}\ln E_F\right\} \tag{3.6}$$

It may be noted here that $\varepsilon_x^{(\mathrm{Breit})} < \varepsilon_x^{(\mathrm{tr})}$ because the neglect of the energy denominator in the expression for $\varepsilon_x^{(\mathrm{tr})}$ decreases the integrand. We plot $\varepsilon_x^{\mathrm{tr}}$ and $\varepsilon_x^{\mathrm{Breit}}$ in Rydbergs in Fig. 5*a* and $v_x^{(\mathrm{tr})}$ and $v_x^{(\mathrm{Breit})}$ in Fig. 5*b*. We may also break up the full transverse interaction into the magnetic part and a part giving corrections to the Coulomb interaction due to retardation effects (Mann and Johnson[26] and Johnson[82]):

$$\varepsilon_m = \frac{3}{4\pi}\beta\left\{-\frac{1}{2} + \frac{E_F}{2\beta}\ln(E_F + \beta) + \frac{9}{8\beta^4}[\beta E_F - \ln(E_F + \beta)]^2\right\} \tag{3.7}$$

$$\varepsilon_{\mathrm{ret}} = \frac{3}{4\pi}\beta\left\{\frac{1}{3} + \frac{1}{3}\beta^2 + \frac{1}{6}\frac{E_F}{\beta}\ln(E_F + \beta) - \frac{2}{3}\frac{E_F^4}{\beta^4}\ln(E_F) - \frac{1}{8\beta^4}[\beta E_F - \ln(E_F + \beta)]^2\right\} \tag{3.8}$$

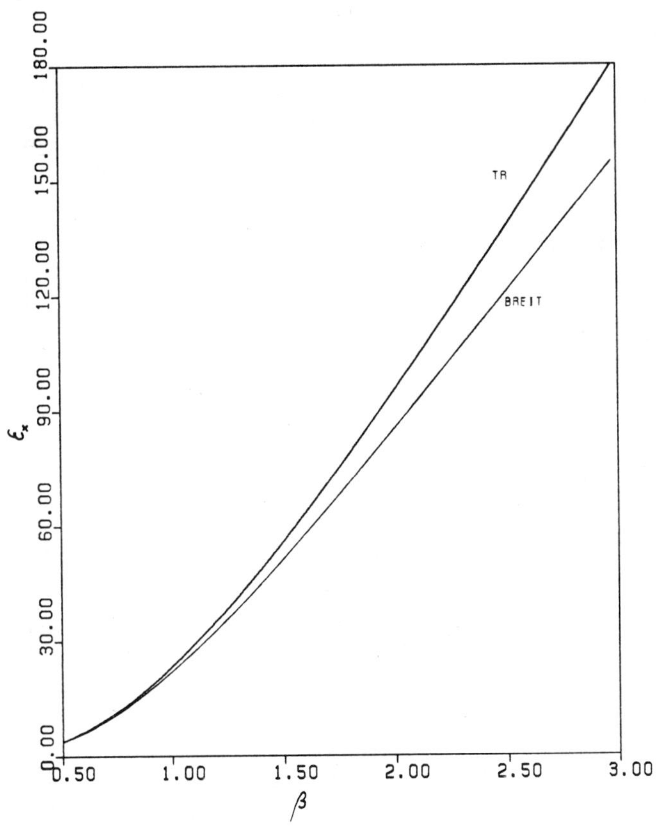

Fig. 5*a*. Plot of $\varepsilon_x^{(\mathrm{tr})}$ and $\varepsilon_x^{(\mathrm{Breit})}$ versus β in Rydbergs.

In the nonrelativistic limit $\beta \to 0$

$$\varepsilon^{\mathrm{tr}} = \frac{3}{4\pi}\left(\frac{5}{9}\beta^2\right) = \varepsilon^{(\mathrm{Breit})} \tag{3.9a}$$

$$\varepsilon_m = \frac{3}{4\pi}\left(\frac{2}{3}\beta^2\right) \tag{3.9b}$$

$$\varepsilon_{\mathrm{ret}} = \frac{3}{4\pi}\left(-\frac{1}{9}\beta^2\right) \tag{3.9c}$$

B. A Self-Consistent Atomic Calculation for Mercury and Fermium

We have modified the relativistic Dirac–Slater program due to Liberman et al.[70] to include the full relativistic Coulomb and transverse exchange as

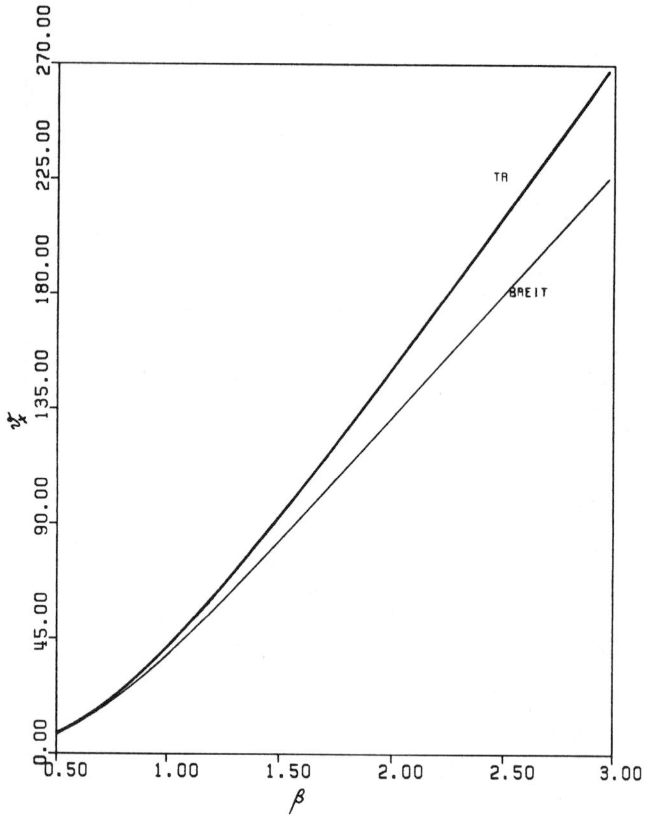

Fig. 5*b*. Plot of $v_x^{(\text{tr})}$ and $v_x^{(\text{Breit})}$ versus β in Rydbergs.

given by Rajagopal[52] and MacDonald and Vosko.[53] Our tabulated values (Table I) for the total relativistic correlation energy per particle and the total correlation potential were fit to a cubic spline as a function of β for calculation at any density.[71] We also integrated numerically the expressions of von Barth and Hedin[48] for the paramagnetic gas and fit the values to a cubic spline as well. The correlation contribution to the total energy of ^{80}Hg and ^{100}Fm and to the 1*s* binding energy in the frozen orbital approximation were evaluated both in the full self-consistent field (SCF) and perturbatively (Ramana et al.[72]). The SCF contribution of the correlation to the 1*s* binding energy is

$$E_{c,\text{SCF}}^{1s} = \{E(N) - E(N-1)\} - \{E^0(N) - E^0(N-1)\} \tag{3.10}$$

The first term in the braces is the frozen orbital binding energy with correlation included in the SCF scheme, while the second term in braces is the binding energy without correlation. The contribution to the total energy is as usual

$$E_c^{\mathrm{SCF}} = \int \varepsilon_c(\rho)\rho(\mathbf{r})\, d^3r \tag{3.11}$$

where ρ is the full self-consistent charge density. The perturbative contribution to the binding energy is given by

$$E_{c;\,\mathrm{pert}}^{1s} = \int \varepsilon_c\left(\rho_N^0\right)\rho_N^0(\mathbf{r})\, d^3r - \int \varepsilon_c\left(\rho_{N-1}^0\right)\rho_{N-1}^0(\mathbf{r})\, d^3r \tag{3.12}$$

where ρ_N^0 is the SCF charge density without correlation for N electrons. The perturbative contribution to the total energy is

$$E_{c;\,\mathrm{pert}} = \int \varepsilon_c\left(\rho^0\right)\rho^0(\mathbf{r})\, d^3r \tag{3.13}$$

Expressions similar to (3.10)–(3.13) were used for the transverse contributions and the Breit contributions.

In Table II we show the results of our correlation calculation for mercury and fermium. In the first three rows we list the correlation contribution to the total energy and in the last three, the contribution to the $1s$ binding energy. There is hardly any difference between the perturbative and SCF calculation for the contribution to the total energy. However, for the $1s$ binding energy SCF changes the perturbative result by about 0.09 a.u. The total correlation contribution in the local density scheme is seen to be much larger than that estimated by Cowan.[29] At this point we must recall the remarks of Cowan[29] and also of Tong.[73] Because of the highly localized nature of atomic electrons, correlation effects calculated using the free electron gas results in local scheme are much overestimated, and we expect this to be more so in the inner electrons of heavy atomic systems.[74] That this may be the case is also seen from the excellent agreement between Dirac–Fock results including Lamb shift and vacuum polarization corrections of Huang et al.[5] and highly accurate experimental data of Bearden and Burr[75] and Deslattes et al.[28] The systematic difference in that case is less than 10 eV.

A comparison of the relativistic and nonrelativistic correlation energies shows that in the case of mercury the relativistic correlation contribution is about -0.2 a.u. to the total energy and about -0.4 a.u. for fermium. This contribution is in the right direction since the relativistic correlation energy

TABLE II
Relativistic (R) and Nonrelativistic (NR) Contributions (in a.u.) to the Total Energy and the 1s Binding Energy Both in SCF and Perturbation (pert) Theory[a]

	Hg	Fm
$-E_{c;\mathrm{R}}^{(\mathrm{pert})}$	10.3502	13.5236
$-E_{c;\mathrm{NR}}^{(\mathrm{pert})}$	10.1597	13.1073
$-E_{c;\mathrm{R}}^{(\mathrm{SCF})}$	10.3537	13.5274
$-E_{c;\mathrm{R}}^{1s(\mathrm{pert})}$	0.292	0.375
$-E_{c;\mathrm{NR}}^{1s(\mathrm{pert})}$	0.215	0.228
$-E_{c;\mathrm{R}}^{1s(\mathrm{SCF})}$	0.203	0.277

[a] The binding energies are obtained using frozen orbitals. (From Ref. 71.)

per particle and potential lie below their nonrelativistic counterparts at high densities for the homogeneous electron gas (Ramana and Rajagopal[71]). In the case of the 1s binding energies (rows 4 and 5 in Table II) the relativistic correlation contribution is $\simeq -0.08$ a.u. for mercury, growing to $\simeq -0.15$ a.u. for fermium. The relativistic contribution to the correlation energy is not large because only a very small number of inner electrons are relativistic.

In Figures 6*a* and *b* we have plotted our correlation contribution to the K_{α_1} and K_{α_2} lines versus the square root of the K_{α_1} and K_{α_2} energies, respectively, expressed in KeV. The straight line is a linear regression fit of differences between the theoretical calculations of Huang et al.[5] and the experimental results of Deslattes et al.[28] The correlation contribution is not large enough to explain the discrepancies or even the trend. The improvements (Deslattes et al.[28]) being carried out in the theoretical calculations of Huang et al.,[5] such as using the transverse interaction instead of the Breit and including it in the SCF, using a more realistic nuclear charge distribution, and the experimental remeasurements, may give a clearer picture of the actual correlation contributions. In Table III we give our values for the correlation contributions to the K_{α_1} and K_{α_2} energies and the total energy for various values of Z. In Fig. 7 we plot the correlation contribution in a.u. to the total energy versus Z. The function $0.058\ Z^{1.18}$ fits this curve within 1.5%.

Table IV presents the Breit and transverse contributions to the 1s binding energy in mercury and fermium. In the first column we show the 1s binding energy; in the next four columns we show the results of the perturbative calculation of the transverse, Breit, magnetic contribution to the transverse, and the retarded contribution to the transverse exchange energy, respec-

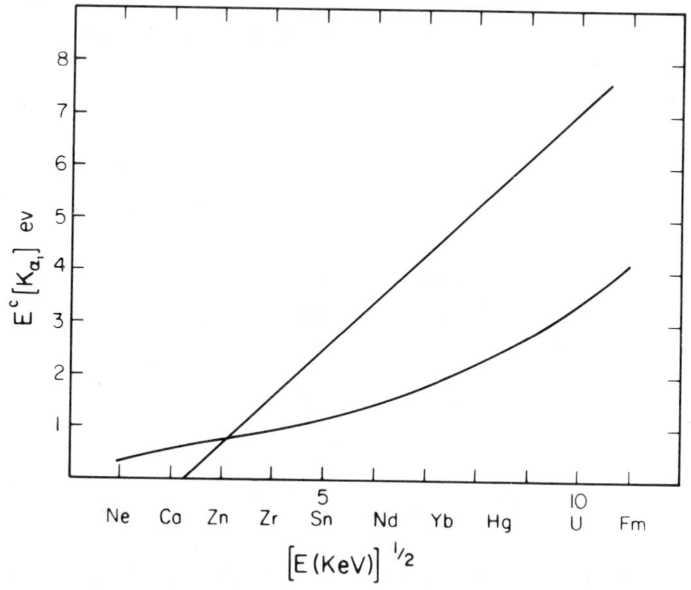

Fig. 6*a*. Correlation contribution in eV to the K_{α_1} line plotted versus the square root of energy in KeV. (From Ref. 72; the straight line is based on Ref. 28.)

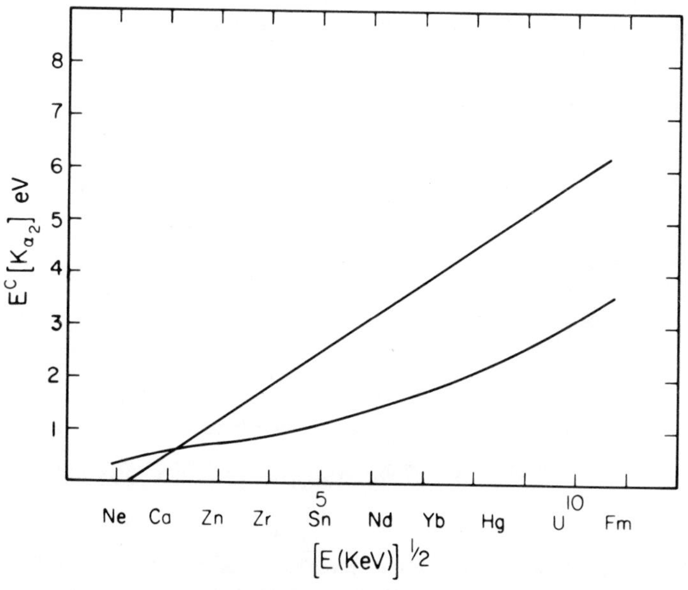

Fig. 6*b*. Same as Fig. 6*a*, for K_{α_2} line. (From Ref. 72.)

TABLE III
Correlation Contribution to the K_α energies (in eV) and to the total energy (in a.u.)[a]

	$E_{cr}[K_\alpha]-E[K_\alpha]$		
	K_{α_2}(eV)	K_{α_1}(eV)	$-E_{cr}$(a.u.)
Ne	0.3102	0.3129	0.8982
Ca	0.5714	0.5742	1.9568
Zn	0.7646	0.7755	3.2819
Zr	0.9034	0.9224	4.4985
Sn	1.1211	1.1565	5.8903
Nd	1.4231	1.4939	7.2465
Yb	1.8014	1.9211	8.8092
Hg	2.2667	2.4545	10.3537
U	3.0259	3.3389	12.1466
Fm	3.6001	4.1362	13.5274

[a](From Ref. 72).

tively. The sixth column is the perturbative result of the Dirac–Fock calculation of Mann and Johnson.[26] The last two columns are the Breit and transverse contributions in the SCF calculation. A comparison of the transverse and Breit contributions in mercury shows that the frequency dependence contributes about 1 a.u. in both perturbation theory and SCF, while SCF reduces the perturbative result by about 0.8 a.u. These effects are larger

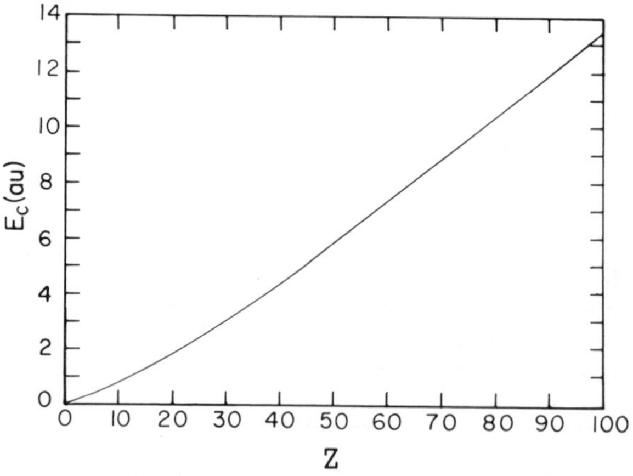

Fig. 7. Plot of the correlation contribution versus the total energy as a function of atomic number Z. The trend is not linear in Z but has an exponent of 1.18 ($\simeq 0.058Z^{1.18}$). (From Ref. 72.)

TABLE IV
Breit and Transverse (Magnetic plus Retarded) Contributions in Atomic Units to the $1s$ Binding Energy[a]

	E^{1s}	$E_x^{(\mathrm{tr})}$	$E_x^{(\mathrm{Breit})}$	E'_m	E'_{ret}	$E_{(MJ)}^{(\mathrm{tr})}$	$E^{1s;(\mathrm{Breit})}$	$E^{1s;(\mathrm{tr})}$
Hg	−3058.8097	15.02	14.03	18.93	−3.91	11.3742	13.26	14.21
Fm	−5236.9692	31.58	28.52	41.24	−9.66	—	26.98	29.89

[a]The first column is the $1s$ binding energy including the transverse term; the next four columns are from perturbative calculations; the last two are SCF contributions. (From Ref. 72.)

in fermium. The major contribution to the transverse energy is from the magnetic part, of which a relatively small amount is cancelled by the negative retardation correction to the Coulomb interaction. A comparison of our result with that of Mann and Johnson[26] shows a difference of almost $\simeq 3.6$ a.u. This difference is much larger in the case of the transverse contribution to the total energy which is displayed in Table V. The frequency dependence contributes about 1.6 a.u. in mercury and about 5.2 a.u. in fermium. The SCF contributes very little. A comparison of the results of Mann and Johnson[26] for magnetic and retardation energy contributions to the total energy in perturbation theory (shown in the last two columns) with our perturbative theory results (shown in the sixth and seventh columns) shows our result to be almost 1.7 times as large. This difference persists for all the heavy atoms investigated. We believe that the difference is due to nonlocal effects. However, the total energies in LDS lie above those of Dirac–Fock values as they should be. In the nonrelativistic limit the ratio $E'_{\mathrm{ret}}/E'_m \simeq -\frac{1}{6}$, as noted by Mann and Johnson;[26] however, their calculations give $\simeq -\frac{1}{10}$. This ratio goes to -1 for $\beta \gg 1$. We get for our two cases a value close to $\simeq -\frac{1}{5}$. In Mann

TABLE V
Breit and Transverse (tr) Contributions in Atomic Units to the Total Energy (first column) in SCF and Perturbation (pert) Theory[a]

	E_{LD}	$E_x^{\mathrm{tr;\,SCF}}$	$E_x^{\mathrm{Breit;\,SCF}}$	$E_x^{\mathrm{tr;\,pert}}$	$E_x^{\mathrm{Breit;\,pert}}$	E'_m(pert)	E'_{ret}(pert)	E'_m(MJ)	E'_{ret}(MJ)
Hg (full SCF)	−19602.26	34.00	—	—	—	—	—	—	—
Hg (tr or Breit in pert theory)	−19636.39	—	32.48	34.25	32.69	42.47	−8.23	24.50	−2.33
Em (full SCF)	−34801.74	78.32	—	—	—	—	—	—	—
Fm (tr or Breit in pert theory)	−34880.41	—	73.12	79.02	73.69	100.33	−21.31	—	—

[a]Perturbative contribution is $\int \varepsilon_x^{\mathrm{tr}} \rho^0(r)\, d^3r$, where ρ^0 is calculated without the transverse potential. For Hg and Fm the ratio of $E'_{\mathrm{ret}}/E'_m \sim -1/5$. (From Ref. 72; the last two columns are from Ref. 26.)

and Johnson's[26] Dirac–Fock calculations the Breit energy is larger than their transverse interaction energy, in contrast to our local results, as remarked earlier after Eq. (3.6).

(i) The correlation contribution. We have calculated the correlation contribution for mercury and fermium using our previous results (Ramana and Rajagopal[71]). We find that the correlation contributions are much larger than previously estimated. However, we may have overestimated the effect of correlation since our relativistic local-density (LD) calculation shares the same objections of the nonrelativistic ones. We find that the contributions to correlation due to relativistic effects are quite small in an actual atom even though the correlation energies and potentials for the relativistic and nonrelativistic electron gas show dramatic differences. We suggest that this is due to the small number of electrons that are relativistic even in an atom such as Fm.

We have plotted the trends, across the period table, of the correlation contributions to the K_α energies and to the total energy. The correlation contribution to the K_α lines is roughly half the discrepancy between accurate theoretical and experimental results. Even the trends are different.

(ii) Breit versus transverse interaction. We calculated the contributions of Breit and transverse interaction energies in mercury and fermium and compared our results with those of Mann and Johnson's[26] Dirac–Fock calculations. We found the LD estimate for the transverse term about $\simeq 1.7$ times larger than the Dirac–Fock result. We attribute this difference to nonlocal effects. This may be seen even in the nonrelativistic case for the exchange term (Perdew and Zunger[75]). Huang et al.,[5] using a single determinant constructed out of the self-consistent wave functions from a local-density calculation (using $\rho^{1/3}$ potential), evaluated the expectation value of the full Hamiltonian and got results in closer agreement to those of Mann and Johnson[26] than ours. We find that the Breit term gives the major part of the transverse contribution but that the frequency dependence is important if comparison is to be made with experiment. Even though this calculation is not variational it points to the fact that nonlocal effects may indeed be significant.

(iii) SCF versus perturbation theory. We have also assessed the importance of SCF for the transverse and correlation contributions and found that SCF makes a difference in the contributions to the binding energy, but practically no difference was found between SCF and perturbation theory results for the contributions to the total energy. This is because the levels are more sensitive to the changes in potential in SCF.

(iv) Other effects. Our calculation has not included relaxation effects which are important for the binding energies and effects due to the finite nuclear size which are important for the inner electrons. Lamb shift and vacuum polarization contributions are much larger than correlation contributions for the inner shells and have been tabulated by Huang et al.[5] All these effects have to be included if a comparison is to be made with experiment. We expect a local spin-density-functional formalism to give a better description as in the nonrelativistic case.

IV. SPIN-DENSITY THEORY

A. Relativistic Spin-Density-Functional Formalism

Consider again the system given by the Hamiltonian (2.1). We are now interested in the effects of spin polarization of the electron gas (Ramana and Rajagopal[76]). In the relativistic theory, however, spin is closely coupled to the kinetic motion of the particle and is not a good quantum number (see, for example, Sakurai[77]). To study the effects of the spin we take the viewpoint discussed by MacDonald and Vosko.[53] Instead of an external interaction term of the form $j^{\mu}A_{\mu}^{(\text{ext})}$, which contains all the electrodynamics of electronic systems, the nonrelativistic viewpoint is taken in that the external fields are those which couple to the particle and spin densities only:

$$\begin{aligned} H_{\text{ext}} = e\int d^3r \quad &:\bar{\psi}(x)\gamma_0\psi(x): A^0_{\text{ext}}(x) \\ &- :\bar{\psi}(x)\sigma_{\mu\nu}\psi(x): F^{\mu\nu}_{\text{ext}}(x) \end{aligned}$$

The field $F^{\mu\nu}_{\text{ext}}$ is a purely fictitious field which couples to the spin only and serves to lift the spin degeneracy.

In principle, specification of A^{μ}_{ext} fixes $F^{\mu\nu}_{\text{ext}}$. But we would like to deal with them as given classical objects. As examples one may cite the $A_{\mu}^{(\text{ext})}$ arising from the nuclei leading to the hyperfine interaction, the Coulomb interaction, and so on (Harriman[78]). In this case the nondegenerate ground state energy is a functional of the expectation values of the four-current density and the magnetization density, and the correct densities minimize the ground state energy for a given external potential. The result of Kohn and Sham[12] can also be generalized and an effective one-particle Dirac equation for an inhomogeneous electron system may be written as follows (for details see Rajagopal[14]):

$$\left[-i\boldsymbol{\alpha}\cdot\nabla - m(1-\beta) + V_{\text{eff}}(\mathbf{r}, J_{\mu}, \mathbf{m}) + \boldsymbol{\Sigma}\cdot\mathbf{W}_{\text{eff}}(\mathbf{r}, J_{\mu}, \mathbf{m})\right]\phi_i(\mathbf{r}) = \varepsilon_i\phi_i(\mathbf{r}) \tag{4.1}$$

where

$$V_{\text{eff}} = -\left(V_{\text{ext}}(\mathbf{r}) + e^2 \int \frac{n(\mathbf{r}')}{|\mathbf{r}-\mathbf{r}'|} d^3\mathbf{r}' + \frac{\delta E_{xc}}{\delta n(\mathbf{r})}\right)$$

$$- e\boldsymbol{\alpha}\cdot\left(\mathbf{A}_{\text{ext}} + \int \frac{\mathbf{J}(\mathbf{r}')\, d^3\mathbf{r}'}{|\mathbf{r}-\mathbf{r}'|} + \frac{\delta E_{xc}}{\delta \mathbf{J}(\mathbf{r})}\right) \tag{4.2}$$

$$W_{\text{eff}} = -e\left(\mathbf{B}_{\text{ext}} + \frac{\delta E_{xc}}{\delta \mathbf{m}(\mathbf{r})}\right) \tag{4.3}$$

where $\boldsymbol{\alpha}$, β and $\boldsymbol{\Sigma}$ matrices are related to the Dirac matrices by $\boldsymbol{\gamma} = \beta\boldsymbol{\alpha}$, $\gamma_0 = \beta$, and $\Sigma^k \equiv \sigma^{ij} = \frac{1}{2}(\gamma^i\gamma^j - \gamma^j\gamma^i)$, with i, j, and k cyclic; and in the ground state the particle density

$$n(\mathbf{r}) = \sum_{i(\text{occ})} \text{tr}(\phi_i^*(\mathbf{r})\phi_i(\mathbf{r})) \tag{4.4a}$$

the current density

$$J_k(\mathbf{r}) = \sum_{i(\text{occ})} \text{tr}(\phi_i^*(\mathbf{r})\alpha_k\phi_i(\mathbf{r})) \tag{4.4b}$$

and the magnetization density

$$m_k(\mathbf{r}) = \sum_{i(\text{occ})} \text{tr}(\phi_i^*(\mathbf{r})\Sigma_k\phi_i(\mathbf{r})) \tag{4.4c}$$

and E_{xc} is the exchange-correlation contribution to the ground state energy of the system. It is a functional, in general, of J_μ and $\mathbf{m}$. As a convenient scheme, one often evaluates E_{xc} for an interacting homogeneous electron gas, which then gives rise to a "local" effective potential. In this scheme E_{xc} is a functional of n and $|\mathbf{m}|$ only since the currents $J_k = 0$ in the ground state.

The Green function for this N-electron system is defined in the usual way

$$S_F(x, x') = -i\langle T(\psi(x)\bar{\psi}(x'))\rangle \tag{4.5}$$

where T is the usual time ordering symbol, and $\langle \ldots \rangle$ is the ground state expectation value. We now construct this Green function in momentum space for the noninteracting system, allowing explicitly for the appearance of spin.

In the formulation of the theory for a relativistic spin-polarized gas we are faced with a difficulty not found in the nonrelativistic theory in that the spin of the particle is not decoupled from its kinetic motion.

To resolve this difficulty we assume that each electron in its rest frame has its spin oriented along a fixed unit vector $\hat{n}$. In a frame at rest relative to a uniform positive background this electron has a momentum $\mathbf{p}$ and energy $p_0 = \sqrt{\mathbf{p}^2 + m^2} \equiv E_p$ in units of $\hbar = c = 1$ which we use throughout. The spin vector $(0, \hat{n})$ is then boosted to this frame (Björken and Drell[3]) such that

$$s_\mu s^\mu = -1 \quad \text{and} \quad s_\mu p^\mu = 0 \tag{4.6}$$

which gives

$$s_0 = \frac{\hat{n}\cdot\mathbf{p}}{m}, \qquad \mathbf{s} = \hat{n} + \left(\frac{E_p}{m} - 1\right)\frac{(\mathbf{p}\cdot\hat{n})\mathbf{p}}{|\mathbf{p}|^2} \tag{4.7}$$

Using (4.5) and the standard plane wave expansions of Björken and Drell[3] for $\psi(x)$,

$$\psi(x) = \sum_{\pm s} \int \frac{d^3\mathbf{p}}{(2\pi)^3} \sqrt{\frac{m}{E_p}} \left[b(p,s)u(p,s)e^{-ip\cdot x} + d^+(p,s)v(p,s)e^{ip\cdot x}\right]$$

and that

$$\langle b^+(p's')b(ps)\rangle = (2\pi)^3 \delta^3(\mathbf{p}-\mathbf{p}')\delta_{s,s'} n_{Fs}(p)$$

for electrons and similarly for positrons, we then obtain the Feynman propagator in momentum space

$$S_{F\alpha\beta}(p) = \sum_{\pm s} \left\{\frac{(\not{p}_+ + m)(1+\gamma_5 \not{s}_+)}{4E_p}\right\}_{\alpha\beta} \left\{\frac{n_{Fs}(|\mathbf{p}|)}{p_0 - E_p - i\eta} + \frac{1 - n_{Fs}(|\mathbf{p}|)}{p_0 - E_p + i\eta}\right\}$$
$$- \sum_{\pm s} \left\{\frac{(\not{p}_- + m)(1+\gamma_5 \not{s}_-)}{4E_p}\right\}_{\alpha\beta} \left\{\frac{1 - \bar{n}_{Fs}(|\mathbf{p}|)}{p_0 + E_p - i\eta} \frac{\bar{n}_{Fs}(|\mathbf{p}|)}{p_0 + E_p + i\eta}\right\} \tag{4.8}$$

where

$$p_+^\mu = (E_p, \mathbf{p}), \quad s_+^\mu = (s_0, \mathbf{s}),$$
$$p_-^\mu = (-E_p, \mathbf{p}), \quad s_-^\mu = (-s_0, \mathbf{s}), \quad \not{s} = s^\mu \gamma_\mu, \quad \not{p} = p^\mu \gamma_\mu \tag{4.9}$$

and $n_F(|\mathbf{p}|)$ is the Fermi function for the positive energy electrons and

$\bar{n}_F(|\mathbf{p}|)$ for the positrons. When $k_F = 0$, that is, zero density, Eq. (4.8) reduces to the familiar Feynman propagator. In our zero temperature theory the negative energy states are assumed filled ($\bar{n}_F = 0$), and there are no electrons above the Fermi level, $n_F(|\mathbf{p}|) = \theta(k_F - |\mathbf{p}|)$, where k_F is the Fermi momentum of the positive energy electrons of spin s. The density-independent term in the second sum of Eq. (4.8) merely contributes to renormalizations, and we may safely drop it and use the physical mass and charge (Bowers et al.[66] and Jancovici[60]). By this procedure we take care of the quantum electrodynamical renormalizations only. We still have the familiar density renormalizations of the ordinary many-body theory. The above is a generalization of (2.43) with spin polarization incorporated. In the nonrelativistic limit, $m \gg k_F$, it is seen that the propagator reduces to

$$S_F(p) \simeq \sum_{\pm \hat{n}} \left(\frac{1 + \boldsymbol{\sigma} \cdot \hat{n}}{2} \right) \left\{ \frac{n_{F\hat{n}}(|\mathbf{p}|)}{p_0 - E_p - i\eta} + \frac{1 - n_{F\hat{n}}(|\mathbf{p}|)}{p_0 - E_p + i\eta} \right\} \tag{4.10}$$

which is the usual nonrelativistic limit of Rajagopal et al.[79] Here $\boldsymbol{\sigma}$ is the Pauli (2×2) spinor.

1. The number of particles with momentum $\mathbf{p}$ is

$$N_{\mathbf{p}} = \int \frac{dp_0}{2\pi i} \operatorname{Tr}(S_F^<(p)\gamma_0) \tag{4.11}$$

where Tr stands for the trace over the γ matrices. Using Eq. (4.8) we obtain

$$N_{\mathbf{p}} = n_{F\uparrow}(|\mathbf{p}|) + n_{F\downarrow}(|\mathbf{p}|) \tag{4.12}$$

where $\uparrow$ and $\downarrow$ now refer to $\hat{n}$ and $-\hat{n}$. The total number of particles is then (Ω is the volume of the system)

$$\int \frac{d^3p}{(2\pi)^3} N_{\mathbf{p}} = \frac{N}{\Omega} = n \tag{4.13}$$

2. The spin magnetization of the system is given by

$$\mathbf{M}_{\mathbf{p}} = \int \frac{dp_0}{2\pi i} \operatorname{Tr}(\gamma_0 \boldsymbol{\Sigma} S_F^<(p)) \tag{4.14}$$

Here $\mathbf{\Sigma}$ is the usual spin operator, the kth component of which is $(\gamma_i\gamma_j - \gamma_j\gamma_i)/2$, where i, j, and k are cyclic. Using Eq. (4.8) we obtain, upon performing the trace over the γ matrices,

$$\mathbf{M}_{\mathbf{p}} = \frac{m\mathbf{s}}{E_p}\left(n_{F\uparrow}(|\mathbf{p}|) - n_{F\downarrow}(|\mathbf{p}|)\right) \tag{4.15}$$

The total magnetization is then

$$\frac{\mathbf{m}}{\Omega} = \int \frac{d^3p}{(2\pi)^3}\mathbf{M}_{\mathbf{p}} = \int \frac{d^3p}{(2\pi)^3}\frac{m\mathbf{s}}{E_p}\left(n_{F\uparrow}(|\mathbf{p}|) - n_{F\downarrow}(|\mathbf{p}|)\right) \tag{4.16}$$

This has an interesting consequence: $\mathbf{m}$ can be shown to be in the direction of $\hat{n}$. In the nonrelativistic limit, $m\mathbf{s}/E_p \simeq \mathbf{n}$ and we obtain the well-known result (see Rajagopal et al.[79]) that $m_{\mathrm{NR}} = (\hat{n}/\Omega)(N_\uparrow - N_\downarrow)$, where $N_\uparrow$ and $N_\downarrow$ are the total number of electrons aligned parallel and antiparallel to $\hat{n}$. One often defines, in this limit, a parameter ζ called the relative magnetizâ-tion:

$$\zeta = \frac{|m_{\mathrm{NR}}|}{N} = \frac{n_\uparrow - n_\downarrow}{n_\uparrow + n_\downarrow} \tag{4.17}$$

which varies between 0 and 1. For noninteracting electrons, we can perform the integrations in Eq. (4.16) upon introducing ζ now as a parameter (Fig. 8). Defining

$$\xi = \frac{|m|}{n}$$

we obtain

$$\xi = \frac{1}{2\beta^3}\left\{\left[\frac{1}{3}\beta^3x^3 + \beta x(1+\beta^2x^2)^{1/2} - \sinh^{-1}(\beta x)\right] - [x \to y]\right\} \tag{4.18}$$

where $x = (1+\zeta)^{1/3}$, $y = (1-\zeta)^{1/3}$ and $\beta = k_F/m(\equiv \hbar k_F/mc)$. Thus $\xi_{\mathrm{NR}} \to \zeta$. In the ultrarelativistic limit ($\beta \gg 1$), one obtains

$$\xi_{\mathrm{UR}} \to \tfrac{1}{3}\zeta \tag{4.19}$$

The factor 1/3 is easily understood, if we go back to Eq. (4.16) and note that helicity is a good quantum number and that $p(p \cdot \hat{n})$ integrated over the Fermi sphere leads to the average of $\cos^2\theta$ over a unit sphere, giving the factor 1/3. This is an interesting feature of the relativistic theory.

Equipped with the propagator, we now compute the ground state energy of the many-electron system whose Hamiltonian includes, besides the static Coulomb interaction, the transverse photon contribution, in the Hartree–Fock approximation. We are led to the expression

$$\frac{\langle H \rangle_{\mathrm{HF}}}{\Omega} = T + E^x_{(c)} + E^x_{(\mathrm{tr})} \tag{4.20}$$

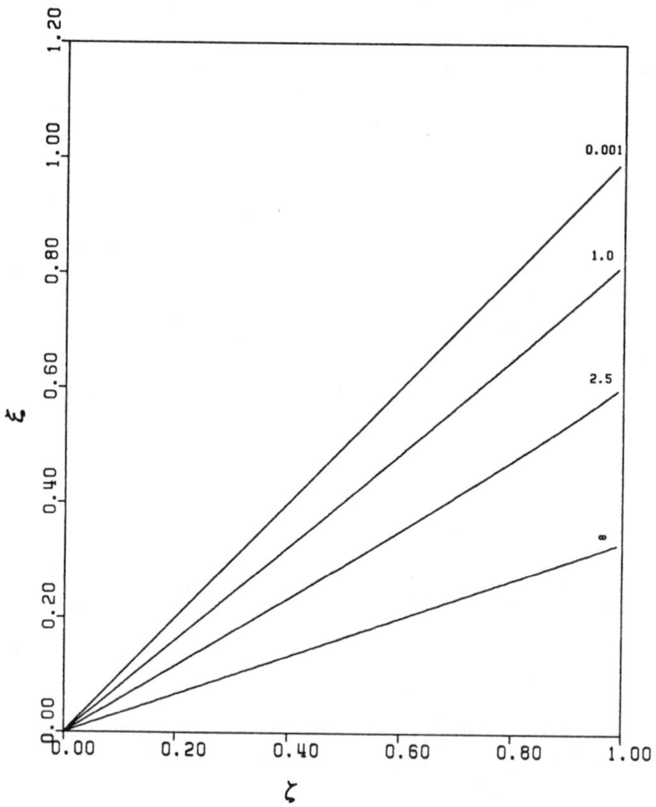

Fig. 8. Plot of the magnetization of the polarized relativistic gas versus the parameter ζ, for $\beta = 0.001$, 2.5, and ∞.

where T is the average kinetic energy, $E^x_{(c)}$ is the exchange energy arising from the bare Coulomb interaction between electrons, and $E^x_{(\mathrm{tr})}$ is the contribution to the exchange energy from the transverse photon–electron interactions. The actual expressions involved contain traces over γ matrices and integrals over momenta. We can evaluate T explicitly and it is given by

$$T=\frac{3}{\beta^5}nE_F\left\{\frac{\beta x}{8}(2x^2\beta^2+1)(\beta^2x^2+1)^{1/2}\right.$$
$$\left.-\frac{1}{3}\beta^3x^3-\frac{1}{8}\sinh^{-1}(\beta x)+(x\to y)\right\} \tag{4.21}$$

where $E_F=k_F^2/2m$ and $n=k_F^3/3\pi^2$.

B. Exchange Energy in a Homogeneous Spin-Polarized Electron Gas*

The contribution to the Coulomb part of the exchange energy is

$$E_x^{(c)}=-2\pi e^2\iint\frac{d^3p\,d^3p'}{(2\pi)^6}\frac{1}{|p-p'|^2}\iint\frac{dp_0\,dp_0'}{(2\pi)^2}$$
$$\cdot\mathrm{Tr}\left[\gamma_0S_F^<(p)\gamma_0S_F^<(p')\right] \tag{4.22}$$
$$=-2\pi e^2\sum_{\hat{n},\hat{n}'}\iint\frac{d^3p\,d^3p'}{(2\pi)^6}\frac{1}{|p-p'|^2}\frac{1}{16E_pE_{p'}}n_{F\hat{n}}(|\mathbf{p}|)n_{F\hat{n}'}(|\mathbf{p}'|)$$
$$\cdot\mathrm{Tr}\left[\gamma_0(\not{p}+m)(1+\gamma_5\not{s})\gamma_0(\not{p}'+m)(1+\gamma_5\not{s}')\right]$$

The trace may be carried out to give

$$E_x^{(c)}=\frac{-e^2}{(2\pi)^3}k_F^4\sum_{\hat{n},\hat{n}'}\iint\frac{d^3x\,d^3x'}{(4\pi)^2}\frac{1}{|\mathbf{x}-\mathbf{x}'|^2}\frac{1}{E_xE_{x'}}$$
$$\cdot\{(1+E_xE_{x'}+\mathbf{x}\cdot\mathbf{x}')(1+s_0s_0'+\mathbf{s}\cdot\mathbf{s}')-s_0s_0'E_xE_{x'}$$
$$-s_0E_{x'}(\mathbf{s}'\mathbf{x})-s_0'E_x(\mathbf{s}\cdot\mathbf{x}')-(\mathbf{x}\cdot\mathbf{s}')(\mathbf{s}\cdot\mathbf{x}')\} \tag{4.23}$$

Similarly for the transverse term we have

$$E_x^{(\mathrm{tr})}=-2\pi e^2\iint\frac{d^3p\,d^3p'}{(2\pi)^6}\iint\frac{dp_0\,dp_0'}{(2\pi)^2}\mathfrak{D}_T^{ij}(p-p')$$
$$\cdot\mathrm{tr}\left[\gamma_iS_F^<(p)\gamma_jS_F^<(p')\right] \tag{4.24}$$

*See Addendum to this chapter.

where $\mathfrak{D}_T^{ij}(p-p')$ is the transverse photon propagator given by Björken and Drell[3] and p, p' are energy momentum four vectors

$$E_x^{(\text{tr})} = -2\pi e^2 \sum_{\hat{n},\hat{n}'} \iint \frac{d^3p\, d^3p'}{(2\pi)^6} \frac{\left[\delta_{ij} - q_i q_j/q^2\right]}{\left[(E_p - E_{p'})^2 - (\mathbf{p}-\mathbf{p}')^2\right]} \frac{1}{16E_p E_{p'}}$$

$$\cdot n_{F\hat{n}}(|\mathbf{p}|)\, n_{F\hat{n}'}(|\mathbf{p}'|)\, \mathrm{Tr}\left[\gamma_i(\not{p}+m)(1+\gamma_5\not{s})\gamma_j(\not{p}'+m)(1+\gamma_5\not{s}')\right] \tag{4.25}$$

The traces over the γ matrices were evaluated and the resulting expressions are given in Appendix D. A look at the various terms in Eqs. (D.1) and (D.2) shows that their structure is similar to various spin–spin, spin–own orbit, spin–other orbit, and other interactions which appear in two-electron theory (Harriman[78]).

In the nonrelativistic limit

$$E_{x\text{NR}}^{(c)} \simeq -\frac{e^2}{(2\pi)^3} k_F^4 \left[x^{4/3} + y^{4/3} - \frac{1}{9}\beta^2(x^6 + y^6)\right] \tag{4.26}$$

and

$$E_{x\text{NR}}^{(\text{tr})} \simeq \frac{e^2}{(2\pi)^3} k_F^4 \left(\frac{2}{3}\beta^2\right) \left\{ \frac{5}{18}(x^6 + y^6) + \frac{1}{9}x^3y^3 + \frac{1}{2}(xy^5 + x^5y) + \frac{1}{4}(x^2y^4 + x^4y^2 - x^6 - y^6)\ln\left|\frac{x+y}{x-y}\right| \right\} \tag{4.27}$$

The transverse term is of order β^2 in the low β limit as it should be. Also, when we set $\alpha = 0$ we recover the old nonmagnetic results.

We have plotted in Fig. 9 the total energy per particle, which includes the kinetic and Coulomb and transverse exchange as a function of the magnetization parameter ζ for various values of β. The energy per particle is scaled to the relativistic Fermi energy $(k_F^2 + m^2)^{1/2} - m$. At the very low density of $\beta = 0.001$ we see that the ground state is fully polarized in this approximation, which is the nonrelativistic result.

At higher densities the kinetic energy is the dominant term and the ground state is a state of zero polarization in the absence of a magnetic field. In the

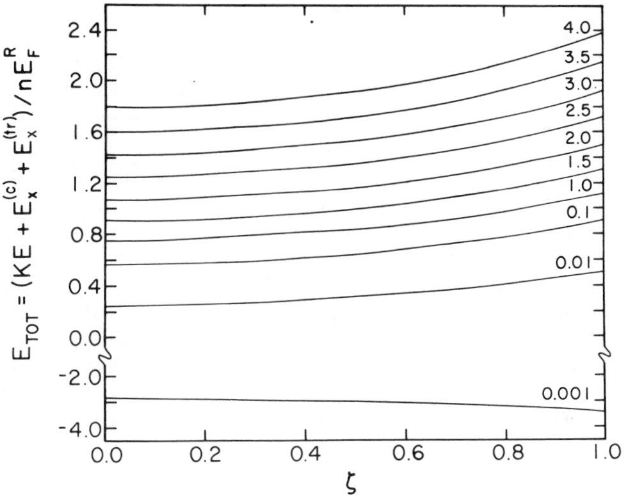

Fig. 9. Curves for the total energy/particle scaled to the relativistic Fermi energy versus the polarization parameter ζ for various values of $\beta(=k_F/m = 1/71.4r_s)$. (From Ref. 76.)

unpolarized case the transverse and Coulomb parts of the exchange cancel each other at $\beta = 2.533$. We reproduce this cancellation with our numerical integration. This cancellation persists for nonzero values of ζ at larger densities. Figure 10 is a plot of this critical density and polarization parameter. The exchange is positive above this curve and negative below it. In Table VI we present a list of values for the total exchange energy as a function of density and the magnetization parameter ζ.

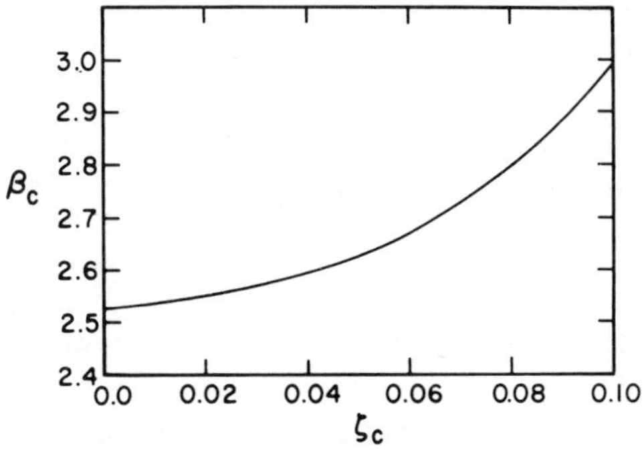

Fig. 10. Plot of β_c versus ζ_c such that $E_x^{(c)}(\beta_c, \zeta_c) + E_x^{(\text{tr})}(\beta_c, \zeta_c) = 0$. The total exchange energy is positive above this curve and negative below it. (From Ref. 76.)

TABLE VI
Representative Values of the Total Exchange Energy/Particle $E_x(\beta,\zeta) = E_x^{(c)}(\beta,\zeta) + E_x^{(\mathrm{tr})}(\beta,\zeta)$ Scaled to the Relativistic Kinetic Energy $(k_F^2 + m^2)^{1/2} - m$ at the Fermi Level[a]

	ζ					
β	0.0	0.2	0.4	0.6	0.8	1.0
0.01	−0.348	−0.351	−0.361	−0.377	−0.402	−0.439
0.1	−0.0347	−0.0350	−0.0360	−0.03776	−0.0401	−0.0438
0.5	−0.00631	−0.00642	−0.00672	−0.00716	−0.00775	−0.00856
1.0	−0.00241	−0.00259	−0.00298	−0.00350	−0.00410	−0.00475
1.5	−0.00106	−0.00110	−0.00191	−0.00263	−0.00342	−0.00423
2.0	−0.000392	−0.000790	−0.00157	−0.00249	−0.00346	−0.00440
2.533	0.0	−0.000522	−0.00150	−0.00259	−0.00365	−0.00461
3.0	+0.000215	−0.000382	−0.00162	−0.00278	−0.00362	−0.00412

[a]From Ref. 76.

C. Effective Exchange Potentials of the Relativistic Spin-Density-Functional Theory*

In a local spin-density approximation, the potentials V_{xc} and W_{xc} may be computed from our expressions for E_{xc} in the usual way:

$$V_{xc} = \frac{\delta E_{xc}}{\delta n} = \frac{\delta E_{xc}^{(c)}}{\delta n} + \frac{\delta E_{xc}^{(\mathrm{tr})}}{\delta n} \equiv V_{xc}^{(c)} + V_{xc}^{(\mathrm{tr})} \tag{4.28}$$

$$W_{xc} = \frac{\delta E_{xc}}{\delta |\mathbf{m}|} = W_{xc}^{(c)} + W_{xc}^{(\mathrm{tr})} \tag{4.29}$$

However, since E_{xc} is most simply expressed in terms of $k_{F\uparrow}$ and $k_{F\downarrow}$, which are the Fermi momenta of "up" and "down" spin electrons related to the particle and magnetization densities, we have

$$\frac{\delta E_{xc}}{\delta n} = \frac{\delta E_{xc}}{\delta k_{F\uparrow}}\left(\frac{\partial k_{F\uparrow}}{\partial n}\right)_{k_{F\downarrow}} + \frac{\delta E_{xc}}{\delta k_{F\downarrow}}\left(\frac{\partial k_{F\downarrow}}{\partial n}\right)_{k_{F\uparrow}} \tag{4.30}$$

and

$$\frac{\delta E_{xc}}{\delta |\mathbf{m}|} = \frac{\delta E_{xc}}{\delta k_{F\uparrow}}\left(\frac{\partial k_{F\uparrow}}{\partial |\mathbf{m}|}\right)_{k_{F\downarrow}} + \frac{\delta E_{xc}}{\delta k_{F_\downarrow}}\left(\frac{\partial k_{F_\downarrow}}{\partial |\mathbf{m}|}\right)_{k_{F\uparrow}} \tag{4.31}$$

so that we may use the expressions for n and $|\mathbf{m}|$ to find the potentials. Since the magnitude of the magnetization is no longer simply $(k_{F\uparrow}^3 - k_{F\downarrow}^3)/6\pi^2$

*See Addendum to this chapter.

as in nonrelativistic theory, we must use

$$\frac{\partial k_{F\uparrow,\downarrow}}{\partial n} = \frac{\pi^2}{k^2_{F\uparrow,\downarrow}} \frac{E_{x\uparrow,\downarrow}(2+E_{x\downarrow,\uparrow})}{(E_{x\uparrow}+E_{x\downarrow}+E_{x\uparrow}E_{x\downarrow})} \tag{4.32a}$$

$$\frac{\partial k_{F\uparrow,\downarrow}}{\partial|\mathbf{m}|} = \frac{(+,-)3\pi^2}{k^2_{F\uparrow,\downarrow}} \frac{E_{x\uparrow}E_{x\downarrow}}{(E_{x\uparrow}+E_{x\downarrow}+E_{x\uparrow}E_{x\downarrow})} \tag{4.32b}$$

In the following we present results based on exchange contributions to E_{xc}, which we denote by E_x. Since the Fermi momentum for "up" and "down" spin electrons $k_{F\uparrow}$ and $k_{F\downarrow}$ appear as upper limits in the integrals for E_x and not in the integrands, the required differentiations may be done simply and the double integrals reduce to a single one which is then evaluated numerically.

In Fig. 11 we present a plot of V_x as a function of β and ζ. It may be seen that, over the range of β, ζ values, cancellation is slightly larger than that of

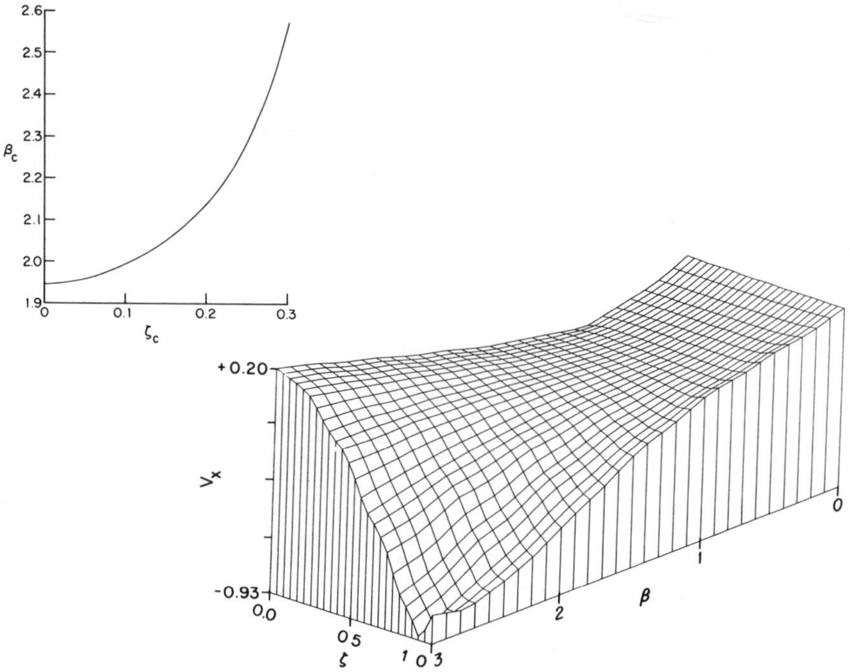

Fig. 11. Plot of the relativistic exchange potential V_x as a function of density ($\beta = k_F/m$) and the spin polarization parameter ζ. The inset is a plot of critical values of β_c versus ζ_c for which transverse exchange potential cancels the Coulomb exchange potential. The value of V_x is negative below and positive above the curve shown. (From Ref. 76.)

the exchange energy. This is displayed as an inset in Fig. 11. Over most of the range of β, ζ values investigated, the exchange potential is negative. Our numerical calculation reproduces known results in the following cases:

(i) For $\zeta = 0$, we obtain the result of Rajagopal.[52]

(ii) For $\beta \ll 1$, $$V^{c}_{x(\mathrm{NR})} = -\frac{\beta}{2\pi}[x_\uparrow + x_\downarrow + O(\beta^2)]. \tag{4.33}$$

and

$$V^{(\mathrm{tr})}_{x\mathrm{NR}} = \frac{\beta}{4\pi}\beta^2\left[\frac{2}{3}\left(x_\uparrow^3 + x_\downarrow^3\right) + x_\uparrow x_\downarrow (x_\uparrow + x_\downarrow) + \frac{1}{6x_\uparrow x_\downarrow}(x_\uparrow - x_\downarrow)^4 (x_\uparrow + x_\downarrow)\ln\left|\frac{x_\uparrow + x_\downarrow}{x_\uparrow - x_\downarrow}\right|\right] \tag{4.34}$$

where $x_\uparrow = k_{F\uparrow}/k_F = (1+\zeta)^{1/3}$ and $x_\downarrow = k_{F\downarrow}/k_F = (1-\zeta)^{1/3}$. In Table VII, the values of V_x for a representative set of (β, ζ) are given. In Fig. 12 we present W_x as a function of β and ζ.

From the plot it is seen that W_x is nowhere positive. Its magnitude remains smaller than V_x for $\beta < 1$ as known from nonrelativistic expressions. However, for values of $\beta > 1$ the magnitude of W_x rapidly becomes much larger than V_x. For heavy elements such as actinides this feature of W_x may play an important role in determining their magnetic properties. We again

TABLE VII
Representative Values of the Total Exchange Potential $V_x = V_x^{(c)} + V_x^{(\mathrm{tr})}$ [a]

β	ζ 0.0	0.2	0.4	0.6	0.8	1.0
0.01	−0.0032	−0.0032	−0.0031	−0.0030	−0.0029	−0.0020
0.1	−0.0315	−0.0314	−0.0309	−0.0301	−0.0284	−0.0199
0.5	−0.1259	−0.1258	−0.1252	−0.1236	−0.1198	−0.0873
1.0	−0.1385	−0.1421	−0.1513	−0.1653	−0.1827	−0.1690
1.5	−0.0777	−0.0917	−0.1266	−0.1801	−0.2529	−0.3186
2.0	0.0101	−0.0209	−0.0970	−0.2113	−0.3753	−0.5331
2.5	0.1058	0.0513	−0.0795	−0..2710	−0.5178	−0.7255
3.0	0.2029	0.1186	−0.0794	−0.3602	−0.6966	−0.7538

[a] From Ref. 76.

have the following limiting cases:

(i) For $\zeta = 0, W_x = 0$ as it must.

(ii) For $\beta \ll 1, W^c_{x(\mathrm{NR})} = -\frac{\beta}{2\pi}[x_\uparrow - x_\downarrow + O(\beta^2)]$. (4.35)

$$W^{(\mathrm{tr})}_{x\mathrm{NR}} = \frac{\beta}{4\pi}\beta^2\left[\frac{4}{9}\left(x^3_\uparrow - x^3_\downarrow\right) + x_\uparrow x_\downarrow (x_\uparrow - x_\downarrow)\right.$$

$$\left. - \frac{1}{6x_\uparrow x_\downarrow}(x_\uparrow - x_\downarrow)(x_\uparrow + x_\downarrow)^4 \ln\left|\frac{x_\uparrow + x_\downarrow}{x_\uparrow - x_\downarrow}\right|\right] \tag{4.36}$$

In Table VIII, the values of W_x for the same set of (β, ζ) values as those in Table VII are given.

For $\zeta = 1$ we note that $|V_x| \neq |W_x|$ but

$$|V^{(c)}_x| = |W^{(c)}_x| \tag{4.37a}$$

and

$$|V^{(\mathrm{tr})}_x| \neq |W^{(\mathrm{tr})}_x| \qquad \text{for all} \quad \beta \tag{4.37b}$$

and, in particular,

$$|V^{(\mathrm{tr})}_x(\beta \ll 1)| \neq |W^{(\mathrm{tr})}_x(\beta \ll 1)| \tag{4.37c}$$

This is unlike the nonrelativistic case where the potentials felt by the charge and magnetization densities are the same for the fully ferromagnetic case. Here the potentials seen by the two densities are highly asymmetric. We trace this to the fact, as remarked earlier, that in the ultrarelativistic limit fully saturated ferromagnetism is not possible. It should also be noted that if the spin unit vector $\hat{n}$ is parallel or antiparallel to the direction of momentum, that is, eigenstates of helicity, then the net magnetization of the noninteracting gas is zero from isotropy.

From Figs. 11 and 12 it may be noted that for $\beta > 2$, V_x and W_x display a valley for ζ values close to unity. We believe that this feature is a manifestation of the same relativistic effect on the magnetized electron gas.

The behavior of the potential V_x is quite different from its nonrelativistic counterpart as given by Eq. (4.33). At low values of ζ, V_x changes sign while $V^{(\mathrm{NR})}_x$ is monotonic and of one sign. For $\zeta \simeq 1$, $V_x \sim V^{(\mathrm{NR})}_x$ and they are of the same sign. The departure of W_x is more drastic from its nonrelativistic value given by Eq. (4.35). For $\beta > 1$ and $\zeta > 0$, the magnitude of W_x is much larger than that given by its nonrelativistic expression (4.35). As noted earlier,

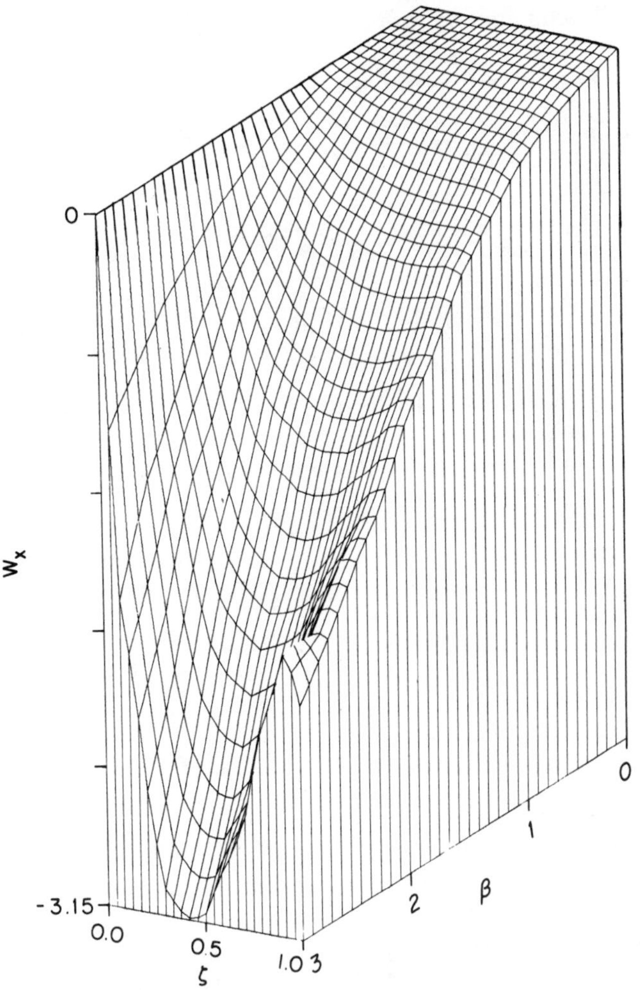

Fig. 12. Plot of the relativistic exchange potential W_x as a function of ζ and β; W_x is nowhere positive. (From Ref. 76.)

W_x is much larger than V_x for $\beta > 1$. The application of these potentials in a high Z solid state system should therefore be of considerable interest as we expect these potentials may help explain the magnetic properties of some actinides and some heavy rare earth elements.

It is of interest to note the following concerning correlation effects in the relativistic context. The correlation contribution does not seem to be as important in the relativistic case as in the nonrelativistic case. While no calculation of correlation effects exist for the polarized electron gas, one may get

TABLE VIII
Representative Values of the Total Exchange Potential $W_x = W_x^{(c)} + W_x^{(\mathrm{tr})}$ [a]

β	ζ 0.2	0.4	0.6	0.8	1.0
0.01	−0.0002	−0.0004	−0.0007	−0.0010	−0.0020
0.1	−0.0022	−0.0044	−0.0069	−0.0101	−0.0200
0.5	−0.0185	−0.03206	−0.0446	−0.0584	−0.1030
1.0	−0.1058	−0.1596	−0.1936	−0.2146	−0.2721
1.5	−0.3599	−0.5132	−0.5840	−0.5944	−0.5993
2.0	−0.8553	−1.1616	−1.2537	−1.1969	−1.0819
2.5	−1.6270	−2.0887	−2.1131	−1.8622	−1.6318
3.0	−2.6736	−3.2029	−2.9569	−2.2852	−2.1048

[a] From Ref. 76.

some idea of it from its behavior in the case of unpolarized gas (Ramana and Rajagopal[71]). In the nonrelativistic case in range $\beta \sim 0.002$ to 0.014 (corresponding to $r_s \sim 6-1$) the correlation potential drops from 44% of exchange potential to 14% and proceeding up in density at $\beta = 0.5$ where it has dropped to 1.3%. However, for larger β the transverse contribution being positive begins to dominate the exchange potential, and the correlation contribution is the major one to the total potentials between $\beta = 1.75$ and 2.0, where the exchange potential goes through zero and drops again to only 8% of the exchange at $\beta = 2.25$. Thus the contribution of correlation to the total exchange-correlation potential is small at relativistic densities except in a narrow region where the Coulomb and transverse parts of the exchange cancel each other. The behavior of the energy is similar. We expect the correlation contribution to be small in the polarized case as well. It is therefore reasonable to use the exchange-only scheme in the relativistic case as a reasonable approximation.

In order to derive the effects of spin-orbit interaction properly in a magnetic system such as Fe, Ni, and Co, in principle one must begin with a relativistic theory such as the one given here, say, by Eq. (4.1), and take its nonrelativistic limit. This procedure leads to additional contributions to the spin-orbit coupling. We give here this equation, without derivation:

$$\left[\frac{p^2}{2m} + V + \boldsymbol{\sigma}\cdot\mathbf{W} - \frac{p^4}{8m^3} - \frac{p^2V}{8m^2} + \frac{i\boldsymbol{\sigma}\cdot[(\mathbf{p}V)\times\mathbf{p}]}{4m^2} - \frac{\boldsymbol{\sigma}\cdot\hat{n}(p^2W)}{8m^2} \right.$$
$$+ \frac{i(\boldsymbol{\sigma}\cdot\hat{n})(\nabla W\cdot\mathbf{p})}{4m^2} - \frac{i(\boldsymbol{\sigma}\cdot\nabla W)(\hat{n}\cdot\mathbf{p})}{4m^2} - \frac{\hat{n}\cdot(\nabla W\times\mathbf{p})}{4m^2} + \frac{(\mathbf{W}\cdot\mathbf{p})(\boldsymbol{\sigma}\cdot\mathbf{p})}{2m^2}$$
$$\left. - \frac{(\boldsymbol{\sigma}\cdot\mathbf{W})p^2}{2m^2} + \frac{i(\boldsymbol{\sigma}\cdot\hat{n})(\nabla W\cdot\mathbf{p})}{4m^2} + \frac{(\mathbf{p}\cdot\mathbf{W})(\boldsymbol{\sigma}\cdot\mathbf{p})}{4m^2}\right]\Psi = E^{(\mathrm{NR})}\Psi$$

where $\hat{n}$ is a unit vector in the direction of magnetization, $\boldsymbol{\sigma}$ is the usual Pauli matrix, and $\mathbf{W} = \hat{n}W$. The first, second, fourth, fifth, and sixth terms are the usual nonrelativistic terms with spin-orbit coupling. The third term is the analog of the second, the seventh of the fifth, and the ninth of the sixth. The others are additional terms due to spin-polarization. It appears from this and the observations in Section IV that the additional contribution besides those arising from V_x are not small in comparison to it, even when one employs the nonrelativistic expressions for V_{xc} and W_{xc} of the von Barth–Hedin form, for example. In the case of the unpolarized gas we obtain the result of MacDonald.[61]

V. SUMMARY, CONCLUSIONS, AND REMARKS

The need for relativistic calculations were felt very early, and such calculations have been carried out with increasing sophistication since the 1940s. The recent band structure calculations of MacDonald et al.[38] show that relativistic effects are important even for valence bands in platinum. The most complete and sophisticated of such calculations for atoms, including Lamb shift and vacuum polarization corrections of Huang et al.,[5] show good agreement with the experimental results of Deslattes et al.[28] However there is a systematic disagreement of a few eV. Further the Dirac–Fock scheme is complicated, for more complex systems, to be implemented in present-day computers. It is also not easy to take correlation into account in the Dirac–Fock scheme. Since the density-functional formalism, and especially the spin-density-functional version, has enjoyed great success in nonrelativistic theory, its extension to the relativistic case seems appropriate (Rajagopal[52] and MacDonald and Vosko[53]). Their relativistic exchange potential, including the transverse term, when applied to atoms (Das et al.[54]) leads to energies lying above the Dirac–Fock values, as they should, unlike the $\rho^{1/3}$ exchange potential used in the Dirac equation.

It seems possible the few eV discrepancy between theory and experiment may be due to correlation contributions. To this end we have calculated the correlation contribution in the relativistic electron gas in the spirit of von Barth and Hedin.[48] To our knowledge the calculation of this correlation potential in a local approximation in the relativistic density-functional theory is new. The relativistic results differ significantly from the nonrelativistic ones at high densities. The approximations of Jancovici,[60] while close to our results at high densities, do not go over to the proper nonrelativistic values.

Using the program of Liberman et al.[70] we calculated the correlation contribution in several heavy atomic systems. The relativistic contributions to correlation, though quite marked in the gas, are not significant in the atoms we investigated. However, the contribution of correlation to the K_α, x-ray energies serves to explain some of the discrepancy between theory and experiment. It should be of some interest to examine the effects of relativistic correlation on the knight shift of heavy systems.

We have also assessed the role of the Breit interaction vis-á-vis the full transverse interaction. This was done perturbatively within the Dirac–Fock scheme but not in the local-density one. The Breit interaction is no simpler to use in the local scheme than the full transverse interaction. Therefore, the transverse interaction can be used in SCF calculations though the Breit interaction is the major part of the transverse interaction. When examining the transverse contributions a comparison with the results of Mann and Johnson[26] showed that nonlocal effects may indeed be significant.

In the second part of the review we have generalized the spin-density-functional formalism of the nonrelativistic theory to the relativistic case. We have not seen such a treatment in the literature. We have calculated the exchange energy and the exchange potentials that appear in an effective one-particle Dirac equation. A calculation of the magnetization of the polarized relativistic electron gas shows that in the ultrarelativistic limit full polarization is not possible. We attribute this to the fact that spin alone is not a good quantum number, but only the projection of spin along the direction of motion. A calculation of the total energy reveals that the ground state of the relativistic electron gas is always unpolarized. This is to be expected since the kinetic energy of the relativistic gas is very large. The improvement of the spin-density-functional over the density-functional formalism in nonrelativistic theory encourages us to believe that this would be the case in relativistic theory too. An application of our potentials should reveal if this is the case. Such an application would be of special interest in the case of rare earth elements and actinides.

In our method for the polarized electron gas we have allowed a fictitious magnetic field to polarize the electrons in the rest frame provided by the uniform positive background and then boosted the electrons to their final velocities. The magnetic field was no longer kept. This is the simplest procedure that can be followed. Any calculation in the explicit presence of the magnetic field is prohibitively complicated. However, we believe that such a procedure would recover our results in the limit of vanishing field. The use of a fixed reference frame should also not pose any difficulties as such a frame is physically present in any atomic or solid state system.

A nonrelativistic limit of the effective one-particle Dirac equation in the relativistic local-density theory yields contributions in addition to the usual spin-orbit coupling arising from the Coulomb field. There is an additional contribution to the spin-orbit interaction due to the use of V_{eff} instead of the Coulomb potential $V(r)$ (MacDonald[61]). Further, there are new contributions which are analogs of the spin-orbit term and the like. And of the same order in v/c arising from the effective potential W (Ramana and

Rajagopal[76]). An application of this in some physical system should be useful to gauge its importance.

It may be worth pointing out that the formalism developed here may be used in the calculation of the energy levels of muonic atoms. The electron screening effects in heavy atoms are important to be taken into account, and one usually does so within an HF or HFS scheme or more sophisticatedly employs a relativistic HFS taking into account the interaction between the muon and the electrons when computing the electronic density. The use of our relativistic density-functional scheme ought to be more accurate than the HFS potential because we have incorporated the proper exchange-correlation contributions in our scheme. For a review of this exciting field of muonic atoms see Borie and Rinker.[80]

The finite temperature version of our relativistic density-functional theory may be of interest in the context of astrophysical and nuclear matter theories. The formalism for this extension exists, and such a calculation of finite temperature QED in a relativistic electron gas has recently been published by Bechler.[81]

We thus conclude that the methods and results obtained in the present review may have wide applications in several areas of physics.

APPENDIX A: EQUIVALENCE OF THE EXPRESSIONS FOR THE DIELECTRIC CONSTANT GIVEN BY VARIOUS AUTHORS

Starting from Eq. (2.48) we derive explicitly the expressions for the polarizabilities Q_{00}, Q_{ij}, Q_{L}, and Q_{T}. We then show the equivalence of the expressions for the dielectric constant given by Jancovici[60] and Tsytovich.[67]

We have for Q_{00}, from (2.48):

$$
\begin{aligned}
Q_{00}(q) = 4\pi e^2 \int \frac{d^3p}{(2\pi)^3} \frac{n_F(p)}{E_p E_{\mathbf{p}+\mathbf{q}}} \\
\cdot \frac{\left[\begin{array}{l}\left(E_{\mathbf{p}+\mathbf{q}}E_p + E_p^2 + \mathbf{p}\cdot\mathbf{q}\right)\left(\omega + E_{\mathbf{p}+\mathbf{q}} - E_p\right) \\ \quad -\left(E_p E_{\mathbf{p}+\mathbf{q}} + E_{\mathbf{p}+\mathbf{q}}^2 - (\mathbf{p}+\mathbf{q})\cdot\mathbf{q}\right)\left(\omega + E_p - E_{\mathbf{p}+\mathbf{q}}\right)\end{array}\right]}{\omega^2 - \left(E_{\mathbf{p}+\mathbf{q}} - E_p\right)^2} \\
+4\pi e^2 \int \frac{d^3p}{(2\pi)^3} \frac{(1 - n_F(p))}{E_p E_{\mathbf{p}+\mathbf{q}}} \\
\cdot \frac{\left[\begin{array}{l}\left(E_p^2 - E_p E_{\mathbf{p}+\mathbf{q}} + \mathbf{p}\cdot\mathbf{q}\right)\left(\omega - E_p - E_{\mathbf{p}+\mathbf{q}}\right) \\ \quad -\left(E_{\mathbf{p}+\mathbf{q}}^2 - E_p E_{\mathbf{p}+\mathbf{q}} - (\mathbf{p}+\mathbf{q})\cdot\mathbf{q}\right)\left(\omega + E_p + E_{\mathbf{p}+\mathbf{q}}\right)\end{array}\right]}{\omega^2 - \left(E_{\mathbf{p}+\mathbf{q}} + E_p\right)^2}
\end{aligned}
\tag{A.1}
$$

In deriving this expression from (2.48) we have used the substitution $-p-q \to -p$ in terms occurring with $n_F(p+q)$ and then combined them. The numerators in the two integrands may be simplified by using $E_{\mathbf{p+q}}^2 - E_p^2 = 2\mathbf{p}\cdot\mathbf{q} + q^2$ and some algebra. The result is (2.62*a*). The term independent of density in the second integral in (A.1) may be dropped after renormalizing the charge and mass.

The longitudinal polarizability may be similarly calculated:

$$Q_L(q) = \frac{q_i q_j}{|\mathbf{q}|^2} Q_{ij}(q) \tag{A.2}$$

This gives

$$Q_L(q) = \frac{4\pi e^2}{|\mathbf{q}|^2} \int \frac{d^3p}{(2\pi)^3} \left\{ \frac{n_F(p)(E_{\mathbf{p+q}} - E_p)^3\left[-q^2 + (E_{\mathbf{p+q}} + E_p)^2\right]}{E_p E_{\mathbf{p+q}}\left[\omega^2 - (E_{\mathbf{p+q}} - E_p)^2\right]} + \frac{(1 - n_F(p))(E_{\mathbf{p+q}} + E_p)^3\left[q^2 - (E_{\mathbf{p+q}} - E_p)^2\right]}{E_p E_{\mathbf{p+q}}\left[\omega^2 - (E_{\mathbf{p+q}} + E_p)^2\right]} \right\} \tag{A.3}$$

By rewriting $(E_{\mathbf{p+q}} - E_p)^3$ as $(E_{\mathbf{p+q}} - E_p)\ \{(E_{\mathbf{p+q}} - E_p)^2 - \omega^2 + \omega^2\}$ and separating out the terms proportional to ω^2 we have:

$$\begin{aligned} Q_L(q) \equiv {} & 4\pi e^2 \frac{\omega^2}{q^2} \int \frac{d^3p}{(2\pi)^3} \left\{ \frac{n_F(p)(E_{\mathbf{p+q}} - E_p)\left[(E_{\mathbf{p+q}} + E_p)^2 - q^2\right]}{E_p E_{\mathbf{p+q}}\left[\omega^2 - (E_{\mathbf{p+q}} - E_p)^2\right]} \right. \\ & \left. + \frac{(1 - n_F(p))(E_{\mathbf{p+q}} + E_p)\left[q^2 - (E_{\mathbf{p+q}} - E_p)^2\right]}{E_p E_{\mathbf{p+q}}\left[\omega^2 - (E_{\mathbf{p+q}} + E_p)^2\right]} \right\} \\ & + \frac{4\pi e^2}{q^2} \int \frac{d^3p}{(2\pi)^3} \left\{ n_F(p) \frac{E_{\mathbf{p+q}} - E_p}{E_p E_{\mathbf{p+q}}} \left[q^2 - (E_{\mathbf{p+q}} + E_p)^2\right] \right. \\ & \left. + (1 - n_F(p)) \frac{E_{\mathbf{p+q}} + E_p}{E_p E_{\mathbf{p+q}}} \left[(E_{\mathbf{p+q}} - E_p)^2 - q^2\right] \right\} \end{aligned} \tag{A.4}$$

The last term in (A.4) can be shown to be zero by using algebra and the final substitution $\mathbf{p} \to \mathbf{p} - \mathbf{q}$ when the angular integral vanishes. The term independent of density may again be dropped.

Tsytovitch defines

$$\varepsilon_L(q) = 1 - \frac{1}{\omega^2} Q_L(q) \tag{A.5}$$

Jancovici has

$$\varepsilon_L(q) = 1 + \frac{4\pi e^2}{q^2}\left\{\int \frac{d^3k}{(2\pi)^3}\frac{n_F(\mathbf{k})(1-n_F(\mathbf{k+q}))}{E_k E_{\mathbf{k+q}}}\right.$$

$$\cdot\frac{\left[(E_{\mathbf{k+q}}+E_k)^2-q^2\right](E_{\mathbf{k+q}}-E_k)}{\left[(E_{\mathbf{k+q}}-E_k)^2-\omega^2\right]}$$

$$\left.+\int\frac{d^3k}{(2\pi)^3}\frac{n_F(\mathbf{k+q})\left[(E_{\mathbf{k+q}}-E_k)^2-q^2\right]\left[E_{\mathbf{k+q}}+E_k\right]}{E_kE_{\mathbf{k+q}}\left[(E_{\mathbf{k+q}}+E_k)^2-\omega^2\right]}\right\} \tag{A.6}$$

In the first term of (A.6) we change $\mathbf{k} \to -\mathbf{k}-\mathbf{q}$ and add dividing by two, and in the second term we simply left $\mathbf{k} \to -\mathbf{k}-\mathbf{q}$:

$$\varepsilon_L(q) = 1 + \frac{4\pi e^2}{q^2}\left\{\frac{1}{2}\int\frac{d^3k}{(2\pi)^3}\left[n_F(\mathbf{k})(1-n_F(\mathbf{k+q}))\right.\right.$$

$$\left.-n_F(\mathbf{k+q})(1-n_F(\mathbf{k}))\right]$$

$$\cdot\left[\left(\frac{(E_{\mathbf{k+q}}+E_k)^2-q^2}{(E_{\mathbf{k+q}}-E_k)^2-\omega^2}\right)\frac{(E_{\mathbf{k+q}}-E_k)}{E_kE_{\mathbf{k+q}}}\right]$$

$$\left.+\int\frac{d^3k}{(2\pi)^3}n_F(\mathbf{k})\left(\frac{(E_{\mathbf{k+q}}-E_k)^2-q^2}{(E_{\mathbf{k+q}}+E_k)^2-\omega^2}\right)\frac{E_{\mathbf{k+q}}+E_k}{E_kE_{\mathbf{k+q}}}\right\} \tag{A.7}$$

$$= 1 + \frac{4\pi e^2}{q^2}\int\frac{d^3k}{(2\pi)^3}\left\{\left(\frac{(E_{\mathbf{k+q}}+E_k)^2-q^2}{(E_{\mathbf{k+q}}-E_k)^2-\omega^2}\right)\frac{E_{\mathbf{k+q}}-E_k}{E_kE_{\mathbf{k+q}}}n_F(k)\right.$$

$$\left.+\left(\frac{(E_{\mathbf{k+q}}-E_k)^2-q^2}{(E_{\mathbf{k+q}}+E_k)^2-\omega^2}\right)\frac{E_{\mathbf{k+q}}+E_k}{E_kE_{\mathbf{k+q}}}n_F(k)\right\} \tag{A.8}$$

which is (A.5).

For the transverse part we have

$$Q_T(q) = \frac{1}{2}\left(\delta_{ij} - \frac{q_i q_j}{q^2}\right) Q_{ij}(q) \tag{A.9}$$

$$Q_T(q) = 4\pi e^2 \int \frac{d^3p}{(2\pi)^3} \left(\frac{n_F(\mathbf{p}) - n_F(\mathbf{p}+\mathbf{q})}{\omega + E_p - E_{\mathbf{p}+\mathbf{q}}} \right) \frac{1}{E_p E_{\mathbf{p}+\mathbf{q}}}$$

$$\cdot \left[E_p E_{\mathbf{p}+\mathbf{q}} - m^2 - \frac{[(\mathbf{p}+\mathbf{q})\cdot\mathbf{q}](\mathbf{p}\cdot\mathbf{q})}{q^2} \right]$$

$$-4\pi e^2 \int \frac{d^3p}{(2\pi)^3} \left(\frac{1 - n_F(\mathbf{p})}{\omega + E_p + E_{\mathbf{p}+\mathbf{q}}} \right) \frac{1}{E_p E_{\mathbf{p}+\mathbf{q}}}$$

$$\cdot \left[E_p E_{\mathbf{p}+\mathbf{q}} + m^2 + \frac{[(\mathbf{p}+\mathbf{q})\cdot\mathbf{q}](\mathbf{p}\cdot\mathbf{q})}{q^2} \right]$$

$$+4\pi e^2 \int \frac{d^3p}{(2\pi)^3} \left(\frac{1 - n_F(\mathbf{p}+\mathbf{q})}{\omega - E_p - E_{\mathbf{p}+\mathbf{q}}} \right) \frac{1}{E_p E_{\mathbf{p}+\mathbf{q}}}$$

$$\cdot \left[E_p E_{\mathbf{p}+\mathbf{q}} + m^2 + \frac{[(\mathbf{p}+\mathbf{q})\cdot\mathbf{q}](\mathbf{p}\cdot\mathbf{q})}{q^2} \right]$$

Manipulating as in the longitudinal case gives

$$Q_T(q) = 8\pi e^2 \int \frac{d^3p}{(2\pi)^3} \frac{n_F(\mathbf{p})}{E_p E_{\mathbf{p}+\mathbf{q}}} \left[\frac{E_p E_{\mathbf{p}+\mathbf{q}} - m^2 - \dfrac{[(\mathbf{p}+\mathbf{q})\cdot\mathbf{q}](\mathbf{p}\cdot\mathbf{q})}{q^2}}{\omega^2 - (E_{\mathbf{p}+\mathbf{q}} - E_p)^2} \right]$$

$$\cdot (E_{\mathbf{p}+\mathbf{q}} - E_p)$$

$$+8\pi e^2 \int \frac{d^3p}{(2\pi)^3} \frac{(1 - n_F(\mathbf{p}))}{E_p E_{\mathbf{p}+\mathbf{q}}} \left[\frac{E_p E_{\mathbf{p}+\mathbf{q}} + m^2 + \dfrac{[(\mathbf{p}+\mathbf{q})\cdot\mathbf{q}](\mathbf{p}\cdot\mathbf{q})}{q^2}}{\omega^2 - (E_{\mathbf{p}+\mathbf{q}} + E_p)^2} \right]$$

$$\cdot (E_{\mathbf{p}+\mathbf{q}} + E_p) \tag{A.10}$$

This can be seen to match Tsytovich's result by rewriting the numerators in (A.10)

$$\frac{1}{E_p E_{\mathbf{p+q}}}\left[E_p E_{\mathbf{p+q}} - m^2 - \frac{(\mathbf{p\cdot q})^2}{q^2} - \mathbf{p\cdot q}\right]$$

$$= 1 + \frac{m^2 + (\mathbf{p\cdot q})^2/q^2 + (\mathbf{p\cdot q})}{E_p E_{\mathbf{p+q}}} \equiv \Lambda_t^{(-)} \text{ of Tsytovich}$$

and in the longitudinal case

$$\frac{1}{q^2 E_p E_{\mathbf{p+q}}}\left[\left(E_{\mathbf{p+q}}^2 - E_p^2\right)^2 - q^2\left(E_{\mathbf{p+q}} - E_p\right)^2\right]$$

$$= \frac{1}{q^2 E_p E_{\mathbf{p+q}}}\left[(2\mathbf{p\cdot q} + q^2)^2 - q^2\left(E_{\mathbf{p+q}}^2 + E_p - 2E_p E_{\mathbf{p+q}}\right)\right]$$

$$= \frac{2}{q^2 E_p E_{\mathbf{p+q}}}\left[q^2(\mathbf{p\cdot q}) + 2(\mathbf{p\cdot q})^2 - q^2 E_p^2 - q^2 E_p E_{\mathbf{p+q}}\right]$$

$$= 2\left[1 + \frac{\mathbf{p\cdot q} + 2(\mathbf{p\cdot q})^2/q^2 - E_p^2}{E_p E_{\mathbf{p+q}}}\right]$$

$$\equiv \Lambda_l^{(-)} \text{ of Tsytovich}$$

We have thus demonstrated the equivalence.

APPENDIX B: APPROXIMATIONS FOR CORRELATION ENERGY

Here we shall indicate how to get to Eq. (3.1) from (2.61). The approximation $q \ll k_F$ and $\omega \ll E_F$ is made. Then the second term in the integrand of (2.62*a*) is negligible, and the remaining integral simplifies and may be done (see Jancovici,[60] especially his Appendix) to give (using $\omega \to i\omega$)

$$Q_{00}(q,\omega) \simeq 4E_F\left\{1 - \frac{\omega E_F}{\beta q}\tan^{-1}\frac{\beta q}{\omega E_F}\right\}$$

This may be used in (2.61) in which the q integral is now cut off at k_F. The further substitutions $\omega/q = u$ and subsequently

$$\frac{uE_f}{\beta} = v \quad \text{and} \quad q = k_F x$$

enable us to perform the x integral resulting in (3.1). We have

$$\Delta E_{\mathrm{corr}}^{(c)} \simeq \frac{1}{8\pi^3}\frac{k_F^5}{\sqrt{k_F^2+m^2}}\int_0^1 x^3\,dx\int_{-\infty}^{\infty} dv$$

$$\cdot \ln\left\{\left(1+\frac{4e^2\sqrt{k_F^2+m^2}}{\pi k_F x^2}R\right)-\frac{4e^2\sqrt{k_F^2+m^2}}{\pi k_F x^2}R\right\}$$

Note that R does not depend on x and is a function of v only. Integrating by parts with respect to x:

$$\simeq \frac{1}{8\pi^3}\frac{k_F^5}{\sqrt{k_F^2+m^2}}\int_{-\infty}^{\infty} dv\left\{\frac{x^4}{4}\left[\ln\left(1+\frac{4e^2\sqrt{k_F^2+m^2}}{\pi k_F x^2}R\right)\right.\right.$$

$$\left.-\frac{4e^2\sqrt{k_F^2+m^2}}{\pi k_F x^2}R\right]_0^1+\int_0^1\frac{x^4}{4}dx$$

$$\left.\cdot\left[\frac{1}{1+4e^2\sqrt{k_F^2+m^2}\,R/\pi k_F x^2}-1\right]\frac{8e^2\sqrt{k_F^2+m^2}\,R}{\pi k_F x^3}\right\}$$

$$E_{\mathrm{corr}}^{(c)} \simeq \frac{1}{8\pi^3}\frac{k_F^5}{\sqrt{k_F^2+m^2}}\int_{-\infty}^{\infty} dv\left\{\frac{1}{4}\left[\ln\left(\frac{4e^2\sqrt{k_F^2+m^2}\,R}{\pi k_F}\right)\right.\right.$$

$$\left.-\frac{4e^2\sqrt{k_F^2+m^2}\,R}{\pi k_F}\right]+\int_0^1 2x\,dx\left[\frac{-1}{x^2+4e^2\sqrt{k_F^2+m^2}\,R/\pi f}\right]$$

$$\left.\cdot 4\left(\frac{e^2\sqrt{k_F^2+m^2}\,R}{\pi k_F}\right)\right\}$$

which leads to (3.1).

The transverse part may be done along similar lines.

APPENDIX C: LOCAL APPROXIMATION FOR THE BREIT INTERACTION

The transverse exchange energy is

$$E_x^{\text{tr}} = \iint \frac{d^3p\, d^3p'}{(2\pi)^6} \frac{4\pi e^2}{\left[-(E_{p'} - E_p)^2 + |\mathbf{p}' - \mathbf{p}|^2\right]} \frac{n_F(p) n_F(p')}{E_p E_{p'}}$$

$$\cdot \left[E_p E_{p'} - m^2 - \frac{[\mathbf{p}\cdot(\mathbf{p}'-\mathbf{p})][\mathbf{p}'\cdot(\mathbf{p}'-\mathbf{p})]}{|\mathbf{p}'-\mathbf{p}|^2} \right] \tag{C.1}$$

The Breit interaction is the approximation obtained by dropping the energy transfer term $(E_{p'} - E_p)^2$ in the denominator (Brown and Ravenhall[41]). The integral over the third term in the brackets is zero. This may be shown by doing the angular integrals:

$$J \equiv \int_{-1}^{+1} d\mu \frac{1}{(p^2 + p'^2 - 2pp'\mu)^2} (p'\mu - p)(p' - p\mu)$$

$$= \int_{-1}^{+1} d\mu \frac{1}{(p^2 + p'^2 - 2pp'\mu)^2} [(p'^2 + p^2)\mu - pp'\mu^2 - pp']$$

Each of the integrals over μ may be separately done and their sum leads to $J = 0$. Thus we have

$$E_x^{(\text{Breit})} \simeq \frac{1}{2} \iint \frac{d^3p\, d^3p'}{(2\pi)^6} \frac{4\pi e^2}{|\mathbf{p} - \mathbf{p}'|^2} \frac{1}{E_p E_{p'}} (2E_p E_{p'} - 2m^2) \tag{C.2}$$

Doing the angular integrals gives

$$E_x^{(\text{Breit})} \simeq \frac{e^2}{2\pi^3} \int_0^{k_F} \frac{p\, dp}{E_p} \int_0^{k_F} \frac{p'\, dp'}{E_{p'}} (E_p E_{p'} - m^2) \ln\left| \frac{p + p'}{p - p'} \right| \tag{C.3}$$

Scaling $p = k_F x$, $p' = k_F x'$, $E_p = \sqrt{p^2 + m^2} = m\sqrt{1 + \beta^2 x^2} = mE_x$, and E_β

$=\sqrt{1+\beta^2}$ gives

$$E_x^{(\text{Breit})} = \frac{3e^2k_F}{2\pi} N \iint_0^1 x\,dx\,x'\,dx' \left(1 - \frac{1}{E_x E_{x'}}\right) \ln\left|\frac{x+x'}{x-x'}\right| \tag{C.4}$$

$$= \frac{3e^2k_F}{2\pi} N \iint_0^1 \left[x\,dx\,x'\,dx \ln\left|\frac{x+x'}{x-x'}\right| - \frac{xx'\,dx\,dx'}{E_x E_{x'}} \ln\left|\frac{x+x'}{x-x'}\right| \right]$$

The first integral gives 1/2.

The second integral has to be integrated by parts and symmetrized to give

$$\iint_0^1 \frac{xx'\,dx\,dx'}{E_x E_{x'}} \ln\left|\frac{x+x'}{x-x'}\right| = \frac{E_\beta}{\beta^2} \int_0^1 \frac{x\,dx}{E_x} \ln\left|\frac{1+x}{1-x}\right| - \frac{1}{\beta^2}\left(\int_0^1 \frac{dx}{E_x}\right)^2 \tag{C.5}$$

Now

$$\int_0^1 \frac{dx}{E_x} = \frac{\ln(\beta + E_\beta)}{\beta}$$

using $\beta x = \sinh\theta$. This same substitution may be used in the first integral of (C.5) and the integral may be carried out. The limits have to be taken rather carefully. Alternatively the first integral in (C.5) may be done by splitting the log into

$$\ln\left|\frac{1+x}{1-x}\right| = 2\ln(1+x) - \ln(1-x^2)$$

Each of the resulting integrals may be separately done.

APPENDIX D: EXPLICIT EXPRESSIONS FOR $E_x^{(c)}$ AND $E_x^{(tr)}$ FOR THE POLARIZED CASE

We present here the expressions for the Coulomb and transverse contributions to the exchange energy after the traces in Eqs. (4.22) and (4.25) have been carried out:

$$E_x^{(c)} = -2\pi e^2 \sum_{\hat{n},\hat{n}'} \iint \frac{d^3p\,d^3p'}{(2\pi)^6} \frac{n_{F\hat{n}}(p)\,n_{F\hat{n}'}(p')}{|\mathbf{p}-\mathbf{p}'|^2} \frac{1}{E_p E_{p'}}$$
$$\cdot\{(1+E_pE_{p'}+\mathbf{p}\cdot\mathbf{p}')(1+s_0s_0'+\mathbf{s}\cdot\mathbf{s}') - s_0s_0'E_pE_{p'}$$
$$- s_0E_{p'}(\mathbf{s}'\cdot\mathbf{p}) - s_0'E_p(\mathbf{s}\cdot\mathbf{p}') - (\mathbf{p}\cdot\mathbf{s}')(\mathbf{s}\cdot\mathbf{p}')\} \tag{D.1}$$

and

$$
\begin{aligned}
E_x^{(\mathrm{tr})} = -\pi e^2 \sum_{\hat{n},\hat{n}'} \iint \frac{d^3p\, d^3p'}{(2\pi)^6} & \frac{n_{F\hat{n}}(p) n_{F\hat{n}'}(p')}{\left[(E_p - E_{p'})^2 - |\mathbf{p}-\mathbf{p}'|^2\right]} \frac{1}{E_p E_{p'}} \\
& \cdot \left\{ E_p E_{p'} - m^2 - \frac{(\mathbf{q}\cdot\mathbf{p})(\mathbf{q}\cdot\mathbf{p}')}{q^2} - s_0 s_0' \left(E_p E_{p'} + m^2 - \frac{(\mathbf{q}\cdot\mathbf{p})(\mathbf{q}\cdot\mathbf{p}')}{q^2} \right) \right. \\
& + \frac{(\mathbf{q}\cdot\mathbf{s})(\mathbf{q}\cdot\mathbf{s}')}{q^2} E_p E_{p'} - m^2 - 3\mathbf{p}\cdot\mathbf{p}' + 3\mathbf{s}\cdot\mathbf{s}' \left(\mathbf{p}\cdot\mathbf{p}' - \frac{(\mathbf{q}\cdot\mathbf{p})(\mathbf{q}\cdot\mathbf{p}')}{q^2} \right) \\
& - E_{p'} s_0 \frac{(\mathbf{s}'\cdot\mathbf{q})(\mathbf{p}\cdot\mathbf{q})}{q^2} - E_p s_0' \frac{(\mathbf{s}\cdot\mathbf{q})(\mathbf{p}'\cdot\mathbf{q})}{q^2} \\
& + 3\frac{(\mathbf{p}'\cdot\mathbf{s})(\mathbf{s}'\cdot\mathbf{q})(\mathbf{p}\cdot\mathbf{q})}{q^2} + 3\frac{(\mathbf{p}\cdot\mathbf{s}')(\mathbf{s}\cdot\mathbf{q})(\mathbf{p}'\cdot\mathbf{q})}{q^2} \\
& \left. + 3(\mathbf{p}\cdot\mathbf{s})(\mathbf{p}'\cdot\mathbf{s}') - 3(\mathbf{p}\cdot\mathbf{s}')(\mathbf{p}'\cdot\mathbf{s}) \right\}
\end{aligned}
\tag{D.2}
$$

The angular integrals may be done and the resulting integrals over p, p' can be done numerically after scaling out the Fermi momentum k_F. The form of the integrals is

$$
E_x^{(c)} = -\frac{e^2 k_F^4}{(2\pi)^3} \sum_{\hat{n},\hat{n}'} \int_0^{x_{\hat{n}}} \frac{x^2\, dx}{E_x} \int_0^{x_{\hat{n}'}} \frac{x'^2\, dx'}{E_x'} \{ I_1 + \hat{n}\cdot\hat{n}' I_2 \} \tag{D.3}
$$

where I_1, I_2 are functions of x, x', E_x, E_x', and β:

$$
E_x^{(\mathrm{tr})} = -\frac{e^2 k_F^4}{(2\pi)^3} 2 \sum_{\hat{n},\hat{n}'} \int_0^{x_{\hat{n}}} \frac{x^2\, dx}{E_x} \int_0^{x_{\hat{n}'}} \frac{x'^2\, dx'}{E_{x'}} \{ T_1 + \hat{n}\cdot\hat{n}' T_2 \} \tag{D.4}
$$

where T_1, T_2 are functions of x, x', E_x, $E_{x'}$ and β, and $k_F x_{\hat{n}} = k_{F\hat{n}}$ is the Fermi energy for up/down spin electrons. In fact, $x_\uparrow = (1+\zeta)^{1/3}$ and $x_\uparrow = (1-\zeta)^{1/2}$ in terms of the ζ parameter.

ADDENDUM

Dr. A. H. MacDonald has pointed out that there may be an error in $E_x^{(\mathrm{tr})}$ and therefore $V_x^{(\mathrm{tr})}$ and $W_x^{(\mathrm{tr})}$. We have since found this error and the expressions (4.27), (4.34), (4.36) and (D.2) in Appendix D should read as follows:

$$E_{x,NR}^{(\mathrm{tr})} = \frac{10}{9}\frac{e^2 k_F^4 \beta^2}{(2\pi)^3}\left(1+\tfrac{7}{15}\zeta^2\right) \tag{4.27}$$

$$V_{x,NR}^{(\mathrm{tr})} = \frac{5}{3}\frac{e^2}{\pi}\beta^2 k_F \tag{4.34}$$

$$W_{x,NR}^{(\mathrm{tr})} = \frac{7}{18}\frac{e^2}{\pi}\beta^2 k_F \zeta \tag{4.36}$$

$$\begin{aligned} E_x^{(\mathrm{tr})}(n_1\zeta) = &-\pi e^2 \int \frac{d^3p}{(2\pi)^3}\int\frac{d^3p}{(2\pi)^3} \sum_{\pm\hat{n},\pm\hat{n}'} \frac{n_{F\hat{n}}(\mathbf{p})\, n_{F\hat{n}'}(\mathbf{p}')}{E_p E_{p'}\left[(E_p - E_{p'})^2 - |\mathbf{p}-\mathbf{p}'|^2\right]} \\ &\cdot\left\{ E_p E_{p'} - m^2 - \frac{(\mathbf{p}\cdot\mathbf{q})(\mathbf{p}'\cdot\mathbf{q})}{|\mathbf{q}|^2} + \mathbf{s}\cdot\mathbf{s}'\left(\mathbf{p}\cdot\mathbf{p}' - \frac{(\mathbf{p}\cdot\mathbf{q})(\mathbf{p}'\cdot\mathbf{q})}{|\mathbf{q}|^2}\right)\right. \\ &+ s_o s_o'\left(m^2 + \frac{(\mathbf{p}\cdot\mathbf{q})(\mathbf{p}'\cdot\mathbf{q})}{|\mathbf{q}|^2}\right) + \frac{(\mathbf{s}\cdot\mathbf{q})(\mathbf{s}'\cdot\mathbf{q})}{|\mathbf{q}|^2}\left(E_p E_{p'} - m^2 - \mathbf{p}\cdot\mathbf{p}'\right) \\ &- \frac{(\mathbf{s}\cdot\mathbf{q})(\mathbf{q}\cdot\mathbf{p}')}{|\mathbf{q}|^2} E_p s_o' - \frac{(\mathbf{s}'\cdot\mathbf{q})(\mathbf{q}\cdot\mathbf{p})}{|\mathbf{q}|^2} E_{p'} s_o + \frac{(\mathbf{s}\cdot\mathbf{p}')(\mathbf{p}\cdot\mathbf{q})(\mathbf{s}'\cdot\mathbf{q})}{|\mathbf{q}|^2} \\ &\left. + \frac{(\mathbf{s}'\cdot\mathbf{p})(\mathbf{p}'\cdot\mathbf{q})(\mathbf{s}\cdot\mathbf{q})}{|\mathbf{q}|^2} - (\mathbf{s}\cdot\mathbf{p}')(\mathbf{s}'\cdot\mathbf{p})\right\} \end{aligned} \tag{D.2}$$

We are therefore revising the corresponding figures and the conclusions based on these expressions in collaboration with Wu-Xing Xu and these will be reported very soon elsewhere.

We thank Dr. A. H. MacDonald for sending us his manuscript before publication and for pointing out the possibility of an error in our work.

Acknowledgments

We thank Professor Walter R. Johnson for many useful discussions and for his suggestions that were incorporated in Section III. We thank Professor A. R. P. Rau for reading an early version of this manuscript and for making several useful comments.

One of us (M.V.R.) expresses his appreciation to his wife for continuing love and support during the progress of this work. He also places on record his appreciation of the good will of the Atomic and Solid State group at LSU, particularly Professor John Kimball and Drs. S. P. Singhal and Kalidas Bhadra.

Thanks are due to Ms. Martha Prather, Ms. Linda Gauthier, and Ms. Julie Bradley for their friendly and gracious typing services and to Ms. Hortensia Delgado for help with the computer system. Financial assistance for the preparing of the doctoral dissertation of M.V.R. was provided by the "Dr. Charles E. Coates Memorial Fund of the LSU Foundation donated by George H. Coates."

References

1. A. Sommerfeld, *Ann. Phys.* (Leipzig) [4] **51**, 1, 125 (1916).
2. P. Pyykkö, *Adv. Quantum Chem.* **11**, 353 (1978).
3. J. D. Björken and S. D. Drell, *Relativistic Quantum Fields*, McGraw-Hill, New York, 1965.
4. V. L. Moruzzi, J. F. Janak, and A. R. Williams, *Calculated Electronic Properties of Metals*, Pergamon Press, Elmsford, N.Y., 1978.
5. K.-N. Huang, M. Aoyagi, M. H. Chen, and B. Craseman, *Atomic Data Nucl. Data* **18**, 243 (1976).
6. P. Pyykkö and J. P. Desclaux, *Acc. Chem. Res.* **12**, 276 (1979).
7. K. S. Pitzer, *Acc. Chem. Res.* **12**, 271 (1979).
8. J. P. Desclaux, *Phys. Scripta* **21**, 436 (1980).
9. O. K. Andersen, B. Johansson, and H. L. Skriver, *Physica* **102B**, 103 (1980).
10. I. P. Grant, *Adv. Phys.* **82**, 747 (1970).
11. P. Hohenberg and W. Kohn, *Phys. Rev.* **B136**, 864 (1964).
12. W. Kohn and L. J. Sham, *Phys. Rev.* **A140**, 1133 (1965).
13. L. J. Sham and W. Kohn, *Phys. Rev.* **A145**, 561 (1966).
14. A. K. Rajagopal, *Adv. Chem. Phys.* **41**, 59 (1980).
15. W. Kohn and P. Vashishta, in *Physics of Solids and Liquids*, S. Lundqvist and N. H. March, eds. Plenum Press, New York, 1982.
16. G. Breit, *Phys. Rev.* **34**, 553 (1929); **36**, 383 (1930); **39**, 616 (1932).
17. B. Swirles, *Proc. Roy. Soc.* (London) **A152**, 625 (1935).
18. A. O. Williams, Jr., *Phys. Rev.* **58**, 723 (1940).
19. D. F. Mayers, *Proc. Roy. Soc.* (London) **A241**, 93 (1957).
20. S. Cohen, *Phys. Rev.* **118**, 489 (1960).
21. C. C. J. Roothaan, *Rev. Mod. Phys.* **23**, 69 (1951); **32**, 179 (1960); C. C. J. Roothaan and P. S. Bagus, *Methods Comp. Phys.* **2**, 47 (1963).
22. T. Kagawa, *Phys. Rev.* **A22**, 2340 (1980), and references therein.
23. I. P. Grant, *Proc. Roy. Soc.* (London), **A262**, 555 (1961).
24. J. B. Mann, Los Alamos Scientific Laboratory Report (1967).
25. J. P. Desclaux, *Comp. Phys. Comm.* **9**, 31 (1975).
26. J. B. Mann and W. R. Johnson, *Phys. Rev.* **A4**, 41 (1971).
27. H. A. Bethe and E. E. Salpeter, *Quantum Mechanics of One and Two Electron Atoms*, Academic Press, New York, 1957.
28. R. D. Deslattes, E. G. Kessler, W. C. Sauder, and A. Henins, *Ann. Phys.* (NY) **129**, 378 (1980).
29. R. D. Cowan, *Phys. Rev.* **163**, 54 (1967).
30. D. Liberman, J. T. Waber, and D. T. Cromer, *Phys. Rev.* **A137**, 27 (1965).
31. T. Ziegler, J. G. Snijders, and E. J. Baerends, *J. Chem. Phys.* **74**, 1271 (1981), and references cited therein.

32. D. Guenzenburger and D. E. Ellis, *Phys. Rev.* **B22**, 4203 (1980).
33. J. Callaway, R. D. Woods, and V. Sirounian, *Phys. Rev.* **107**, 934 (1957).
34. H. L. Skriver, O. K. Andersen, and B. Johansson, *Phys. Rev. Lett.* **41**, 42 (1978).
35. H. L. Skriver, O. K. Andersen, and B. Johansson, *Phys. Rev. Lett.* **44**, 1230 (1980).
36. D. Glötzel, *J. Phys. F: Metal Phys.* **8**, L163 (1978).
37. K. Schwarz and P. Herzig, *J. Phys. C: Solid State Phys.* **12**, 2277 (1979).
38. A. H. MacDonald, J. M. Daams, S. H. Vosko, and D. D. Koelling, *Phys. Rev.* **B23**, 6377 (1981).
39. *Proceedings of the International Conference on the Physics of Actinides and Related 4f Materials*, Institute of Physics, London (1980).
40. N. C. Pyper, and I. P. Grant, *Proc. Roy. Soc.* (London) **A376**, 483 (1981), and references therein.
41. G. E. Brown and D. G. Ravenhall, *Proc. Roy. Soc.* (London), **A208**, 552 (1951).
42. M. H. Mittleman, *Phys. Rev.* **A24**, 1167 (1981).
43. J. Sucher, *Phys. Rev.* **A22**, 348 (1980).
44. G. W. F. Drake and S. P. Goldman, *Phys. Rev.* **A23**, 2093 (1981).
45. D. Bagayoko, *Phys. Rev.* A, submitted for publication (1982).
46. N. D. Mermin, *Phys. Rev.* **A137**, 1441 (1965).
47. U. Gupta and A. K. Rajagopal, *Phys. Rev.* **A22**, 2792 (1980).
48. U. von Barth and L. Hedin, *J. Phys. C: Solid State Phys.* **5**, 1629 (1972).
49. M. M. Pant and A. K. Rajagopal, *Solid State Commun.* **10**, 1157 (1972).
50. A. K. Rajagopal, and J. Callaway, *Phys. Rev.* **B7**, 1912 (1973).
51. O. Gunnarsson, and B. I. Lundqvist, *Phys. Rev.* **B13**, 4274 (1976).
52. A. K. Rajagopal, *J. Phys. C: Solid State Phys.* **11**, L943 (1978).
53. A. H. MacDonald, and S. H. Vosko, *J. Phys. C: Solid State Phys.* **12**, 2977 (1979).
54. M. P. Das, M. V. Ramana, and A. K. Rajagopal, *Phys. Rev.* **A22**, 9 (1980).
55. M. P. Das, *Int. J. Quantum Chem. Symp.* **14**, 67 (1980).
56. D. E. Ellis, *J. Phys. B: Atom Molec. Phys.* **10**, 1 (1978).
57. M. Gell-Mann, and K. A. Brueckner, *Phys. Rev.* **106**, 364 (1957).
58. M. S. Freedman, F. T. Porter, and J. B. Mann, *Phys. Rev. Lett.* **28**, 711 (1972).
59. B. Fricke, J. P. Desclaux, and J. T. Waber, *Phys. Rev. Lett.* **28**, 714 (1972).
60. B. Jancovici, *Nuovo Cim.* **25**, 428 (1962).
61. A. H. MacDonald, Ph.D. Thesis, Univ. of Toronto (1978).
62. J. Sapirstein, *Phys. Rev. Lett.* **47**, 1723 (1981).
63. S. A. Chin, *Ann. Phys.* (NY) **108**, 301 (1977).
64. I. A. Akhieser, and S. V. Peletminskii, *Soviet Phys. JETP* **11**, 1316 (1960).
65. E. S. Fradkin, *Proc. (Trudy) Lebedev Phys. Inst.* **29**, 1 (1967).
66. R. L. Bowers, J. A. Campbell, and R. L. Zimmerman, *Phys. Rev.* **D7**, 2278 (1973).
67. V. N. Tsytovich, *Soviet Phys.—JETP* **13** 1249 (1961).
68. B. Bezzerides and D. F. Dubois, *Ann. Phys.* (NY) **70** 10 (1972).
69. J. W. Connolly, in *Semi-empirical Methods of Electronic Structure Calculation*, Pt. A: *Techniques*, G. A. Segal ed., Plenum, New York, 1977.

70. D. Liberman, J. T. Waber, and D. T. Cromer, *Comp. Phys. Commun.* **2**, 107 (1971).
71. M. V. Ramana and A. K. Rajagopal, *Phys. Rev.* **A24**, 1689 (1981).
72. M. V. Ramana, A. K. Rajagopal, and W. Johnson, *Phys. Rev.* **A25**, 96 (1982).
73. B. Y. Tong, *Phys. Rev.* **A4** 1375 (1971).
74. B. Y. Tong and L. J. Sham, *Phys. Rev.* **144**, 1 (1966).
75. J. P. Perdew and A. Zunger, *Phys. Rev.* **B23**, 5048 (1981).
76. M. V. Ramana and A. K. Rajagopal, *J. Phys. C: Solid State Phys.* **12**, L845 (1979) and Corrigendum, ibid. **14**, L111 (1981). See also, ibid. **14**, 4291 (1981) for a detailed version.
77. J. J. Sakurai, *Advanced Quantum Mechanics*, Addison-Wesley, Reading, Mass. 1973.
78. J. E. Harriman, *Theoretical Foundations of Electron Spin Resonance*, Academic Press, 1978 New York, Appendix F.
79. A. K. Rajagopal, H. Brooks, and N. R. Ranganathan, *Nuovo Cimento Suppl.* **5**, 807 (1967).
80. E. Borie and G. A. Rinker, *Rev. Mod. Phys.* **54**, 67 (1982).
81. A. Bechler, *Ann. Phys.* (NY) **135**, 19 (1981).
82. W. R. Johnson, private communication (1981).

POLARITON AND SURFACE EXCITON STATE EFFECTS IN THE PHOTODYNAMICS OF ORGANIC MOLECULAR CRYSTALS

J. M. TURLET AND PH. KOTTIS

Centre de Physique Moléculaire Optique et Hertzienne
Université de Bordeaux I and CNRS Talence, France

AND

M. R. PHILPOTT

IBM Research Laboratory San Jose, California

CONTENTS

I. INTRODUCTION

Since the development in the 1950s of the theory of molecular excitons by Davydov and others, the study of organic molecular crystals, which had until then been largely neglected because of the low symmetry of these systems and the complexity of their constituent molecules, has been the focus of increasing and intense interest. Among the fundamental reasons for this may be the "vital" urge of researchers to draw closer to biological systems, or perhaps the attraction exerted by the diversity of organic compounds. There are, however, more immediate motives: While the eigenstates of crystalline systems, like those of any periodic structure, are collective eigenstates (excitons) related to the equivalence of the sites, they are always closely related to those of the free molecule because the intermolecular actions are very weak. Hence, the term *molecular crystal.* Research on these systems is, therefore, of twofold interest: (1) to exhibit collective optical, electrical, magnetic, and other properties related to the crystalline structure and often very different from those found in other types of crystal; and (2) to deduce from these properties those of the free molecule, which is always hard to study as such.

Indeed, the results of Davydov and others explaining the complementary light which molecule and crystal shed on each other, removed one of the main difficulties in the study of organic compounds and led immediately to fresh developments. Visible and ultraviolet spectroscopy, which has proved well adapted to these crystals, has confirmed overall in absorption (by transmission) and emission the exciton representation predicted by Frenkel and Davydov. It has been shown that in crystals containing two molecules per unit cell, such as naphthalene and anthracene, molecular transitions split into two components (Davydov splitting) with different energies and polarizations from those of the free molecule.

However, as experimental techniques improved (better resolution, lower temperatures), disagreements appeared for most systems between the measurements and theory. In the case of anthracene, for example, it was not understood in the exciton framework why the absorption bandwidths of several hundred wave numbers at liquid nitrogen temperatures (77 K) persisted at 4 K (150 cm^{-1}). As we shall show later, most of the difficulties arose from too simple a picture of the real crystal and the fact that from the start coupling between crystal elementary excitations and light was neglected. It was assumed for some time that the optical response of a crystal would contain only the eigenstates of an isolated, strictly periodic, model crystal.

In fact, in close to moderate or strong transitions, such as the first singlet transition ($S_0 \rightarrow S_1^*$) in anthracene, the optical properties of the crystal are not simply those of its exciton modes, but those of the mixed (polariton) state due to coupling between crystal modes and radiation. Moreover, the finite boundaries of the crystal, and the associated breakdown of periodicity, may give rise to special modes (surface and subsurface excitons). Finally, it may be necessary to take into account coupling between the zero-order modes and other elementary crystal excitations, like intra- and intermolecular vibrations (phonons) which may be created at the same time. It is advantageous to study these modes and their relaxation by the reflected light which cannot get into the crystal, rather than by traditional transmission spectra, since, theoretically at least, exciton–photon coupling forbids propagation in the neighborhood of the resonance.

In the present work we have taken account of these relatively recent general ideas to improve our knowledge of organic molecular crystals close to their resonances through a theoretical and experimental description of different bulk and surface modes. The precise problems discussed below have origins in many theoretical and experimental results. We shall first describe the history of the above main ideas and the associated first principles. The major goals of our work are to be found at the end of Section II, which is in fact the main introduction.

II. BULK AND SURFACE EXCITONS IN MOLECULAR CRYSTALS

Since the systematic studies undertaken at the turn of the century,[1] solid state spectroscopy has made considerable progress, simultaneously with solid state physics and chemistry. While its developments have had well-known effects on everyday life, it is unfortunately true that different branches have become increasingly segregated as new ones appear. It is clear at large meetings that spectroscopists do not all speak the same language and that despite the generality of the problems studied, methods vary widely. Hence, we shall begin by discussing the properties of organic molecular crystals and the theory of excitons in them. The history of the theory and of its main applications will illustrate its major limitations and the object of the present work.

A. Molecular Crystals and Frenkel Excitons

Contrary to the situation in, say, metallic or ionic crystals, the intermolecular forces of cohension, often Van der Waals forces, are much weaker than the intramolecular binding forces. Molecular crystals therefore have rather low melting points, and their heats of sublimation, of the order of a kcal/mol, are an order of magnitude smaller than the dissociation energy of the molecule. Examples are the crystals of the inert gases and most organic compounds including the aromatic hydrocarbons [benzene, naphthalene, anthracene (Fig. 1), etc.], in which we shall be interested here.

The exciton, nowadays understood as a collective electronic or vibronic crystal excitation, was first introduced by Frenkel[2] in the form of "excitation waves," as a first stage in the degradation of absorbed photons into "heat." The theory, developed for cubic system crystals of the inert monoatomic gases, contains the first account of the localization of the excitation by lattice vibration and its migration from site to site. Subsequent work[3,4] soon showed that the Frenkel exciton, in which site excitation plays a special role, was not applicable to all crystals and led to a second model, the

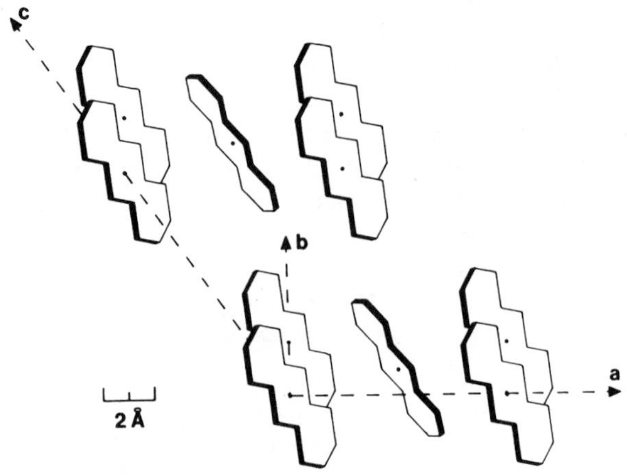

Fig. 1. Crystalline structure of anthracene.

Wannier exciton, better suited to describing the optical properties of inorganic insulators like CdS and ZnO or ionic crystals like NaCl, where excited electronic orbitals cover many sites. The Frenkel exciton remains the better model for the first singlet states of crystals of the aromatic hydrocarbons, owing to the small overlap between π or π^* orbitals on different molecules.

Well-known experiments by Prinsheim and Kronenberger[5] and Obreimov and Prikhot'ko[6-10] in the 1920s and 1930s showed that intermolecular actions, weak as they are, cause crystal transitions to differ in energy and above all in polarization from those of the free molecule. The crystal could no longer be treated as an oriented gas, and the need became clear for a theory connecting crystal states to those of the molecule. Such a theory of molecular crystals was proposed by Davydov.[11] His theory is outlined below and is discussed in another form in the following chapter. Numerous books[12-16] and review articles[17-28] have been written on it.

1. *First-Order Energy Levels of the Crystal*

This theory describes the eigenstates of a perfect, rigid (absolute zero temperature) crystal of $N\sigma$ molecules, N being the number of unit cells and σ the number of (translationally inequivalent) molecules per unit cell. By the cyclic Born–von Karman conditions, generally assumed to hold (though this is not always true—more on this later), the sites are made indiscernible by closing the crystal on itself in all three directions. The Hamiltonian of such a system is

$$H_{ex} = \sum_{n\alpha}^{N\sigma} H_{n\alpha} + \tfrac{1}{2} \sum_{n\alpha m\beta}{}' V_{n\alpha m\beta} \qquad \left[\textstyle\sum' \equiv (n\alpha \neq m\beta)\right] \tag{1}$$

where $V_{n\alpha m\beta}$, the potential energy of the Coulomb interaction between molecules $n\alpha$ and $m\beta$, is sufficiently small in molecular crystals to be treated as a perturbation ("tight binding" approximation) of the Hamiltonian ($\Sigma_{n\alpha} H_{n\alpha}$) of the "oriented gas" of $N\sigma$ noninteracting molecules.

Neglecting overlap between molecular orbitals, the crystal eigenstates are properly antisymmetrized tensor products of the eigenfunctions $\varphi^r_{n\alpha}$ (energy ε^r) of the free molecule (Hamiltonians $H_{n\alpha}$.)

a. Ground State. Disregarding contributions of excited molecular states to the crystal ground state, its zeroth-order wave function is the antisymmetrized product of the molecular ground state wave functions $\varphi^0_{n\alpha}$:

$$\psi^0 = \prod_n \varphi^0_{n\alpha}$$

First-order perturbation theory leads to the ground state energy E^0:

$$E^0 = \sum_{n\alpha} \varepsilon^0 + \tfrac{1}{2} \sum_{n\alpha,\, m\beta}{}' \langle 00 | V_{n\alpha m\beta} | 00 \rangle \tag{2}$$

b. Excited States. If $\varphi^f_{n\alpha}$ is the wave function of excited state f (supposed for simplicity to be a pure electronic state) on free molecule $n\alpha$, with energy ε^f, then exciton wave functions can be written, keeping only resonance interactions, as linear combinations of the $N\sigma$-fold degenerate functions:

$$\phi^f_{n\alpha} = \varphi^0_{11} \cdot \varphi^0_{12} \cdot \cdots \cdot \varphi^f_{n\alpha} \cdot \cdots \cdot \varphi^0_{N\sigma} \tag{3}$$

N.B. Since this model associates the same excitation energy ε^f with all the sites, they must have the same weights in the exciton wave function. The corresponding idea of collective excitation of the crystal, delocalized over all the sites, is incompatible, at least at low temperatures, with the model of site excitation followed by energy migration from site to site[29-32].

Crystal periodicity is accounted for by dividing the $N\sigma$ molecules into σ groups of N and by introducing a wave vector $\mathbf{k}$ with N values in the first Brillouin zone to define sublattice wave functions:

$$\phi^f_\alpha(\mathbf{k}) = \frac{1}{\sqrt{N}} \sum_{n=1}^{N} \phi^f_{n\alpha} \exp i(\mathbf{k} \cdot \mathbf{r}_{n\alpha}) \tag{4}$$

and writing the exciton wave functions as:

$$\psi^\nu(\mathbf{k}) = \sum_{\alpha=1}^{\sigma} c^\nu_\alpha(\mathbf{k}) \phi^f_\alpha(\mathbf{k}) \tag{5}$$

Owing to the intermolecular terms $V_{n\alpha m\beta}$, solution of Schrödinger's equation for this type of wave function and Hamiltonian (1) lifts the degeneracy and yields eigenenergies $E^\nu(\mathbf{k})$ and wave functions $\psi^\nu(\mathbf{k})$ [characterized by the $c^\nu(\mathbf{k})$ obtained for each energy $E^\nu(\mathbf{k})$] for each of the $N\sigma$ levels in the exciton quasi continuum corresponding to molecular state f:

$$E^\nu(\mathbf{k}) - E^0 = \Delta E^\nu(\mathbf{k}) = \Delta E^f_m + D + I^\nu(\mathbf{k}) \tag{6}$$

where $\quad \Delta E^f_m = \varepsilon^f - \varepsilon^0$ is the molecular excitation energy (7)

and

$$D = \sum_{m\beta}^{N\sigma}{}' \{ \langle 0f | V_{n\alpha m\beta} | 0f \rangle - \langle 00 | V_{n\alpha m\beta} | 00 \rangle \} \tag{8}$$

is the difference between the interaction of an excited and a ground state molecule $n\alpha$, with the rest of the crystal in the ground state. Since molecules generally interact more strongly in the excited state, this Van der Waals term is negative, causing the whole band to be lowered relative to the excited molecular level.

The exciton exchange or transfer interaction between site $n\alpha$ and the rest of the crystal is

$$I^{\nu}(\mathbf{k}) = \sum_{m\beta}^{N\sigma}{}' \{\langle 0f | V_{n\alpha m\beta} | f0\rangle \exp(i\mathbf{k}\cdot(\mathbf{r}_{n\alpha} - \mathbf{r}_{m\beta}))\} \tag{9}$$

It is thus the quantum exchange term which causes the dispersion of the exciton band (Fig. 2).

2. *Transition Probabilities — Selection Rules*

Time-dependent perturbation theory is applicable to the calculation of transition probabilities between the crystal ground state E^0 and the exciton states $E^{\nu}(\mathbf{k})$,[12] provided we can neglect the perturbation of the exciton states by the field. This is approximately true for weak transitions. Consider an applied electromagnetic field of frequency ω and wave vector $\mathbf{q}$:

$$\boldsymbol{\varepsilon}(\mathbf{r}, t) = \boldsymbol{\varepsilon}_0 \exp i(\mathbf{q}\cdot\mathbf{r} - \omega t)$$

and let $\mathbf{d}^{0\mathrm{f}}$ be the transition dipole moment between states $|0\rangle$ and $|f\rangle$ on site α of the unit cell:

$$\mathbf{d}_{\alpha}^{0f} = \langle \varphi_n^f | \mathbf{d}_{\alpha}^{0f} | \varphi_{n\alpha}^0 \rangle$$

while $D_{\mathbf{k},\mathbf{q}}^{0\nu}$ is the electric dipolar interaction of the exciton transition:

$$D_{\mathbf{k},\mathbf{q}}^{0\nu} = \langle \psi^{\nu}(\mathbf{k}) | D_{\mathbf{k},\mathbf{q}} | \psi^0 \rangle = (\sigma N)^{-1/2} \boldsymbol{\varepsilon}_0 \left[\sum_{\alpha} (c_{\alpha}^{\nu} \mathbf{d}_{\alpha}^{0f}) \right] \left[\sum_{n=1}^{N} \exp i(\mathbf{q} - \mathbf{k})\cdot\mathbf{r}_n \right] \tag{10}$$

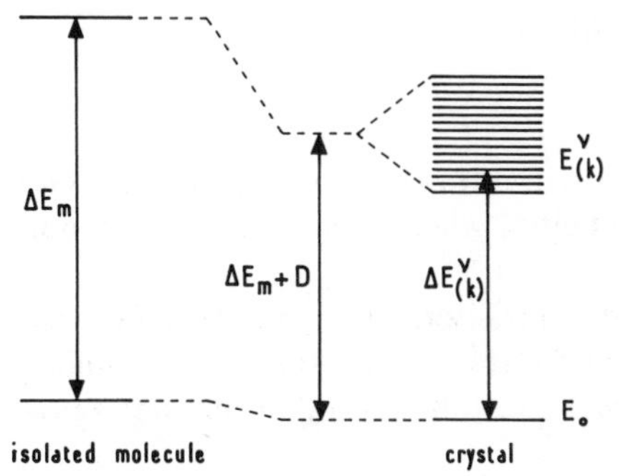

Fig. 2. Relation between molecular and crystal levels.

Then, the transition probability is

$$P^{0\nu}(\mathbf{k}) = \frac{2\pi}{\hbar}|\mathbf{D}^{0\nu}_{\mathbf{k},\mathbf{q}}|^2\,\delta(\Delta E^{\nu}(\mathbf{k}) - \hbar\omega) \tag{11}$$

This expression implies selection rules for exciton transitions.

a. Conservation of Energy. The frequency ω of the applied field must be resonant with the transition frequency $\Delta E^{\nu}(\mathbf{k})/\hbar$.

b. Allowed Molecular Transition. The magnitude of the dipole moment $\mathbf{d}^{0f}_{\alpha}$, being the same for all sites α, factorizes in $P^{0\nu}(\mathbf{k})$ and must be different from zero. This is a first-order rule, and in fact a dipole forbidden molecular state may become an allowed exciton transition through state mixing. Inversely, the cooperation of dipoles in (10) may forbid an exciton transition by symmetry, even when there is a strong molecular transition.

c. Polarization of the Unit Cell. Let $\mathbf{P}^{\nu}$ be the polarization of the unit cell for each exciton subband:

$$\mathbf{P}^{\nu} = \sum_{\alpha=1}^{\sigma} c^{\nu}_{\alpha}\mathbf{d}^{0f}_{\alpha} \tag{12}$$

It must have a nonzero projection on the direction of the electric field vibration.

d. Conservation of Photon Momentum. Since

$$\sum_{n} \exp i(\mathbf{q}-\mathbf{k})\cdot\mathbf{r}_n = \delta_{\mathbf{q},\mathbf{k}}$$

the wave vector $\mathbf{k}$ must be the same as the wave vector $\mathbf{q}$ of the photon, modulo $\pm 2\pi\mathbf{G}$, $\mathbf{G}$ being a vector in the reciprocal lattice. At optical frequencies ($q = 2\pi/\lambda \sim 0$), we must have:

$$\begin{cases} |\mathbf{k}| = |\mathbf{q}| \sim 0 \\ \hat{k} = \hat{q} \end{cases} \tag{13}$$

This double condition explains why exciton transition energies depend on the direction of the photon (nonanalyticity), whereas the long-used condition $k = 0$ alone does not.

Thus in the one-particle event, photon → exciton, only one level $E^{\nu}(\mathbf{k} = \mathbf{q})$ is optically accessible, provided the first three conditions are fulfilled. Therefore, this very much simplified model of total crystal "excitation" predicts at most σ transitions.

3. *Crystals with Two Molecules Per Unit Cell*

Applied to crystals with two molecules per unit cell such as anthracene, ($\sigma = 2$), this model yields two subbands ($\nu = +$ and $\nu = -$) for each molecular excited state f, with first-order energy relative to the crystal ground state

$$\Delta E^{\pm}(\mathbf{k}) = \Delta E_m + D + I_{11}(\mathbf{k}) \pm I_{12}(\mathbf{k}) \tag{14}$$

where $I(\mathbf{k})$, the sum of excitation exchange terms defined in (9), has been split into a sum over equivalent molecules, $I_{11}(\mathbf{k})$, and one over inequivalent molecules, $I_{12}(\mathbf{k})$. Carrying these energies, $E^{\pm}(\mathbf{k})$, into the equations for the c_α^ν leads to the corresponding wave functions:

$$\Psi^{\pm}(\mathbf{k}) = \frac{1}{\sqrt{2}}\left(\phi_1^f(\mathbf{k}) \pm \phi_2^f(\mathbf{k})\right) \tag{15}$$

There are two allowed transitions $\Delta E^{\pm}(\mathbf{k} = \mathbf{q})$, separated by a Davydov splitting Δ (see Fig. 3):

$$\Delta = \Delta E^{+}(\mathbf{k} \sim \mathbf{q}) - \Delta E^{-}(\mathbf{k} \sim \mathbf{q}) = 2I_{12}(\mathbf{k}) \tag{16}$$

polarized along the polarization directions $\mathbf{P}^\nu$ of the unit cell:

$$\mathbf{P}^{\pm} = \frac{1}{\sqrt{2}}\left(\mathbf{d}_1^{0f} \pm \mathbf{d}_2^{0f}\right) \tag{17}$$

In the first singlet transition $S_0 \rightarrow S_1^*$ of anthracene, at roughly $\omega \simeq 26{,}000$ cm^{-1}, the $\mathbf{P}^-$ transition is polarized along the $\mathbf{b}$ axis and the P^+ one in the $(\mathbf{a}, \mathbf{c})$ plane (see Fig. 4).

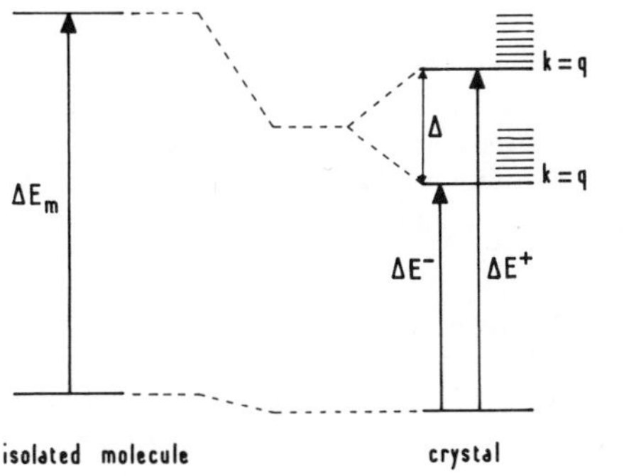

Fig. 3. Exciton component and Davydov splitting of a crystal with two molecules per unit cell.

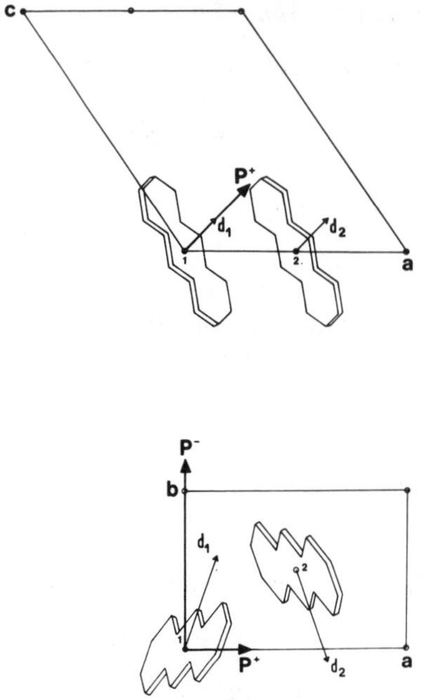

Fig. 4. Polarization $\mathbf{P}^+$ and $\mathbf{P}^-$ of the exciton transitions ΔE^+ and ΔE^-.

N.B. While for the sake of simplicity we have discussed only pure electronic transitions, the theory is readily generalized to cover vibronic transitions as well. Corresponding to a vibronic transition $\Delta E_m^f + \Delta e_V$ of a free molecule in the Born–Oppenheimer approximation, one has:

$$\Delta E^{\pm,v}(\mathbf{k}) = \Delta E_m + \Delta e_v + D + I_{11}(\mathbf{k}) \prod_u |\langle \chi_u^{f_1} | \chi_u^{01} \rangle|^2$$

$$\pm I_{12}(\mathbf{k}) \prod_u \langle \chi_u^{f_1} | \chi_u^{01} \rangle \langle \chi_u^{f_2} | \chi_u^{02} \rangle \tag{18}$$

where χ_u is the vibrational wave function of the uth normal mode of the totally symmetric vibration v.

The Davydov splitting of vibronic states is thus less than that of the pure electronic state by a Franck—Condon factor $F_v^f = \Pi_u |\langle \chi_u^f | \chi_u^0 \rangle|^2$ which must be between 0 and 1 since $\Sigma_v F_v^f = 1$.

Besides vibronic excitons (vibrons) built from electronic and vibrational states on the same molecule, one can also build collective two-particle states in which the electronic and vibrational excitations are on different molecules.[241]

4. *Calculation of Exciton Levels*

Calculation of the first-order exciton energy levels depends on the electrostatic interactions $V_{n\alpha m\beta}$ between molecules $n\alpha$ and $m\beta$ [cf. D defined in (8) and the sums $I^{\nu}(\mathbf{k})$ defined in (9)]. Detailed knowledge of the molecular wave functions would produce rigorous results, but despite great efforts, the molecular orbitals of "relatively large" systems like the aromatic hydrocarbons are still too imprecise compared to the sensitivity of the expressions evaluated. The charge distribution of each molecule is commonly approximated by a multipolar development centered on the molecule, where the potential $V_{n\alpha m\beta}$ appears as a series of interaction terms between point multipoles, separated by $\mathbf{R} = \mathbf{r}_{n\alpha} - \mathbf{r}_{m\beta}$. Depending on symmetry and the strength of the transition, one has dipole–dipole (R^{-3}), dipole–quadrupole (R^{-4}), quadrupole–quadrupole (R^{-5}), dipole–octupole (R^{-5}), octupole–octupole (R^{-7}), and similar terms.[15,23]

The different terms in D depend on the mean values of the multipole moments of the molecule in the ground and excited states. In molecules with a center of symmetry the first nonzero terms, quadrupole–quadrupole interactions, are not large and necessitate a better, second-order estimation of the energy.[33] The difficulty of this calculation and the fact that D lowers all the exciton bands en masse have discouraged theoretical work on this topic. However, as we shall see below, the D terms are of fundamental importance in the theory of surface and subsurface excitons, so it is to be hoped that experimental results in this fields will stimulate new interest. However, much work has been done on the calculation of the excitation exchange terms $I^{\nu}(\mathbf{k})$, since they can be checked by comparison with the Davydov splitting and the polarization ratios. Although there is no general procedure, methods and approximations abound, each depending on the systems and transitions investigated. If one divides transitions into three classes of oscillator strength, one can comment as follows.

a. Weak Transitions ($f \sim 0.01$ or less). Since these transitions are dipole forbidden, the short-range multipole interactions are readily summed. Unfortunately the smallness of these terms obliges one to take account of second-order effects, such as mixing with neutral states or charge transfer states and electron exchange.

b. Medium Transitions ($f \sim 0.1$). These will be discussed at greater length below, since the $S_0 \rightarrow S_1^*$ transition of anthracene ($\sim 27{,}000$ cm^{-1}) falls in this class and is the main object of this work. Suffice it to say here that the main terms are dipole—dipole interactions. The problem is the summation, since these terms drop off in R^{-3} whereas their number increases in R^3 and the long-range terms, for k small but nonzero, are always important.

c. Strong Transitions. In this case dipole representation should work, but few systems have been definitively studied.

5. *Limitations of Exciton Representation*

The principal merit of the preceding elementary theory of excitons is that it explains globally and qualitatively the absorption spectra of many molecular crystals. However, detailed analysis of experimental spectra reveals many discrepancies, such as that between theoretical and experimental linewidths.

The reason for this is mainly that the exciton described above is the eigenstate of a very idealized and isolated crystal. A better theory must include couplings between the exciton and other elementary crystal excitations, such as exciton–phonon, exciton–photon, and exciton–exciton coupling. It must discuss crystal defects and the finite bounds of real crystals (edge effects). Only by these interactions can a crystal absorb or emit light. The theory must also ultimately include small-scale roughness of real crystal surfaces.

B. Effect of Exciton–Photon Coupling

Consider the dispersion relation $\omega = f(\mathbf{k})$ in a crystal irradiated with a field $(\omega,\mathbf{k})$ with polarization such that it is coupled to only one exciton mode $E^{\nu}(\mathbf{k})$. Let us suppose that close to $\mathbf{k}=0$, $E^{\nu}(\mathbf{k}) = E^{\nu}(0) = \hbar\omega_T$.

If there were no coupling, a photon of energy $\hbar\omega = \hbar ck/n_{\infty}$ would propagate unperturbed through the crystal, except when $\omega = \omega_T$, that is, when an exciton $(\omega_T,\mathbf{k})$ would be created. The crystal–field system is described either by a photon or by an exciton, the role of the field being to pick the exciton mode at resonance. This is not an unreasonable representation in the case of weak transitions.

In medium to strong transitions exciton–photon coupling is no longer negligible, and one must rediagonalize the crystal–field system. This leads to mixed (polariton) states in which the weight of the exciton varies between 0 and 1. The dispersion curve is very different from that of the uncoupled systems (Fig. 5) and strongly influences the optical properties of the crystal. Briefly we note here:

1. There are now two distinct branches separated by the gap $(\omega_L - \omega_T)$, which is a function of the exciton–photon coupling strength. In this interval, if there are no relaxation phenomena, all propagation modes $(\omega,\mathbf{k})$ are forbidden. In this region the refractive index $n(\omega)$ is a pure imaginary and the reflectivity of the crystal

$$R(\omega) = \left| \frac{n(\omega)-1}{n(\omega)+1} \right|^2$$

is equal to 1. The total reflection may be attenuated, especially for $\omega > \omega_T$ at

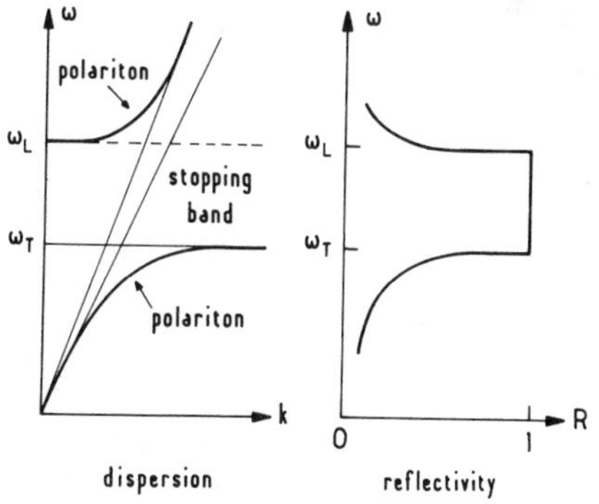

Fig. 5. Relation between the dispersion curve $\omega(k)$ of the polariton and the reflectivity $R(\omega)$ of the crystal.

very low temperatures, if one takes account of various relaxation channels in the crystal (e.g., phonons, traps or surface modes) in which a polariton and another elementary excitation are simultaneously created.

The nonpenetration of the light into the crystal in the stopping band region even for crystals as thin as 0.1 μm explains the large bandwidths observed by transmission even at very low temperature (4 K) in the case of medium and strong transitions. This is the main reason for the development of absorption study through reflection spectroscopy in such cases.

2. While at high and low frequencies the weight of the photon is practically 1, in the bends of the dispersion curve the wave vector of the polariton differs from that of the photon of same energy. Hence, in an infinite, perfect crystal, collision between the polariton and another particle is necessary, just as is the case for the exciton, before emission can occur. This is one reason for the missing phononless fluorescence origin in the emission spectra of some crystals, notably anthracene.

C. Surface and Subsurface Excitons

1. Definitions

The eigenstates described above are those of a molecular crystal with cyclic Born–Von Karman conditions to make all the unit cells identical in all three directions. We shall call these modes "bulk excitons."

Consider now a "semi-infinite" crystal composed of a pile of infinite planes laid down perpendicular to one direction, in which the cyclic conditions can no longer apply. Molecules in the first few planes will have missing interactions (if the $N=0$ plane faces onto a vacuum) or different interactions (if the surface is covered with some impurity) from those of the molecules situated sufficiently far into the bulk of the crystal for the convergence of the interactions to have been reached. If exchange coupling is much stronger in the planes than between them, surface and subsurface zero-order modes will appear. The localization of these modes in the first planes will increase with the energy difference between them and the bulk exciton.

N.B. (1) Surface excitons, which originate essentially in the different environment of surface resonances ("Site Shift Surface Excitons"), may be likened, though with certain fundamental differences of course, to the single-electron surface states exhibited in some semiconductors and introduced in the band model by Tamm.[34] (2) The properties and localization of the surface exciton completely distinguish it from the surface polariton, which is an electromagnetic mode due to coupling between a photon and an elementary surface excitation, occurring at the boundary of two media, 1 and 2, when their dielectric constants satisfy $\varepsilon_1(\omega)+\varepsilon_2(\omega)<0$. The latter propagates in the surface, with an electric field polarized in the plane of incidence (p polarized) and an evanescent wave decaying exponentially in direction perpendicular to the surface[35] (See Section VIII).

2. *Simple Theory of Surface Excitons*

A complete theory of surface modes is possible only if we know the variation in energy Δe of the center of gravity of the zero-order exciton with the distance to the surface. This is not being the case, we must use a model with perfect, "unreconstructed" surfaces and suppose Δe to be nonzero only in the first few planes.

As we saw earlier in this section, the first-order energy of a bulk exciton mode is

$$E_V^\nu(\mathbf{k}) = \Delta E_m + D + I^\nu(\mathbf{k}) \tag{19}$$

where D and $I(\mathbf{k})$ defined in (8) and (9) are sums over all molecules $m\beta \neq n\alpha$ of bimolecular $(n\alpha, m\beta)$ matrix elements for the change in interaction when molecule $n\alpha$ is excited (D) and for excitation exchange $[I(\mathbf{k})]$. If the crystal is divided into planes the summation may be done plane by plane. When $\mathbf{k}$ is perpendicular to the planes

$$D = D^{(0)} + 2D^{(1)} + 2D^{(2)} + \cdots \tag{20}$$

$$I^\nu(\mathbf{k}) = I^{\nu(0)}(\mathbf{k}) + 2I^{\nu(1)}(\mathbf{k}) + 2I^{\nu(2)}(\mathbf{k}) + \cdots \tag{21}$$

where index (0) refers to sums over molecules $m\beta$ in the same plane as $n\alpha$, (1) refers to molecules in the neighboring plane, and so on.

Assumptions by which the series may be truncated. Given present knowledge of the variation of D and $I(\mathbf{k})$ with the distance between sites, we may truncate series (20) and (21). Regarding D, it is only known that the terms are short range. We shall keep the first three $[D^{(0)}, D^{(1)}$, and $D^{(2)}]$ and bear in mind that the number of terms retained defines a model with as many nondegenerate modes: surface, subsurface, and bulk. There being no theory to estimate these terms, only experimental evidence of a certain number of surface modes can tell where to truncate and how many planes are necessary to define a bulk exciton.

The $I^{\nu(i)}(\mathbf{k})$ terms have been calculated plane by plane by De Wette and Schacker's method,[36] for $|\mathbf{k}|=0$, normal to the (001) face of some aromatic hydrocarbons.[37]

In the $S_0 \rightarrow S_1^*$ transition of anthracene polarized along the short axis M,

$$\begin{aligned} I^{\pm(0)}(k=0) &= (-211{,}4 \pm 114{,}4)\mathrm{cm}^{-1} = I_{11}^{(0)} \pm I_{12}^{(0)} \\ I^{\pm(1)}(k=0) &= (-0{,}44 \pm 0{,}8)\mathrm{cm}^{-1} = I_{11}^{(1)} \pm I_{12}^{(1)} \\ I^{\pm(2)}(k=0) &= (1{,}6.10^{-3} \pm 3{,}7.10^{-5})\mathrm{cm}^{-1} = I_{11}^{(2)} \pm I_{12}^{(2)} \end{aligned} \tag{22}$$

Thus, in the configuration $\mathbf{k}$ normal to the (001) plane $I^{\pm}(\mathbf{k})$ can, with a very good approximation, be terminated at the first term—at least when $|\mathbf{k}| = 0$. Since dispersion in anthracene is very weak in this direction,[38-40] this result can be extended to $I(k)$ for all optically prepared excitons with wave vector $q = 2\pi/\lambda$.

The final expression for the bulk exciton energy is

$$\boxed{\Delta E_V^{\pm}(k) = \Delta E_m + D^{(0)} + 2D^{(1)} + 2D^{(2)} + I^{\pm(0)}(k)} \tag{23}$$

The above assumptions imply two extra exciton bands for the surface and the subsurface (Fig. 6), with energies

$$\boxed{\Delta E_S^{\pm}(k) = \Delta E_m + D^{(0)} + D^{(1)} + D^{(2)} + I^{\pm(0)}(k) = \Delta E_V^{\pm}(k) + \delta_S} \tag{24}$$

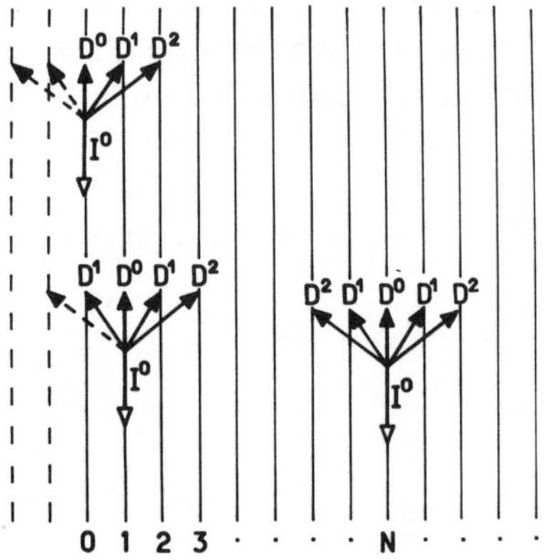

Fig. 6. Representation of the different molecule–plane interactions for molecules in the surface (0), the subsurface (1), or the volume (N).

and

$$\boxed{\Delta E_{sS}^{\pm}(k) = \Delta E_m + D^{(0)} + 2D^{(1)} + D^{(2)} + I^{\pm(0)}(k) = \Delta E_V^{\pm}(k) + \delta_{sS}} \tag{25}$$

shifted relative to the bulk exciton band by an amount independent of k:

$$\delta_S = -(D^{(1)} + D^{(2)}) \quad \text{and} \quad \delta_{sS} = -D^{(2)} \tag{26}$$

Since D is negative in the aromatic hydrocarbons, surface excitons have higher energies than bulk ones.

This result calls for some comments:

(a) Since the spatial dispersion of $E(k)$ is given by the same term, $I^{\pm(0)}(k)$, in (23), (24), and (25), bulk, surface, and subsurface $k=0$ excitons will all have the same polarizations and Davydov splittings.

(b) It is to be expected that the $D^{(l)}$ terms tend rapidly to zero with l, so the center of gravity of the surface and subsurface modes will tend equally quickly towards the energy of the bulk exciton ($\delta_S \gg \delta_{sS}$).

(c) The energy contribution of coupling between nonresonant levels is proportional to the square of the interaction matrix element $I^{\pm(1)}(k)$ and to the inverse of the energy separation. So long as

$$\delta_S \gg \delta_{sS} > I^{\pm(1)}(k)$$

the surface, subsurface, and bulk modes will perturb one another very little. The two-dimensional localization of subsurface excitons must disappear gradually as one gets further away from the surface.

(d) Since there is no evidence that the termination of $I^{\pm}(k)$ after the first term works for $k \gg q$, we cannot assume that Eqs. (24) and (26) hold for all wave vectors of the two-dimensional Brillouin zone for surface states. The frequency and wave vector dispersion of surface states are problems of fundamental interest.

N.B. If one keeps $I^{\pm(1)}(k)$, it can be shown[41] that for **k** perpendicular to the plane, the energy and the Davydov splitting of the surface exciton are

$$\Delta E_S^{\pm} = \Delta E_V^{\pm} + \frac{(-\delta_S + I^{\pm(1)})^2}{\delta_S}$$

$$\Delta_S = \Delta E_S^{+} - \Delta E_S^{-} = 2I_{12}^{(0)} + \frac{(I^{+(1)})^2 - (I^{-(1)})^2}{\delta_S} \tag{27}$$

3. *Experimental Evidence of Surface Excitons*

Given that the above assumptions are justified, particularly that $\delta > I^{(1)}(k)$, some ways of exhibiting surface modes are the analysis of transmission and reflection and fluorescence spectra at very low temperatures so that thermally induced transitions between planes and other dispersion phenomena are as small as possible.

a. Transmission. Exhibiting surface modes by this technique is hard because it is insensitive to surface phenomena and has the disadvantage of involving two boundaries. Medium or strong transitions could be studied only in extremely thin crystals, with all the concomitant inconveniences, and then only for modes shifted well outside the broad bulk band.

b. Reflection. Various reflection spectra–ordinary, modulated, ellipsometry, or attenuated total reflection—have proved to be very sensitive to surface phenomena.[42,43]

The circumstances are particularly favorable in the case of surface excitons associated with medium to strong transitions, because the exciton—photon coupling gives rise to a strong reflection band $(\omega_T - \omega_L)$. A surface mode excited in this band, no matter how close to ω_T, will cause a distinct structure in the band. The various methods[44,45] used to find reflectivity as a function of energy are described later. Qualitatively, however, these structures are due to constructive and destructive interference between the wave reflected by the bulk and "those" (artificially distinguished from it) re-

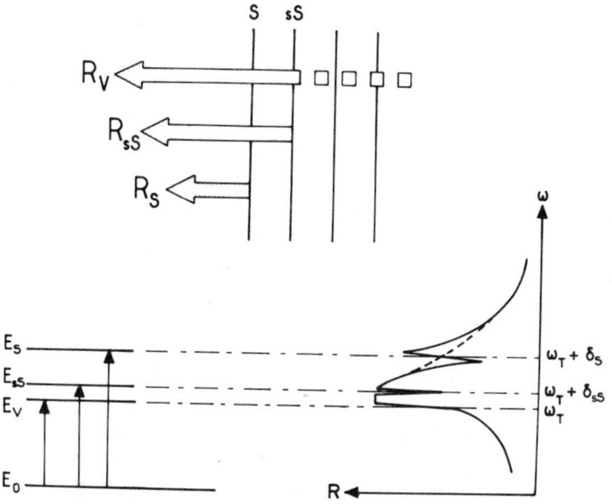

Fig. 7. Influence of surface and subsurface exciton transitions on the reflectivity of the crystal.

flected by the first few planes close to their resonance $\omega_T + \delta$, owing to the rapid phase change around the resonance. Close to ω_T the reflectivity approaches unity, and only destructive interference will be seen as a narrow minimum $(\omega_T + \delta_{sS})$ in the reflection spectrum (see Fig. 7).

When excitation occurs outside the stopping band, then owing to the weak bulk reflectivity the effect is much less pronounced and one finds a small peak (see Section VI, Fig. 51).

c. **Fluorescence.** In molecular crystals for which the surface and subsurface exciton states fall above those of the bulk, direct emission from them at energies $\omega_T + \delta$ should be unobservable owing to various fast, nonradiative processes towards the bulk states.

If, on the other hand, their radiative lifetime is of the order of or less than the relaxation time, they should be observable in fluorescence.

D. Principal Theoretical and Experimental Results Obtained on the First Singlet Transition of Anthracene

The above extensions of Davydov's initial theory came about only through constant comparison between theory and experiment, insofar as both did not contain fundamental misconceptions (not always the case). It is difficult to separate these complementary aspects of the problem. It would be impossible to describe here all the work on molecular crystals, or even that on anthracene; thus we shall discuss below only the first singlet transition of anthracene, since it is a significant example and consequently became the main subject of our work.

1. *Free Molecule*

The first singlet transition of anthracene is $A_g \rightarrow {}^1B_{2u}$ polarized along the short molecular axis M and has been observed in vapor at 27,688 cm^{-1} with an oscillator strength of 0.3.[46] Its absorption and fluorescence spectra in solution in Shpolskii matrices[47] show vibronic structure composed of broad bands at intervals of roughly 1400 cm^{-1}, even at 4 K. This is due to the presence, besides the intramolecular vibration at 390 cm^{-1}, of intense modes around 1400 cm^{-1} (1028, 1159, 1397, 1464, 1503, and 1550 cm^{-1}).

2. *Monocrystal*

Excepting a few developments ahead of (or behind) their time, the experimental (here only transmission, reflection, and fluorescence) and theoretical study of the crystal may be divided into several stages.

a. **Transmission and Davydov Splitting: Context of the Exciton.** Most absorption and transmission spectra have been made for a wave vector normal to the (001) face of the crystal, since it is the only well-developed one obtainable. Very thin crystals up to 0.1 μm thick and well-polarized light must be used because of the large oscillator strength of the transition. As noted in Refs. 48–49, these exacting conditions may largely explain the measured Davydov splittings ranging from 0 to 383 cm^{-1}[50–55]. Confusion reigned for a long time because of wrong estimates of the dipole sums, due to a mathematically unjustifiable truncation (see, for example, Ref. 56) of the dipole–dipole interaction sum.[57,58] Because of these difficulties it was also suggested that the Davydov splitting depended on the thickness of the crystal.[54,59] This was disproved by the calculations of Silbey et al.[60] Though they were imprecise owing to the use of molecular orbitals, they did show at least what approximations were justifiable. Lest our brief description of the history of the theory of the Davydov splitting of anthracene be judged too harsh, we remind the reader that at the time, the confusion was sustained by wrong estimates of the dipole sum. The truncated spherical sums were chosen because they best fitted the available data.

Even at 4 K the $A_g \rightarrow A_u$ (state $-$) and $A_g \rightarrow B_u$ (state $+$) Davydov components are broad (~ 150 cm^{-1}) so that the energies of their maxima are not very significant, but the Davydov splitting must be around the values (~ 200 cm^{-1}) given by Wolf[61] and later by Brodin and Marisova[48] and Claxton et al.[62] Their values agree well with theoretical estimates by Ewald's methods and summation plane by plane by Mahan,[63] Davydov and Sheka,[38] (excepting the second part of the article), and Philpott.[37,39,64] The last two authors and, more recently, Homa[40] have calculated the exciton band structure for the three principal directions of the wave vector. The density of states of the band has been given by Schroeder and Silbey[65] and more recently by Sceats and Rice.[66]

b. Reflection and Anisotropy: Context of the Polariton. Rapid development between 1958 and 1960 of the theory of exciton–photon coupling, described in a later chapter, showed up the difficulties of absorption and transmission work and the advantages of using reflectivity instead. Attention to the importance of the reflection spectrum and to the variation of refractive index close to resonance was first drawn in the case of anthracene by, among others, Borisov[67] and Wolf.[61] The first reflection studies of molecular crystals were made by Anex and Simpson[68] on a dye best referred to mnemonically as BDP.

Since reflection spectra could be made on even very small crystal facets, it was possible to check the variation of the Davydov splitting with the direction of the wave vector. The main results in this field are due to Clark and Philpott[69,71] at room temperature, and Marisova et al.[72,74] and, later, Morris et al.[75,77] at low temperature.

Models of reflectivity based on the transition characteristics have been developed by Sugakov[44] and Philpott.[45,78,79]

c. Fluorescence: How May a Crystal Emit? The emission spectrum of the $S_1^* \rightarrow S_0$ transition of a monocrystal of anthracene (total quantum yield nearly unit) comprises a large number of lines between 25,080 and 22,000 cm^{-1}. They are relatively narrow (5 to 15 cm^{-1}) at low temperature. Much has been written about this emission,[80,85] most of it concerning low temperatures (less than 4.2 K) owing to rapid thermal broadening of the lines. One of the most important studies of the "intrinsic" fluorescence of anthracene was made by Glockner and Wolf,[86,87] who used a high-resolution method to obtain the positions, polarizations, and widths of the lines between 1.7 and 77 K. They were very careful to use particularly pure crystals with few defects (very low background).

A curious fact about this emission is that the fluorescence origin corresponding to a pure electronic (0,0) transition from the k state at the bottom of the lower Davydov component has never been "properly" observed. All the lines involve either a vibrationally excited ground state (phonons[84,88] or intramolecular vibrations[47]) or an excited state of lower energy (trap) due to structural defects or to chemical impurities. The lines of highest energy, with phonons up to 80 cm^{-1}, are also the strongest and are completely polarized along **b** (polarization of the lower exciton subband), while in others the (**a**, **c**)-polarized component appears gradually toward lower energy.[86,87]

The absence of a fluorescence origin has caused much discussion and widely differing interpretations,[80] among which the possibility of reabsorption of this line has been the view most tenaciously held.[89] In the exciton context, Glockner and Wolf have put this line at $25{,}097 \pm 0.5$ cm^{-1} (in vacuo) by vibrational analysis and interpreted its absence by the fact that the lowest level of the exciton band is a $k \neq 0$ state.

In fact before asking how a crystal may emit, we should first ask how it may absorb. In a polariton treatment which should prevail for this transition, it has been shown recently by Brodin et al.[90] that "intrinsic" fluorescence can be explained principally by inelastic polariton scattering during excitation by phonons and intramolecular vibrations.

This explanation also appears in the interpretation of the measured lifetime,[91] which in the case of the higher lines (for example, 25,081 cm^{-1}) has progressively shortened from 10 to 0.6 ns[92,94] as more selective and faster measuring apparatus became available.

d. Surface Excitons: Fact or Fiction?. Since 1970, among all the problems posed by the interpretation of absorption, reflection, and emission spectra of anthracene, the question of whether or not surface excitons have been detected has give rise to some of the "hottest" debates. The idea of surface excitons, introduced simultaneously (less, than a month apart) by Pekar[95] and Selivanenko[96] in 1957, during development of the polariton theory, had no immediate effects on experiments in molecular crystals.

The debate began nearly a decade later. Marisova,[97, 98] discussing low temperature (20 K and 4 K) reflection spectra of the first singlet transition of crystalline anthracene [(001) face], mentions occasional structures in the **b**-polarized spectra: (I) very pronounced, at 25,225 cm^{-1} on average, whose position seems to change with the thickness of the crystal (25,245–25,290 cm^{-1}); and (II) close to the maximum reflectivity of the crystal (~ 25,100 cm^{-1}), appearing clearly as a minimum only below 20 K. While this was confirmed by Brodin et al.[99] in 1968, the interpretations set forward [phononless line (0,0) for II and dispersion phenomenon close to $n = 1$ for I] had no basis. Following this, two other groups, in Stuttgart and Chicago, besides that in Kiev, were more or less directly interested in these anomalous structures. They produced three radically different interpretations. For the sake of clearness we shall describe them separately, though in fact they are closely interrelated.

1. In 1969 Glockner and Wolf found two very weak emissions, I (25,298 cm^{-1}) and II (25,103 cm^{-1}), in the fluorescence spectra of the very "best" crystals.

Crystals mounted on quartz have very broad, shifted lines. Detailed study of these emissions[100,101] on a large number of freely mounted crystals showed up two structures at the energies given above in two-thirds of the samples (type A) and emissions shifted to lower energies, 25,227 cm^{-1} (I) and 25,100 cm^{-1} (II), in the rest (type B). Below 20 K, some A-type crystals became like B-type ones. These authors, basing their interpretation on the second part of the article by Davydov and Sheka, suggested that strong dispersion in the exciton band could explain these emissions above the fluorescence origin. Thus:

- Line I could be emission from a bulk $k = 0$, exciton state, which would not therefore be at the bottom of the band, but some 200 cm^{-1} higher up.
- Line II could be emission from a bulk $k \neq 0$, exciton state, 6 cm^{-1} above the bottom of the band and populated by phonons.

2. With Sugakov's 1969 theory of the reflectivity of crystals with two molecules per unit cell,[41] the Kiev group showed that the shape of structure I observed by Marisova can be reproduced by giving the surface of the crystal a different energy from that of the volume.[44] Their theory of surface excitons was later extended to cover any direction of the wave vector.[102] More experiments at 4 K, based on this theory, by Brodin et al.[103,104] exhibited the

coincidence of the two structures in reflection and fluorescence and led to the following interpretation:

- Line I: Transition to (or from) the **b**-polarized surface excition.
- Line II: Transition to (or from) the **b**-polarized bulk exciton component [phononless (0,0) line].

Other lines observed only in fluorescence (25,117 and 25,142 cm^{-1}) were atributed to localized states close to the surface. Recently the same authors[105] have examined the reflectivity of anthracene for different directions of the wave vector [**k** normal to the (001), (111), (110), and (201) faces] and have deduced the position of the second surface exciton Davydov component as 25,524 cm^{-1}. No structures attributable to surface effects have been observed on faces other than (001).

3. As early as 1970 the Chicago group, which had already discussed the width of absorption bands in anthracene,[106] studied the reflection spectrum at 4 K, finding two minima (25,232 and 25,113 cm^{-1}), higher than the fluorescence origin.[107] These structures were associated with the many "low energy" lines also observed and interpreted as due to physical defects around chemical impurities. In 1973, in a very detailed account of the reflectivity of anthracene,[108] Morris and Sceats[109] again questioned the origin of these structures, but unfortunately based their arguments on spectra taken at 80 K which contained a prominent shoulder at 25,290 cm^{-1}. The shoulder seemed to be favored by fast cooling or by rigid mounting of the sample and could be completely removed by keeping the crystal in the dark at 80 K for a few hours (8 h). It was attributed to dislocations and others forms of internal roughness caused by strains during cooling. Spectra published a year later[110] taken at low temperature (10 K) did not have any definite structure except a shoulder at 25,160 cm^{-1}. This was taken as support of their view that stresses and strains were the cause of the low temperature reflection minima.

Thus by 1974 we have three explanations of the anomalous structures in the **b**-polarized reflectivity spectrum of anthracene. While we cannot give a complete discussion of these results here, we may briefly show why none of these theses is entirely satisfactory.

The first of them links these structures with bulk transitions between the ground state and **k** states 200 and 6 cm^{-1} above the bottom of the exciton band. In the exciton picture, this would imply strong dispersion $E(\mathbf{k})$ of the band for **k** normal to the (001) face. Now the dispersion of the band correctly calculated in this direction, by Ewald's method with an infinite radius, or by plane-by-plane summation, is practically nil.[38,40] It has already been shown[56] that the second part of Davydov and Sheka's[38] article, in which the dispersion appears as a dipole sum on spheres of different radii, is wrong. As a simple proof of this, the penetration depth of 500 Å (50 layers) found in this direction is overestimated, since both calculation and the intensity of absorption show it is no more than a few layers.[37,110]

The second thesis (above), despite considerable and admirable theoretical and experimental antecedents, is vulnerable on two counts:

(*a*) The first is the assignment of structure II to the phononless (0,0) bulk transition. This interpretation, introduced without grounds in 1967[73] and reproduced without full discussion in the later publications, is hard to accept for two reasons: The (0,0) line cannot appear as a minimum in the stopping band of the reflectivity spectrum, but corresponds to the steep low energy (ω_T) edge of it. In emission, the fluorescence origin is given by Glockner and Wolf[76,87] by vibrational analysis at 25,097 $\pm$ 0.5 cm^{-1} and not at the energy of structure II (25,103 cm^{-1}), which they also observed.

(*b*) The skepticism which greeted the surface exciton interpretation of structure I, despite the solid theoretical support given by Sugakov, is due to two main points:

1. The unstable and often nonreproducible nature of this structure (presence, form, and energy level) is a barrier to its acceptance as an intrinsic feature of the crystal.
2. The link between this structure and the crystal surface has never been clearly established.

The third point of view (above), based on defects, has no sound experimental basis, since the 80- and 10-K spectra of Morris and Sceats are not really comparable with those at 4 K of Marisova. Supposing that the shoulder at 80 K observed by the former authors corresponded to the narrow structure seen at 4 K by the latter, it is not clear that keeping the crystal 8 h at 80 K in a vacuum of 10^{-4} torr would "cancel out" the defects. To the contrary, in these conditions, in such a cryostat the chances of deteriorating the surface are increased.

Thus, though in 1974 the idea of surface excitons was well launched experimentally and theoretically, much remained to be done in both areas.

E. Subject of the Present Work

Anthracene has been extensively studied not because it has spectacular peculiarities but, on the contrary, because it is a typical, not too simple, not too complicated organic molecular crystal. Further, in the summary given above of the first singlet transition of the crystal, we encountered some of the main unsolved problems of solid state spectroscopy.

As we have seen, many of these problems depend on one another and may be categorized under the following heading: "Determination of the optical properties of real crystals close to their exciton resonances."

This encompasses the two principal questions asked in this field in the 1970s:

First, what are the eigenstates of the crystal–field system when a crystal is irradiated with electromagnetic waves? In other words, by what means do propagation, degradation, scattering, and reemission of the incident light occur? We can at present only begin to answer these questions.

Secondly, in the context of the first question, how may we model a real crystal and what are the consequences of this "reality" for the optical response? We have already mentioned that a real crystal is finite and nonrigid and has physical and chemical defects. The problem is thus to bring the physical sample as close as possible to the model, by careful refining, low temperatures, and so on, and to introduce in the model those aspects of the real crystal, such as its finite boundaries which cannot be eliminated experimentally.

In the first part of this work we shall describe different theoretical tools—classical, semiclassical, and quantum mechanical—for describing exciton–photon coupling and apply them to the "optical constants" of anthracene.

The second part is a new experimental study of the reflectivity and fluorescence of anthracene at very low temperatures (1.6 K) and relatively high resolution (0.3 cm^{-1}), necessary to decide between the different explanations of these spectra in the literature. We shall attempt to answer the following questions:

1. Are there any structures in the **b**-polarized reflection spectrum of anthracene?
2. If so, can they be associated experimentally with the crystal surface?
3. Is it then possible to detect the other (**a**-polarized) Davydov component of the surface exciton, thus strengthening its status?
4. Is there any relation, such as absorption to emission, between the features observed in reflection and fluorescence spectra?

The third section incorporates the results of the first two to propose a model of the reflectivity of finite crystals, with a phenomenological treatment of relaxation mechanisms.

Finally, we reexamine these structures in light of our results and those obtained during (or since) our work; we try to give them a more satisfactory interpretation and to outline generalizations of these effects which might be established experimentally.

III. POLARITONS IN "INFINITE" CRYSTALS

A. Introduction

Several kinds of elementary excitations are found in ordered media: excitons, phonons, plasmons, magnons, and so on, depending on the degrees of freedom involved.[111] In general, when there is strong matter–radiation coupling, a proper account of the optical properties of such media cannot treat the field as an external perturbation of the matter system; for example, the optics of crystals close to their resonances cannot be correctly explained by the presence of either a photon or an exciton. In practice, depending on the nature (frequency, wave vector) of the field, certain degrees of freedom are coupled to it, causing a polarization of the medium.

Hence the electromagnetic wave in the medium is not just the applied field, but the superposition of the applied field and of the induced polarization wave. The corresponding "mixed" wave, or quasi particle, is generally called a polariton, no matter what the excitation coupled to the field. There are, therefore, phonon–polaritons, exciton–polaritons or photo–excitons (Davydov), plasmon–polaritons, and so on, depending on the medium in hand.

N.B. As Burstein[112] has pointed out, the term "polariton" was first introduced by Hopfield[113] to describe a polarization wave (i.e., the generalized exciton), not the "mixed" state. It is entertaining to note that due to later confusion the term "polariton" was attributed to the mixed wave and became so widespread that even Hopfield[114] refers to his article for a definition which he had not in fact given.

We refer the reader to Ref. 112 for a complete account of the development of these ideas, and shall give here a summary of them. Following the appearance of successive dispersion theories due to Lorentz, Drude, Born, and Tosati, in 1957 Huang,[115, 116] in a classical study of coupling between the field and optical modes of diatomic ionic crystals, first foresaw that the "mixed wave" nature should vary continuously as a function of the wave vector, between the photon and the elementary excitation, in his case an optical phonon.

Fano[117] in 1956 and Pekar[118] a year later, gave the first quantum theories of coupling between a "field of oscillators" and the electromagnetic field. Pekar's theory deals specifically with exciton–photon coupling. It is only fair to point out that Pekar introduced the variation of the dielectric constant with the wave vector (spatial dispersion) and two ideas which subsequently became very important, those of "anomalous waves" and "surface excitons."

Hopfield's 1958 article,[113] developing a quantum theory of exciton–photon coupling, is well known for its lucid discussion of absorption in crystals and the necessity of a third "body" such as a phonon or surface defect to explain it.

The above theories involved models of cubic crystals. The first theory of anisotropic crystals, also in second quantization, is due to Agranovich.[119]

The first experimental dispersion curve $\omega(k)$ of the polariton was obtained by Henry and Hopfield[114] in the cubic system GaP. Since then the theory has spread into many areas of physical and chemical research on solids, surfaces, and interfaces. These developments are described in several review articles and anthologies: phonon–polaritons.[120, 121] exciton–polaritons.[122, 125] surface polaritons,[126, 130] and the more general accounts.[131, 133]

As we saw above, the polariton can be studied from two standpoints:

1. Either in a "semiclassical," or phenomenological or dipolar, theory in which the crystal–field system is replaced by a field of dipoles interacting and vibrating in phase, and in which Maxwell's equations yield the optical properties of the system.
2. Or in an entirely quantum (second quantization) theory in which the total Hamiltonian of the field–matter system is diagonalized to find the eigenstates.

In the case of simple two-body coupling, when the dipole approximation is valid, that is, when the dipole–dipole contribution is sufficiently large in the interaction matrix element, the two methods are equivalent and must give identical results. The main advantage of the semiclassical theory is the simplicity with which it describes molecular interactions and justifies certain approximations allowing, for instance, inclusion of the contributions of excited states, which are not considered explicitly in the coupling.

However, it cannot properly account for absorption, that is, polariton scattering on phonons, defects, and so on, except phenomenologically by introducing damping of the oscillators. Hence in more complicated cases, such as that of exciton–photon–phonon coupling, in nonlinear processes, or when the dipolar approximation is inappropriate, the quantum treatment must be used.

We shall examine these different methods applied to the first excited singlet of anthracene.

B. Semi-Classical Theory of the Polariton

1. General

The principle of the "semiclassical" theory of matter–field coupling is to treat macroscopically, by Maxwell's equations, propagation of the electromagnetic wave in a continuous ($\lambda \gg a$) medium.

In a nonmagnetic medium without free charges the equations for the electric field $\mathbf{E}(\mathbf{k},\omega)$ and the magnetic field $\mathbf{H}(\mathbf{k},\omega)$ are

$$\nabla\Lambda\mathbf{E}=-\frac{1}{c}\frac{\partial\mathbf{B}}{\partial t}\qquad \nabla\cdot\mathbf{D}=0$$

$$\nabla\Lambda\mathbf{H}=\frac{1}{c}\frac{\partial\mathbf{D}}{\partial t}\qquad \nabla\cdot\mathbf{H}=0 \tag{28}$$

The electric displacement vector $\mathbf{D}(\mathbf{k},\omega)$ is linked to $\mathbf{E}(\mathbf{k},\omega)$ by

$$\mathbf{D}(\mathbf{k},\omega)=\tilde{\varepsilon}(\mathbf{k},\omega)\cdot\mathbf{E}(\mathbf{k},\omega)=\mathbf{E}(\mathbf{k},\omega)+4\pi\mathbf{P}(\mathbf{k},\omega) \tag{29}$$

where $\tilde{\varepsilon}(\mathbf{k},\omega)$ is the (generally complex) dielectric tensor; $\mathbf{P}(\mathbf{k},\omega)$ is the "macroscopic polarization" vector; ω is the frequency of the wave and $\mathbf{k}$ its wave vector (generally complex quantities). The magnetic induction $\mathbf{B}(\mathbf{k},\omega)$ $=\mathbf{H}(\mathbf{k},\omega)$ since $\tilde{\mu}(\mathbf{k},\omega)=1$.

Solution of system (28) for the propagation of the electromagnetic wave proceeds in two stages:

a. Relation Between $\mathbf{D}(\mathbf{k},\omega)$ and $\mathbf{E}(\mathbf{k},\omega)$. This is, in fact, the problem of determining the dielectric tensor $\tilde{\varepsilon}(\mathbf{k},\omega)$. In general, in "discontinuous" media composed of interacting molecules this is properly done in a microscopic (quantum) approach.[26] In a "semiclassical" theory the macroscopic field $\mathbf{E}(\mathbf{k},\omega)$ of Maxwell's equations differs from the applied field $\mathbf{E}_0(\mathbf{k},\omega)$, so that by (29) determination of $\tilde{\varepsilon}(\mathbf{k},\omega)$ amounts to solving the self-consistent field, in which the polarization $\mathbf{P}(\mathbf{k},\omega)$ depends on $\mathbf{E}(\mathbf{k},\omega)$, which in turn depends on $\mathbf{P}(\mathbf{k},\omega)$.

b. Proper Solution of Maxwell's Equations. Having determined $\tilde{\varepsilon}(\mathbf{k},\omega)$ we shall deduce from (28) Fresnel's relation, involving only the electric field. Thus if we look for plane wave solutions of the form $\mathbf{X}(\mathbf{r})\exp(i(\mathbf{k}\cdot\mathbf{r}-\omega t))$, system (28) reduces to

$$\mathbf{D}(\mathbf{k},\omega)=\frac{c^2}{\omega^2}\left[k^2\mathbf{E}-\mathbf{k}(\mathbf{k}\cdot\mathbf{E})\right]=\tilde{\varepsilon}(\mathbf{k},\omega)\cdot\mathbf{E}(\mathbf{k},\omega) \tag{30}$$

Nontrivial solutions ($\mathbf{E}(\mathbf{k},\omega)\neq 0$) exist if the determinantal relation below holds:

$$\left\|\frac{\omega^2}{c^2}\varepsilon_{ij}(\omega,\mathbf{k})-k^2\delta_{ij}+k_ik_j\right\|=0 \tag{31}$$

The solutions $\omega = f(\mathbf{k})$ for which the determinant vanishes are the eigenmodes of propagation of the electromagnetic wave in the medium, that is, the dispersion relations of the polariton.

N.B. For transverse ($\mathbf{E} \perp \mathbf{k}$) waves in isotropic media, $\tilde{\varepsilon}(\mathbf{k}, \omega)$ is a scalar. In this case (31) reduces to the well-known dispersion formula:

$$\tilde{\varepsilon}(k, \omega) = \varepsilon_1(k, \omega) + i\varepsilon_2(k, \omega) = \frac{c^2k^2}{\omega^2} = (\tilde{n})^2 \tag{32}$$

where $\tilde{n} = n_1 + in_2$ is the complex index of refraction.

Before discussing the polariton we must examine the significance of excitons in this semiclassical approach and what the exciton concept means in terms of electromagnetic propagation.

2. *"Instantaneous Propagation of Interaction" Approximation: Coulomb Excitons*

In Frenkel's definition[134] excitons are crystal modes in which only electrostatic interactions are considered. It is known (see, for example, Ref. 116, p. 329, and Ref. 135, paragraph 6) that this amounts, in propagation terms, to setting the transverse (relative to $\mathbf{k}$) component of the electric field as zero ($\mathbf{E}_\perp = 0$). The approximation is said to be that of instantaneous propagation since it amounts to neglecting the time-dependent part of Maxwell's equations (28) and to letting $c \to \infty$.

Solution of (28) outlined previously leads to:

$$\mathbf{E} = \mathbf{E}_\parallel \quad \text{and} \quad \mathbf{D} = \tilde{\varepsilon}(\mathbf{k}, \omega)\cdot\mathbf{E}_\parallel = \mathbf{E}_\parallel + 4\pi\mathbf{P} \tag{33}$$
$$\nabla\cdot\mathbf{D} = 0$$

There are two kinds of solutions, both called Coulomb excitons because their origins can be traced to Coulomb gauge.

a. Longitudinal Excitons: $\mathbf{D} = 0$. If the electric displacement $\mathbf{D}$ is zero the polarization, of nontrivial solutions ($\mathbf{E} \neq 0$), is by (33) longitudinal, that is, in the same direction as $\mathbf{k}$. The eigenvalues of the longitudinal exciton are, by (33):

$$\tilde{\varepsilon}(\mathbf{k}, \omega) = 0 \tag{34}$$

If the excitation is entirely transverse, for a longitudinal polarization, there can be no coupling between the field and these modes, the dispersion curve

$\omega(k)$ being a straight line parallel to the axis: $\omega(k) = \omega_L$. However, as Hopfield[123] has pointed out, one should beware of indirect couplings, even in completely transverse experiments.

b. Nonlongitudinal Excitions: D·k = 0, D ≠ 0. It follows from (33) that the corresponding polarization must have a nonzero transverse component. Introducing the transverse dielectric tensor $\tilde{\varepsilon}_\perp(\mathbf{k}, \omega)$[136] defined by

$$\mathbf{D}_{(\omega,\mathbf{k})} = \tilde{\varepsilon}_{\perp(\omega,\mathbf{k})} \cdot \mathbf{E}_{\perp(\omega,\mathbf{k})} \tag{35}$$

the eigenvalues of these modes are given by (33) as:

$$|\tilde{\varepsilon}_\perp(\omega,\mathbf{k})|^{-1} = 0 \tag{36}$$

In contrast to longitudinal modes, nonlongitudinal modes are coupled to the electromagnetic field giving a dispersion curve $\omega(k)$ between that of the photon and that of the exciton. This will lead naturally to the idea of the polariton.

N.B. It will be noted that (36) contains one reason for "spatial dispersion," namely, variation of $\tilde{\varepsilon}$ with the magnitude of **k**. We see that this variation is linked to the variation of the excition energy $\omega_T(k)$. A second cause, not appearing here, is the dependence of the exciton–photon coupling on **k**.

3. *Retarded Interactions: The Polariton*

As the speed of light is finite, one must allow for the retardation of contributions from different regions of the medium in defining the electromagnetic wave. We must solve stages *a* and *b* of Section III.B.1. Before proceeding we must establish the relationship between the medium, which is composed of interacting molecules in an external field, and its semiclassical representation as an ensemble of oscillators vibrating under the influence of an effective, or local, field.

a. Contribution of Electronic Excitations to the Dielectric Tensor. Consider a set of neutral atoms which may become excited from their ground state $|\phi_0(\mathbf{r})\rangle$ to excited states $|\phi_j(\mathbf{r})\rangle$. In order not to mix different problems, suppose the molecular interactions are negligible (oriented gas).

If all these molecules are subjected to the same field, polarized along Ox, $(\mathbf{E}(t) = \mathbf{E}_x(e^{i\omega t} + e^{-i\omega t}))$, the response of the system may be found by a simple perturbation treatment. We look for wave functions of the form:

$$\Psi(\mathbf{r}, t) = \phi_{0(\mathbf{r})}\exp(-\omega_0 t) + \sum_j C_{j(t)}\phi_{j(\mathbf{r})}\exp(-\omega_j t)$$

where $\Psi(\mathbf{r},0) = \phi_0(\mathbf{r})$ and the (weak) perturbation is $e\mathbf{E}\cdot\mathbf{r}$. The $C_j(t)$ may be found by carrying $\Psi(\mathbf{r}, t)$ into Schrödinger's equation:

$$C_j(t) = e\mathbf{E}_x\mathbf{X}_{j0}F_{0j}\left\{\frac{1-\exp|i(\omega+\omega_j-\omega_0)t|}{\hbar|\omega+(\omega_j-\omega_0)|} + \frac{1-\exp|i(\omega+\omega_j-\omega_0)t|}{\hbar|\omega-(\omega_j-\omega_0)|}\right\}$$

where F_{0j} is the Franck–Condon factor for the $|0\rangle \to |j\rangle$ transition and

$$e\mathbf{X}_{0j} = \langle \phi_j(\mathbf{r})|e\mathbf{x}|\phi_0(\mathbf{r})\rangle$$

is the matrix element between $|0\rangle$ and $|j\rangle$ of the projection of the dipole moment on the direction of polarization of the electric field: $e(\mathbf{r}\cdot\mathbf{E})$.

The connection with dipole representation is as follows. In the average value of the dipole moment,

$$\mathbf{d} = \langle e\mathbf{x}\rangle = \langle \Psi(\mathbf{r}, t)|e\mathbf{x}|\Psi(r,t)\rangle$$
$$= E_x\sum_j e^2F_{0j}^2|X_{0j}|^2\frac{1}{\hbar}\left\{\frac{1}{\omega_j-\omega}+\frac{1}{\omega_j+\omega}\right\}(e^{i\omega t}+e^{-i\omega t})+\ldots$$

the part linear in E_x can be written:

$$\mathbf{d} = \alpha_{(\omega)}\cdot\mathbf{E} \tag{37}$$

which is the same as the classical relation between the polarization and the electric field.

$$\mathbf{p} = \alpha(\omega)\cdot\mathbf{E}$$

Introducing the oscillator strength f_j of the $|0\rangle \to |j\rangle$ transition

$$f_j = \frac{2m^*}{\hbar}\omega_j F_{0j}^2|X_{0j}|^2 \tag{38}$$

(m^*: effective mass of the electron),

verifying the Thomas–Reich–Kuhn rule $\Sigma f_j = 1$, $\alpha(\omega)$ may be written in the simplified form

$$\alpha(\omega) = \sum_j \frac{e^2F_{0j}^2|X_{0j}|^2}{\hbar}\cdot\frac{2\omega_j}{\omega_j^2-\omega^2} = \frac{e^2}{m^*}\sum_j\left(\frac{f_j}{\omega_j^2-\omega^2}\right) \tag{39}$$

Thus, according to (37) our system of molecules in an electromagnetic field may be replaced classically by a molecular polarizability $\alpha(\omega)$, in which each molecule is replaced by dipoles $\mathbf{d}_j$, corresponding to its j transitions

$$\mathbf{d}_j = \frac{e^2}{m^*} \frac{f_j}{\omega_j^2 - \omega^2} \cdot \mathbf{E} \tag{40}$$

N.B. This treatment to show the basis of the dipolar theory cannot be applied to all matter–field couplings and is in fact justified only for medium-strength transitions.

b. Local Field. The dipole representation set out above leads in principle, and if applicable, to the macroscopic polarization $\mathbf{P}(\mathbf{k}, \omega)$, obtained by summing the contributions of all the molecules in unit volume. From (29) one may then deduce the dielectric tensor, $\tilde{\varepsilon}(\mathbf{k}, \omega)$. However, in the above argument it was supposed that the N molecules of the system were independent and that the electric field acting on each molecule [the field in definition (40) of each dipole] was therefore the applied macroscopic field. Whereas in this model one gets the same answer as the quantum result for an "oriented gas" ($H = \Sigma^N H_n$) with N degenerate modes for each molecular mode, we know that it cannot be applied to a medium in which there are molecular interactions, no matter how weak. We must now determine how these forces affect the dipole theory.

Generally speaking, when interactions cannot be neglected, the electric field in the definition of a dipole (40) is not the applied field, but an effective or local field due to the superposition of the macroscopic field in the medium and of the field due to all the other dipoles.[132, 137]

$$\mathbf{E}^{\text{local}}_{(\omega,\mathbf{k})} = \mathbf{E}^{\text{external}}_{(\omega,\mathbf{k})} - L \cdot \mathbf{P}_{(\omega,\mathbf{k})} + \mathbf{E}^{d}_{(\omega,\mathbf{k})} \tag{41}$$

Macroscopic field in the medium — **Field created by the dipoles**

where L is a depolarization factor.

It has been shown[138] that taking account of the local field acting on the dipoles of the unperturbed molecules in the dipole theory is equivalent to the usual quantum treatment in which the external field is applied to the system perturbed by molecular interactions ($H = \Sigma_n H_n + \Sigma'_{m,m} V_{m,n}$).

c. Some Approximations. Any microscopic treatment of the local field to determine the dielectric tensor, and hence the dispersion curve of the polariton, is in general a difficult task. We shall make certain simplifying assumptions in what follows:

Linear medium. We shall suppose that the light intensities used are weak enough for the polarization to vary linearly with the fields ($\mathbf{P}(\mathbf{k}, \omega) = \alpha(\mathbf{k}, \omega)\mathbf{E}$).

No spatial dispersion. At optical frequencies only a very small part of the first Brillouin zone around $k = 0$ is used. We shall suppose, therefore, that the energy and the oscillator strength of the exciton are independent of the wave vector **k**, so that $\tilde{\varepsilon}(\mathbf{k}, \omega) = \tilde{\varepsilon}(0, \omega) = \tilde{\varepsilon}(\omega)$. Exciton band calculations for anthracene[139,140] show that this is reasonable for **k** normal to the (001) face.

No Umklapp effect. Strictly speaking, an external field $E(\mathbf{k})$ produces, in addition to a polarization $\mathbf{P}(\mathbf{k})$, a polarization $\mathbf{P}(\mathbf{k}+2\pi\mathbf{G})$, where **G** is a vector in the reciprocal lattice. Even for vanishing wave vectors, $\mathbf{k} \sim 0$, this polarization appears by reaction in the definition of the local field $\mathbf{E}(\mathbf{k})$. While this contribution may be introduced in the dipole theory, at the cost of some complications, we shall neglect it here.

Rigid lattice approximation. It is certain that a crystal cannot properly be described unless the nuclear movement is introduced,[113] for it influences the electric dipoles and hence the electric field, which in turn influences the nuclear movement. This interaction leads to nonstationary solutions and hence to absorption and emission by the crystal. The dipole theory accounts for this process phenomenologically by introducing a damping term in the equations for the dipole movements. While this term simulates relaxation in the crystal and its effect on the optical properties, its physical relation to such a relaxation channel is far from clear.

4. *Simple Example of a Polariton Dispersion Curve*

As an illustration of the above ideas, consider an isotropic dispersive medium represented by a set of harmonic oscillators of mass m, charge e, and **k**-independent eigenfrequency ω_0, placed one on each molecular position.

a. Negligible Absorption and Molecular Interactions. In this case the equation of the forced vibration **x** is

$$m\left(\frac{\partial^2 \mathbf{x}}{\partial t^2} + \omega_0^2 \mathbf{x}\right) = e\mathbf{E}^{\mathrm{loc}} = e\mathbf{E}^{\mathrm{ext}} = eE_0 e^{-i(\omega t - \mathbf{kr})} \tag{42}$$

of which the solution is

$$\mathbf{d} = e\mathbf{x} = \frac{e^2}{m}\left(\frac{1}{\omega_0^2 - \omega^2}\right)\mathbf{E}^{\text{ext}}$$

If N is the number of oscillators per unit volume, the macroscopic polarization $\mathbf{P}$ is

$$\mathbf{P}_{(\omega)} = \sum_1^N \mathbf{d} = \frac{Ne^2}{m}\left(\frac{1}{\omega_0^2 - \omega^2}\right)\mathbf{E}^{\text{ext}}$$

From (29), $\varepsilon(\omega)$ is given by:

$$\varepsilon(\omega) = \left(1 + \frac{4\pi Ne^2/m}{\omega_0^2 - \omega^2}\right) = 1 + \frac{\omega_P^2}{\omega_0^2 - \omega^2} \tag{43}$$

where ω_P has the dimension of a frequency and is called the *plasma frequency* (cf. forced movement of gas of free electrons; see Ref. 111, p. 95).

$$\omega_P^2 = \frac{4\pi Ne^2}{m}$$

Expression (43) is generalized to include the effects of other sources of polarization by adding a term ε_∞, independent of $\mathbf{k}$ and ω, to represent the dielectric constant at high frequencies, far from the resonance (Fig. 8).

$$\varepsilon(\omega) = \varepsilon_\infty + \frac{\omega_P^2}{\omega_0^2 - \omega^2} \tag{44}$$

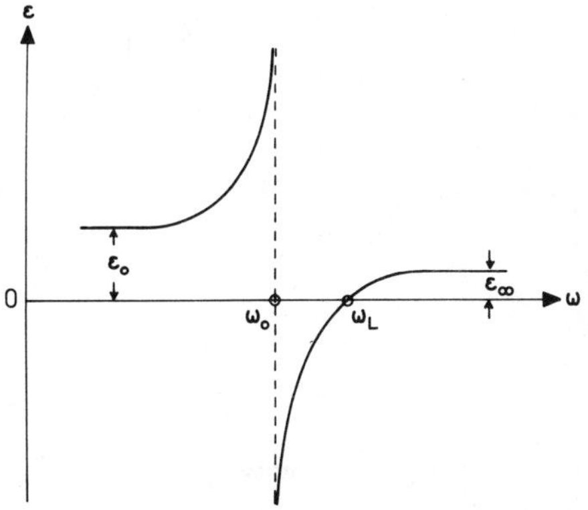

Fig. 8. Dielectric function $\varepsilon(\omega)$ of a set of harmonic, uncoupled, and undamped oscillators of natural frequency ω.

The dispersion relation (31) is then

$$\frac{c^2k^2}{\omega^2} = \varepsilon_\infty + \frac{\omega_P^2}{\omega_0^2 - \omega^2}$$

with solutions

$$\omega_{\pm}(k) = \frac{1}{2}\left(\frac{c^2k^2}{\varepsilon_\infty} + \omega_0^2 + \omega_P^2\right) \pm \frac{1}{2}\left[\left(\frac{c^2k^2}{\varepsilon_\infty} - \omega_0^2 - \omega_P^2\right)^2 + 4\frac{c^2k^2}{\varepsilon_\infty}\omega_P^2\right]^{1/2} \tag{45}$$

illustrated in Fig. 9, where

$$\omega_L^2 = \omega_0^2 + \frac{\omega_P^2}{\varepsilon_\infty}; \qquad \varepsilon_0 = \varepsilon_\infty + \frac{\omega_P^2}{\omega_0^2}; \qquad k = k_1 + ik_2$$

N.B. It follows that

$$\frac{\omega_L^2}{\omega_0^2} = \frac{\varepsilon_0}{\varepsilon_\infty}$$

the Lyddane–Sachs–Teller (LST) relation,[141] from which the strength of the matter–field coupling may be estimated, knowing the dielectric constant at high and low frequencies.

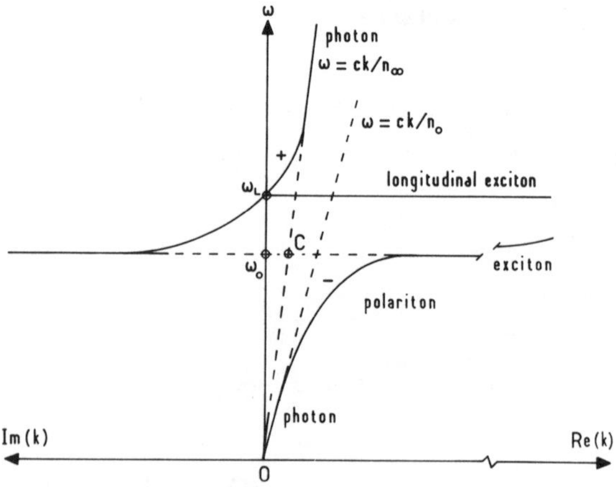

Fig. 9. Dispersion curve $k(\omega)$ of a set of harmonic, uncoupled, and undamped oscillators of natural frequency ω_0.

The dispersion curve calls for several remarks:

1. Close to point C, where coupling is large, neither photon nor exciton are natural modes of the crystal–field system. This is because when $\omega \sim \omega_0$ oscillators vibrating at frequency ω_0 will emit and their emission will tend to be absorbed. These two modes repel each other and lead to $(+)$ and $(-)$ branches, here both transverse (isotropic medium), for the polariton dispersion curve.

2. At high ($\omega \gg \omega_L$) and low ($\omega \ll \omega_0$) frequencies the oscillators do not exchange energy with the field, which may propagate in the crystal. Note, however, that at low frequencies, even far from the resonance, the photon is very much retarded.

3. As $\omega \to \omega_0$, the polariton becomes less and less like a photon and more and more like a pure exciton at ω_0. As we saw in Section III.B.2, the energy of the exciton is indeed a pole of $\varepsilon_\perp(\omega)$; here, $\varepsilon(\omega) = \varepsilon_\perp(\omega)$.

4. Between ω_0 and ω_L the wave vector is a pure imaginary. The only form of propagation in the crystal is a wave decaying in $\exp(-|k_2|x)$ perpendicular to the surface. If there is no damping, therefore, the crystal is totally reflecting in this region, called the "stopping band."

5. Meaning of ω_L. When $k \to 0$, the $(-)$ and $(+)$ branches tend respectively to $\omega^-_{(0)} = 0$ and $\omega^+_{(0)} = \omega_L = (\omega_0^2 + \omega_P^2/\varepsilon_\infty)^{1/2}$.

According to Section III.B.2, the value of the frequency at which the permittivity $\varepsilon(\omega)$ is zero also corresponds to the longitudinal exciton. Hence, for $\omega = \omega_L$, the transverse polariton and the longitudinal exciton are degenerate, which is quite natural given that when $k = 0$ the distinction between transverse and longitudinal modes is meaningless.

b. Damping and the Local Field. The above model may be improved by drawing in the interactions between molecules in the form of the local field, and the global effect of the relaxation channels in the form of damping of the oscillations. The equation of motion of the dipole (15) is then

$$m\left(\frac{\partial^2 \mathbf{x}}{\partial t^2} + \gamma \frac{\partial \mathbf{x}}{\partial t} + \omega_0^2 \mathbf{x}\right) = e \cdot \mathbf{E}^{\text{local}} \tag{46}$$

In a cubic lattice the local field is the sum of the external field and of the Lorentz correction $\mathbf{E}^d = (4\pi/3)\mathbf{P}$. Since $L = 0$, in the present example (41) yields

$$\mathbf{E}^{\text{local}} = \mathbf{E}^{\text{ext}} + \frac{4\pi \mathbf{P}}{3} = \mathbf{E}^{\text{ext}} + \frac{4\pi}{3} Ne\mathbf{x}$$

which carried into (46) gives

$$m\left(\frac{\partial^2 \mathbf{x}}{\partial t^2}+\gamma\frac{\partial \mathbf{x}}{\partial t}+\left(\omega_0^2-\frac{4\pi}{3}\frac{Ne^2}{m}\right)\mathbf{x}\right)=e\cdot\mathbf{E}^{\text{ext}} \tag{47}$$

It should be realized that in this expression the local field shifts the resonance of the dipoles from ω_0 for the oriented gas to ω_T such that

$$\omega_T^2=\omega_0^2-\frac{4\pi}{3}\frac{Ne^2}{m}$$

This confirms the equivalence between the local field in the dipole theory and the molecular interactions in the quantum theory. The solution of (47) is

$$e\mathbf{x}=\mathbf{d}=\frac{e^2/m}{\omega_T^2-\omega^2-i\omega\gamma}\mathbf{E}^{\text{ext}} \quad \text{and} \quad \mathbf{P}=\frac{Ne^2/m}{\omega_T^2-\omega^2-i\omega\gamma}\mathbf{E}^{\text{ext}}$$

Putting these values in (29) we have

$$\tilde{\varepsilon}(\omega)=\varepsilon_\infty+\frac{4\pi Ne^2/m}{\omega_T^2-\omega^2-i\omega\gamma}=\varepsilon_\infty+\frac{\omega_P^2}{\omega_T^2-\omega^2-i\omega\gamma}=\varepsilon_1(\omega)+i\varepsilon_2(\omega) \tag{48}$$

This expression for $\tilde{\varepsilon}(\omega)$ leads to complex wave vectors at some frequencies from which the reflection and absorption spectra follow (see Fig. 10).

5. *Polariton Dispersion Curves in Anisotropic Media*

Up to now we have considered isotropic media, where the dielectric tensor is a simple scalar and there is only one transverse mode in the dispersion

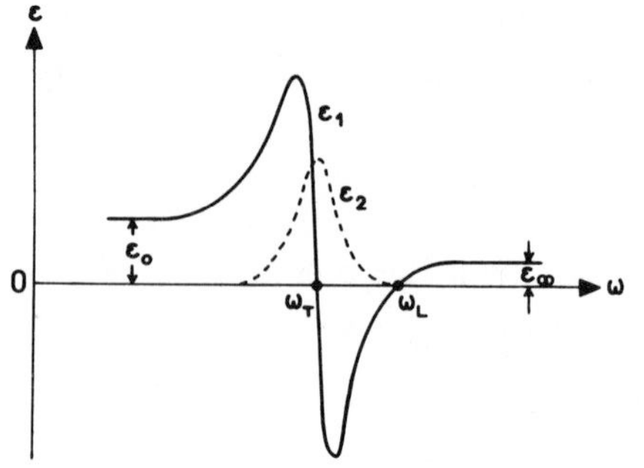

Fig. 10. Real (ε_1) and imaginary (ε_2) parts of the dielectric function $\tilde{\varepsilon}(\omega)$ of a set of harmonic, coupled, and damped oscillators of natural frequency ω_0.

equations. In less symmetrical systems $\tilde{\varepsilon}(\mathbf{k}, \omega)$, or $\tilde{\varepsilon}(\omega)$ for brevity, is a 3×3 complex tensor. Solution of (31) leads, depending on the direction of the wave vector **k**, to the existence of several modes of propagation. They are not entirely transverse, as is the case in isotropic media. Consider, for example, the "relatively simple" case of a uniaxial crystal.

a. Uniaxial Crystals.[142, 143] Although the real and imaginary principal values need not coincide, we shall suppose that $\tilde{\varepsilon}(\omega)$ is of the form

$$\tilde{\varepsilon}(\omega)=\left\|\begin{matrix}\varepsilon_{\perp}(\omega) & 0 & 0\\ 0 & \varepsilon_{\perp}(\omega) & 0\\ 0 & 0 & \varepsilon_{\|}(\omega)\end{matrix}\right\|$$

where the optical axis is along $0z$. Here $\varepsilon_{\|}(\omega)$ and $\varepsilon_{\perp}(\omega)$ are supposed to be known and of the form (44) or (48). If we suppose that **k** is perpendicular to $0y$ (this is no loss of generality),[131] and if θ is the angle between **k** and the optical axis, Eq. (31) becomes:

$$\left\|\begin{matrix}\dfrac{\omega^2}{c^2}\varepsilon_{\perp}-k^2\cos^2\theta & 0 & k^2\sin\theta\cos\theta\\ 0 & \dfrac{\omega^2}{c^2}\varepsilon_{\perp}-k^2 & 0\\ k^2\sin\theta\cos\theta & 0 & \dfrac{\omega^2}{c^2}\varepsilon_{\|}-k^2\sin^2\theta\end{matrix}\right\|=0$$

No matter what the direction of **k**, there is always a completely transverse solution given by

$$\frac{\omega^2}{c^2}\varepsilon_{\perp}(\omega)-k^2=0,$$

exactly as above. This solution is generally called the *ordinary polariton*.

The other solutions depend on the direction of **k** and are not generally transverse. If **k** is perpendicular or parallel to the optical axis they are the solutions of

$$\varepsilon_{\|}(\omega)\left(\frac{c^2k^2}{\omega^2}-\varepsilon_{\perp}(\omega)\right)=0 \qquad (\text{for } \theta=0)$$

$$\varepsilon_{\perp}(\omega)\left(\frac{c^2k^2}{\omega^2}-\varepsilon_{\|}(\omega)\right)=0 \qquad (\text{for } \theta=90°)$$

and are respectively a transverse polariton (better called extraordinary polariton) and a longitudinal exciton for which $\varepsilon_{\|}(\omega)=0$ (for $\theta=0$) or $\varepsilon_{\perp}(\omega)=0$ (for $\theta=90°$).

b. Anthracene. The dispersion formulae of anthracene and other biaxial molecular crystals of the space group C_{2h}^5 are even more complicated. If **b** is the symmetry axis of order 2 in the frame **a**, **b**, **c**′ [**c**′ axis perpendicular to the (**a**, **b**) plane], $\tilde{\varepsilon}(\omega)$ is of the form[144]

$$\varepsilon(\omega)=\begin{vmatrix} \varepsilon_{aa} & 0 & \varepsilon_{ac'} \\ 0 & \varepsilon_{bb} & 0 \\ \varepsilon_{ac'} & 0 & \varepsilon_{c'c'} \end{vmatrix}$$

and the dispersion relation is

$$\begin{vmatrix} \dfrac{\omega^2}{c^2}\varepsilon_{aa}-k^2+k_a^2 & k_ak_b & \dfrac{\omega^2}{c^2}\varepsilon_{ac'}+k_ak_{c'} \\ k_bk_a & \dfrac{\omega^2}{c^2}\varepsilon_{bb}-k^2+k_b^2 & k_bk_{c'} \\ \dfrac{\omega^2}{c^2}\varepsilon_{ac'}+k_ak_{c'} & k_{c'}k_b & \dfrac{\omega^2}{c^2}\varepsilon_{c'c'}-k^2+k_{c'}^2 \end{vmatrix}=0 \quad (49)$$

Realizing that the $\varepsilon_{ij}(\mathbf{k},\omega)$ are generally complex functions of **k** (since spatial dispersion is large in some directions), it is easy to see the complexity of the description of the optical properties of anisotropic crystals for arbitrary directions of the wave vector **k**.[144–147]

According to (49), completely transverse propagation in anthracene can occur only for:

- **k** normal to the (001) face and **E** polarized along **b**.
- **k** normal to the (010) face and for two directions of **E**.[144]

The bulk of our experimental work involves the first and simpler of these cases. We now apply the dipole theory to the dispersion curve of anthracene in the first case.

C. Application of the Semiclassical Theory to Crystalline Anthracene

1. *Dispersion Equation of an Isotropic Crystal*

The above theory, illustrated in the case of an isotropic medium with only one transition, can be extended to the case of an anisotropic medium such as anthracene, or even to a triclinic system. The general principle of a set of dipoles in forced motion due to a local field is the same, but is harder to apply because of the lower symmetry, the greater number of transitions (hence of dipoles), and the slow convergence of the dipole sums in the local field.

Several authors[139, 148–158] have tried to explain the dependence, on the direction of the wave vector, of the exciton energy in anthracene and similar crystals, close to $\mathbf{k} = 0$, and have all to some extent discussed exciton–photon coupling. Most of them solved the problem of the convergence of the dipole sums by using the old, but very good, treatment given by Ewald,[159, 160] who was one of the first to give a microscopic basis to the dispersion theories of Lorentz and Drude. We shall examine the dispersion of the polariton in the formalism due to Mahan[150] and Philpott,[154] who have made a special study of this problem in anthracene.

a. Dynamic Equation of a Dipole. Consider an infinite crystal with σ nonequivalent molecules per unit cell, in which each molecule is indexed by $n\alpha$ (or the radius vector $\mathbf{r}_{n\alpha}$), where n is the number of the cell and α is the place in this cell.

If each molecule may undergo several vibronic transitions $|0\rangle \to |r\rangle$, then by the equivalence established in Section III.B.3, it may be represented in an electromagnetic field by a set of dipoles

$$\sum_r \mathbf{d}_{n\alpha, r}(t)$$

located on the sites $n\alpha$ (point dipole approximation) pointing in the same direction as the transition moments $\mathbf{M}_{n\alpha, r}$. We shall use the following notations:

e and m = electronic charge and mass

ω_r = angular frequency (rad/s) of the $|0\rangle \to |r\rangle$ transition

$f_r = \dfrac{2m\omega_r F_{0r}^2 |M_{0r}|^2}{\hbar}$: oscillator strength of the $|0\rangle \to |r\rangle$ transition.

$\hat{d}_{n\alpha, r}$ = unit vector along dipole $\mathbf{d}_{n\alpha, r}$ (or $\mathbf{M}_{n\alpha, r}$)

$\mathbf{E}^{\text{loc}}_{(\mathbf{r}_{n\alpha}, t)}$ = local electric field at site $n\alpha$.

The equation of motion of the dipole is (see Section III.B.3)

$$m\left(\frac{\partial^2}{\partial t^2} + \omega_r^2\right)\mathbf{d}_{n\alpha, r}(t) = e^2 \cdot f_r \cdot \hat{d}_{\alpha r}\left(\hat{d}_{\alpha r} \cdot \mathbf{E}^{\text{loc}}_{(\mathbf{r}_{n\alpha}, t)}\right) \tag{50}$$

b. Form of the Local Field. The local field $\mathbf{E}^{\text{loc}}_{(\mathbf{r}_{n\alpha}, t)}$ is the sum of the applied field $\mathbf{E}^{0}_{(\mathbf{r}_{n\alpha}, t)}$ and the field $\mathbf{E}^{d}_{(\mathbf{r}_{n\alpha}, t)}$ at $n\alpha$ due to all the other vibrating dipoles on sites $p\beta \neq n\alpha$. The field $\mathbf{E}^d_{(\mathbf{r}_{n\alpha},t)}$ may be found by using Hertz's formalism.

The translational invariance of the crystal may be introduced by writing the dipoles as plane waves:

$$\mathbf{d}_{n\alpha,r}(t) = \mathbf{d}_{\alpha r} e^{i(\mathbf{k}\cdot\mathbf{r}_{n\alpha} - \omega t)} \tag{51}$$

The Hertz vector at site $n\alpha$ due to the dipoles $\mathbf{d}_{p\beta,s}$ is

$$\boldsymbol{\pi}(\mathbf{r}_{n\alpha},t) = \sum_{p\beta}\sum_{s} \frac{\mathbf{d}_{\beta s}}{[\mathbf{r}_{n\alpha} - \mathbf{r}_{p\beta}]} \exp\left[i\left(\mathbf{k}\cdot\mathbf{r}_{p\beta} + \frac{\omega}{c}[\mathbf{r}_{n\alpha} - \mathbf{r}_{p\beta}] - \omega t\right]\right.$$

The associated electric field $\mathbf{E}^d$ is

$$\mathbf{E}^d_{(\mathbf{r}_{n\alpha},t)} = \left[\nabla\nabla - \frac{1}{c^2}\frac{\partial^2}{\partial t^2}\right]\boldsymbol{\pi}(\mathbf{r}_{n\alpha},t)$$

or expanding,

$$\begin{aligned}\mathbf{E}^d_{(\mathbf{r}_{n\alpha},t)} = \sum_{p\beta}{}'\sum_{s} &\left\{\left(\frac{\omega^2}{c^2}R^{-1} + i\frac{\omega}{c}R^{-2} - R^{-3}\right)\mathbf{d}_{\beta s}\right.\\ &\left.+\left(3R^{-5} - 3\frac{\omega}{c}iR^{-4} - \frac{\omega^2}{c^2}R^{-3}\right)(\mathbf{d}_{\beta s}\cdot\mathbf{R})\mathbf{R}\right\}\\ &\cdot\exp\left[i\left(\mathbf{k}\cdot\mathbf{r}_{p\beta} + \frac{\omega}{c}R - \omega t\right)\right]\end{aligned} \tag{52}$$

where $\mathbf{R} = \mathbf{r}_{n\alpha} - \mathbf{r}_{p\beta}$; $R =$ length of $\mathbf{R}$; and $\hat{R} = \mathbf{R}/R$ (unit vector along $\mathbf{R}$).

c. Dispersion Equation. The dispersion relation of the natural polariton modes (no applied field) results from putting (52) for the local field in (50), the equation of motion of the dipoles on site $n\alpha$. Using the time-dependent expression (51) of the dipoles, one arrives at:

$$\sum_{\beta,s}\left[\left(\omega_r^2 - \omega^2\right)\delta_{\alpha\beta}\delta_{rs} + \omega_p^2 f_r \phi_{\alpha r\beta s}(\mathbf{k},\omega)\right]\bar{d}_{\beta s} = 0 \tag{53}$$

where ω_p is the "plasma" frequency of Section III.B.4, given by

$$\omega_p^2 = \frac{4\pi Ne^2}{m} = \frac{4\pi e^2}{mV_c} \quad \text{or} \quad \omega_p^2 = \frac{e^2}{\varepsilon_0 mV_c}$$

(CGS) (S.I.)

with V_c the volume of the unit cell and

$$\phi_{\alpha r \beta s}(\mathbf{k}, \omega) = \hat{d}_{\alpha r} \cdot \tilde{\phi}_{\alpha\beta}(\mathbf{k}, \omega) \cdot \hat{d}_{\beta s}$$

with

$$\tilde{\phi}_{\alpha\beta}(\mathbf{k}, \omega) = \frac{V_c}{4\pi} \sum_p \left\{ (1 - \delta_{np}\delta_{\alpha\beta}) \left[\left(1 - i\frac{\omega}{c}R\right) R^{-3}(\mathbb{1} - 3\hat{R}\hat{R}) - \frac{\omega^2}{c^2} R^{-1}(\mathbb{1} - \hat{R}\hat{R}) \right] \cdot \exp\left[i\left(\mathbf{k}\cdot\mathbf{R} + \frac{\omega}{c}R\right)\right] \right\} \tag{54}$$

Diad (54) is the retarded dipole–dipole interaction between sites α and β.

d. Dipole Sums. The $\tilde{\phi}_{\alpha\beta}(\mathbf{k}, \omega)$ of an infinite crystal are hard to calculate and can only be summed if approximations are made. We shall divide the problem into two parts.

First, the $\tilde{\phi}_{\alpha\beta}(\mathbf{k}, \omega)$ may be expressed as functions of the instantaneous interactions. Mahan has shown in the appendix of Ref. 115 that if $\hat{k}$ is the unit vector along the wave vector $\mathbf{k}$ and $\tilde{T}_{\alpha\beta}(\mathbf{k}, \omega)$ is the diad representing *instantaneous* or static dipole–dipole interaction between α and β:

$$\tilde{T}_{\alpha\beta}(\mathbf{k}, \omega) = \left(\frac{V_c}{4\pi}\right) \sum_p (1 - \delta_{np}\delta_{\alpha\beta}) R^{-3}(\mathbb{1} - 3\mathbf{RR})\exp(i\mathbf{k}\cdot\mathbf{R}) \tag{55}$$

with $n(\mathbf{k}, \omega) = ck/\omega$, the refractive index, then

$$\tilde{\phi}_{\alpha\beta}(\mathbf{k}, \omega) = \tilde{T}_{\alpha\beta} + \frac{\mathbb{1} - \hat{k}\hat{k}}{1 - n^2(\mathbf{k}, \omega)} + O\left(\frac{\omega_a}{c}\right)^2 \tag{56}$$

This result holds for all wave vectors in the first Brillouin zone at frequencies such that $(\omega a/c)^2 \ll 1$. Since $\omega/c \sim 25{,}000\ \text{cm}^{-1}$ and a, the length of the unit cell, is roughly 10^{-7} cm, this condition is fulfilled.

In the second stage we must sum the *instantaneous* dipole terms $\tilde{T}_{\alpha\beta}(\mathbf{k}, \omega)$, for which the same convergence problems occur. Long-range contributions, which are by no means negligible, cause the crystal energy to be a nonanalytic function of $\mathbf{k}$.[161]

Ewald solved the problem by replacing the distribution of dipoles $\mathbf{d}_{p\beta}$ around dipole $\mathbf{d}_{n\alpha}$ by a normal distribution.[162] Convergence is rapid for the right value of the width of the bell-shaped distribution. The dipole sums depend on the form chosen.[157] Modern methods of summation, such as summation by planes, give results identical with Ewald's,[158] differences in the methods being realized in the time to complete the summation.

In Ewald's method, for small k the $\tilde{T}_{\alpha\beta}$ have the form

$$\tilde{T}_{\alpha\beta}(\mathbf{k}, \iota) = \tilde{t}_{\alpha\beta}(\mathbf{k}) + \hat{k}\hat{k} \tag{57}$$

where the first term $\tilde{t}_{\alpha\beta}$ is an analytic function of the wave vector and the second term $\hat{k}\hat{k}$, representing the (macroscopic) long-range interactions, is the nonanalytic part of $\tilde{T}_{\alpha\beta}$. [In (54) we will finally have terms in $(\hat{d}_{\alpha r}\cdot \hat{k})(\hat{k}\cdot \hat{d}_{\beta s})$ whose values for vanishing $\mathbf{k}$ depend on its direction.]

Carrying this form of $\tilde{T}_{\alpha\beta}$ into (56) for $\tilde{\phi}_{\alpha\beta}$ we have

$$\tilde{\phi}_{\alpha\beta}(\mathbf{k}, \omega) = \tilde{t}_{\alpha\beta}(\mathbf{k}) + \frac{\mathbf{kk} - (\omega^2/c^2)\mathbb{1}}{k^2 - \omega^2/c^2} \tag{58}$$

e. Explicit form of the Dispersion Relation. Like the lowest $\pi - \pi^*$ transitions of most plane aromatic hydrocarbons with D_{2h} symmetry, those of anthracene are in-plane polarized along the long (L) or the short (M) axes.

In (53) the sum over all molecular transitions $(\Sigma_s|0\rangle \to |s\rangle)$ can be split into a sum, $\Sigma_n|0\rangle \to |n\rangle$, over M-polarized transitions and another, $\Sigma_l|0\rangle \to |l\rangle$, over L-polarized ones.

Moreover, this expression holds for any initial choice of dipole $\mathbf{d}_{\alpha r}$, so it may be written for all values of α and r.

Let us choose first of all $r = m^\circ$, and M-polarized transition. Equation (53) becomes

$$\sum_\beta \left\{ \sum_m \left[\delta_{\alpha\beta}\delta_{m^\circ m} + \frac{\omega_p^2}{\omega_{m^\circ}^2 - \omega^2} f_{m^\circ}\phi_{\alpha m^\circ \beta m} \right] \bar{d}_{\beta m} \right.$$
$$\left. + \sum_l \left[\frac{\omega_p^2}{\omega_{M^\circ}^2 - \omega^2} f_{m^\circ}\phi_{\alpha m^\circ \beta l} \right] \bar{d}_{\beta l} \right\} = 0$$

or

$$\sum_\beta \left\{ \delta_{\alpha\beta} + \sum_m \left[\frac{\omega_p^2}{\omega_{m^\circ} - \omega^2} f_{m^\circ}\phi_{\alpha m^\circ \beta m}\bar{d}_{\beta m} \right] \right.$$
$$\left. + \sum_l \left[\frac{\omega_p^2}{\omega_{m^\circ}^2 - \omega^2} f_{m^\circ}\phi_{\alpha m^\circ \beta l}\bar{d}_{\beta l} \right] \right\} = 0 \tag{59}$$

Since the $\tilde{\phi}_{\alpha r \beta s}$ depend only on the polarizations $\hat{d}_{\alpha r}$ and $\hat{d}_{\beta s}$ of transitions $|r\rangle$ and $|s\rangle$, we must have, for all m: $\phi_{\alpha m^\circ \beta m} = \phi_{\alpha M \beta M}$ and for all l: $\phi_{\alpha m^\circ \beta l} = \phi_{\alpha M \beta L}$.

Introducing the L and M components of the total molecular polarization, $\mathbf{P}_{\beta M} = \Sigma_m \mathbf{d}_{\beta m}$ and $\mathbf{P}_{\beta L} = \Sigma_l \mathbf{d}_{\beta l}$, (59) becomes:

$$\sum_\beta \left\{ \left[\delta_{\alpha\beta} + \frac{\omega_p^2}{\omega_{m^\circ}^2 - \omega^2} f_{m^\circ} \phi_{\alpha M \beta M} \right] \bar{P}_{\beta M} + \left[\frac{\omega_p^2}{\omega_{m^\circ}^2 - \omega^2} f_{m^\circ} \phi_{\alpha M \beta L} \right] \bar{P}_{\beta L} \right\} = 0$$

Summing this expression for all m° (M polarized), we have

$$\sum_\beta \left\{ \left[\delta_{\alpha\beta} + \sum_m \left(\frac{\omega_p^2}{\omega_m^2 - \omega^2} f_m \right) \phi_{\alpha M \beta M} \right] \bar{P}_{\beta M} + \left[\sum_m \left(\frac{\omega_p^2}{\omega_m^2 - \omega^2} f_m \right) \phi_{\alpha M \beta L} \right] \bar{P}_{\beta L} \right\} = 0 \tag{60}$$

Similarly, for $r = l^\circ$, an L-polarized transition, we have

$$\sum_\beta \left\{ \left[\delta_{\alpha\beta} + \sum_l \left(\frac{\omega_p^2}{\omega_l^2 - \omega^2} f_l \right) \phi_{\alpha L \beta L} \right] \bar{P}_{\beta L} + \left[\sum_l \left(\frac{\omega_p^2}{\omega_l^2 - \omega^2} f_l \right) \phi_{\alpha L \beta M} \right] \bar{P}_{\beta M} \right\} = 0 \tag{61}$$

Letting α_M and α_L be the principal components along L and M of the molecular polarization,

$$4\pi\alpha_M = \sum_m \left(\frac{\omega_p^2}{\omega_m^2 - \omega^2} f_m \right) \quad \text{and } 4\pi\alpha_L = \sum_l \left(\frac{\omega_p^2}{\omega_l^2 - \omega^2} f_l \right) \tag{62}$$

Eqs. (60) and (61) may be written in the condensed form

$$\sum_\beta \left[\left(\delta_{\alpha\beta} + 4\pi\alpha_M \phi_{\alpha M \beta M} \right) \bar{P}_{\beta M} + \left(4\pi\alpha_M \phi_{\alpha M \beta L} \right) \bar{P}_{\beta L} \right] = 0 \tag{63}$$

$$\sum_\beta \left[\left(\delta_{\alpha\beta} + 4\pi\alpha_L \phi_{\alpha L \beta L} \right) \bar{P}_{\beta L} + \left(4\pi\alpha_L \phi_{\alpha L \beta M} \right) \bar{P}_{\beta M} \right] = 0 \tag{64}$$

If there are σ nonequivalent molecules per unit cell, these relations written for α and β between 1 and σ lead to a $2\sigma \times 2\sigma$ determinant for the eigenmodes $\omega(k)$ of the polariton. Organic crystals such as napthalene, anthracene, and tetracene have two molecules per unit cell. We must solve:

$$\begin{vmatrix} (1+4\pi\alpha_M\phi_{1M1M}) & 4\pi\alpha_M\phi_{1M2M} & 4\pi\alpha_M\phi_{1M1L} & 4\pi\alpha_M\phi_{1M2L} \\ 4\pi\alpha_M\phi_{2M1M} & (1+4\pi\alpha_M\phi_{2M2M}) & 4\pi\alpha_M\phi_{2M1L} & 4\pi\alpha_M\phi_{2M2L} \\ 4\pi\alpha_L\phi_{1L1M} & 4\pi\alpha_L\phi_{1L2M} & (1+4\pi\alpha_L\phi_{1L1L}) & 4\pi\alpha_L\phi_{1L2L} \\ 4\pi\alpha_L\phi_{2L1M} & 4\pi\alpha_L\phi_{2L2M} & 4\pi\alpha_L\phi_{2L1L} & 1+4\pi\alpha_L\phi_{2L2L} \end{vmatrix} \begin{vmatrix} \bar{P}_{1M} \\ \bar{P}_{2M} \\ \bar{P}_{1L} \\ \bar{P}_{2L} \end{vmatrix} = 0$$

2. *Application to Anthracene*

a. General Formula. Since anthracene has a center of symmetry,

$$\phi_{1X2Y} = \phi_{2X1Y} \quad \text{where } X, Y \in \{M, L\}$$

This is the only simplification of the determinant which can be made for arbitrary incidence of the field.

If **k** is normal to the (001) face (the experimental situation investigated here), then[139]

$$\phi_{1X1Y} = \phi_{2X2Y} \quad \text{where } X, Y \in \{M, L\}$$

Now, taking the sum and the difference of (63) and (64) and introducing the following combinations for the polarizations of the unit cell,

$$\phi^{\pm}_{MM} = \phi_{1M1M} \pm \phi_{1M2M}; \qquad \phi^{\pm}_{LL} = \phi_{1L1L} \pm \phi_{1L2L}; \qquad \phi^{\pm}_{ML} = \phi_{1M1L}\phi_{1M2L};$$
$$\mathbf{P}^{\pm}_{M} = \mathbf{P}_{1M} \pm \mathbf{P}_{2M}; \qquad \mathbf{P}^{\pm}_{L} = \mathbf{P}_{1L} \pm \mathbf{P}_{2L} \tag{65}$$

we have

$$\left|\begin{array}{cc|cc} 1+4\pi\alpha_M\phi^+_{MM} & 4\pi\alpha_M\phi^+_{ML} & 0 & 0 \\ 4\pi\alpha_L\phi^+_{ML} & 1+4\pi\alpha_L\phi^+_{LL} & 0 & 0 \\ \hline 0 & 0 & 1+4\pi\alpha_M\phi^-_{MM} & 4\pi\alpha_M\phi^-_{ML} \\ 0 & 0 & 4\pi\alpha_L\phi^-_{ML} & 1+4\pi\alpha_L\phi^-_{LL} \end{array}\right| \cdot \begin{vmatrix} \bar{P}^+_M \\ \bar{P}^+_L \\ \bar{P}^-_M \\ \bar{P}^-_L \end{vmatrix} = 0 \tag{66}$$

Thus the polariton modes $\omega(\mathbf{k})$ corresponding to normal incidence on the $(\mathbf{a}, \mathbf{b})$ face of the crystal require that both blocks ($+$ and $-$) in this determinant are set to zero.

Due to the symmetry relations between the nonequivalent molecules 1 and 2 in Fig. 11, the cell polarizations are directed as follows:

Polarization $(+)$: $$\mathbf{P}_M^+ = \sum_m (\mathbf{d}_{1m} + \mathbf{d}_{2m}) \quad \text{and} \quad \mathbf{P}_L^+ = \sum_l (\mathbf{d}_{1l} + \mathbf{d}_{2l})$$
all in the $(\mathbf{a}, \mathbf{c})$ plane

Polarization $(-)$: $$\mathbf{P}_M^- = \sum_m (\mathbf{d}_{1m} - \mathbf{d}_{2m}) \quad \text{and} \quad \mathbf{P}_L^- = \sum_l (\mathbf{d}_{1l} - \mathbf{d}_{2l})$$
all along the monoclinic axis **b**

N.B. Both $\mathbf{P}_M^+$ and $\mathbf{P}_L^+$ are in the $(\mathbf{a}, \mathbf{c})$ plane but are not colinear. For this reason, the m and l polarized contributions vary with ω and change direction in the $(\mathbf{a}, \mathbf{c})$ plane. Hence both the numerical values and the directions of the two principal components of the dielectric tensor, excepting ε_{bb}, change with ω (axial dispersion).[145, 163]

The two exciton transitions corresponding to a molecular transition, polarized respectively along the **b** axis and in the $(\mathbf{a}, \mathbf{c})$ plane, turn up in the polariton (propagation mode), for which $\omega(\mathbf{k})$ satisfies both the $(+)$ and the $(-)$ blocks of (66) set to zero.

By suitable polarization of the external field (hence creation of particular matter polarizations), we may set either $\mathbf{P}^+$ or $\mathbf{P}^-$ to zero and reduce the problem to that of one block.

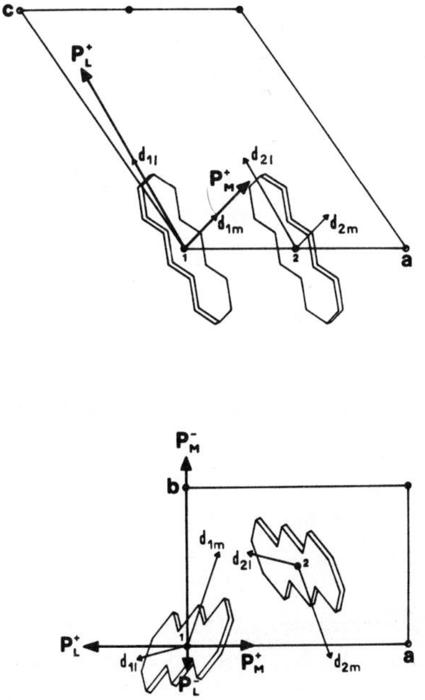

Fig. 11. Projections on the $(\mathbf{a}, \mathbf{b})$ and $(\mathbf{a}, \mathbf{c})$ planes of the $\mathbf{P}^+$ and $\mathbf{P}^-$ unit cell polarizations corresponding to the short axis (M) and long axis (L) polarized transitions $\mathbf{d}_m$ and $\mathbf{d}_l$.

Special case of $\mathbf{k} \perp (001)$ *and* $\mathbf{E}$ *polarized along the* $\mathbf{b}$ *axis.* These conditions are one of the two ways of creating transverse propagation modes in the crystal and lead to linear polarization. The other, $\mathbf{k} \perp (010)$ and $\mathbf{E}$ polarized along either of the principal directions depending on ω, corresponds to elliptically polarized propagation due to the angle between $\mathbf{P}_M^+$ and $\mathbf{P}_L^+$.

If $\mathbf{E}(\mathbf{k}, \omega)$ is $\mathbf{b}$ polarized, $\mathbf{P}_M^+$ and $\mathbf{P}_L^+$ vanish and the dispersion curve corresponds to the vanishing of the $(-)$ block in determinant (66):

$$(1+4\pi\alpha_M\phi_{MM}^-)(1+4\pi\alpha_L\phi_{LL}^-)-(4\pi^2)\alpha_M\alpha_L(\phi_{ML}^-)^2=0 \tag{67}$$

b. Dispersion Due to One Transition. If there is only one transition, (67) reduces to

$$1+4\pi\alpha_M\phi_{MM}^-=0 \tag{68}$$

where

$$4\pi\alpha_M=\frac{\omega_p^2 f_m}{\omega_m^2-\omega^2}$$

and

$$\phi_{MM}^-=\phi_{1M1M}-\phi_{1M2M}=\hat{d}_{1M}\cdot\tilde{\phi}_{11}\cdot\hat{d}_{1M}-\hat{d}_{1M}\cdot\tilde{\phi}_{12}\cdot\hat{d}_{2M}$$

and

$$\tilde{\phi}_{\alpha\beta}=\tilde{t}_{\alpha\beta}+\frac{\mathbf{kk}-(\omega/c)^2\mathbb{1}}{k^2-(\omega/c)^2}$$

expanding,

$$\phi_{MM}^-=t_{1M1M}-t_{1M2M}+\frac{k^2[(\hat{d}_{1M}\cdot\hat{k})(\hat{k}\cdot\hat{d}_{1M})-(\hat{d}_{1M}\cdot\hat{k})(k\cdot\hat{d}_{2M})]-(\omega/c)^2[(\hat{d}_{1M}\cdot\hat{d}_{1M})-(\hat{d}_{1M}\cdot\hat{d}_{2M})]}{k^2-(\omega/c)^2}$$

Let

$$t_{1M1M}-t_{1M2M}=A \quad \text{and} \quad (\hat{d}_{1M}\cdot\hat{d}_{1M})-(\hat{d}_{1M}\cdot\hat{d}_{2M})=B$$

Since

$$(\hat{d}_{1M}\cdot\hat{k})(\hat{k}\cdot\hat{d}_{1M})-(\hat{d}_{1M}\cdot\hat{k})(\hat{k}\cdot\hat{d}_{2M})=0$$

whenever **k** is normal to the (**a**, **b**) face, with **E**||**b**, we have

$$\phi_{MM}^{-}=A-\frac{4\pi^2\omega^2 B}{k^2-4\pi^2\omega^2}\qquad(\omega \text{ in cm}^{-1})$$

Putting the last expression in (68), we have the dispersion curve

$$k^2=4\pi^2\omega^2+\frac{\omega_p^2 f_m B\cdot 4\pi^2\omega^2}{\omega_m^2+\omega_p^2 f_m A-\omega^2}\tag{69}$$

where ω, ω_m, and ω_p are in cm^{-1} and k in radian · cm^{-1}, or in the permittivity form

$$\varepsilon_{(\omega)}=n_{(\omega)}^2=1+\frac{\omega_p^2 f_m B}{\omega_m^2+\omega_p^2 f_m A-\omega^2}\tag{70}$$

The effect of other transitions may be included by adding ε_∞, the high-frequency permittivity, far from the resonance of interest. We have, therefore,

$$k^2=4\pi^2\omega^2\varepsilon_\infty+\frac{\omega_p^2 f_m B\cdot 4\pi^2\omega^2}{\omega_m^2+\omega_p^2 f_m A-\omega^2}\tag{71}$$

and

$$\varepsilon_{(\omega)}=n_{(\omega)}^2=\varepsilon_\infty+\frac{\omega_p^2 f_m B}{\omega_m^2+\omega_p^2 f_m A-\omega^2}\tag{72}$$

Here, $k(\omega)$, $\varepsilon(\omega)$, and $n(\omega)$ have been calculated from these formulae (using the numerical values in Tables A.1 and A.2 in the Appendix) for the Van der Waals shifted molecular transition of anthracene at $\omega_m=25{,}435\ \text{cm}^{-1}$ (Fig. 12). The dispersion curve has a stopping band in which $\varepsilon(\omega)$ is negative and $k(\omega)$ and $n(\omega)$ are both pure imaginary, between $\omega_T(25{,}098\ \text{cm}^{-1})$ and $\omega_L(25{,}383\ \text{cm}^{-1})$, where

$$\omega_T^2=\omega_m^2+\omega_P^2\cdot f_m\cdot A\quad\text{and}\quad\omega_L^2=\omega_T^2+\frac{\omega_p^2 f_m B}{\varepsilon_\infty}$$

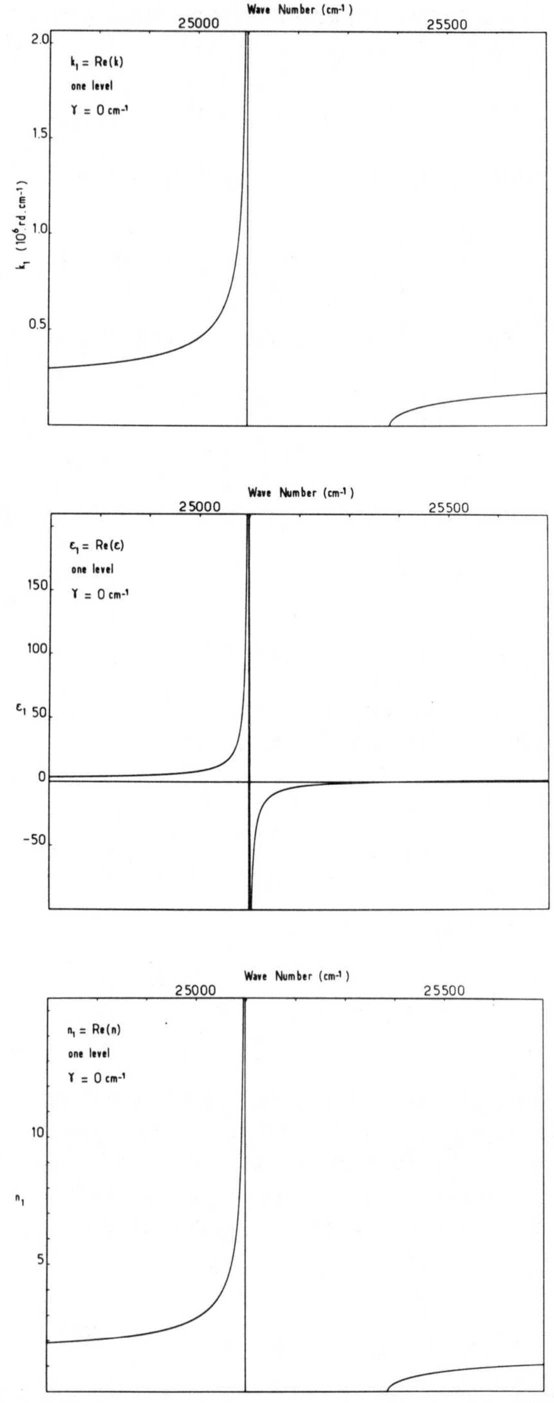

Fig. 12. Real (1) and imaginary (2) parts of the wave vector k, the permittivity $\tilde{\varepsilon}$, and the refractive index $\tilde{n}$ versus energy in anthracene, due to the transition at $\omega_m = 25{,}435\ \mathrm{cm}^{-1}$; no damping ($\gamma = 0\ \mathrm{cm}^{-1}$).

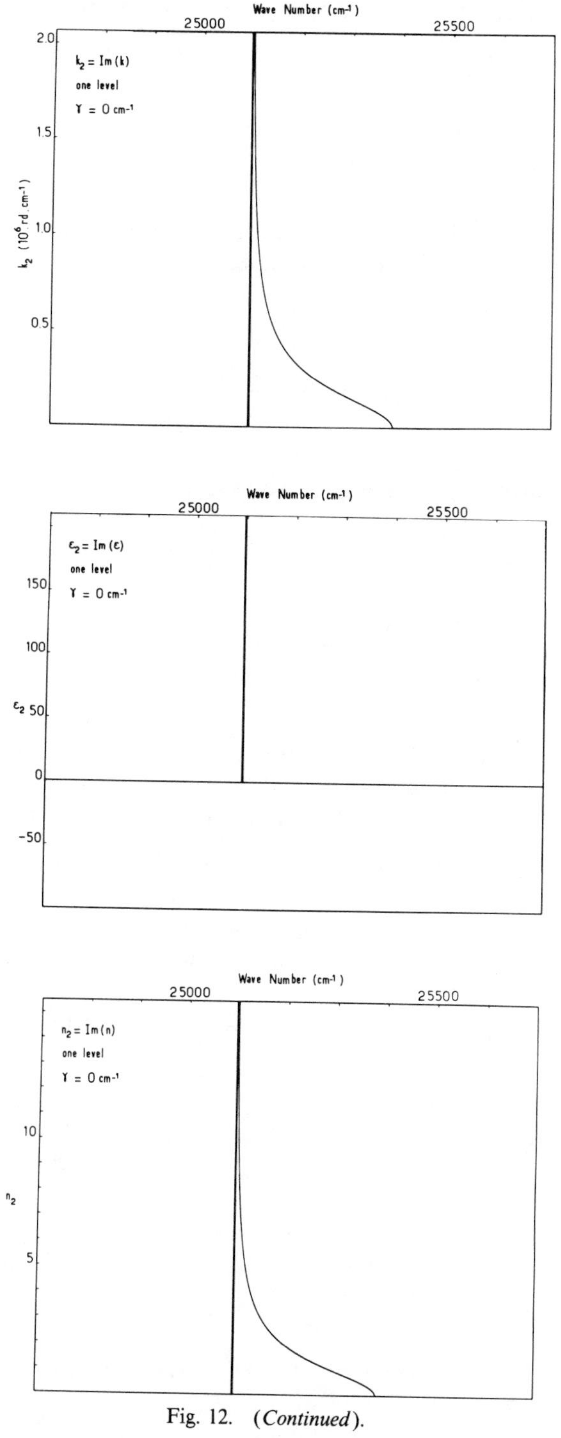

Fig. 12. (*Continued*).

N.B. If $\Delta\omega = \omega_L - \omega_T$ is the width of the stopping band and $\omega_{cG} = \frac{1}{2}(\omega_L - \omega_T)$ is its center, then

$$\omega_{cG} \cdot \Delta\omega = \frac{1}{2} \frac{\omega_p^2 f_m B}{\varepsilon_\infty}$$

Effect of damping. The overall effect of dissipative processes in (50) can be simulated by adding a damping term, $m\gamma\, \partial/\partial t$, which implies a molecular polarizability of the form

$$4\pi\alpha_M = \frac{\omega_p^2 f_m}{\omega_m^2 - \omega^2 - i\omega\gamma}$$

Equation (68) then yields a complex wave vector, $\tilde{k} = k_1 + ik_2$, obeying the dispersion relation

$$\tilde{k}^2 = 4\pi^2\omega^2\varepsilon_\infty + \frac{\omega_p^2 f_m B 4\pi^2\omega^2}{\omega_m^2 + \omega_p^2 f_m A - \omega^2 - i\omega\gamma} \qquad (71_{\text{bis}})$$

Figure 13 illustrates the real and imaginary parts of the wave vector $\tilde{k} = k_1 + ik_2$, the permittivity $\tilde{\varepsilon} = \varepsilon_1 + i\varepsilon_2$, and the refractive index $\tilde{n} = n_1 + in_2$ of anthracene, with $\gamma = 15\ \text{cm}^{-1}$. The real and imaginary parts of these three quantities are no longer strictly zero in and outside the stopping band.

c. Effect of Several Transitions. If other states of the crystal before coupling to the field have large transition probabilities, even at some distance from the corresponding resonances, then the additional term ε_∞ is insufficient to account for them and one must solve the complete equation (67).

If we express ϕ_{LL}^- and ϕ_{LM}^- in the same way as ϕ_{MM}^- in the preceding paragraph,

$$\phi_{LL}^- = C - \frac{4\pi\omega^2 \cdot D}{k^2 - 4\pi^2\omega^2}; \qquad \phi_{LM}^- = E - \frac{4\pi^2\omega^2 \cdot F}{k^2 - 4\pi^2\omega^2}$$

where

$$C = t_{1L1L} - t_{1L2L}; \qquad D = (\hat{d}_{1L} \cdot \hat{d}_{1L}) - (\hat{d}_{1L} \cdot \hat{d}_{2L})$$

and

$$E = t_{1L1M} - t_{1L2M}; \qquad F = (\hat{d}_{1L} \cdot \hat{d}_{1M}) - (\hat{d}_{1L} \cdot \hat{d}_{2M})$$

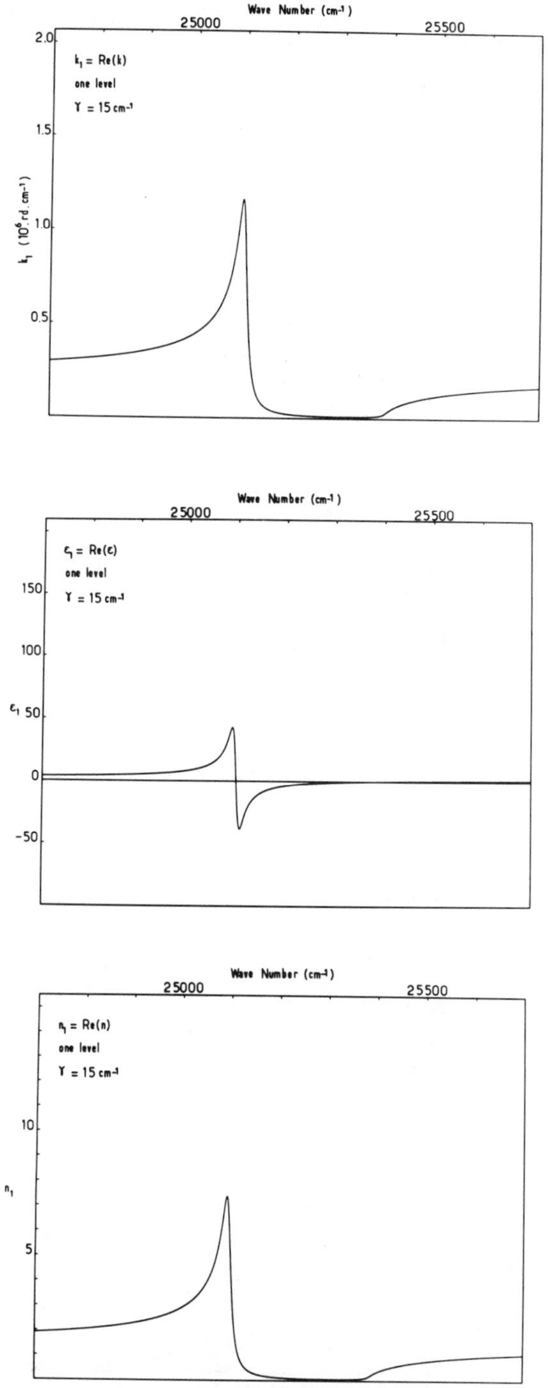

Fig. 13. Real (1) and imaginary (2) parts of the wave vector $\tilde{k}$, the permittivity $\tilde{\varepsilon}$, and the refractive index $\tilde{n}$ versus energy in anthracene, due to the transition at $\omega_m = 25{,}435\ \text{cm}^{-1}$; damping ($\gamma = 15\ \text{cm}^{-1}$).

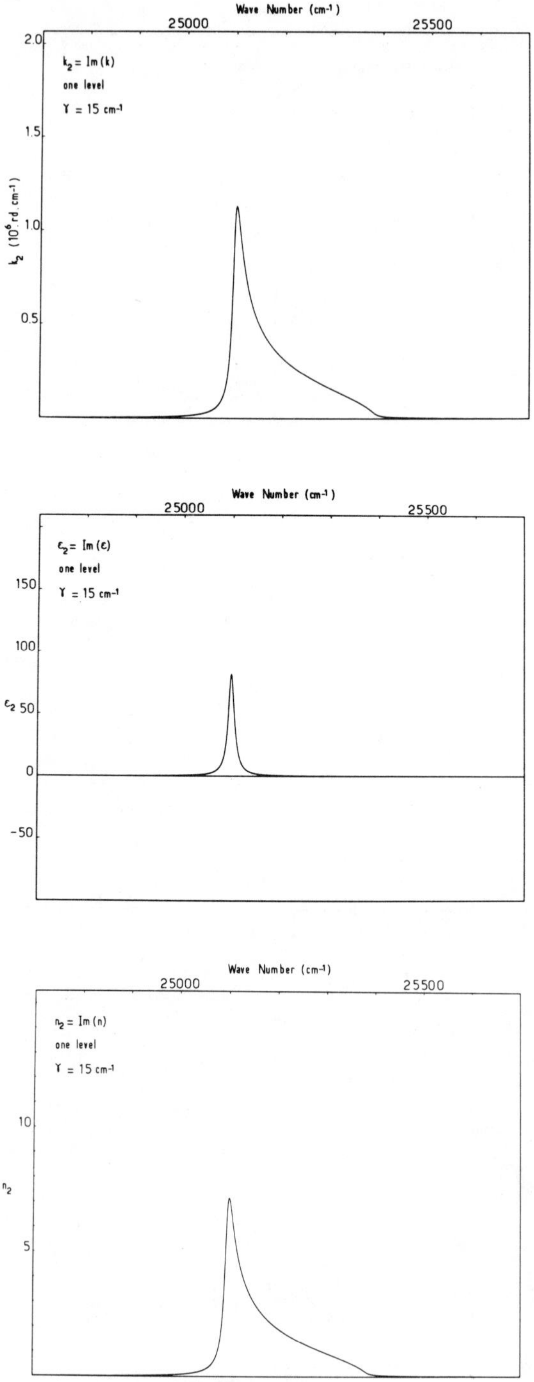

Fig. 13. (*Continued*).

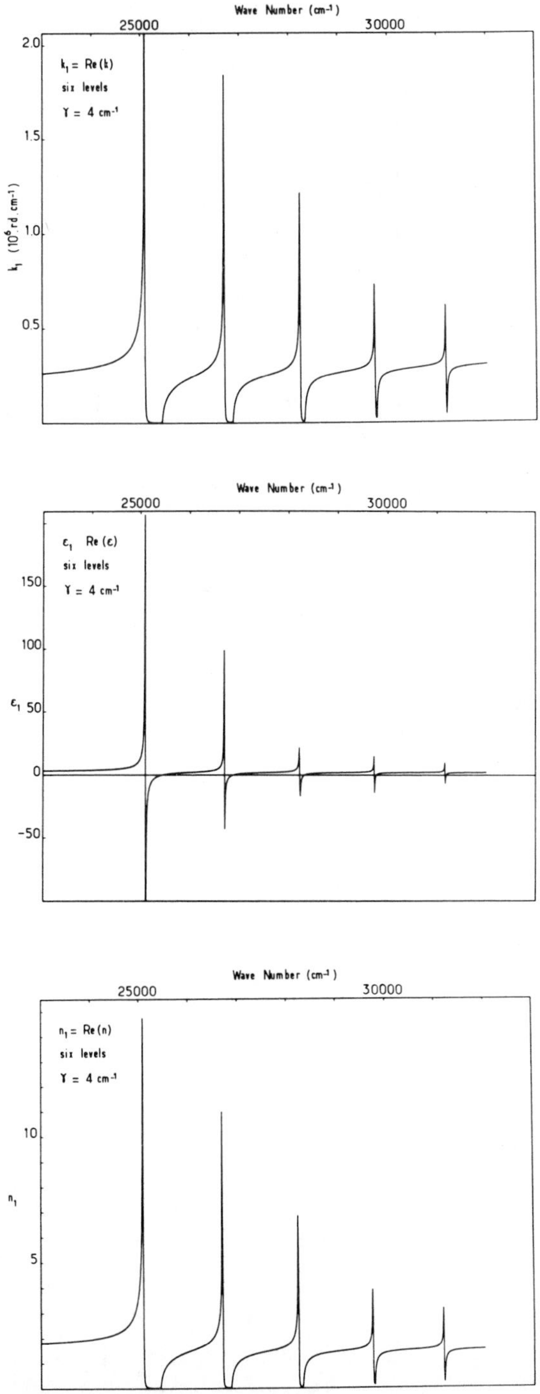

Fig. 14. Real (1) and imaginary (2) parts of the wave vector $\tilde{k}$, the permittivity $\tilde{\epsilon}$, and the refractive index $\tilde{n}$ of anthracene, calculated from six vibronic levels (five for S_1 and one for S_2) and a damping coefficient $\gamma = 4\ \mathrm{cm}^{-1}$ throughout.

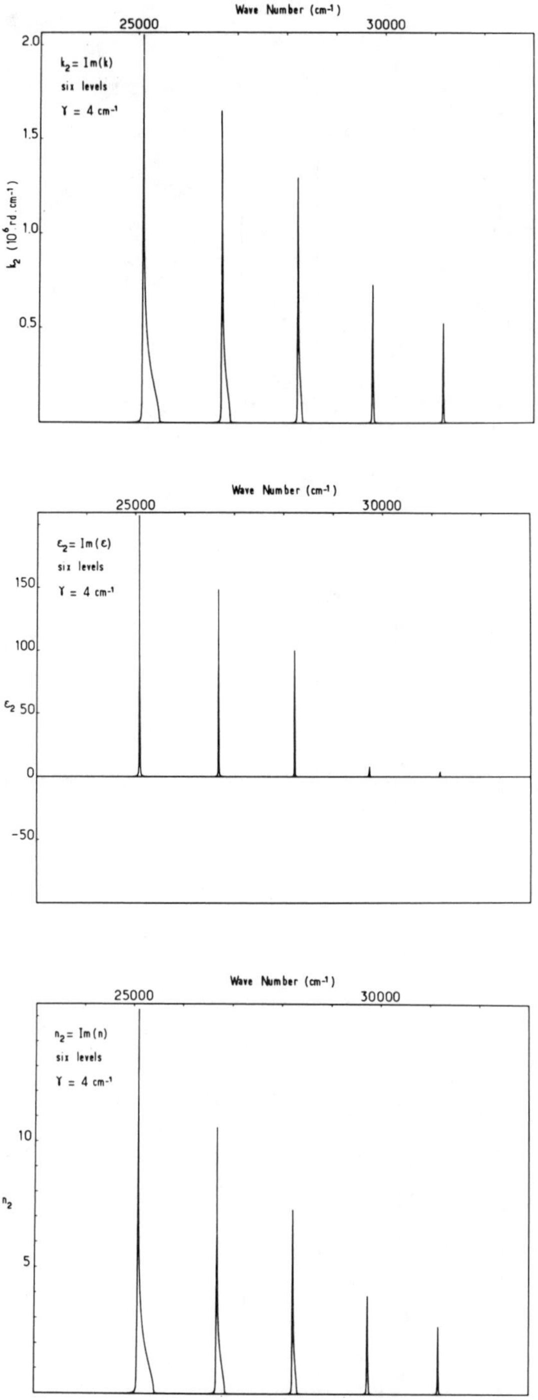

Fig. 14. (*Continued*).

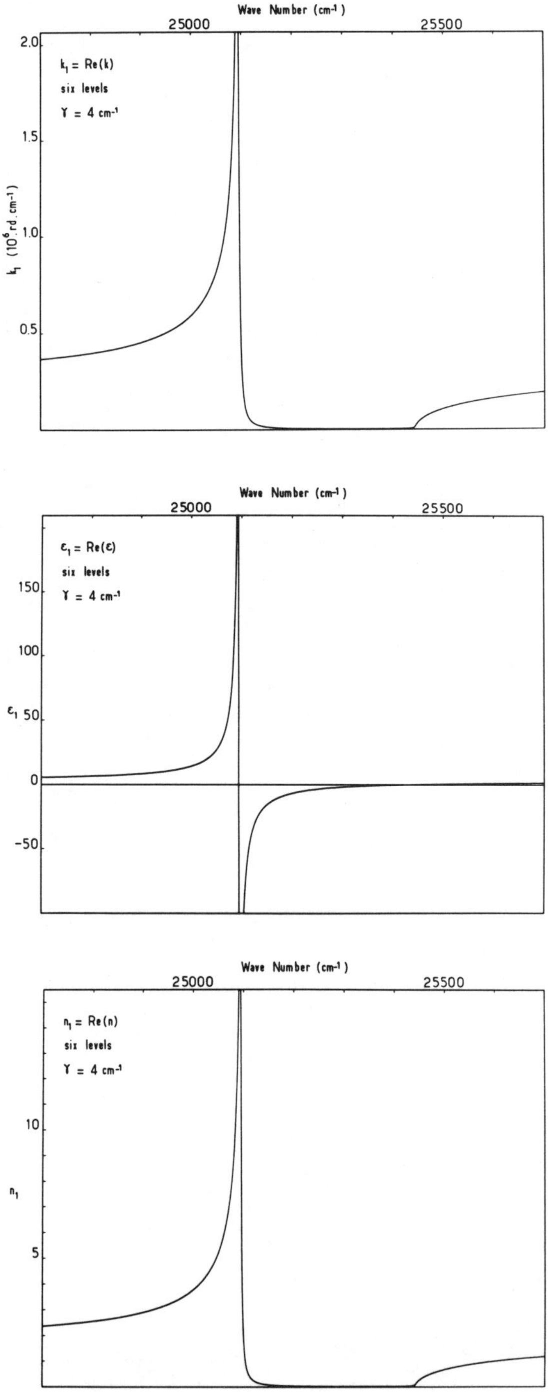

Fig. 15. Same as Fig. 14. Detailed illustration of the (0,0) region of the $S_0 \rightarrow S_1$ transition.

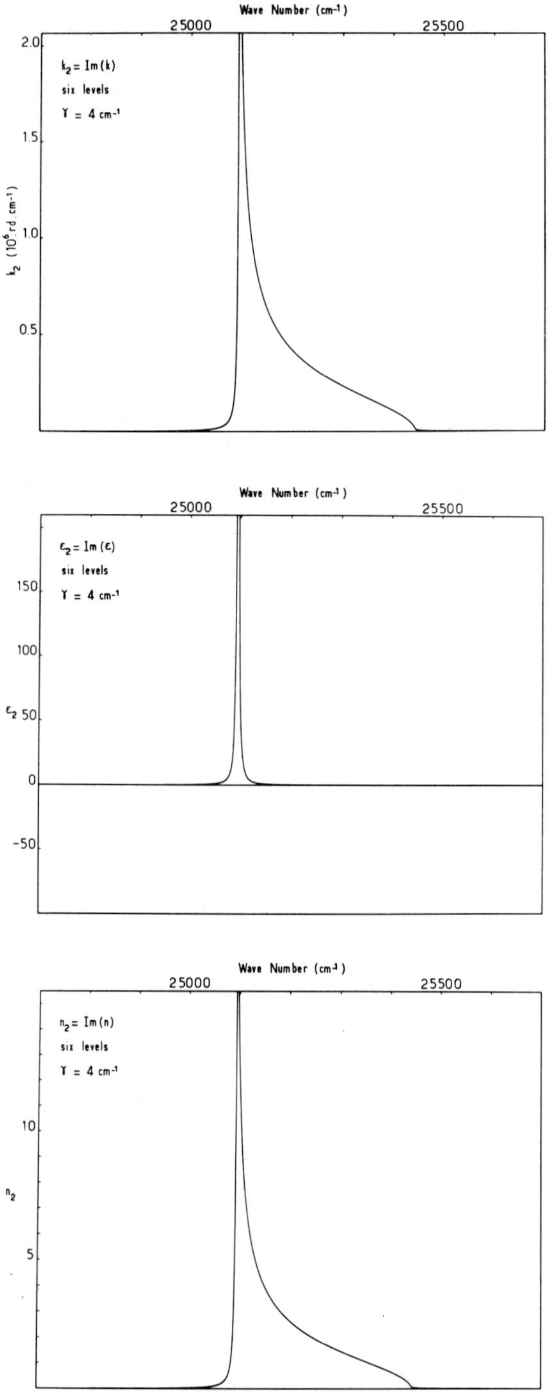

Fig. 15. (*Continued*).

Let

$$P=\omega_p^2\sum_m \frac{f_m}{\omega_m^2-\omega^2} \quad \text{and} \quad Q=\omega_p^2\sum_l \frac{f_l}{\omega_l^2-\omega^2}$$

Then (67) becomes

$$\left[P\cdot Q(BD-F^2)\right]\cdot X^2-\left[PB+PQ(BC+AD-2EF)+QD\right]\cdot X + \left[1+PA+PQ(AC-E^2)+QC\right]=0$$

in $X=(\tilde{\varepsilon}(\omega)-1)^{-1}$

The solution of this equation is the polariton dispersion curve (the other root being a free photon). As above we may introduce the permittivity at high frequency ε_∞ and a set of damping coefficients γ_m, γ_l for the various molecular levels involved.

We have applied this method to anthracene, using Philpott's level system comprising five vibronic levels of S_1 [$\omega_m = 25{,}614\ \text{cm}^{-1} + n\cdot 1400\ \text{cm}^{-1}$; $n \in (0,4)$] and a straight electronic level S_2 (at $\omega_l = 39{,}000\ \text{cm}^{-1}$), all with the same damping $\gamma=\gamma_l=\gamma_m = 4\ \text{cm}^{-1}$, and $\varepsilon_\infty = 2.56$. The real and imaginary parts of the wave vector $\tilde{k}(\omega)$, the permittivity $\tilde{\varepsilon}(\omega)$, and the refractive index $\tilde{n}(\omega)$ are illustrated in Fig. 14, which covers all the vibronic band of S_1. Figure 15 is a more detailed illustration of the S_1 origin (0,0).

D. Quantum Theory of the Polariton

1. *Introduction*

For some years now it has been possible to give a completely quantum description of complex systems involving many interacting parts by the techniques of second quantization. These methods, which have few restrictions (in principle, at least), were first applied to the polariton by Fano[117] and Hopfield,[113] and more specifically to polaritons in molecular crystals by Agranovich.[119] We shall draw on the analysis of Agranovich here.

Although, as we shall see later, the quantum method inevitably leads to a polariton dispersion curve equivalent to that predicted by the semiclassical theory, it does have the advantage of treating the polariton more as a particle, a mixed quasiparticle in fact, in which the weights of its observable aspects, the photon and the exciton, may be calculated.

Moreover, this method is better suited to descriptions of real crystals (finite and nonrigid) since the photon–exciton treatment given here may be extended by the addition of a phonon subspace, or of polariton–polariton interaction, when the field is very strong.

2. *Hamiltonian of the Crystal–Field System*

The Hamiltonian of a crystal in an electromagnetic field may be written as

$$H = H_{\text{ex}} + H_f + H_{\text{int}}$$

where

H_{ex} = the crystal Hamiltonian including electrostatic interaction between charges.
H_f = the Hamiltonian of the electromagnetic field, corresponding to free transverse photons.
H_{int} = the interaction operator between the crystal charges and the transverse photon field.

We shall derive below the second quantization forms of each of these terms.

a. Unperturbed Crystal Hamiltonian: Exciton Theory in Second Quantization

(i) Principle. We shall briefly describe this well-known theory in order to form a plan of the method of diagonalization chosen.

In space representation the Hamiltonian of a crystal of $N\sigma$ molecules (N number of cells, σ number of molecules per unit cell) is

$$H_{\text{ex}} = \sum_{n} H_n + \tfrac{1}{2} \sum_{n,m}{}' V_{\text{nm}} \qquad \left(\sum_{n,m}{}' \equiv \sum_{n,m\neq n} \right) \tag{75}$$

where

H_n = the Hamiltonian of the "free" molecule on site n [n will be developed later as $n\alpha$, $n \in (1, N)$, and $\alpha \in (1, \sigma)$]; and
V_{nm} = the electrostatic interaction between sites n and m.

The eigenfunctions φ_{nr} of H_n corresponding to states $|r\rangle$ of each molecule n, with eigenvalues ε_r, form a basis of singly excited crystal states on which to determine the crystal eigenstates.

In second quantization, on the other hand, the crystal states are characterised by functions $\psi(\ldots, N_{nr}, \ldots)$ of the occupation numbers of the eigenfunction φ_{nr} of each molecular Hamiltonian H_n. The occupation number N_{nr} of state $|r\rangle$ is 1 or 0 as the molecule is or is not in state $|r\rangle$. All the possible N_{nr} are eigenvalues of the $\psi(\ldots, N_{nr}, \ldots)$ of the operator $\hat{N}_{nr}$.

The operator $\hat{N}_{nr}$, expressed as a function of the creation (b_{nr}^+) and annihilation (b_{nr}) operators of state $|r\rangle$ on molecule n is

$$\hat{N}_{nr} = b_{nr}^+ b_{nr}$$

The definitions and commutation rules of b^+ and b depend on the statistics of the particles (bosons or fermions). If $[X, Y] = XY - YX$ and $[X, Y]_+ = XY + YX$ are the commutator and anticommutator of operators X and Y, then:

(ii) For bosons:

$$a_i^+ \psi(\ldots, n_i, \ldots) = \sqrt{n_i + 1}\, \psi(\ldots, n_i + 1, \ldots)$$
$$a_i \psi(\ldots, n_i, \ldots) = \sqrt{n_i}\, \psi(\ldots, n_i - 1, \ldots)$$

with

$$[a_i, a_j] = [a_i^+, a_j^+] = 0 \quad \text{and} \quad [a_i, a_j^+] = \delta_{ij}$$

(iii) For fermions:

$$c_i^+ \psi(\ldots, n_i, \ldots) = A\sqrt{1 - n_i}\, \psi(\ldots, n_i + 1, \ldots)$$
$$c_i \psi(\ldots, n_i, \ldots) = A\sqrt{n_i}\, \psi(\ldots, n_i - 1, \ldots)$$

where $A = (-1)\Sigma_{j<i} n_j$ is a factor ensuring the fermion wave functions are antisymmetric, with

$$[c_i, c_j]_+ = [c_i^+, c_j^+]_+ = 0 \quad \text{and} \quad [c_i, c_j^+]_+ = \delta_{ij}$$

In the case at hand, the Pauli exclusion principle is satisfied if b_{nr}^+ and b_{nr} corresponding to state $|r\rangle$ on molecule n anticommute like fermion operators, but operators corresponding to the same state on different molecules, or to different states on the same molecules, commute as for bosons. The b^+ and b operators which are neither fermion nor boson operators, therefore, are called Pauli operators.

The second quantization of the crystal Hamiltonian (75) is

$$H_{\text{ex}} = \sum_{n,r} \varepsilon_r b_{nr}^+ b_{nr} + \frac{1}{2} \sum_{\substack{n,m \\ r,r' \\ s,s'}} b_{nr'}^+ b_{ms'}^+ b_{ms} b_{nr} \langle r's' | V_{nm} | rs \rangle \qquad (76)$$

where

$$\langle r's'|V_{nm}|rs\rangle = \int \phi_{nr'}\phi_{ms'}V_{nm}\phi_{nr}\phi_{ms}\,d\xi\,d\eta$$

Because H_{ex} does not commute with the occupation number operator $\hat{N}_{nr}$, the N_{nr} cannot be eigenstates of the crystal and the excitation cannot be localized on one site. The eigenstates can be found by canonical transformation of the b^+ and b terms to diagonalize H_{ex}.

(iv) Heitler–London approximation for diagonalizing H_{ex}. In this approximation, the following simplifications are made:
—Only one excited molecular state, f, is used. The indices r, r', s, and s' can be only 0 or f.
—The density of excited molecules is assumed to be low enough to neglect matrix elements between excited molecules of the form

$$\langle ff|V_{nm}|ff\rangle,\quad \langle ff|V_{nm}|00\rangle \quad \text{and} \quad \langle 00|V_{nm}|ff\rangle$$

These simplifications are not fundamental and may be removed if they are not justified.[164]

We shall keep only the following matrix elements:

$\langle 00|V_{mn}|00\rangle$: interaction between molecules in the ground state
$\langle 0f|V_{mn}|0f\rangle$: interaction between a molecule in the ground state and another in the excited state
$\langle 0f|V_{mn}|f0\rangle$: excitation exchange term

Let

$\Delta\varepsilon^f = \varepsilon_f - \varepsilon_0$ be the excitation energy of the free molecule,
$E_0 = \sum_n \varepsilon_0 + \frac{1}{2}\sum'_{n,m}\langle 00|V_{nm}|00\rangle$ the ground state energy of the crystal, .
$D^f_{nm} = \langle 0f|V_{nm}|0f\rangle - \langle 00|V_{nm}|00\rangle$ the difference between the interactions of molecules m and n when one of them is excited and when both are in their ground state (a site shift Van der Walls term), and
$M^f_{nm} = \langle 0f|V_{nm}|f0\rangle$ the exchange term.

The crystal Hamiltonian is then

$$H_{\text{ex}} = E_0 + \sum_n \Big(\Delta\varepsilon_f + \sum_m{}' D^f_{mn}\Big) b^+_{nf} b_{nf} + \sum_{n,m}{}' M^f_{nm} b^+_{n0} b^+_{mf} b_{m0} b_{nf}$$

which may be simplified by introducing the operators β^+ and β for transitions of molecule n between the states 0 and f:

$$\beta_{nf}^{+} = b_{nf}^{+} b_{n0}, \qquad |0\rangle \to |f\rangle$$
$$\beta_{nf} = b_{n0}^{+} b_{nf}, \qquad |f\rangle \to |0\rangle$$

The commutation rules of these new operators are easily verified to be

$$\beta_{nr}\beta_{ms}^{+} - \beta_{ms}^{+}\beta_{nr} = \begin{cases} 0 & \text{if } n \neq m \quad \text{or} \quad r \neq s \\ 1 - 2\hat{N}_{nr} & \text{if } n = m \quad \text{and} \quad r = s \end{cases}$$

It is important in what follows that for low light intensities the number of excited molecules is small, so that the average $\langle \hat{N}_{nr} \rangle$ of $\hat{N}_{nr}$ is of the order of 10^{-6}. We may suppose, therefore, that these new operators commute like boson operators:

$$\beta_{nr}\beta_{mS}^{+} - \beta_{mS}^{+}\beta_{nr} = \delta_{nm}\delta_{rS}$$

Then

$$H_{\text{ex}} = E_0 + \sum_{n} \Delta E_g \beta_{nf}^{+}\beta_{nf} + \sum_{n,m}' M_{nm}^{f} \beta_{nf}^{+}\beta_{mf} \tag{77}$$

where

$$\Delta E_g = \left(\Delta\varepsilon^{f} + \sum_{m}' D_{mn}^{f} \right)$$

The $\beta+$ and β operators being treated as boson operators, Hamiltonian H_{ex} can be diagonalized by Tyablikov's method[165]

$$H_{\text{ex}} = E_0 + \Delta E_0 + \sum_{\nu} E_\nu B_\nu^{+} B_\nu \tag{78}$$

where the operators B_ν^{+} and B_ν are defined by

$$\beta_{nf} = \sum_{\nu} u_{n\nu} B_\nu + v_{n\nu}^{*} B_\nu^{+}$$
$$\beta_{nf}^{+} = \sum_{\nu} u_{n\nu}^{*} B_\nu^{+} + v_{n\nu} B_\nu \tag{79}$$

with

$$\Delta E_0 = -\Sigma_{n\nu} E_\nu |v_{n\nu}|^2.$$

We must now introduce the explicit form $n\alpha$ of the index n, n referring to the cell and α to the site in the cell. Translational periodicity can be accounted for by introducing the usual wave representation with wave vectors $\mathbf{k}$ in the first Brillouin zone. Using the cyclic Born–von Karman conditions, the coefficients in the above transformations are:

$$u_{n\nu} \equiv u_{n\alpha,\nu} = N^{-1/2} \cdot u_{\alpha,\nu}(\mathbf{k})\, e^{i(\mathbf{k}\cdot\mathbf{r}_{n\alpha})}$$
$$v_{n\nu} \equiv v_{n\alpha,\nu} = N^{-1/2} \cdot v_{\alpha,\nu}(\mathbf{k})\, e^{i(\mathbf{k}\cdot\mathbf{r}_{n\alpha})} \tag{80}$$

Comparison of (77) and (78) for H_{ex} after transformation (79) leads to the system

$$I_{\alpha\beta}(\mathbf{k}) = \sum_m{}' M^f_{n\alpha,\, m\beta} e^{i\mathbf{k}\cdot(\mathbf{r}_{m\beta}-\mathbf{r}_{n\alpha})} \tag{81}$$

where

$$(E_\nu - \Delta E_g)\, u_{\alpha\nu} = \sum_{\beta=1}^{\sigma} I_{\alpha\beta}(\mathbf{k}) \cdot u_{\beta\nu}$$
$$-(E_\nu + \Delta E_g)\, v_{\alpha\nu} = \sum_{\beta=1}^{\sigma} I_{\alpha\beta}(\mathbf{k}) \cdot v_{\beta\nu} \tag{82}$$

These equations, for all values of α from 1 to σ, form a system of size 2σ. Setting its determinant equal to zero yields an equation of degree σ in E_ν^2. The σ positive solutions are the energies of the σ exciton branches.

In anthracene, with two molecules per unit cell, the $\nu = +$ and $\nu = -$ components are

$$E_\pm(\mathbf{k}) = E_0 + \Delta E_g + I_{11}(\mathbf{k}) \pm I_{12}(\mathbf{k})$$

The $E_\nu(\mathbf{k})$ substituted into (82) give 2σ equations for the $u_{\alpha\nu}(\mathbf{k})$ and the $v_{\alpha\nu}(\mathbf{k})$, which are completely determined by the system and the normalization conditions.

For positive energies in anthracene,

$$u_{1+} = u_{1-} = u_{2+} = -u_{2-} = \frac{1}{\sqrt{2}}$$

and

$$v_{1+} = v_{1-} = v_{2+} = v_{2-} = 0$$

The $u_{\alpha\nu}(\mathbf{k})$ and $v_{\alpha\nu}(\mathbf{k})$ completely define the canonical transformations (79), linking the β^+ and β to the $B^+_{\mathbf{k}\nu}$ and $B_{k\nu}$ creation and annihilation operators of an exciton of wave vector $\mathbf{k}$ and energy $E_\nu(\mathbf{k})$.

The crystal Hamiltonian is then given by (78), where by the Heitler–London approximation:

$$\Delta E_0 = -\sum_{\alpha,k,\nu} E_\nu(\mathbf{k})|v_{\alpha\nu}(\mathbf{k})|^2 = 0 \quad \text{since} \quad v_{\alpha\nu}(\mathbf{k}) = 0$$

Taking $E_0 = 0$ as the origin of energies, the diagonalized crystal Hamiltonian is

$$H_{\text{ex}} = \sum_{k,\nu} E_\nu(\mathbf{k}) B^+_{k\nu} B_{k\nu} \tag{83}$$

b. Hamiltonian of the Electromagnetic Field. The second quantized Hamiltonian of a field of free transverse photons of wave vectors $\mathbf{q}$ and polarizations $\hat{\varepsilon}_{qj}$ $[j \in (1,2)]$ is:[135]

$$H_f = \sum_{q,j} \hbar qc a^+_{qj} a_{qj} \tag{84}$$

where a^+_{qj} and a_{qj} are the (bose) creation and annihilation operators of a transverse photon $(\mathbf{q}, \hat{\varepsilon}_j)$.

N.B. Clearly, in the above analysis the excitons are bosons and both they and the photons obey the same boundary conditions. This is a necessary condition for an exciton of wave vector $\mathbf{k}$ to be coupled only to a photon of wave vector $\mathbf{q} = \mathbf{k}$ to form a stable particle, the polariton. However, in a finite crystal the exciton interacts with a photon continuum whose dimension depends on that of the crystal; for example, one-dimensional excitons will be coupled to a two-dimensional photon continuum whereas two-dimensional excitons will be coupled to a one-dimensional photon continuum. In both cases the particles are radiatively very unstable. Work on this instability is currently being done in our group.[223]

We shall also neglect the "Umklap" effect corresponding to coupling between a photon $\mathbf{q}$ and an exciton $\mathbf{k} = \mathbf{q} + \mathbf{G}$, where $\mathbf{G}$ is a reciprocal lattice vector. At optical frequencies far from x- and γ-rays, the Umklap effect is a negligible perturbation and in any case could be introduced later.

c. Matter–field Interaction. Neglecting relativistic effects, crystal–field interaction can be described by that of the "optical electrons," that is, those able to interact effectively and the field:

$$H_{\text{int}} = -\sum_{n\alpha} \frac{e}{mc} \mathbf{A}_{(\mathbf{r}_{n\alpha})} \cdot \mathbf{J}_{n\alpha} + \frac{e^2 S}{2mc^2} \sum_{n\alpha} \mathbf{A}^2_{(\mathbf{r}_{n\alpha})} = H^{(1)}_{\text{int}} + H^{(2)}_{\text{int}} \tag{85}$$

where $\mathbf{A}(\mathbf{r}_{n\alpha})$ = vector potential at site $n\alpha$; $\mathbf{J}_{n\alpha}$ = total momentum of the electrons of molecule $n\alpha$; and S = total number of optical electrons on site $n\alpha$.

In the Coulomb gauge ($\nabla\cdot\mathbf{A}=0$) the second quantized form of the vector potential operator is[135]

$$\mathbf{A}_{(\mathbf{r})}=\sum_{q,j}\left(2\pi c^2\hbar/Vqc\right)^{1/2}\cdot\hat{\varepsilon}_{qj}\left(a_{qj}e^{i\mathbf{q}\cdot\mathbf{r}}+\underline{a}^{+}_{qj}e^{-i\mathbf{q}\cdot\mathbf{r}}\right) \tag{86}$$

where V is the volume of quantification (Here the crystal, NV_c).

The $\mathbf{J}_{n\alpha}$ are readily determined if we take advantage of the periodicity of the crystal. Let

$$J_{\pm k}=\sum_{n\alpha}e^{\pm i(\mathbf{k}\cdot\mathbf{r}_{n\alpha})}\hat{\varepsilon}_{qj}\cdot\mathbf{J}_{n\alpha} \tag{87}$$

The $J_{\pm k}$ operators, being sums of molecular operators, can be expressed[166] in terms of the exciton operators $B^{+}_{k\nu}$ and $B_{k\nu}$ [see Eq. (83)]:

$$J_{\pm k}=\frac{im}{\hbar e}\sum_{\nu}E_{\nu}(\mathbf{k})\left\{\sum_{\alpha}\hat{\varepsilon}_{qj}\cdot\mathbf{d}^{0f}_{\alpha}\left(u_{\alpha\nu}(\mathbf{k})+v_{\alpha\nu}(\mathbf{k})\right)\right\}\cdot\left(B_{\mp k}-B_{\pm k}\right) \tag{88}$$

where $\mathbf{d}^{0f}_{\alpha}$ is the $|0\rangle\to|f\rangle$ transition dipole moment of the molecule on site α. The final form of H_{int}, taking account of $\mathbf{A}(\mathbf{r})$ and $\mathbf{J}_{\pm k}$, is:

$$H^{(1)}_{\text{int}}=\sum_{\nu,k,j}T(j,\mathbf{k},\nu)\left\{a_{kj}\left(B_{-k\nu}-B^{+}_{k\nu}\right)+a^{+}_{kj}\left(B_{k\nu}-B^{+}_{-k\nu}\right)\right\} \tag{89}$$

where

$$T(j,\mathbf{k},\nu)=i(2\pi N/kcV\hbar)\sum_{\alpha}\left\{\left(\hat{\varepsilon}_{kj}\cdot\mathbf{d}^{0f}_{\alpha}\right)\left\{u_{\alpha\nu}(\mathbf{k})+v_{\alpha\nu}(\mathbf{k})\right\}\right\}E_{\nu}(\mathbf{k}) \tag{90}$$

describes exciton–photon coupling.

N.B. In calculating $H^{(1)}_{\text{int}}$ the sums

$$\sum_{n}e^{i(\mathbf{q}-\mathbf{k})\cdot\mathbf{r}_{n\alpha}}=N\delta(\mathbf{q}-\mathbf{k})\quad\text{and}\quad\sum_{n}e^{i(\mathbf{q}+\mathbf{k})\cdot\mathbf{r}_{n\alpha}}=N\delta(\mathbf{q}+\mathbf{k})$$

appear. It follows that only excitons of wave vector $\mathbf{k}=\pm\mathbf{q}$ can be coupled to a photon of wave vector $\mathbf{q}$. Below, we shall write $\mathbf{k}$ for both $\mathbf{k}$ and $\mathbf{q}$.

The two-photon interaction term is

$$H_{\text{int}}^{(2)}=\frac{\hbar\omega_p^2 S}{4}\sum_{kj}\frac{1}{kc}\{2a_{kj}^{+}a_{kj}+a_{kj}^{+}a_{-kj}^{+}+a_{kj}a_{-kj}\} \tag{91}$$

where $\omega_p^2=4\pi e^2/mV_c$ is the squared plasma frequency introduced earlier.

3. *Diagonalization of the Crystal—Field Hamiltonian*

We are now in a position to write the full crystal—field Hamiltonian. For given **k**, summation is over $\pm\mathbf{k}$. Hence

$$\begin{aligned}H(\mathbf{k})=&\sum_{\nu}E_{\nu}(\mathbf{k})(B_{k\nu}^{+}B_{k\nu}+B_{-k\nu}^{+}B_{-k\nu})+\sum_{j}\hbar kc(a_{kj}^{+}a_{kj}+a_{-kj}^{+}a_{-kj})\\&+\sum_{\nu j}T(j,\mathbf{k},\nu)\{(a_{kj}+a_{-kj}^{+})(B_{-k\nu}-B_{k\nu}^{+})\\&+(a_{-kj}+a_{kj}^{+})(B_{k\nu}-B_{-k\nu}^{+})\}\\&+\frac{\hbar\omega_p^2}{2}\{a_{kj}^{+}a_{kj}+a_{-kj}^{+}a_{-kj}+a_{kj}a_{-kj}+a_{kj}^{+}a_{-kj}^{+}\}\end{aligned} \tag{92}$$

This Hamiltonian is diagonal neither on the photon occupation number basis $a_{kj}^{+}a_{kj}$ nor on the exciton occupation number basis $B_{k\nu}^{+}B_{k\nu}$, so that neither photon nor exciton are crystal–field eigenstates.

Let ξ_l^{+} and ξ_l be new bose operators such that:

$$\begin{aligned}B_{k\nu}&=\sum_{l}[\xi_l U_{k\nu}(l)+\xi_l^{+}V_{k\nu}^{*}(l)]\\B_{k\nu}^{+}&=\sum_{l}[\xi_l^{+}U_{k\nu}^{*}(l)+\xi_l V_{k\nu}(l)]\\a_{kj}&=\sum_{l}[\xi_l U_{kj}(l)+\xi_l^{+}V_{kj}^{*}(l)]\\a_{kj}^{+}&=\sum_{l}[\xi_l^{+}U_{kj}^{*}(l)+\xi_l V_{kj}(l)]\end{aligned} \tag{93}$$

H is diagonal in the form

$$H=\sum_{l}(\Delta E_l+E_l\xi_l^{+}\xi_l) \tag{94}$$

where

$$\Delta E_l=-E_l\left\{\sum_{j}|V_{kj}(l)|^2+\sum_{j}|V_{-kj}(l)|^2+\sum_{\nu}|V_{k\nu}(l)|^2+\sum_{\nu}|V_{-k\nu}(l)|^2\right\}$$

This transformation introduces a new quasi particle, the "polariton," whose creation and annihilation operators are ξ^+ and ξ.

Comparing (92) with (94) we see by (93) that the U and V must obey:

$$\{E_\nu(\mathbf{k})-E_l\}U_{k\nu}-\sum_j T(j,\mathbf{k},\nu)\{U_{kj}+V_{-kj}\}=0$$

$$\{E_\nu(\mathbf{k})+E_l\}V_{-k\nu}+\sum_j T(j,\mathbf{k},\nu)\{U_{kj}+V_{-kj}\}=0$$

$$\{\hbar kc-E_l\}U_{kj}+\sum_\nu T(j,\mathbf{k},\nu)\{U_{k\nu}-V_{-k\nu}\}+\frac{\hbar\omega_p^2 S}{2kc}\{U_{kj}+V_{-kj}\}=0$$

$$\{\hbar kc+E_l\}V_{-kj}+\sum_\nu T(j,\mathbf{k},\nu)\{U_{k\nu}-V_{-k\nu}\}+\frac{\hbar\omega_p^2 S}{2kc}\{U_{kj}+V_{-kj}\}=0$$

(95)

Combining these equations one can find an equation in the U_{kj} only.[166] Written for the polarizations $j=1$ and $j=2$, it yields a system for U_{k1} and U_{k2} with determinant

$$\left\|\left\{E_l^2-(\hbar ck)^2-S\left(\hbar\omega_p^2\right)\right\}\delta_{jj'}+\sum_\nu T(j,\mathbf{k},\nu)T(j',\mathbf{k},\nu)\frac{4\hbar kcE_\nu(\mathbf{k})}{E_l^2-E_\nu(\mathbf{k})}\right\| \tag{96}$$

where $(j,j')\in(1,2)$.

The zeros of this determinant provide the dispersion curve $E_l=f(\mathbf{k})$ of the polariton.

Introducing the $T(j,\mathbf{k},\nu)$ of (90) and considering the total transition moment of the unit cell,

$$\mathbf{P}_\nu(\mathbf{k})=\sum_{\alpha=1}^{\sigma}\mathbf{d}_\alpha^{0f}\left[u_{\alpha\nu}(\mathbf{k})+v_{\alpha\nu}(\mathbf{k})\right] \tag{97}$$

and the oscillator strength of the electronic transition,

$$F_\nu(\mathbf{k})=\frac{2m}{\hbar^2e^2}|\mathbf{P}_\nu(\mathbf{k})|^2E_\nu(\mathbf{k}) \tag{98}$$

where $\Sigma_\nu F_\nu(\mathbf{k})=\sigma S$, number of electrons per cell, one arrives at the general dispersion equation $\omega_l = E_l/\hbar$ and $\omega_\nu(\mathbf{k}) = E_\nu(\mathbf{k})/\hbar$

$$k^2 = 4\pi^2\omega^2 - \frac{1}{2}\sum_\nu \frac{\omega_p^2 F_\nu(\mathbf{k})\sin^2\phi(\nu,k)}{\omega^2(\mathbf{k}) - \omega_\nu^2(\mathbf{k})}\cdot 4\pi^2\omega^2$$
$$\pm\frac{1}{2}\left[\left\{\sum_\nu \frac{\omega_p^2 F_\nu(\mathbf{k})\left[\cos^2\phi_1(\nu,k) - \cos^2\phi_2(\nu,k)\right]}{\omega^2(\mathbf{k}) - \omega_\nu^2(\mathbf{k})}\right\}^2\right.$$
$$\left.+4\left\{\sum_\nu \frac{\omega_p^2 F_\nu(\mathbf{k})\cos\phi_1(\nu,k)\cos\phi_2(\nu,k)}{\omega^2(\mathbf{k}) - \omega_\nu^2(\mathbf{k})}\right\}^2\right]^{1/2}\cdot 4\pi^2\omega^2 \qquad (99)$$

where $\phi(\nu,k)$, $\phi_1(\nu,k)$, and $\phi_2(\nu,k)$ are the angles between $\mathbf{P}_\nu(k)$ and $\mathbf{k}$, $\hat{\varepsilon}_{1k}$ and $\hat{\varepsilon}_{2k}$ (see Fig. 16).

In anthracene and other crystals with two molecules per cell, there are, as we saw above, two orthogonal transitions $E_-(k)$ and $E_+(k)$ for each molecular transition $|0\rangle \to |f\rangle$. In addition, $\mathbf{P}_-(k)$ is parallel to the $\mathbf{b}$ axis and $\mathbf{P}_+(k)$ lies in the $(\mathbf{a},\mathbf{c})$ plane. In the simple case in which we are interested here, $\mathbf{k}$ is perpendicular to face $(\mathbf{a},\mathbf{b})$ and the photon is $\mathbf{b}$ polarized, so that

$$\phi = \frac{\pi}{2}; \qquad \phi_{1-} = 0; \qquad \phi_{2-} = \frac{\pi}{2}; \qquad \phi_{1+} = \frac{\pi}{2}; \qquad F_+(k) = 0$$

Then (99) simplifies to

$$k^2 = 4\pi^2\omega^2 - \frac{\omega_p^2 F_-(k) 4\pi^2\omega^2}{\omega^2 - \omega_-^2(k)} \qquad (100)$$

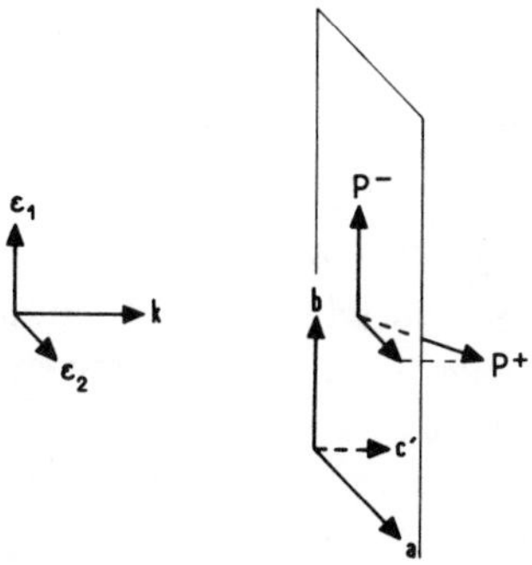

Fig. 16. Directions of wave vector and polarizations $\hat{\varepsilon}_1$ and $\hat{\varepsilon}_2$ of an incident photon, relative to the crystal polarization $\mathbf{P}^+$ and $\mathbf{P}^-$.

which in terms of the permittivity is

$$\varepsilon_-(\omega) = n_-^2(\omega) = 1 - \frac{\omega_p^2 F_-(k)}{\omega^2 - \omega_-^2(k)} \tag{101}$$

These expressions are exactly those found in the semiclassical theory, (69) and (70), involving molecular data ω_m and f_m, whereas here the data are exciton values $\omega_-(k)$ and $F_-(k)$ [$\omega_-(k) \equiv \omega_T(k)$: energy (in cm^{-1}) of the Coulomb exciton].

Spatial dispersion is negligible here and we shall set

$$\omega_-(k) \equiv \omega_-(0)$$

4. Energy Dependence of the Exciton and Photon Components of the Polariton

Given the polariton energy $E(k)$, system (95) can be used[166] to find the coefficients of the canonical transformations $U_{k\nu}$, U_{kj}, $V_{k\nu}$, and V_{kj}. They must also obey a normalization condition.

In the present case of **b**-polarized photons, there is coupling only with the exciton branch ($j \equiv b, \nu \equiv -$). We find:

$$|U_{kj}|^2 \equiv |U_{kb}|^2 = \frac{(\hbar kc + E)^2 (E_-^2 - E^2)}{8\hbar kcE(E_-^2 + \hbar^2 k^2 c^2 - 2E^2)}$$

$$|V_{kj}|^2 \equiv |V_{kb}|^2 = \frac{(\hbar kc - E)^2}{(\hbar kc + E)^2} |U_{kb}|^2$$

$$|U_{k\nu}|^2 \equiv |U_{k-}|^2 = \frac{(\hbar^2 k^2 c^2 - E^2)(E_-^2 - E^2)\hbar kc}{E_-(E_- - E)^2(\hbar kc + E)^2}$$

$$|V_{k\nu}|^2 \equiv |V_{k-}|^2 = \frac{(E - E_-)^2}{(E + E_-)^2} |U_{k-}|^2. \tag{102}$$

Using the polariton dispersion curve $E(k)$, the coefficients can be expressed in terms of the energy only, and the weights of the photon ($|U_{kj}|^2 - |V_{kj}|^2$) and of the exciton ($|U_{k\nu}|^2 - |V_{k\nu}|^2$) can be deduced. Figure 17 shows the weights of each component, for numerical values typical of anthracene and for strong and weak coupling between the exciton and the photon.

Once again, we see that far from resonance the polariton is very much like a photon, whereas close to resonance it is nearly an exciton.

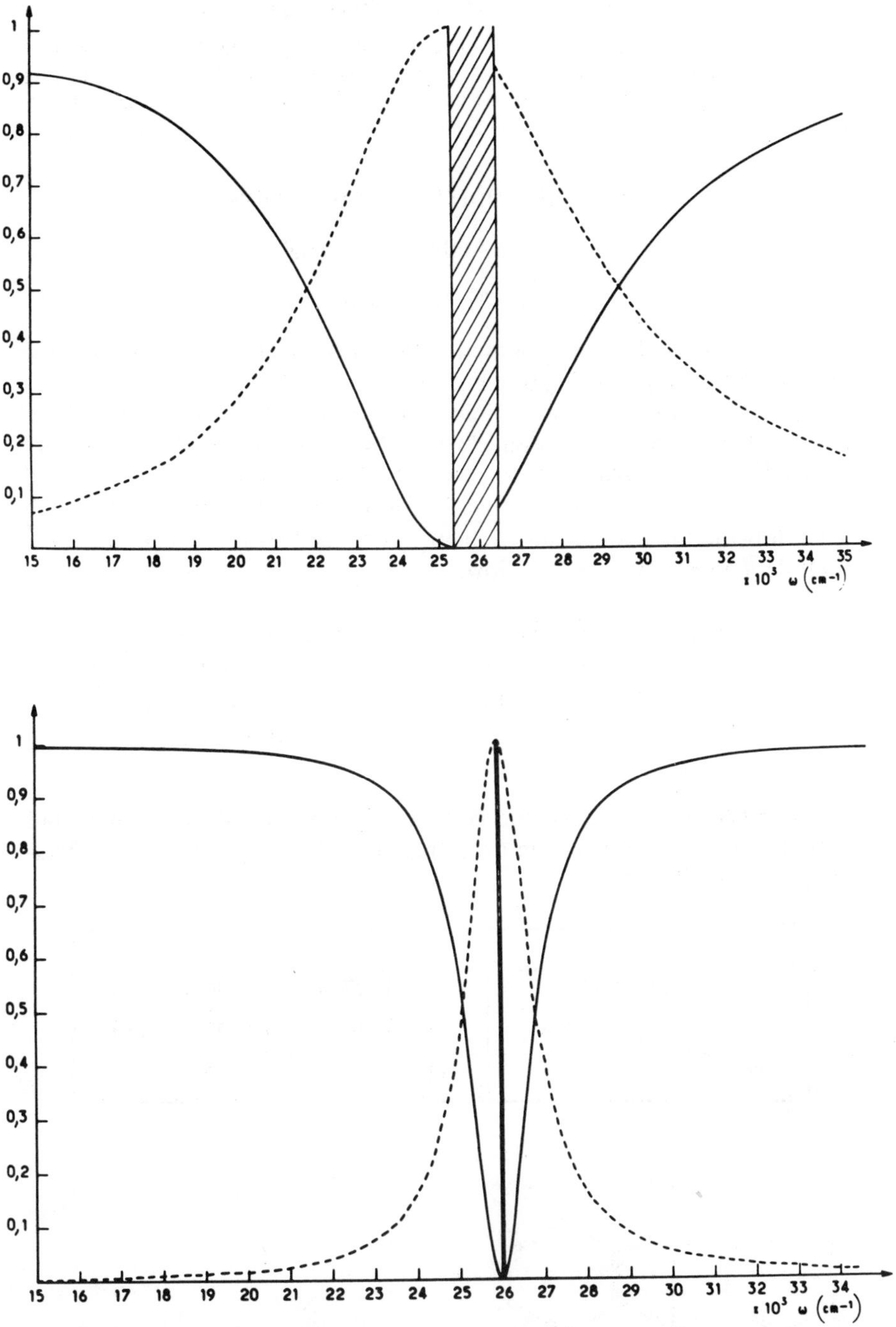

Fig. 17. Weights of the photon (—) and exciton (---) components of the polariton versus energy ω for two different exciton–photon couplings: strong coupling (upper curve), weak coupling (lower curve).

N.B. These curves are generally similar to those published by Hopfield[113] in his first article. It is interesting to compare them with those published by Huang[115] showing the variation of the "mechanical energy percentage" in transverse vibration modes in ionic crystals (classical study of phonon–photon coupling).

IV. EXPERIMENTAL METHODS

A. The Anthracene Crystal

1. *Crystal Structure*

Anthracene, $C_{10}H_{14}$, is composed of three benzene rings. It is flat and belongs to the point group D_{2h} (see Fig. 18).

The crystal, which is monoclinic, belongs to the space group $P2_1/a$ (C_{2h}^5 in Schöenflies's notation.)[167, 168] The operators in this group are $\{E, i, C_2^{(b)}, \sigma_{ac}\}$, whereas the site group in which both molecule and crystal are invariant is $\{E, i\}$. The unit cell is formed by vectors **a**, **b**, and **c**, with $\mathbf{a} \perp \mathbf{b}, \mathbf{b}$ being the monoclinic symmetry axis (see Fig. 19). The latest structural data, due to Mason,[169] follow:

At 290 K	$a = 8.562$ Å	At 90 K	$a = 8.443$ Å
	$b = 6.038$ Å		$b = 6.002$ Å
	$c = 11.184$ Å		$c = 11.124$ Å
	$\beta = 124°42'$		$\beta = 125°36'$

The molecular axes **L**, **M**, and **N** relative to the crystalline axes **a**, **b**, and **c**′ ($\mathbf{c}' \perp (\mathbf{a}, \mathbf{b})$) are given in the accompanying tabulation for two temperatures (290 and 95 K).

	a	b	c'
L	119°6/120°7	97°3/97°5	30°6/31°9
M	71°5/72°3	26°6/26°2	71°6/71°3
N	36°0/36°5	115°3/115°0	66°3/65°2

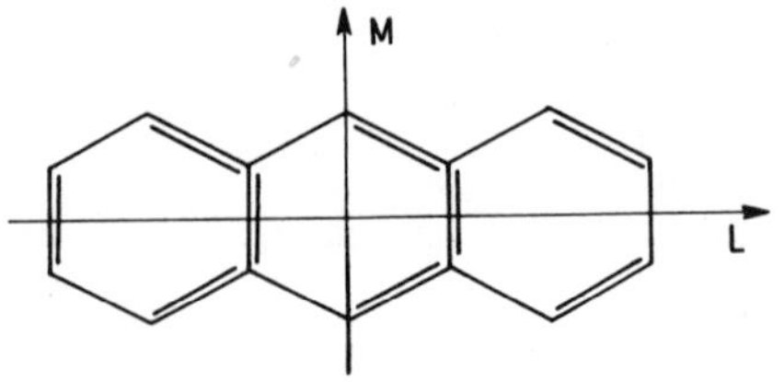

Fig. 18. The anthracene molecule.

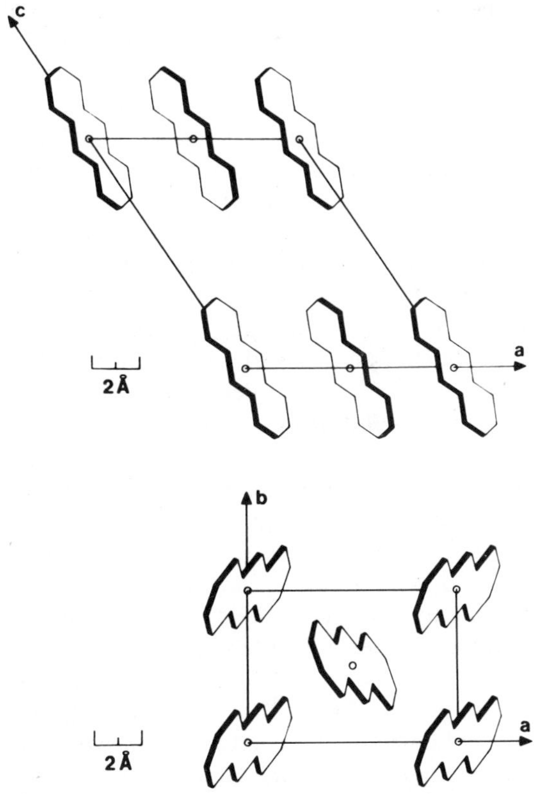

Fig. 19. Projections of the unit cell of anthracene on the (**a**, **b**) and (**a**, **c**) planes.

The unit cell contains two nonequivalent molecules: 1 at (0,0,0), and 2 at (1/2,1/2,0). They are related by reflection in the plane at $b/4$ followed by translation of $a/2$, or by rotation about the C_2^b axis at $a/4$ followed by translation of $b/2$. These symmetry operations and the site wave functions lead to group factor wave functions corresponding to the two exciton components of the crystal.[170,171]

2. *Refining and Crystal Growth*

The purest commercial anthracene, containing roughly 10^{-3} M per molar of impurities, is unsuitable for spectroscopic purposes. In order to reduce these impurities to less than those detectable chemically (i.e., less than 1 ppm), which only certain experiments (e.g., fluorescence, lifetime of the triplet) can qualitatively appreciate, it must be extensively purified by methods amply described elsewhere.[172–183]

On exposure to near ultraviolet light, anthracene may photodimerize in the form of stable dipara-anthracene or change into peroxides and anthraquinone.[184-188] These reactions seem to occur preferentially along faults and dislocations.[189] In general we have avoided these effects, except once when they were useful. All the purification and growing stages were carried out in yellow light and under an inert atmosphere of very pure argon. Most of the experiments carried out at the IBM laboratories were on anthracene kindly provided by D. M. Burland. The purification process is described in Ref. 190. Monocrystals were obtained by methods derived from the classical techniques of Bridgman fusion, saturated solutions and sublimation.[191-196] In Bordeaux the raw anthracene (Fluka "puriss," Eastman–Kodak "scintillation grade," and "synthetic" anthracene) is placed in sealed pyrex tubes and zone refined 200 times at 1 cm/h, after degassing at 10^{-6} torr; normal, slow fusion, and cooling under 400 torr of argon; degassing again; and refilling with argon. The middle section of each tube is removed and the whole process repeated. Only a few centimeters from the middle of this tube are used for final purification and crystal growth.

In fluorescence studies we used crystals with as few structural defects as possible. Our sublimator (Fig. 20) is based on those of Lipsett[191] and Glockner.[197] Better joints have been added, able to withstand a vacuum of 10^{-6} torr; a silicon oil bath heater and tapered shape have also been added, the latter increasing the area of product formed. The pressure of the argon at room temperature is 165 torr, and the crystals are grown at 170°C.[198] Seventeen cycles of sublimation of the anthracene powder, in each of which the crystals formed are removed and the pumping and filling with argon are repeated, have shown that the best crystals form after 10 or so cycles, the first ones serving only as purifications.

The resulting crystals are thin, on average a micron thick, roughly $1/2\ \text{cm}^2$ in area, with a well-developed (001) face. They can be handled by sucking on their edges with a pipette.[191]

3. *Measurement of Crystal Thickness*

Anthracene has a monoclinic structure and is therefore optically a biaxial crystal. The principal direction, corresponding to the middle index n_β, is along the **b** axis. Values of the principal indices and directions at different visible wavelengths have been published.[199-201]

When light falls normally on the $(\mathbf{a},\mathbf{b})$ face, the thickness e can be deduced from the known birefringence and the measured path difference from the relation

$$\Gamma = (n_b - n_a)e$$

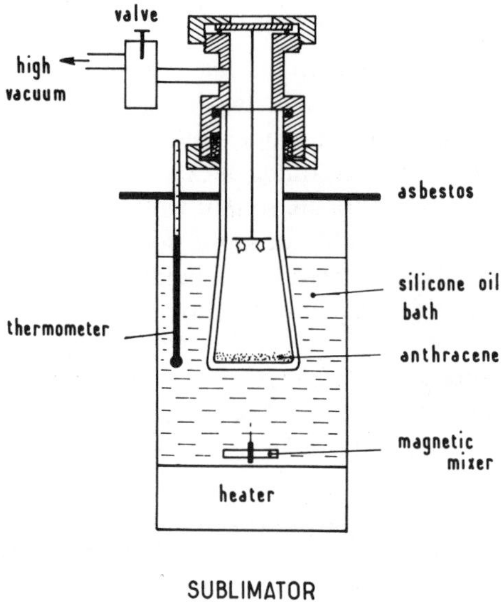

Fig. 20. Principle of the sublimator.

The commonly used[202] value of $n_b - n_a$ is

$$(n_b - n_a) = (1.79 - 1.63) = 0.16$$

The value of Γ can be estimated by comparison of the interference color of the crystal placed diagonally between crossed polarizers, with the Michel Levy table. Allowance must be made for the anomalous colors[203] in anthracene, due to strong dispersion.

Although this method does not yield the exact thickness, it is, due to the sensitivity of the eye to shading, a very good test of very small changes in thickness and of the uniformity of the surface. We used only those crystals which were of absolutely uniform color (constant thickness) and which reflected light simultaneously from all regions (a test of flatness).

Better estimates of the thickness were made with a rotating de Senarmont compensator[204,205] in monochromatic light at 5460 Å. Bree and Lyons's interference method[206] could not be used because of the glass–crystal contact. The former method easily measures path differences in the range 0 to λ due to anthracene crystals a few microns thick. Accuracy (~ 10%) is limited by the error in the birefringence.

4. *Mounting and Orientation*

Strains at very low temperatures may cause structural defects, so we mounted our crystals freely between two black paper diaphragms with 3 mm holes, placed in the recess of the sample holder, which also had a hole through which light could pass. A small frame of exactly the same shape as the crystal was glued to the back diaphragm, so that while the paper did not exert any pressure on the crystal it could not move sideways. The front face, pressed against the frame only, was fixed to the holder with four spots of low temperature putty (see Fig. 21). These operations were carried out under a binocular microscope.

The crystalline axes were found relative to the holder by noting the positions of extinction between crossed polarizers. The holder could be placed directly on the microscope stage. The **a** and **b** axes can be determined by observing the isogyre along **a** in convergent light.[191,201,207] However, due to the difference in the brightness at extinction, observation in parallel light was sufficient. The polarization can be finely adjusted at low temperatures by rotating the crystal in its plane with a mechanism controlled from the outside (see (3) on Fig. 22).

B. Low Temperature Techniques

The crystal is cooled to very low temperatures by placing the sample holder in a liquid helium—filled cryostat. Besides rotation (3), above, of the crystal relative to the holder, two other controls can be used to turn (through 360°) or slide (through 3 cm) the holder relative to the cryostat (Fig. 22). By this means the angle of incidence and the region examined can be varied, and in absolute reflectivity measurements, a reference mirror covered by an identi-

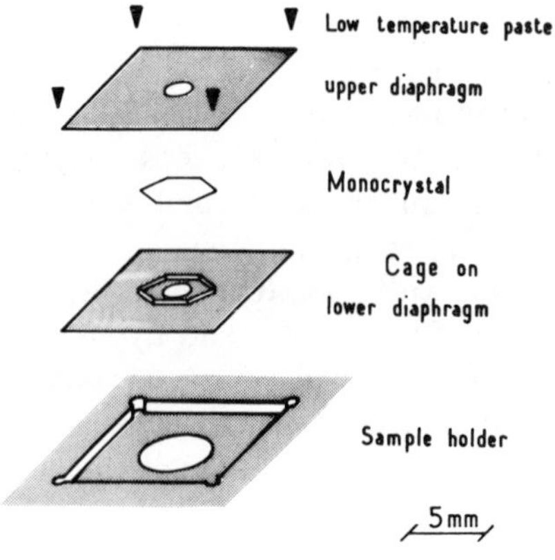

Fig. 21. A "free" mounting.

cal diaphragm to that of the crystal, and situated just below it, can be raised into place. The cryostat itself rests on a stable support, rotatable about a vertical axis through 360° (useful, among other things, for eliminating reflections from the windows) and movable through 10 cm perpendicular to the optical axis, so all the crystal can be scanned.

A Germanium resistance thermometer placed close to the crystal on the holder is used to measure its temperature. The temperature of the resistance, calibrated to 1/100 K between 1.5 and 100 K is continuously monitored by a four-wire bridge connected to a current generator (1, 10, and 100 μA) and a digital microvoltmeter.

The cryostat (Fig. 23) has a 4-liter liquid helium tank joined to the sample chamber via a "cold" needle valve, a capillary tube, and a heat exchanger. The sample is constantly surrounded by gaseous or liquid helium. The lowest temperatures (1.6 K) are reached by continuous pumping over a pool of liquid helium kept at a constant level in the chamber. Temperatures around 1.6 K can be maintained for 4 h, whereas at 5 K the full tank lasts more than a day.

We have found a very simple way of coating the crystal with a thin coat of gas. The crystal is immersed in liquid helium which is all evaporated off by pumping so that the chamber containing helium at low pressure sucks in the gas as soon as the bottle is opened very slightly. Condensation on the crystal (at roughly 10 K) is monitored by the decline of the reflectivity peak

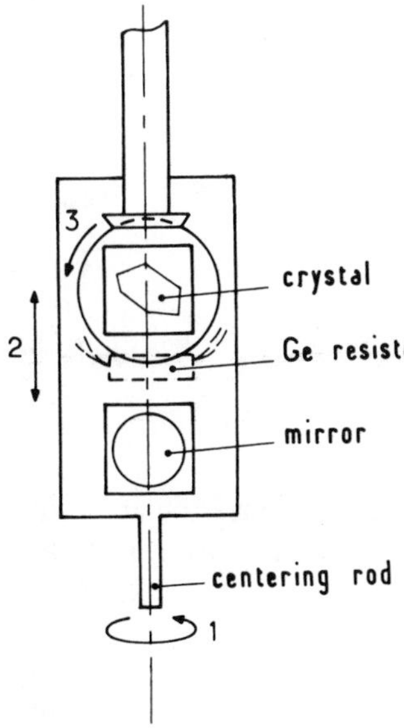

Fig. 22. The sample holder.

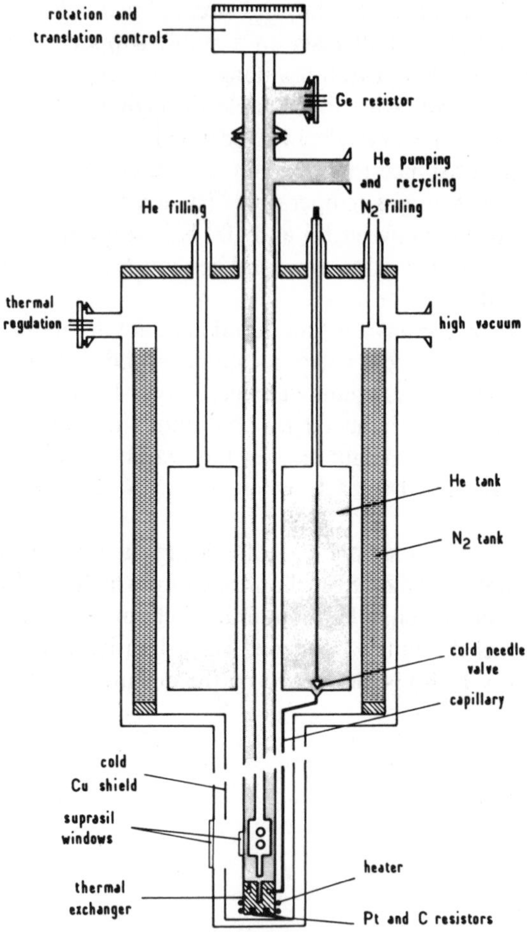

Fig. 23. Cryostat design.

of system I at 25,310 cm^{-1}. The gas coat can be removed by warming the crystal in a current of gaseous helium.

C. Apparatus

We shall describe only the set-up made in Bordeaux, since the apparatus used at the IBM laboratories was nearly the same as the reflectivity part (Fig. 24).

1. Reflectivity

The light source used is a 450-W xenon lamp with a continuous spectrum between 3100 and 4100 Å, varying linearly in intensity with the wavelength.

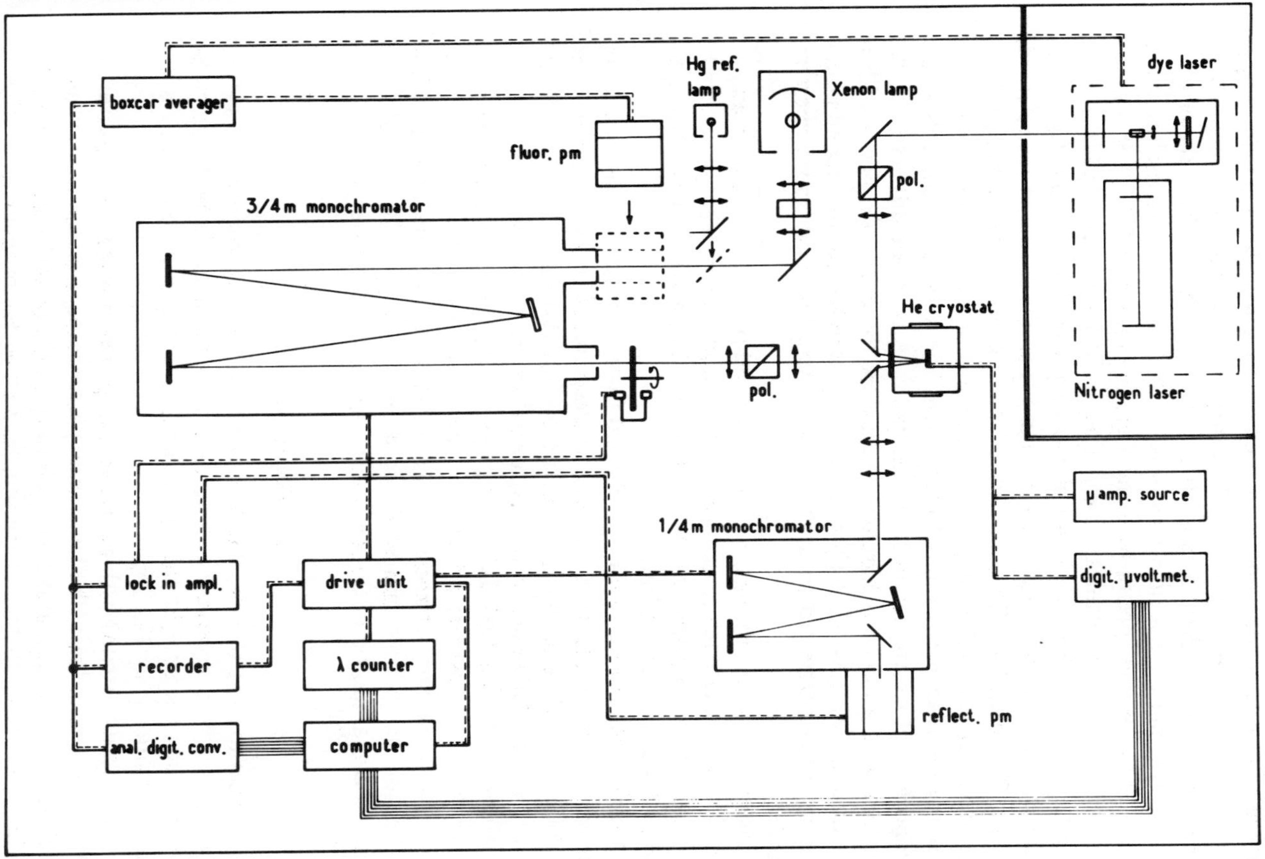

Fig. 24. Reflectivity and fluorescence apparatus.

There are no superposed lines. The light passes through a collimator, and infrared filter (water cell), and is focused on the entry slit of a 0.75-m Spex 1702 monochromator used at $f/6.8$. Using a B. L. engraved grating of 1200 lines/mm blazed at 3000 Å in the first order, with slit widths of the order of 12 μm, we had an exit linewidth of 0.1 Å (0.6 cm^{-1}). This resolution was doubled for certain spectra by using a grating blazed at 5000 Å in the second order. After mechanical modulation at 30 Hz the exit beam is made parallel, polarized by an 18-mm glan-laser prism (Fichou) and focused at nearly normal incidence (always less than 5°) on the crystal or the reference mirror. The specular reflection, viewed by a beveled mirror on an x, y, θ_1, θ_2 stand, is focused on the entry slit of a 0.25-m J. A. monochromator to filter out any fluorescence in the reflected beam. The wavelength controls of the monochromators are carefully synchronized. Any slight difference at the beginning of a sweep is harmless because of the difference between their bandwidths. An EMI 6256 photomultiplier placed behind the second monochromator is connected to a synchronous detection unit and a chart recorder.

After each sweep, the xenon lamp is replaced by a mercury vapor lamp to provide an accurate reference (4046.56 Å) relative to which any shifts can be detected. We used sweep speeds of 0.1 to 0.01 Å/s and a time constant typically of the order of 300 ms.

Using as reference signal light reflected from the cryostat window and detected through an identical optical and electronic channel (not represented in Fig. 24), a data acquisition unit normalizes (for lamp fluctuations and spectral and polarization variations), digitalizes the reflectivity spectrum, and finally sends it on a 4051 Tektronix graphic computer where it is stocked on magnetic tape. Absolute reflectivity spectra are obtained by replacing the crystal with the mirror. Experimental and theoretical absolute reflectivity and dispersion curves were calculated and drawn on the CII.HB computer and Benson digital plotter of the Bordeaux I university computer center.

2. *Fluorescence*

In order to show that the details in the reflection and fluorescence spectra coincided, it was necessary to keep resolution the same in both and to be able to study one straight after the other so that they were both made in the same conditions (see Fig. 25).

The crystal is excited with a tunable laser pumped at the lowest power (12 kV) by a nitrogen laser. The laser beam passes through a glan-laser polarizer and falls, out of focus, at normal incidence, on the crystal. Fluorescence, detected from the front face of the sample, follows the reverse path to that in reflectivity experiments. It passes through a second glan-laser polarizer into

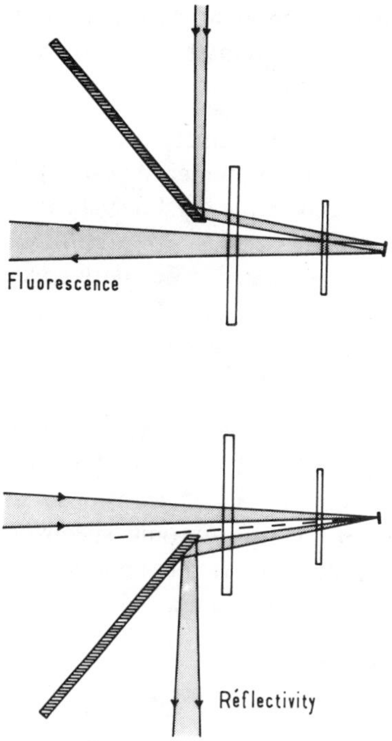

Fig. 25. Beam paths in reflection and fluorescence experiments.

the Spex monochromator. The pulsed signal is detected by an EMI 6256 photomultiplier and a sampler-averager (PAR.) which delivers a continuous signal to the data processing unit and the chart recorder.

Excitation spectra between 26,300 and 24,500 cm^{-1} were taken with the dye BBQ (in a 50/50 toluene–ethanol solvent) by varying the laser wavelength and setting the Spex on the emission of interest. The spectra were normalized to take account of the variation of laser power with wavelength by directing the signals from a direct reference beam and the fluorescence to the entry channels of a boxcar ratiometer.

V. EXPERIMENTAL REFLECTIVITY AND FLUORESCENCE STUDIES OF THE PHOTODYNAMIC PROPERTIES OF CRYSTALLINE ANTHRACENE

A. Introduction

The apparatus and methods described above were gradually developed to answer the four questions we asked at the end of the first chapter concerning the "anomalous" structures observed (or not) in the optical reflection and fluorescence spectra of crystalline anthracene [face (001)] in the region of the first singlet transition (25,100 cm^{-1}). This experimental work, carried out first in the IBM laboratories at San Jose, California, and later at the University of Bordeaux I, can be divided into four main parts:

1. The **b**-polarized reflectivity spectrum at very low temperatures (1.6 K) and high resolution (0.4 cm^{-1}) at nearly normal incidence ($i < 5°$).
2. The influence on the spectra of modifications of the crystal surface in reversible and irreversible experiments.
3. A similar investigation (1 and 2) of the **a**-polarized reflectivity spectrum.
4. Comparative study of the reflectivity and fluorescence spectra of the same sample in the same experimental conditions.

The **a**- and **b**-polarized normal incidence spectra of the (001) face are quite different. Room temperature spectra made by Clark and Philpott[71] between 20,000 and 55,000 cm^{-1} (Fig. 26) show, besides the $\pi-\pi^*$ transitions at higher energy (37,000, 45,000, and 52,000 cm^{-1}), the vibronic spectrum (25,000 to 30,000 cm^{-1}) of the first singlet transition. At this temperature the vibronic structure is like that of the free molecule, composed of broad

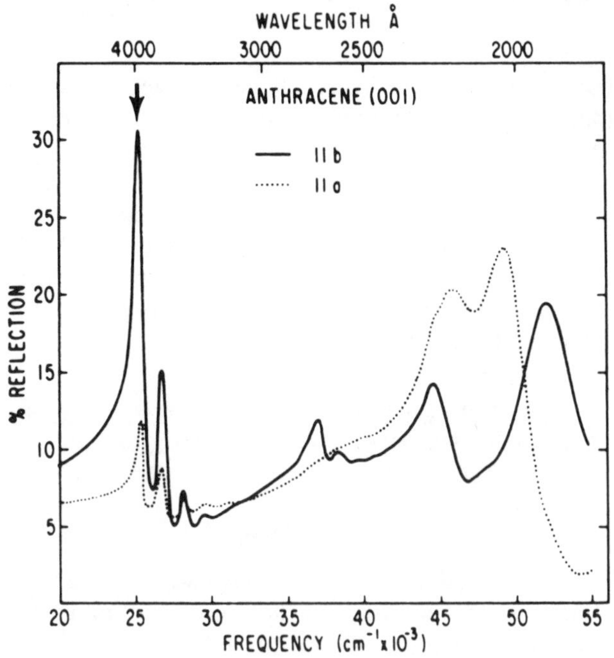

Fig. 26. Absolute, polarized reflection spectra of the (001) face of anthracene at 300 K. After Clark and Philpott.[71]

bands [denoted (0,0), (0,1), (0,2), etc.] at regular intervals of roughly 1400 cm^{-1} due to increasing vibrational quantum numbers.

At low temperatures (6 K), in addition to the high reflectivity [for $\mathbf{E}\|\mathbf{b}$ the maximum of the (0,0) line increases from 30 to 90% reflection], the vibrational bands are structured to varying degrees (see Fig. 27). The (0,1) region from 25,500 to 27,500 cm^{-1} is a broad band, quite strong if $\mathbf{E}\|\mathbf{b}$, on which

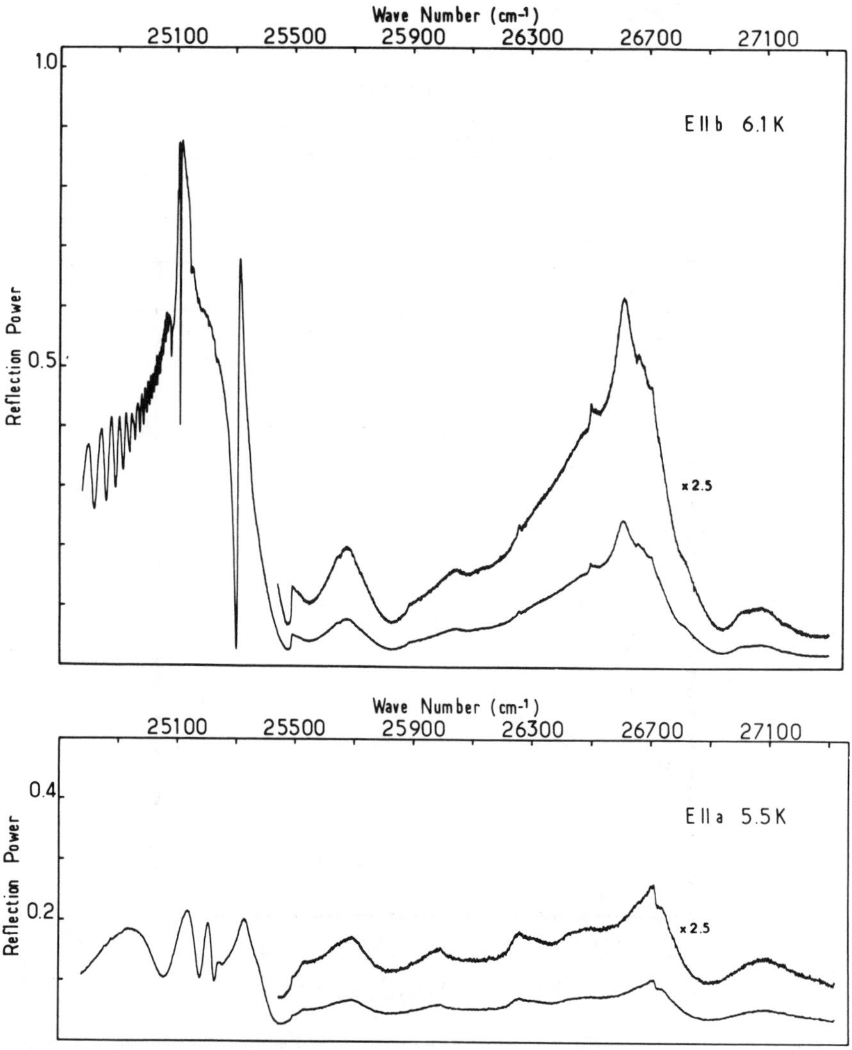

Fig. 27. Polarized, absolute reflectivity spectra of the (001) face of anthracene (sublimation grown) at 6 K, in the (0,0) and (0,1) vibronic regions of the first singlet transition.

many narrow or fairly narrow lines are superposed. The large number of intramolecular vibrational modes possibly present in these spectra makes vibrational analysis very complicated and necessitates comparison between different spectra. A tentative interpretation of most of the structures between 25,500 and 28,000 cm^{-1} has been made.[33] It invokes ground state molecular vibrations (two-particle systems) and molecular vibrations of the electronic excited states (vibrons). This analysis will not be developed here as it would not be germane.

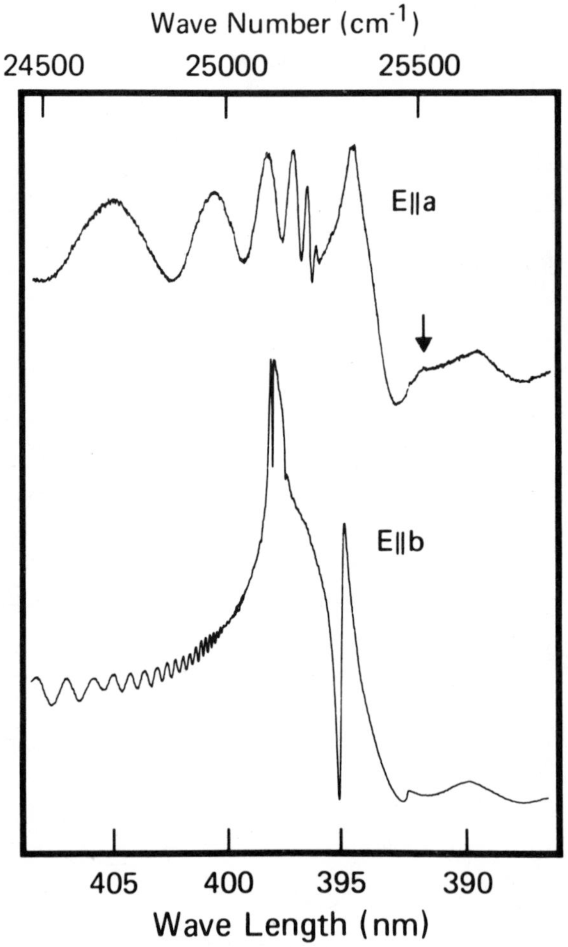

Fig. 28. Polarized reflectivity spectrum of the (001) face of anthracene (sublimation grown) at 2 K, in the (0,0) region of the first singlet transition. The **E**||**a** spectrum has been multiplied by 4.

The description below is limited therefore to the (0,0) region corresponding to pure electronic transitions and to phonon modes (0 to ~150 cm^{-1}). Note that on the high energy side of Fig. 28 there can also be seen a transition involving the 390 cm^{-1} intramolecular vibration: asymmetric line at 25,484 cm^{-1} (two-particle system) and broad maximum at 25,600 cm^{-1} (vibron).

B. b-Polarized Reflectivity Spectrum

1. Description

Figure 29 shows the characteristic features in the (0,0) region of the **b**-polarized, normal incidence reflectivity spectrum of the (001) face of a "sufficiently" pure crystal at low temperature (6 tol. 7 K)

The gross shape of this spectrum is similar to the theoretical rectangular stopping band described above, except that it is progressively attenuated toward high energies. This may be explained by the various relaxation processes coupling the polariton to dissipative channels (e.g., phonons), which in a real crystal may allow propagation and or absorption (lowering the reflectivity) in the stopping band. The sum of these processes has greater probability on the high side of ω_T as the density of phonons increases. Despite the gradual

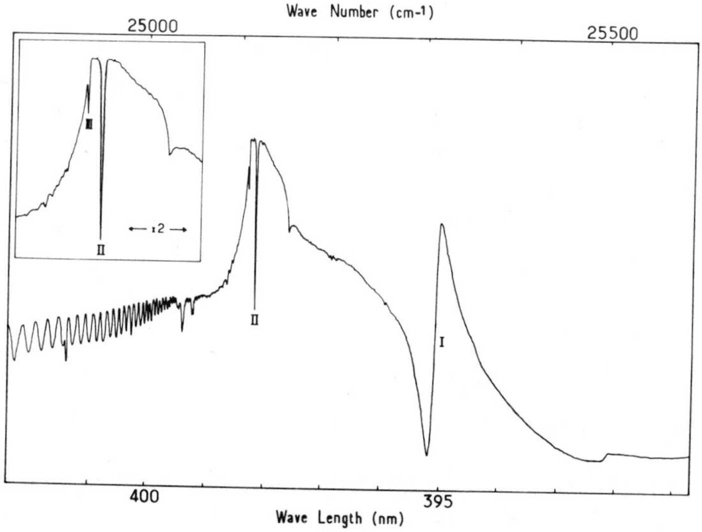

Fig. 29. The **E**||**b**-polarized reflectivity spectrum of the (001) face of a (sublimation prepared) anthracene crystal, of a few microns thickness, at 1.7 K. Besides structure I (25,294, 25,307 cm^{-1}), the magnified view in the corner shows structures II (25,102 cm^{-1}) and III (25,095 cm^{-1}), as well as the dip at 25,142 cm^{-1}.

drop in reflectivity the stopping bandwidth may be estimated at roughly $(\omega_L - \omega_T) = 400\ \text{cm}^{-1}$. Close examination of the spectrum reveals:

(*a*) Up to 25,095 cm^{-1} (ω_T) the reflectivity increases, slowly at first, and rapidly close to ω_T. This corresponds to passing around the knee region in the $(-)$ branch of the dispersion curve of the polariton (see Fig. 9). Thin crystals, say, 1 μm thick, exhibit oscillations. In the transparent frequency range (polariton $\equiv$ photon) of the crystal, they are easily explained by the interference of reflected waves off the front and back faces. The strong dispersion of the refractive index accounts for the shortening of the interfringe separation. As ω_T is approached from below there is a closer and closer spacing of the polariton modes.

(*b*) Between 25,095 and 25,450 cm^{-1} lies the stopping band. Besides diffuse but reproducible structures perhaps due to weakly active phonon modes, convoluted with their density of states, one can pick out four definite features:

1. The reflectivity minimum ($\sim$ 0%) at 25,294 cm^{-1} in system I and the maximum ($\sim$ 70%) at 25,307 cm^{-1}
2. A very narrow minimum (II) at 25,102 cm^{-1}
3. A minimum (III) at 25,095 cm^{-1} (see inset in upper left-hand corner)
4. A noticeable dip at 25,142 cm^{-1}, between I and II

Using the monochromator in the second order (resolution 0.3 cm^{-1}) and sweeping very slowly at very low temperature (1.7 K) we have been able to examine II and III more closely (Fig. 30). Note that structure III, which occurs only when the minima of I and II are close to zero, most probably has not been completely resolved, since the measured width is of the order of the resolution of the equipment.

(*c*) Beyond 25,450 cm^{-1} the reflectivity falls off to very low values ($\sim$ 3%) and rises at the 390 cm^{-1} vibration mentioned in the introduction to this section.

2. *Temperature Dependence*

Figure 31 shows the variation of structure I (II and III in inset) between 2 and 120 K.

Above 100 K the reflectivity spectrum is shapeless, without any structure. Increasing the temperature activates various intra- or intermolecular vibrational modes and so depresses the reflectivity as more polariton relaxation channels are created.

At this temperature the crystal is nothing like our model, and it is not surprising that the reflectivity drops and the structures disappear.

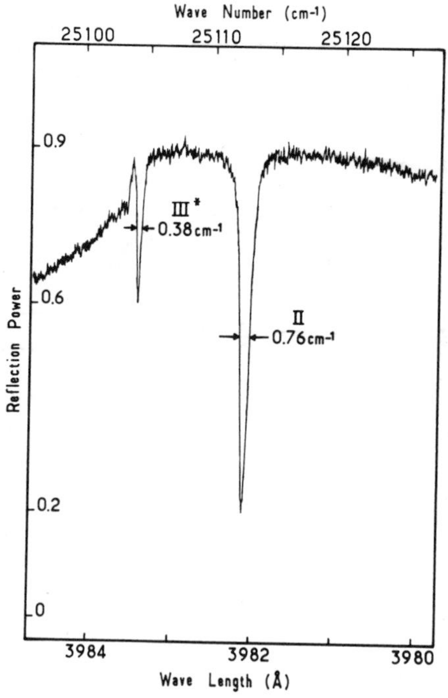

Fig. 30. High resolution sweep (0.3 cm^{-1}) of structures II and III of the **E**‖**b** reflectivity spectrum of Fig. 29. Structure III apparently is not completely resolved. The minimum probably is narrower and deeper than shown.

On cooling, system I begins to appear around 80 K as a shoulder progressively deepening and narrowing until at 10 K there is little further change. (There is hardly any difference between the forms at 5 and 1.7 K.)

These variations occur in reverse order on warming the sample, which may be repeatedly cooled and warmed with the same results. Their positions are constant at all temperatures and their intensities are unaffected by very slow cooling or by prolonged storage (even several weeks) at low temperature.

3. *Influence of Crystal Thickness*

Spectra of Bridgman-grown crystals several millimeters thick (Fig. 32) are very similar to those of this sublimation prepared crystals (Fig. 29). They also exhibit structures I, II, and III. The only difference is that the oscillations on the low energy side are absent.

4. *Influence of Crystal Purity and Surface Quality*

While the general shape of the reflectivity spectrum depends little on the sample (excepting a general drop in reflectivity), the presence of structures I, II, and above all III depends strongly on the purity and the surface quality,

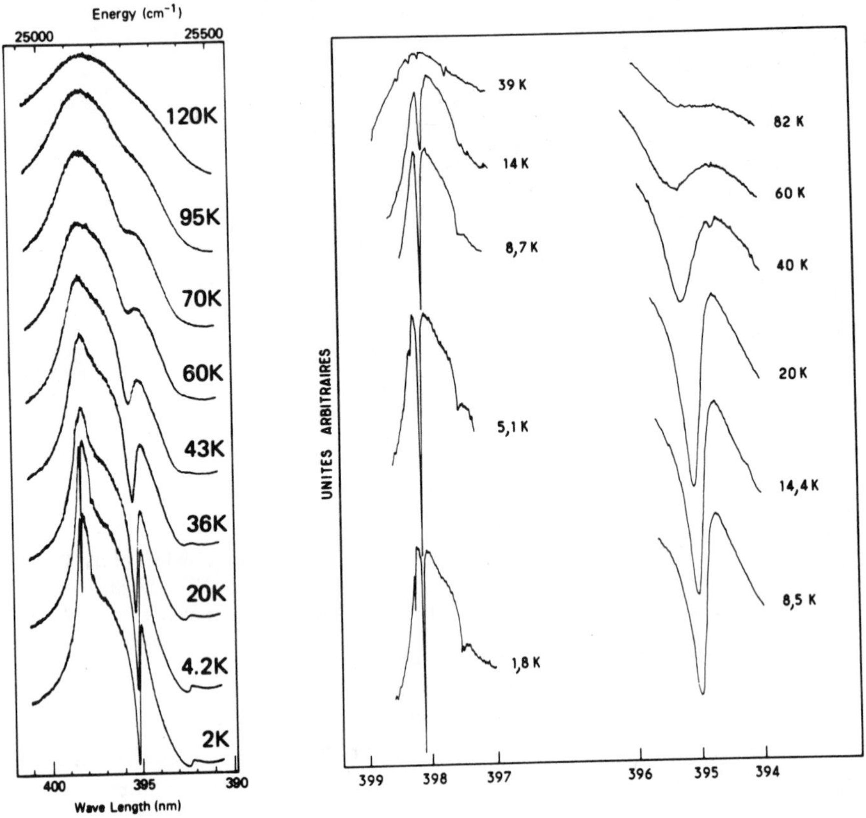

Fig. 31. Temperature dependence of the **E**||**b** reflectivity spectrum of the (001) face of anthracene. Inset on right shows an enlarged view of structures I, II, and III.

requiring thorough purification. This can be shown by comparing the spectra of successive sublimitations.

Figure 33 compares the spectra of a thoroughly purified sample (1) and of a crystal obtained during the first sublimitations (2). The minima I and II are hardly visible in (2). However, (2) shows a broad minimum at 25,106 cm^{-1} not unlike that announced at 10 K.[106] Note that the general shape of the spectrum, excepting structures I, II, and III, is the same in both cases.

This remark is illustrated also be Fig. 34, showing the reflectivity spectra of three samples prepared by evaporation of a saturated solution. Such crystals generally have poor surfaces with defects visible to the naked eye, in which inclusion of solvent molecules cannot be ruled out. Structures I and II (III was not observed) vary from one sample to another.

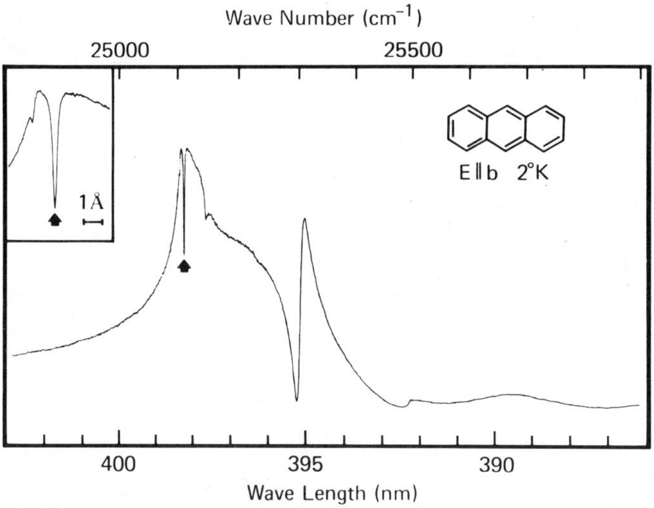

Fig. 32. The 2 K, **E**||**b** reflectivity spectrum of the (001) face of an anthracene crystal several millimeters thick, grown by Bridgman fusion. The inset shows structures II and III.

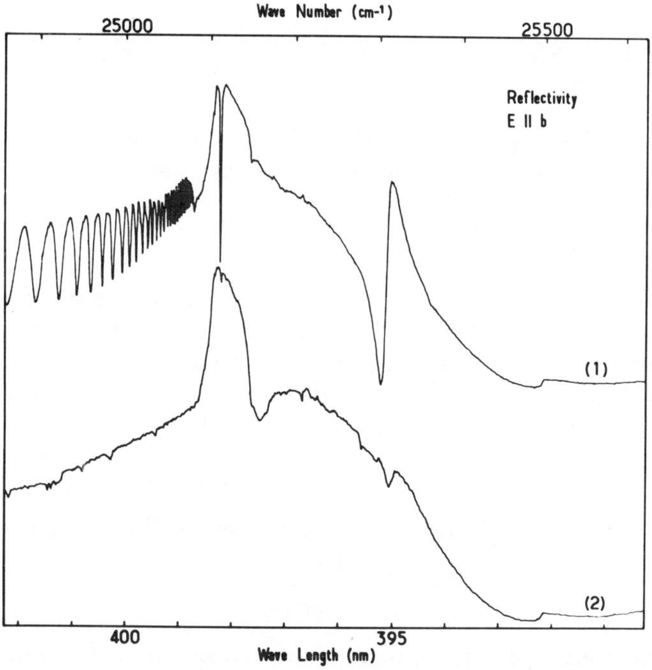

Fig. 33. Comparison of the 5 K, **E**||**b** reflectivity spectra of the (001) faces of very pure (1) and less pure (2) anthracene.

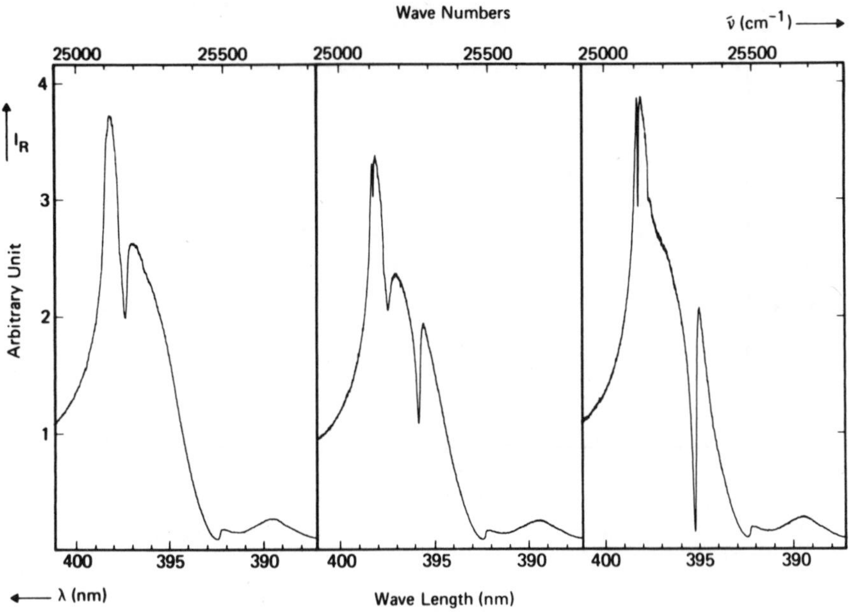

Fig. 34. The 2 K, **E**||**b** reflectivity spectra of the (001) faces of three crystals grown from the same purified raw product by evaporation of a saturated solution.

It is remarkable that from I to III the depth of a given minimum depends on that of the lower numbered minima. Thus, minimum III is seen only if II and I are very deep.

5. *Conclusion*

The main point established here is that in sufficiently pure samples we have indeed seen the structures I and II recorded by Marisova and in the "best cases" have found a third one (III). Besides the last structure, the new point made here is that structures I, II, and III seem correlated. Their behavior as a function of purity and temperature differs markedly from that of the rest of reflectivity spectrum.

C. Influence of Surface Modifications on the E||b Reflectivity Spectrum

1. *Principle*

The basic idea here is to distinguish between surface and subsurface features and the rest of the spectrum related to the bulk of the sample. It was necessary to find ways of perturbing (reversibly if possible), the first few crystal planes to observe the effects on the spectra. We have carried out two kinds of experiments. In the first (irreversible), advantage is taken of the photochemical reactions which may occur when anthracene is exposed to near ultraviolet sources. In the second (reversible), a layer of material, transparent in the region studied, is deposited on the surface. These two experiments modify molecular interactions at the surface—the first rather violently since a new chemical species is formed, the second softly since only the molecular interactions are altered.

2. *Surface Photochemistry*

As we mentioned when describing the properties of anthracene (see Section IV), it may, under certain circumstances, become photo-oxydized when

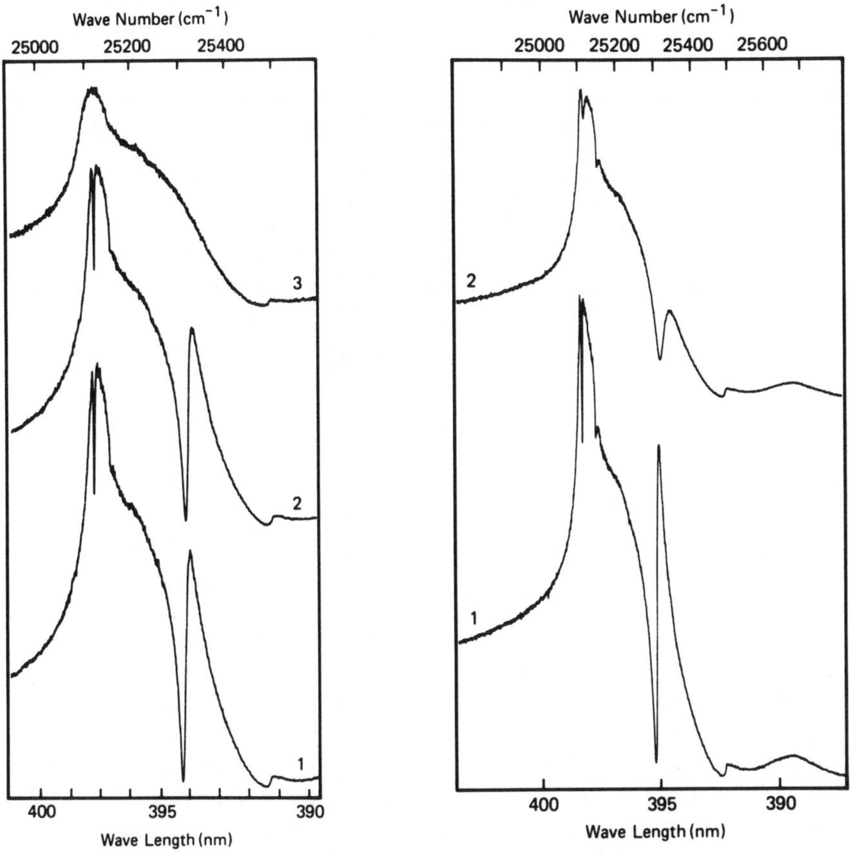

Fig. 35. Effect on the **E**||**b** reflectivity spectrum [(001) face at 2 K] of illuminating the sample with ultraviolet light. *Left*: (1) before irradiation; (2) after 30 min of treatment, the sample being in liquid helium at 4.2 K; (3) after 1 min at 300 K in air. *Right*: Same experiment on another sample (1) before and (2) after 15-s illumination in air at 300 K.

illuminated with a near ultraviolet source. The result of this effect, brought about by shining a small pencil-type mercury vapor lamp on the sample, is shown in Fig. 35; all the spectra are at 2 K.

On the left we see that spectrum (2) obtained after 30-min exposure in liquid helium is little different from (1) before exposure. This is normal since photodimerization requires some movement of the molecule (very small at 4 K) and photo-oxidation cannot occur in a liquid helium atmosphere. However, after even a short exposure (a minute or so) in air at 300 K [spectrum (3)], structures I, II, and III [hardly visible in (1) and (2)] disappear entirely, and the reflectivity generally diminishes, though keeping the same form.

The right-hand side shows the same experiment on another sample illuminated very briefly (about 15 s) in air. We have here an intermediate situation in which structures I and II are partially destroyed. It is reasonable to suppose that during such a brief exposure only the very first few layers are affected. These spectra show that, unlike the rest of the spectrum, structures I, II, and III are particularly sensitive to surface perturbations.

3. Condensing and Evaporating a Transparent Layer of Gas on the Surface

This new experiment is directly related to the surface exciton described in Section II. For if one observes a surface exciton state above those of the bulk, due to the different interactions affecting molecules in the surface and subsurface, then by covering the surface with a transparent substance one could hope to recreate the "missing" interactions and make the environment of surface molecules more like that of bulk ones. The surface exciton should therefore be stabilized, and the reflectivity structures associated with the first layers should be shifted to lower energies (i.e., toward the bulk exciton energy ω_T).

In order to be conclusive, the experiment must be carried out "in situ" and reversibly, to be sure that one has the same sample in the same experimental conditions. [It is out of the question to deposit a thin coat (e.g., Au, Ag,...) in an evaporator.] Since our observations are made at very low temperatures, we condensed a layer of gas on the crystal, which was all the time kept between 4 and 20 K in the cryostat. Different stages of the coating can be followed by successive additions of small quantities of the gas, and on heating above the boiling point of the condensed gas, the original surface is restored. All this occurs in a neutral atmosphere of helium (see Section IV).

We used various gases, such as air, oxygen, nitrogen, argon, krypton, and xenon, and organic substances like methane and ethylene.

Figure 36 shows the effect of successive coatings with small amounts of air. All the spectra were taken at 2 K. On the left we show the effect of

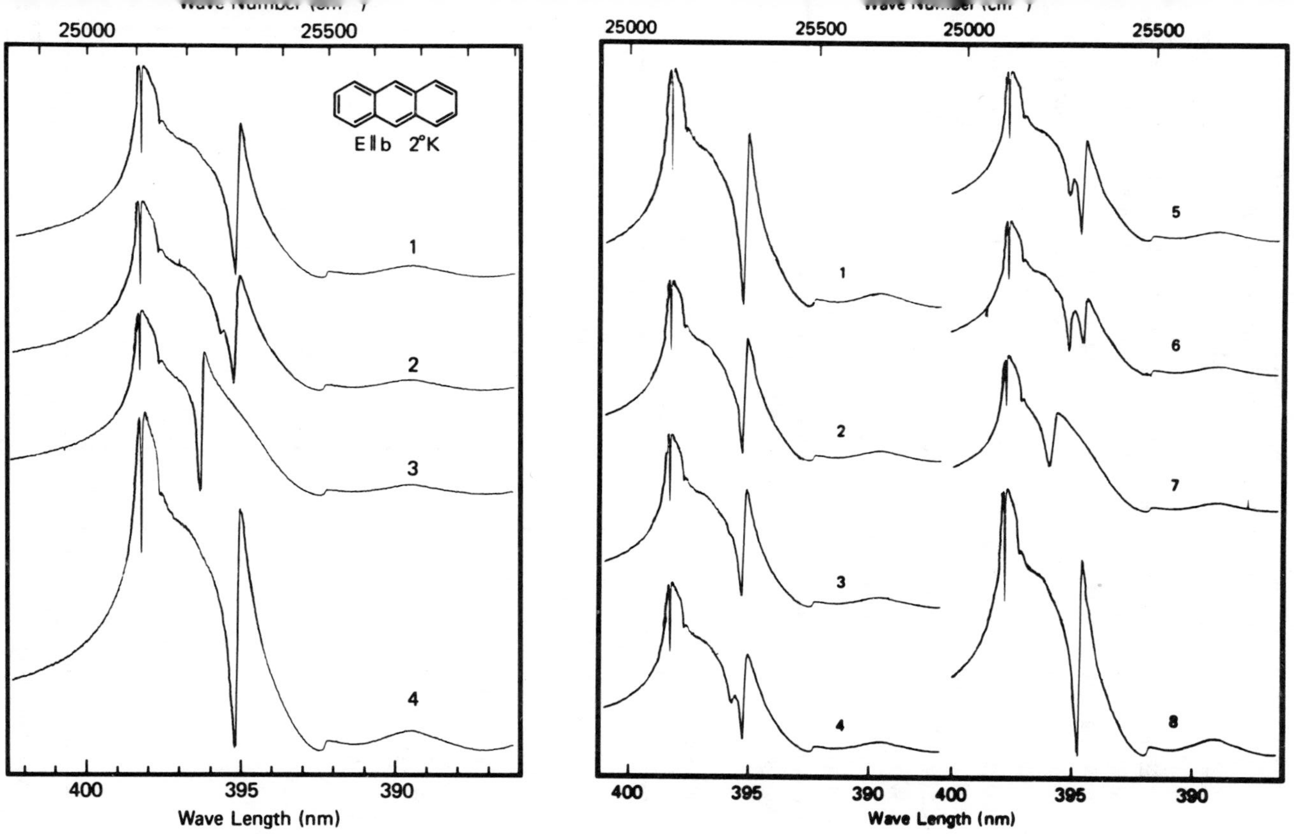

Fig. 36. **E**‖**b**, 2 K reflectivity spectrum of the (001) face of anthracene. Effect of condensing air on the surface. *Left*: (1) before; (2) and (3) after increasing condensation; (4) after evaporating the layer of gas deposited. *Right*: Details of various stages of this process following successive [(2) to (7)] condensations of very small amounts of air.

successive coats of air, (2) to (3), shifting structure I 70 cm^{-1} and structure II (much harder to see the change) 3 cm^{-1} relative to their initial positions in (2). Note that the rest of the spectrum, including the dip at 25,412 cm^{-1} is unchanged. After evaporating (4) the gas layer, the structures shift in the other direction and regain their initial positions (1). On the right we show various stages of the shift of structure I due to coating with air. Spectra (2) to (7) show that the shift is not abrupt, following the first deposit, but has intermediate values, with some oscillation, probably due to partial coverage of the surface. The final shift observed may be due to annealing of the film.

Most other gases, such as argon or nitrogen, cause nearly the same shifts of structures I and II as those caused by a coat of air (see Fig. 37). However, methane (see Fig. 38) causes a much greater shift of structure I (roughly 85 cm^{-1}). While several reasons may be advanced to explain the special effect of this gas, no firm conclusion could be drawn from the observations. However, we took advantage of the large shift due to methane to check that only interface interactions—not the thickness of the coat deposited—affect the shift. If methane is deposited on a surface already covered with air, the final

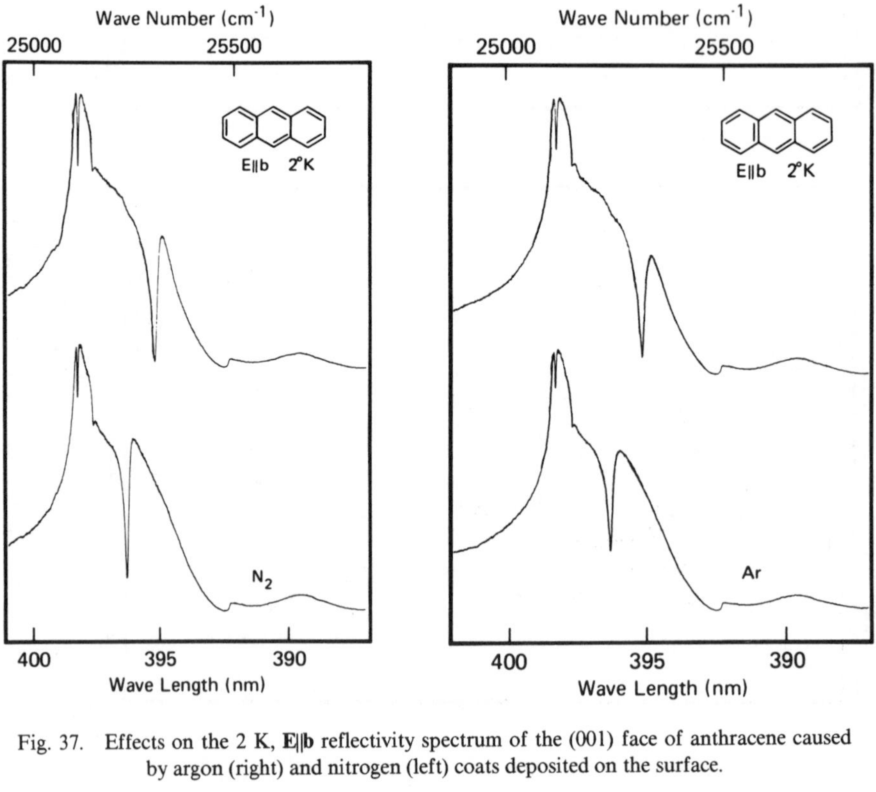

Fig. 37. Effects on the 2 K, **E**||**b** reflectivity spectrum of the (001) face of anthracene caused by argon (right) and nitrogen (left) coats deposited on the surface.

shift of structure I due to the air (70 cm^{-1}) does not increase. The right-hand side of Fig. 38 shows that ethylene, which causes a smaller shift (77 cm^{-1}) than does methane, singularly and almost irreversibly perturbs the surface since structures appear on coating and persist after evaporation.

4. *Conclusion*

These experiments are very important, since they distinguish structures I, II, and III from the rest of the **b**-polarized reflectivity spectrum. They alone

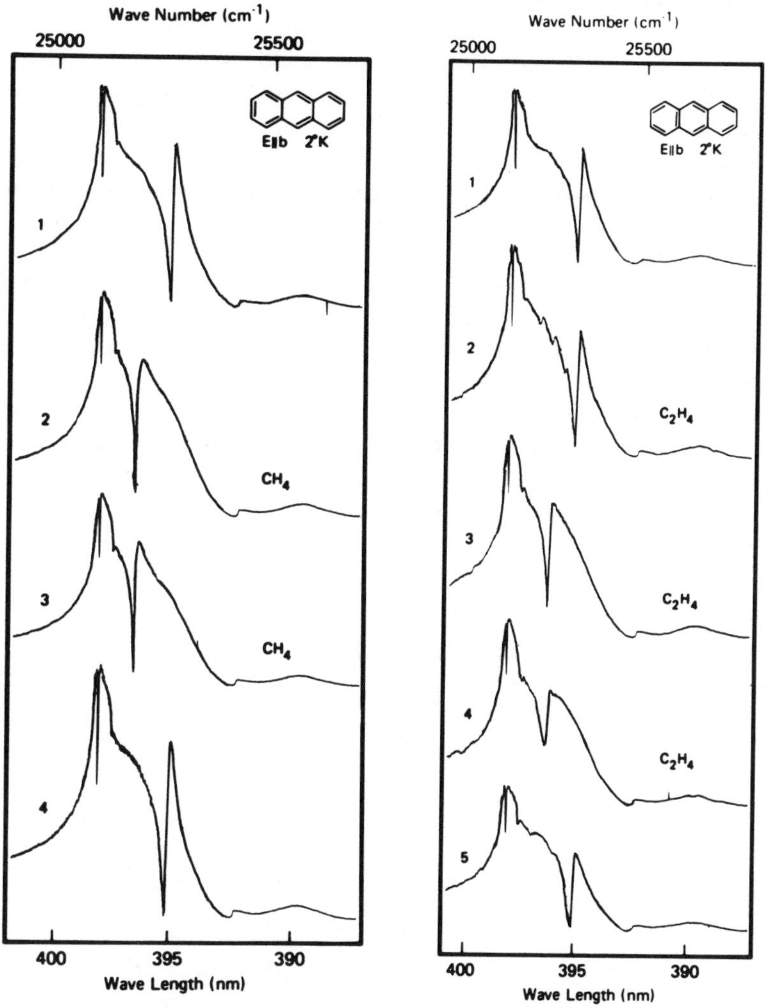

Fig. 38. Effects on the 2 K, **E**∥**b** reflectivity of the (001) face of anthracene caused by methane (left) and ethylene (right) coats deposited on the surface.

are sensitive to perturbation of the crystal surface. We conclude that not only system I, but also (though less obviously) minimum II and minimum III are related to the surface and the first few planes of the crystal. The gas-coating experiments are entirely compatible with the idea of excitons localized in the first few layers of the crystal.

D. a-Polarized Reflectivity Spectrum

An important point in the theory of surface excitons which must be examined now is whether or not "analogous" structures in the **a**-polarized spectrum can be exhibited. As we saw in the Section II, one can expect the surface exciton to have about the same Davydov splitting as the bulk exciton, so that an **a**-polarized surface exciton should also exist. The reflectivity of the (001) face in **a**-polarized light is much smaller than in **b**-polarized light, so that no "stopping band" is seen. The main reason for this is that this transition has a much smaller oscillator strength and that propagation of this polarization is not entirely transverse. Since the **E**||**a** reflectivity is so weak compared to the **E**||**b** reflectivity, particular care must be taken when adjusting the polarization.

In Fig. 39 the maximum reflectivity for **E**||**a** at 25,319 cm^{-1} is about 20%. Spectra of thin crystals exhibit oscillations at low energies like those of the **E**||**b** spectrum but are much more widely space in energy owing to the much smaller refractive index.

On the high energy side we observe at roughly the same positions as before the structures (asymmetric peak at 25,480 cm^{-1} and broad maximum at

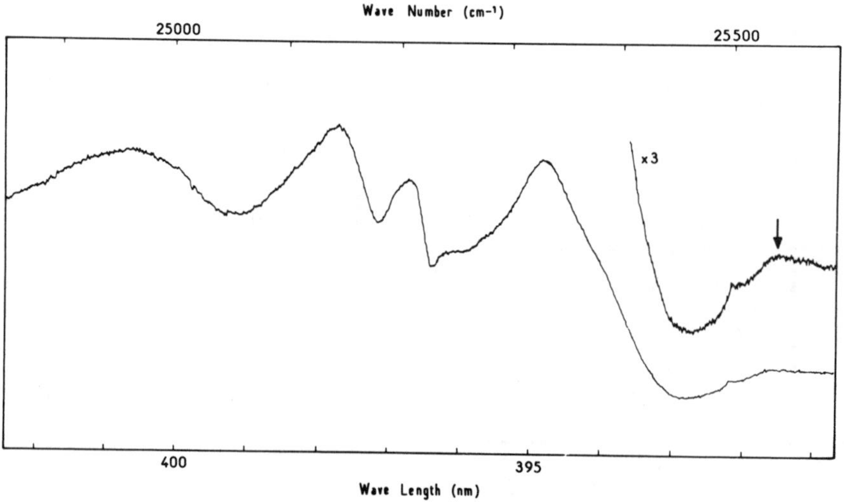

Fig. 39. The 1.7 K, **E**||**a** reflectivity spectrum of the (001) face of a sublimation grown anthracene crystal. The arrow shows the maximum seen at 25,523 cm^{-1}.

25,673 cm^{-1}) related to the 390-cm^{-1} intramolecular vibration, seen in the **E**||**b** spectra. There is also a fairly broad maximum at 25,523 cm^{-1} (arrow on Fig. 39).

This maximum is shifted relative to the overall maximum by about the same amount (204 cm^{-1}) as structure I relative to the maximum of the **E**||**b** spectrum (203 cm^{-1}). Hence we concentrated our attention on it, despite its low intensity, and repeated the experiments above. Two points result from our findings:

1. In Fig. 40 we note that coarser crystals from the first sublimations [spectrum (2)] do not exhibit this maximum in the **E**||**a**-polarized spectrum, just as in the **E**||**b** spectrum structures I, II, and III did not appear. [In both polarizations spectra (1) and (2) are taken on the same samples.] This seems to establish the same criteria for the presence of this maximum as for the structures I, II, and III.
2. In gas-coating experiments (see Fig. 41) the maximum, indicated by the arrow, disappears. This establishes the sensitivity of this "structure" to perturbations of the surface.

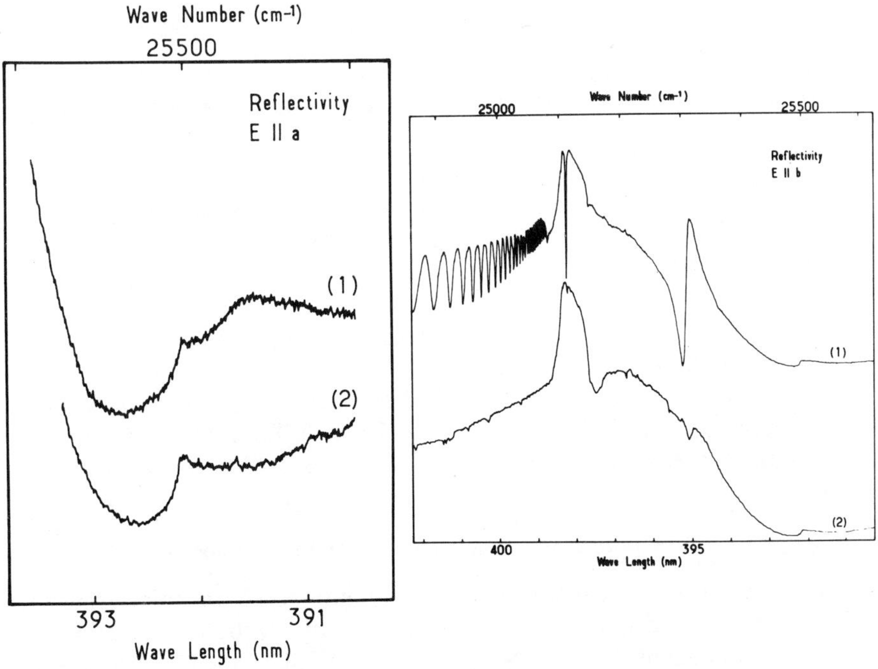

Fig. 40. Detail of the 1.7 K, **E**||**a** reflectivity of the (001) face of very pure anthracene (1) and of a coarser sample (2). On the right the corresponding **E**||**b** spectra, for comparison.

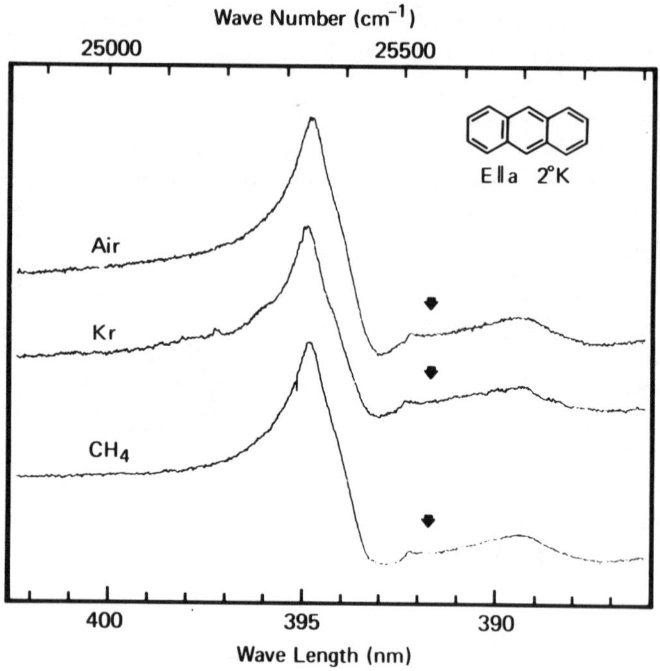

Fig. 41. Effect of various gas layers on the **E**∥**a** reflectivity of anthracene at 2 K. Note the absence in every case of the maximum at 25,523 cm^{-1} (arrow) observed on uncoated surfaces.

N.B. Recent experimental results that we will discuss in a forthcoming paper show clearly that under surface coating this structure moves to lower energy of approximately the same (although smaller) energy shift.[242]

In conclusion, despite the much less favorable conditions for the **E**∥**a** spectra, a special "structure" can be related to the crystal surface and has similar behavior to that of structure I in the **E**∥**b** spectrum. These results seem to confirm more "tangibly" the observation (at 25,524 cm^{-1}) and interpretation of the **a**-polarized component of the surface exciton reported by Brodin et al.[105]

E. Fluorescence Studies in the Region of I and II

The last part of this experimental study of anthracene was to determine if the structures I and II seen above in the reflectivity spectra are related to the very weak emissions detected by Glockner and Wolf on the high energy side of the fluorescence spectrum of the crystal (see Section II). This is of great importance in understanding the process by which a surface exciton, if it exists can be deactivated. Since it is of greater energy than the quasicontinuum of bulk states, one might expect it to relax nonradiatively towards them. If, on the other hand, the emissions found by Glockner and Wolf can be linked to the first planes of the crystal, then this confirms Sugakov's idea that surface states should have a very short radiative lifetime. It is easy to appreciate the importance of connecting the emissions with the reflectivity structures in order to find out if these two processes involve the same set of states associated with the first few planes of the crystals. In establishing this connection we had to make, almost simultaneously, reflectivity (**E**||**b**) and fluorescence measurements in the vicinity of these structures on the same samples in the same experimental conditions.

1. Description

Figure 42 shows the region of the fluorescence origin of an anthracene monocrystal at 1.7 K. The strongest part of the spectrum, stretching toward

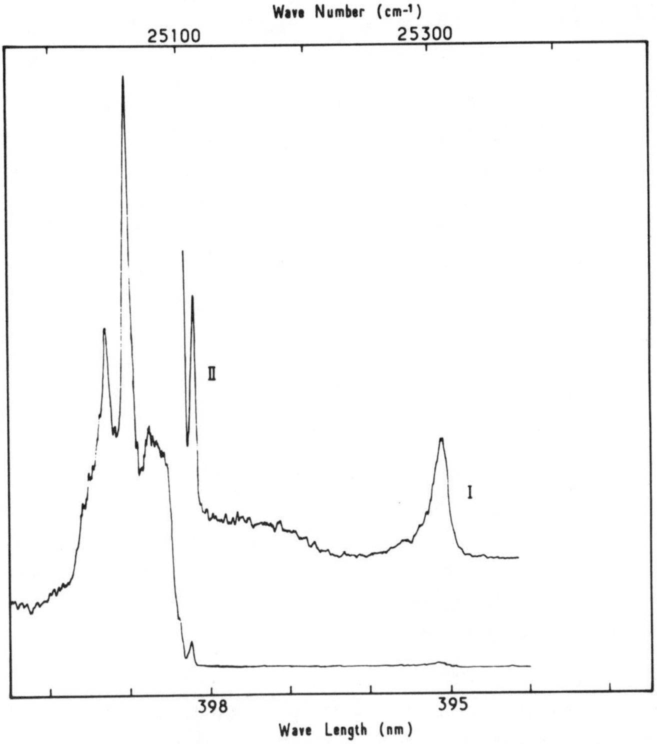

Fig. 42. Fluorescence spectrum at 1.7 K of the (0,0) region of the first singlet transition of sublimation grown anthracene. Monochromator slitwidths: 20 μm in the lower spectrum; 50 μm in the upper one.

low frequencies from 25,085 cm^{-1}, is very like that reported by Glockner and Wolf.[86,87] This is the "normal" fluorescence of the bulk, involving phonon modes and ground state intramolecular vibrations. For reasons discussed in the Section II, the fluorescence origin [(0, 0) phononless line] is not observed. The first, strongest lines, emerging from the background whose strength depends on the purity of the crystal, are entirely **b** polarized (for all exciting polarizations) and can be associated with the principal phonon modes of anthracene.[84,88] Their energies (25,085, 25,073, 25,056, and 25,047 cm^{-1}) are in good agreement with published values. Note, however, that depending on the sample and the traps in it, other lines of various intensities can appear and form new vibronic series.

In fact, our work concerned not the low energy side, which has been widely studied, but the high energy side of the (0, 0) position, where (Fig. 42) there are two emissions I and II, first reported by Glockner and Wolf. This spectrum shows the weak intensity of these emissions compared to the "normal" fluorescence. Figure 43 shows the reflectivity spectrum (**E**||**b**) above the fluorescence of the same sample at 1.8 K. The structures I and II, and even III, in reflectivity coincide nearly perfectly with the "high energy" structures in the fluorescence spectrum. We will discuss these emissions successively.

Line I. Line I, centered on 25,302 cm^{-1}, is completely **b** polarized. An important point of Fig. 43 is that the maximum of this emission coincides with the middle of system I in reflectivity, and not with its maximum or minimum. The line is asymmetric and has a width of 18 cm^{-1} at half-height. The smaller half-width (high energy side) corresponds to a width of 12 cm^{-1} at half-height, this being roughly the difference between the maximum and the minimum of system I in reflectivity. A diffuse though reproducible structure is visible on the low energy tail. It will be noted that there is no continuous background between emission I and emission II when a **b**-oriented polarizer is placed in front of the detector (this was not yet done for the spectrum of Fig. 42).

Line II. This emission, centered at 25,103 cm^{-1}, is also completely **b** polarized. In our first experiments, as in those of Glockner and Wolf, line II was superposed on the front of the intense bulk fluorescence. At very low temperature (1.7 K) and using large time constants and the narrowest slits possible, we were able to completely resolve this line. Its width at half-height is 1.8 cm^{-1} (see left-hand insert of Fig. 43).

Line III. Again on the left-hand insert there is a very reproducible structure at 25,095 cm^{-1} in the fluorescence spectrum, coinciding closely with structure III in reflectivity. Note that the description of lines I and II given above agrees well with that given by Glockner and Wolf[100, 101] for their type-A crystals.

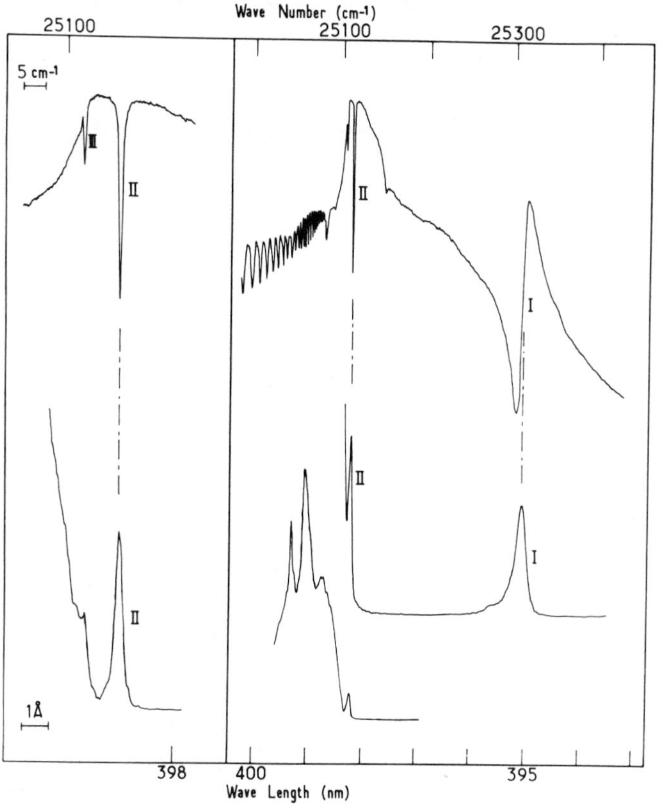

Fig. 43. Reflectivity **E**||**b** (above) and fluorescence (below) of the same crystal of anthracene (sublimation grown) at 1.7 K. On left a sweep at higher resolution of the region of structures II and III.

2. *Temperature Dependence*

Structures I and II exhibit the same temperature dependence as the corresponding features of the reflectivity spectrum (Fig. 44).

Note that line II disappears in both cases at around 15 K; whereas line I persists at much higher temperatures. The broadening of the bulk fluorescence prevents observation of its final disappearance that occurs in reflectivity around 100 K.

3. *Dependence on the Purity of the Sample*

There is perfect agreement between the dependences of intensity of the fluorescence and reflectivity structures on the purity of the crystal. Rela-

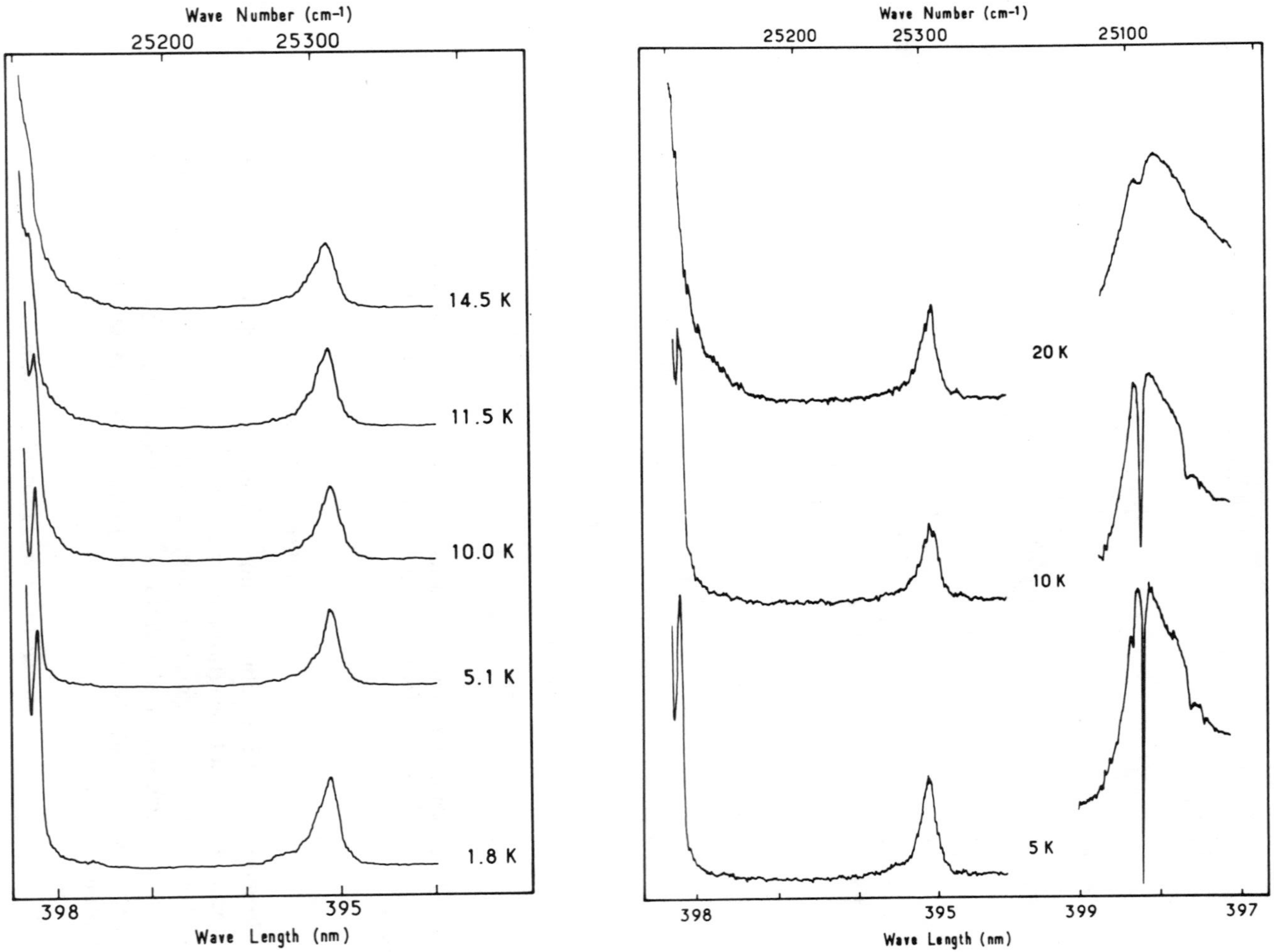

Fig. 44. *Left*: Temperature dependence of the fluorescence spectrum of anthracene in the region of structures I and II. *Right*: Insert shows the identical variation of structure II in fluorescence and in reflec-

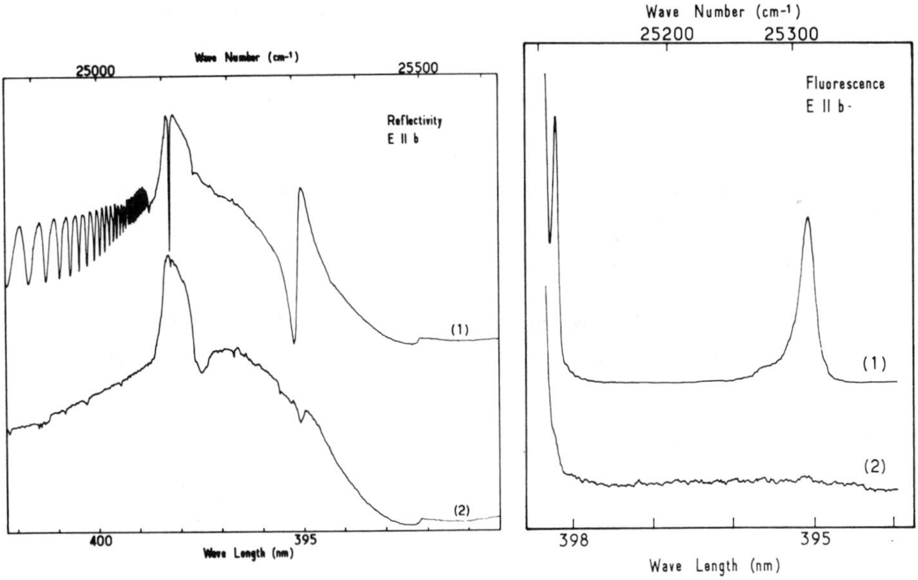

Fig. 45. *Right*: Fluorescence spectra at 4 K of the region of structures I and II of a carefully purified (1) and a less well-purified (2) anthracene crystal. *Left*: Comparison with the reflectivity spectra ($E \| b$) of the same samples.

tively impure crystals spectra (2) exhibit structures I and II neither in the reflection nor in the fluorescence spectra (Fig. 45).

4. *Influence of Gas Layers Condensed on the Surface*

In order to completely check the similarities between structures I and II in fluorescence and reflectivity, but above all to establish the origin of these emissions, we repeated the gas-coating experiments and observed almost simultaneously the fluorescence and the reflectivity of the same sample.

This experiment is summarized in Fig. 46, where beside the fluorescence spectrum we show also the reflectivity for comparison. The gas used in this case was nitrogen. All the spectra were recorded at 5 K.

It will be seen that relative to the spectra before condensation (1) or after evaporation (3) of the gas, the spectra (2) with the gas coating exhibit the same shifts (78 cm^{-1} for I and 3 cm^{-1} for II) in fluorescence and reflectivity. It is important to note that the shifted positions of the emissions (25,224 cm^{-1} for and 25,100 cm^{-1} for II) coincide rather well with the values given by Glockner and Wolf[100,101] for type-B crystals (25,227 cm^{-1} for I and 25,100 cm^{-1} for II; see Section II). One can find in this experiment the probable explanation of the two types of crystal and of the transition on warming from type A to type B, reported by the same authors. It also explains rather nicely the "unstable" position of structure I noted by Marisova et al.[97,99] It should also be noted from Fig. 46 that the intense "normal" fluorescence lines are undisturbed by the gas coat.

N.B. There are large variations of intensity between spectra (1), (2), and (3) of lines I and II. These differences are due principally to fluctuations in the power of the dye laser used to excite the crystal. (These spectra are not normalized.) Power varies widely from day to day (there is a day between each spectrum and the one after) depending on the "mood" of the nitrogen laser used to pump the dye laser.

The spectra in Fig. 46:

1. Distinguish emissions I and II, sensitive to surface perturbations from the intense "normal" spectrum, which can now be called bulk fluorescence.

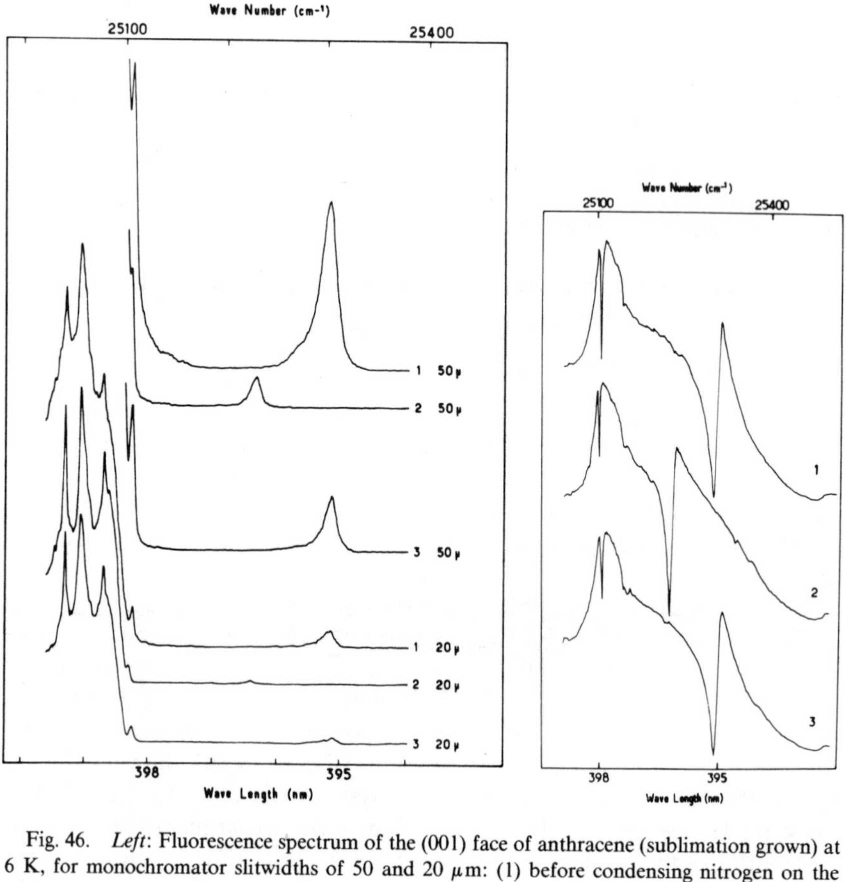

Fig. 46. *Left*: Fluorescence spectrum of the (001) face of anthracene (sublimation grown) at 6 K, for monochromator slitwidths of 50 and 20 μm: (1) before condensing nitrogen on the surface; (2) after condensing nitrogen on the surface; and (3) after evaporating the coat of solid nitrogen. *Right*: Corresponding reflectivity (**E**||**b**) spectra recorded during the same experiment.

2. Show clearly the correlation between structures I and II in reflectivity and fluorescence.
3. Establish for the first time the link between these emissions and the first planes of the crystal.

5. *Excitation Spectra of Structure I and a Bulk Line*

Whereas in Section IV we describe the optical arrangement for exciting and detecting fluorescence, we did not specify the spectral region used to excite fluorescence from lines I and II. In fact, in the range of dyes used (20,000 to 27,000 cm^{-1}), observation of lines I and II is possible (given correct temperature and purity) at any excitation energy higher than that of structure I (25,302 cm^{-1}). However, the strength of the emission varies greatly with the exciting wavelength.

Further, we have compared the excitation spectrum (**E**||**b** and **E**||**a**) of line I (line II is too close to the bulk emission) with that of the strongest bulk line (25,056 cm^{-1}) in order to distinguish once more between emissions I and II and the bulk fluorescence and to determine from what states of higher en-

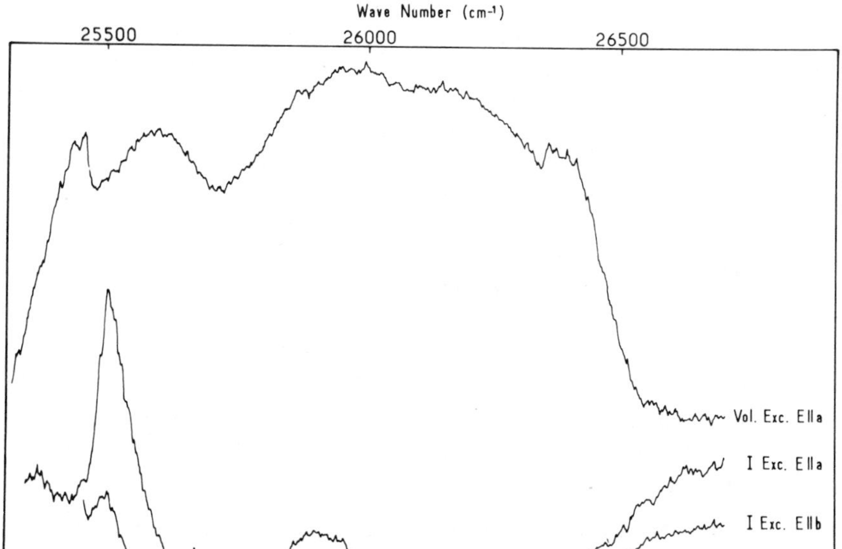

Fig. 47. Normalized excitation spectrum of line I (25,302 cm^{-1}) for **E**||**a** and **E**||**b**, and of the most intense line of the bulk fluorescence (25,056 cm^{-1}).

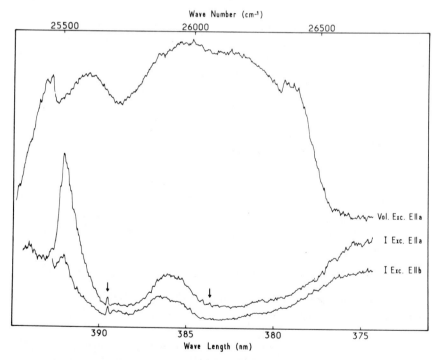

ergy these emissions may be excited. Note that an excitation spectrum is similar, but distinct from, an absorption spectrum.

The spectra, normalized relative to the excitation power, are shown in Fig. 47. The lower two spectra, corresponding to line I, show very reproducible structures (small arrows) which may be vibronic states associated with this emission. Confirmation would require a much more detailed study and somewhat more sophisticated apparatus.[242] An important fact revealed here is that emission I (which is completely **b** polarized) is very much enhanced by an **a**-polarized excitation centered on 25,525 cm^{-1}. This peak is absent in the **b**-polarized excitation spectrum of I and in the bulk excitation spectrum (upper curve in Fig. 47), which is very different from, and even complementary to, that of emission I. While it would be premature to give a definite interpretation of this peak, it is tempting, due to its polarization ($\mathbf{E}\|\mathbf{a}$) and its position (~ 25,525 cm^{-1}), to associate it with the weak maximum seen in the $\mathbf{E}\|\mathbf{a}$ reflectivity spectrum, since we have also shown that this maximum was correlated with structure I in the $\mathbf{E}\|\mathbf{b}$ reflectivity spectrum.[242]

VI. THEORETICAL MODEL OF THE REFLECTIVITY OF A SEMI-INFINITE AND A FINITE CRYSTAL

It would seem from the experimental results described in the preceding sections that the first few planes of the crystal play an important part in the optical properties after excitations in the configuration studied [**k** normal to the (001) face]. Hence the theoretical description of these properties must be based on a finite, or at least semi-infinite, crystal, contrary to the polariton theory developed in Section III. We must now discover how the optical properties of a finite crystal can be deduced from this theory. Obviously any theoretical treatment needs information on the structure of the surface (which, unfortunately, we do not possess), so we must make a simple model of the surface and its coupling to the bulk. While the main point of our model is to show that the introduction of special electronic excitations in the first few planes explains the structures described above, so strengthening the hypothesis of surface excitons, it is also hoped to lay the foundations of a better understanding of the optical properties of real crystals, (finite and absorbing), near their resonances.

A. Theory of the Reflectivity of a Semi-infinite Crystal

Rapid development of the polariton theory around 1960 led naturally to a theoretical study of the reflectivity of semi-infinite crystals. Much theoretical work has been done on the problem since then.[208-219]

Mahan and Obermair[212] contributed largely, with a classical theory of the reflectivity of a semi-infinite cubic crystal: If there are N nondegenerate planes before the bulk of the crystal (N being very small compared to the total number of planes in the crystal), then such a semi-infinite crystal has in an electromagnetic field $N+1$ nondegenerate propagation (polariton) modes. Each mode has its own refractive index $n_j(\ell)$ being a root of an $(N+1)^{\text{th}}$-degree equation involving the polarization $\mathbf{P}(\ell)$ and the transverse part of the vector potential $\mathbf{A}(\ell)$ of the ℓth nondegenerate plane.

They conclude that the reflectivity is the product of the reflectivities of each mode:

$$R(\omega)=\prod_{j=1}^{N+1}\left|\frac{n_j(\omega)-1}{n_j(\omega)+1}\right|^2$$

This theory has recently been extended, by Philpott[216-218] and others, to crystals of lower symmetry retaining the spatial dispersion and to the case of oblique incidence.

The narrower problem of the reflectivity of semi-infinite crystal with a surface exciton, such as that of anthracene, has been studied principally by Sugakov[214-215] and Philpott.[217] Their methods are identical and lead to the same expression for the reflectivity.

Relative to an incident wave vector normal to the (001) face and a **b**-polarized electric field, the crystal is considered to be a pile of (**a**, **b**) planes (Fig. 48) connected by excitation exchange terms $I^{(1)}_{(k)}$, $I^{(2)}_{(k)}$, and so on, negligible compared to the intraplane exchange term $I^{(0)}_{(k)}$. As we saw in Section II, this leads to a surface exciton shifted to a higher energy than that of the volume by an amount δ_S and essentially localized in the first molecular layer owing to the smallness of the exchange terms. Because of the localization, Sugakov and Philpott consider the crystal, from the point of view of electromagnetic propagation, as two continuous media (surface and bulk) placed side by side and characterized by indices of the form given in Section III [Eq. (72)].

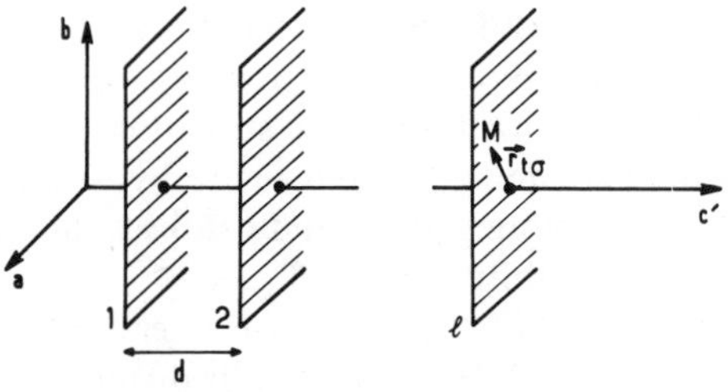

Fig. 48. Model of a semi-infinite crystal of anthracene in the present experimental conditions.

N.B. In view of the relative magnitudes of ω_m, ω, δ_S, and γ ($\omega_m \sim \omega \gg \delta_S$ and γ), Sugakov and Philpott assume that n_S can be deduced from n_V by a simple energy translation δ_S:

$$n_S(\omega) \simeq n_V(\omega - \delta_S)$$

The final form of the reflection power is

$$R(\omega) = \left| \frac{1 - n_V(\omega) + i\dfrac{\omega d}{c} n_S^2(\omega)}{1 + n_V(\omega) - i\dfrac{\omega d}{c} n_S^2(\omega)} \right|^2 \tag{103}$$

where d is the distance between planes.

If a damping term γ is introduced (n_V would otherwise be a pure imaginary in all the stopping band), this expression reproduces the minimum or maximum reflectivity due to the surface exciton, according to whether the latter falls inside or outside the stopping band.

In our model of the reflectivity of a finite anthracene crystal (**E**‖**b**) in which surface and subsurface excitons may be created, we took up[166, 198] the basic idea of Sugakov and Philpott to divide the crystal into several regions of polariton propagation. As we saw in Section III, it is entirely justified to neglect interplane excitation exchange in this direction. In each region (i), the propagation of the polariton is described by a dispersion $k_i(\omega)$ found by the methods of Section III:

$$k_i^2 = 4\pi^2\omega^2\varepsilon_\infty + \frac{\omega_p^2 f_m B 4\pi^2\omega^2}{(\omega_m + \delta_i) + \omega_p^2 f_m A - i\omega\gamma_i - \omega^2} \tag{104}$$

where $\omega_m(i)$ is the molecular excitation energy lowered by the Van der Waals term $D(i)$ of region (i), determined from our experimental observations of δ_S and δ_{sS}. It is difficult to justify (or to refute) the assumption that the same form of dispersion formula holds for all regions (surface, subsurface and bulk), because we lack information on the structure of the first few planes. [While this hypothesis is similar to that made by Sugakov and Philpott, $n_S(\omega) = n_V(\omega - \delta_S)$, it is more general and can be applied for all values of the surface and subsurface shifts, δ_S and δ_{sS}.]

We first give an account of the model of a semi-infinite crystal before passing on to the finite crystal. We suppose it is divided into a surface and a volume (degenerate or not as the case may be).

B. Reflectivity of a Semi-infinite Crystal

1. *Principle*

The semi-infinite crystal, in which a surface exciton may be created, consists of a surface region (II) of thickness $d = OA$ and a semi-infinite volume (Fig. 49).

An incident photon with wave vector $\mathbf{k}_1$ in vacuo (region I) normal to the $(\mathbf{a},\mathbf{b})$ plane will propagate in regions II and III as a polariton with wave vectors $\mathbf{k}_2$ and $\mathbf{k}_3$, respectively. The following are the corresponding wave functions:

$$\begin{aligned}&\text{Region I} && \phi_1(x) = A_1 e^{ik_1x} + A_1' e^{-ik_1x}\\ &\text{Region II} && \phi_2(x) = A_2 e^{ik_1x} + A_2' e^{-ik_2x}\\ &\text{Region III} && \phi_3(x) = A_3^{ik_3x}\end{aligned} \tag{105}$$

The continuity conditions in O and A imply:

$$\text{Point } O \quad \begin{cases} A_1 + A_1' = A_2 + A_2' \\ ik_1(A_1 - A_1') = ik_2(A_2 - A_2') \end{cases} \tag{106}$$

$$\text{Point } A \quad \begin{cases} A_2 e^{ik_2d} + A_2' e^{-ik_2d} = A_3 e^{ik_3d} \\ ik_2\left(A_2 e^{ik_2d} - A_2' e^{-ik_2d}\right) = ik_3 A_3 e^{ik_3d} \end{cases}$$

The reflection coefficient of the incident photon is

$$\rho = \frac{A_1'}{A_1} = \frac{k_2(k_1 - k_3)\cos k_2d + i\left(k_2^2 - k_1k_3\right)\sin k_2d}{k_2(k_1 + k_3)\cos k_2d + i\left(k_2^2 + k_1k_3\right)\sin k_2d} \tag{107}$$

N.B. The large wavelength approximation ($k_2d \ll 1$) is valid at the wavelengths considered, since $k_1n_s = k_2$ and $k_1n_V = k_3$. The first-order approximation of ρ is then the same as the

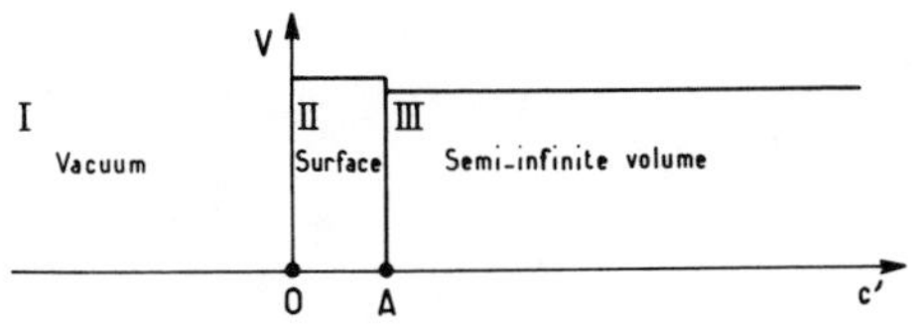

Fig. 49. Model of a semi-infinite crystal with surface exciton.

expression found by Sugakov and Philpott [formula (103)] by neglecting $n_V(\omega)$ relative to $n_S(\omega)$. This limits its use to the case of large energy shifts δ_S of the surface exciton.

2. *Results*

We used expression (107) to determine the reflectivity $R = |\rho|^2$ as a function of ω. If we define

$$F(x) = \frac{1+x}{1-x} \quad \text{and} \quad F^{-1}(x) = \frac{x-1}{x+1} \tag{108}$$

and the ratio of the reflected and incident amplitudes $\rho_i = A_i'/A_i$ (107) becomes

$$\rho_i = \frac{A_i'}{A_1} = F^{-1}\left(\frac{k_1}{k_2} F\left(e^{2ik_2d} \cdot F^{-1}\left(\frac{k_2}{k_2}\right)\right)\right) \tag{109}$$

Given the dispersions $k(\omega)$ of k_1, k_2, and k_3, this form is amenable to easy machine calculation of the reflectivity $R = |\rho_1|^2$

a. Degenerate Surface and Volume. We first calculated the reflectivity of a crystal for which ω_m (molecular excitation energy lowered by the appropriate van der Waals term) is the same for both the surface and the volume ($\omega_m = 25{,}435$ cm^{-1}). The dispersions of k_2 and k_3 are the same and were calculated without damping ($\gamma = 0$) and for various values of the damping. The reflectivity curves (Fig. 50) show how the volume reflectivity for $\gamma = 0$ becomes asymmetric when a damping term representing different relaxation processes is introduced.

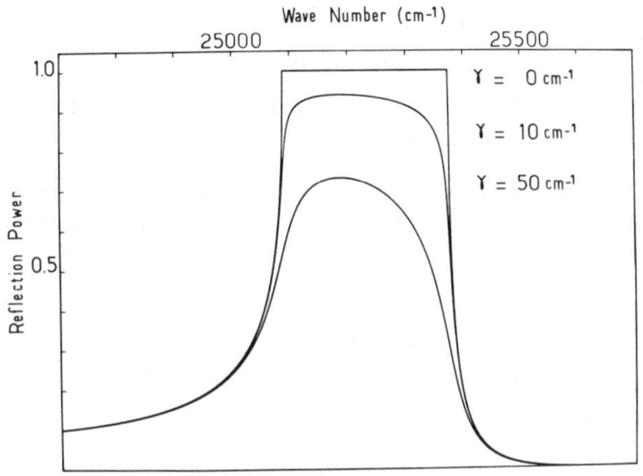

Fig. 50. Absolute reflectivity of a semi-infinite anthracene crystal with degenerate surface and volume, calculated for various values of the damping γ.

b. Introduction of the Surface Exciton. If now we suppose $\omega_{m(\text{surface})} = \omega_{m(\text{volume})} + \delta$, then $k_2(\omega)$ and $k_3(\omega)$ have different dispersion curves. Reflectivity curves (Fig. 51) calculated for the same damping ($\gamma = 4\ \text{cm}^{-1}$) and different values of the energy shift δ exhibit a structure due to the surface resonance. Depending on whether the surface resonance is inside or outside the stopping band, it gives either a minimum or a maximum. These curves are equivalent to those given by Philpott.[217]

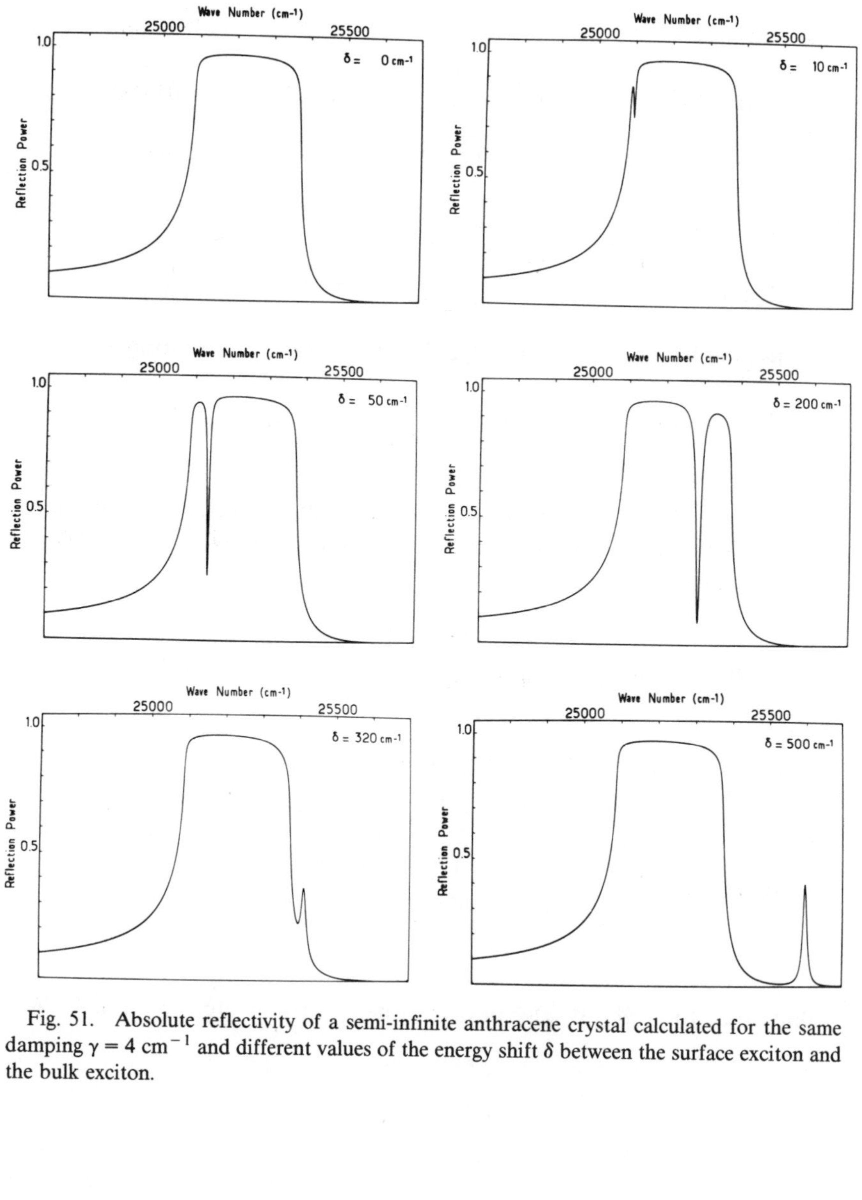

Fig. 51. Absolute reflectivity of a semi-infinite anthracene crystal calculated for the same damping $\gamma = 4\ \text{cm}^{-1}$ and different values of the energy shift δ between the surface exciton and the bulk exciton.

C. Reflectivity of a Finite Crystal

1. *Principle*

Consider now a finite crystal [along **c**′ perpendicular to the (**a**, **b**) plane] in which the first two planes are not degenerate with the bulk (Fig. 52).

As before, the crystal is separated into different regions (II to VI) in each of which the polariton has a different dispersion curve. The wave functions for each propagation mode are:

Region I	$\phi_1(x) = A_1 e^{ik_1x} + A_1' e^{-ik_1x}$	(vacuum)
Region II	$\phi_2(x) = A_2 e^{ik_2x} + A_2' e^{-ik_2x}$	(surface)
Region III	$\phi_3(x) = A_3 e^{ik_3x} + A_3' e^{-ik_3x}$	(subsurface)
Region IV	$\phi_4(x) = A_4 e^{ik_4x} + A_4' e^{-ik_4x}$	(volume)
Region V	$\phi_5(x) = A_5 e^{ik_3x} + A_5' e^{-ik_3x}$	(subsurface)
Region VI	$\phi_6(x) = A_6 e^{ik_2x} + A_6' e^{-ik_2x}$	(surface)
Region VII	$\phi_7(x) = A_7 e^{ik_1x}$	(vacuum)

If region IV corresponding to the bulk contains N planes of separation $\alpha = OA = AB = CD = DO'$ the continuity conditions in O, A, B, C, D, O' lead to the systems:

Point O
$$\begin{cases} A_1 + A_1' = A_2 + A_2' \\ ik_1(A_1 - A_1') = ik_2(A_2 - A_2') \end{cases}$$

Point A
$$\begin{cases} A_2 e^{ik_2d} + A_2' e^{-ik_2d} = A_3 e^{ik_3d} + A_3' e^{-ik_3d} \\ ik_2\left(A_2 e^{ik_2d} - A_2' e^{-ik_2d}\right) = ik_3\left(A_3 e^{ik_3d} - A_3' e^{-ik_3d}\right) \end{cases}$$

. . . .

Point O'
$$\begin{cases} A_6 e^{ik_2(N+4)d} + A_6' e^{-ik_2(N+4)d} = A_7 e^{ik_1(N+4)d} \\ ik_2\left(A_6 e^{ik_2(N+4)d} - A_6' e^{-ik_2(N+4)d} = ik_1 A_7 e^{ik_1(N+4)d}\right. \end{cases}$$

Taking the ratio of both sides of each system, introducing $F(x)$ and $F^{-1}(x)$, and letting the $\rho_i = A'_i/A_i$, we have, for the coefficient of reflection:

$$\rho_1 = F^{-1}\left(\frac{k_1}{k_2}\cdot F\left(e^{2ik_2d}\cdot F^{-1}\left(\frac{k_2}{k_3}\cdot F\left(e^{2ik_3d}\right.\right.\right.\right.$$

$$\cdot F^{-1}\left(\frac{k_3}{k_4}\cdot F\left(e^{2ik_4Nd}\cdot F^{-1}\left(\frac{k_4}{k_3}\cdot F\left(e^{2ik_3d}\right.\right.\right.\right.$$

$$\left.\left.\left.\left.\left.\left.\left.\left.\left.\left.\left.\cdot F^{-1}\left(\frac{k_3}{k_2}\cdot F\left(e^{2ik_2d}\cdot F^{-1}\left(\frac{k_2}{k_1}\right)\right)\right)\right)\right)\right)\right)\right)\right)\right)\right)\right) \qquad (110)$$

2. *Results*

The wave vectors in the surface (k_2), subsurface (k_3), and volume (k_4) were calculated from Eq. (104), using in each case the value of ω_m found from the experimental **E**||**b** reflectivity spectra. The values are:

$$\omega_{m(\text{volume})} = 25{,}435 \text{ cm}^{-1}$$

$$\omega_{m(\text{subsurface})} = 25{,}445 \text{ cm}^{-1} \qquad (\delta_{sS} = 10 \text{ cm}^{-1})$$

$$\omega_{m(\text{surface})} = 25{,}638 \text{ cm}^{-1} \qquad (\delta_S = 203 \text{ cm}^{-1})$$

The dispersions $k(\omega)$ were calculated with a constant damping of 4 cm^{-1} for the surface and 2 cm^{-1} for the subsurface.

Figure 53 shows various stages of our efforts to simulate the experimental **E**||**b** reflectivity spectrum and to understand the different processes governing its shape and structures. All the spectra were calculated for a crystal of 1.7 μm thickness (1700 planes), corresponding to a sample whose absolute reflectivity we had measured.

a. Effect of the Finite Thickness. We note first on comparing spectra (1) and (2) that the finite thickness of the crystal [perpendicular to (**a**, **b**)]

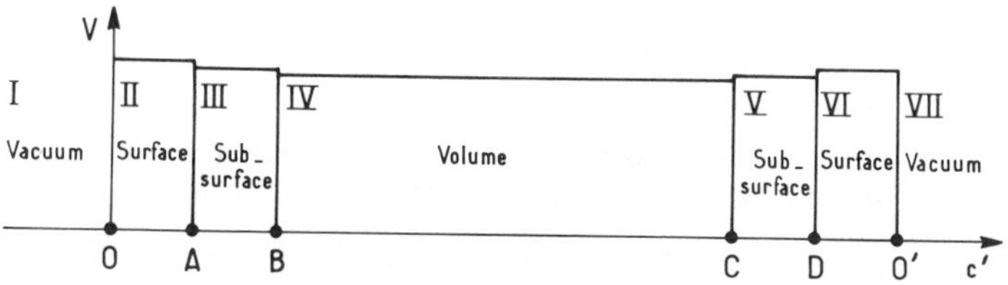

Fig. 52. Model of a finite crystal with surface and sub-surface excitons.

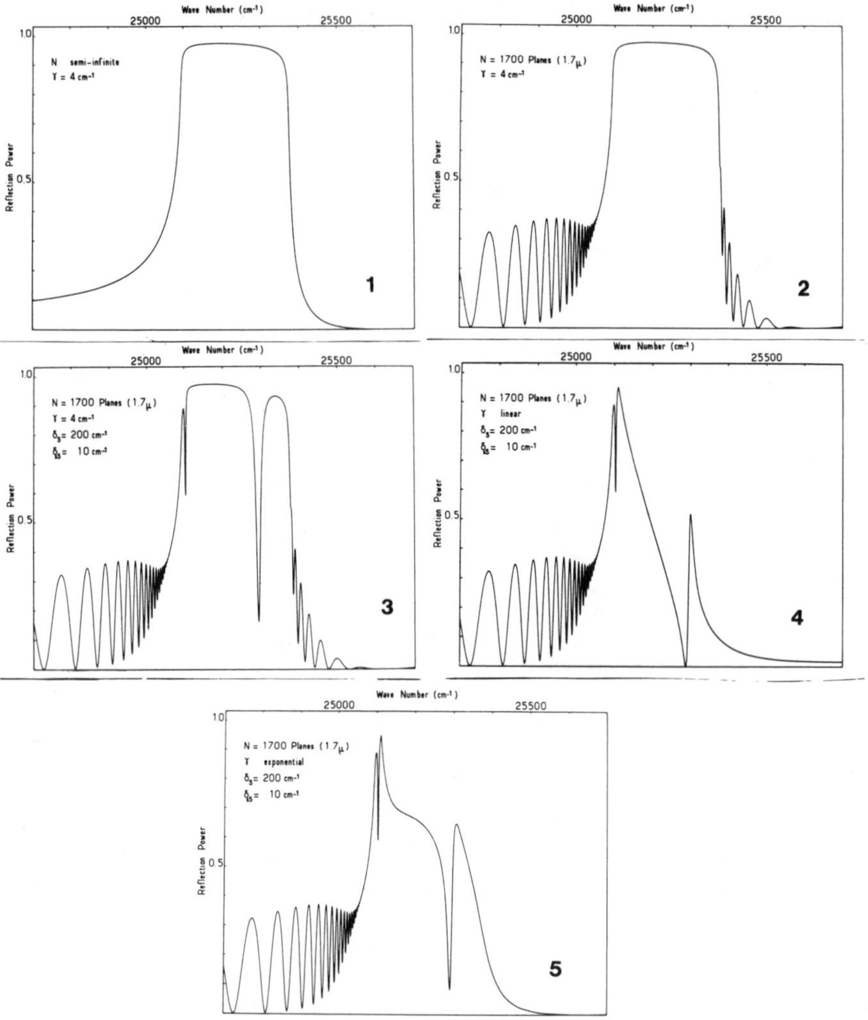

Fig. 53. Various stages of the model of the reflectivity of a finite crystal.

introduces oscillations in the wings of the stopping band like those we observed on the low energy side of the experimental spectra (Figs. 27–29) of crystals of roughly the same thickness. The oscillations, which are faster close to ω_T (25,095 cm^{-1}), can be interpreted classically as interference between waves reflected off the front and the back faces of the crystal (cf. Perot–Fabry interferometer), in which the index rises rapidly close to ω_T. From the quantum point of view, these oscillations correspond to resonances of polariton modes in finite crystals. Their finite lifetimes are due to coupling to a photon continuum, and their density increases rapidly close to ω_T.[220–223]

b. Influence of Surface and Subsurface Excitons. Spectrum (3) shows that introduction of a subsurface exciton at $\omega_T + \omega_{sS}$ and of a surface exciton at $\omega_T + \omega_S$ reproduces structures II and I of the experimental spectrum, with comparable linewidths. Note, however, that structure I falls in a region of high bulk reflectivity, so that we do not find the typical minimum–maximum form of the experimental spectra.

c. Introduction of a Variable Bulk Damping $\gamma(\omega)$. As we saw above, spectrum (3) calculated with a constant damping term $\gamma = 4$ cm^{-1} still has a nearly rectangular stopping band, whereas the experimental reflectivity falls off at energies above ω_T. In this region, ($\omega_T < \omega < \omega_L$) it is possible (among other things) to create a polariton and a phonon, by which process the electromagnetic wave can get into the crystal and propagate in it. Since the probability of this process increases with the number of phonons, this may explain the sharp drop in the reflectivity at 25,142 cm^{-1} (corresponding to the 47 cm^{-1} phonon mode of anthracene) followed by the continuous fall towards higher energies, observed experimentally.

We have attempted to give a phenomenological representation of the effect of these relaxation processes by introducing in the bulk dispersion formula $k_4(\omega)$, a damping term increasing with ω from ω_T. Obviously this is not meant to represent analytically the complicated coupling between the polariton and the phonon; it is simply a qualitative simulation of the processes responsible for the experimental picture.

Spectrum (4) shows that a linear damping term

$$\omega \leqslant 25{,}110 \text{ cm}^{-1}; \qquad \gamma = 4 \text{ cm}^{-1}$$
$$\omega \geqslant 25{,}110 \text{ cm}^{-1}; \qquad \gamma = 13(\omega - 25{,}110) + 4 \text{ cm}^{-1}$$

does not model well the experimental curve, whereas curve (5) shows that an exponential damping term

$$\omega \leqslant 25{,}110 \text{ cm}^{-1}; \qquad \gamma = 4 \text{ cm}^{-1}$$
$$\omega \geqslant 25{,}110 \text{ cm}^{-1}; \qquad \gamma = 60\left[1 - \exp\left(-\frac{\omega - 25{,}110}{20}\right)\right] + 4 \text{ cm}^{-1}$$

gives better results.

Figure 54 shows in more detail the agreement between experiment and theory. Comparison of the two spectra calls for some comments:

1. On the low energy side of $\omega_T = 25{,}095$ cm^{-1} the frequency of the oscillations is roughly correct on the theoretical spectrum, but their ampli-

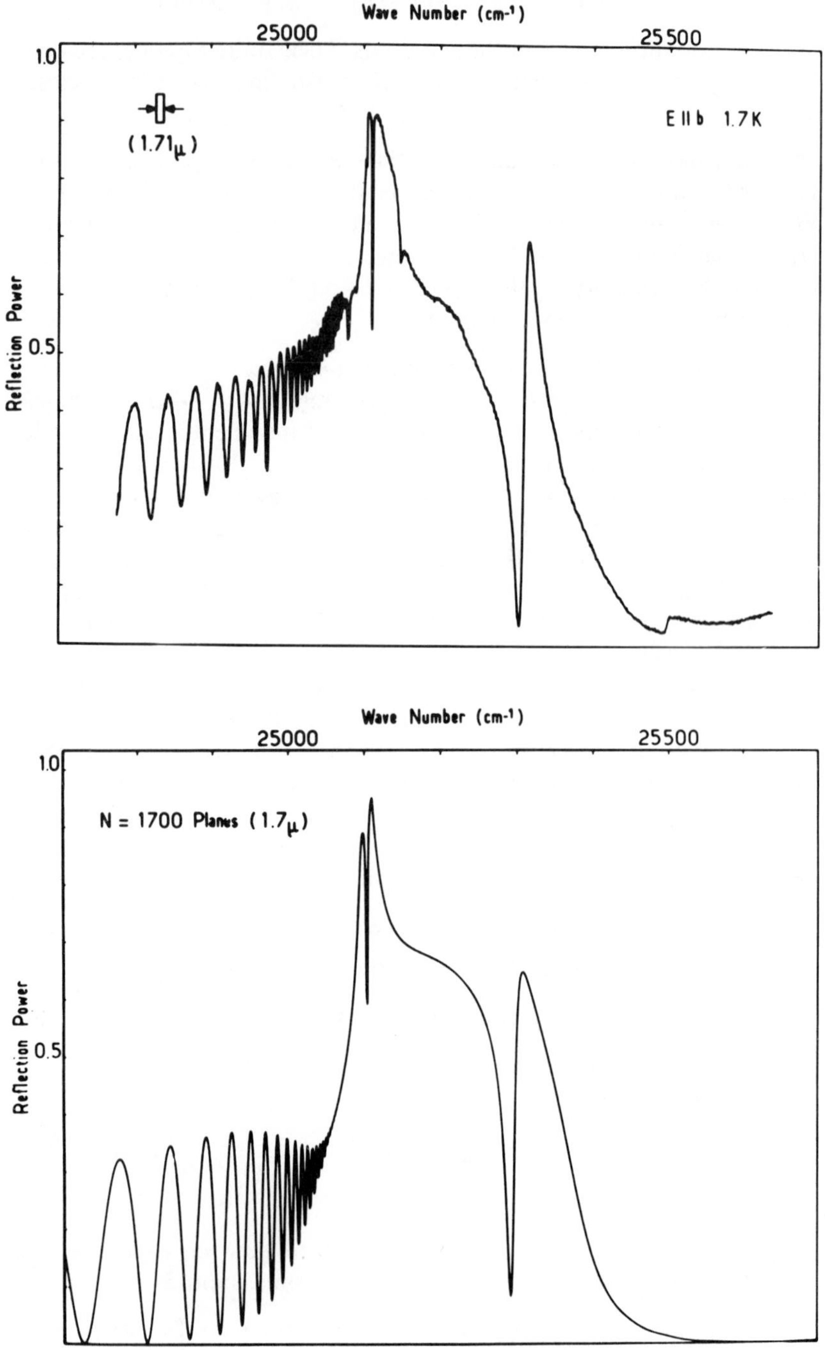

Fig. 54. Comparison of the observed reflectivity (upper curve) and the theoretical curve given by the model.

tude is greater and their average value lower than for the experimental spectrum. The first discrepancy is explained by the fact that in this region constant damping ($\gamma = 4\ \text{cm}^{-1}$) does not account properly for the decay of the crystal, which depends strongly on the wavelength. The low average reflectivity may be explained by the fact that our model contains only one molecular transition ($\omega_m = 25{,}435\ \text{cm}^{-1}$) and the contribution of other levels is described[2,125] by the constant ε_∞, which cannot represent correctly the influence of S_1 vibronic and other $\pi-\pi^*$ transitions, known to be important in anthracene. Introduction of these levels and a quantitative analysis of the damping mechanisms of the polaritons should improve the model.

2. In the stopping band the main difference between the spectra is the absence of structures showing a drop in reflectivity, attributable to the simultaneous creation of a polariton and of a particular phonon with a well-defined energy. While in introducing $\gamma(\omega)$ we were not expecting to reproduce these structures, it is to be hoped that a more quantitative form of $\gamma(\omega)$ would reproduce them as a result of an explicit polariton–phonon coupling.

3. On the high energy side the width of the stopping band is correct, but the vibronic structure of the experimental spectra are absent from the model curve since vibronic levels were not included in it. A better model involving these vibronic levels can be elaborated on the same principles.

We used the model with exponential damping to calculate the $\mathbf{E}\|\mathbf{b}$ reflectivity as a function of the number N of $(\mathbf{a},\mathbf{b})$ planes (Fig. 55). This is interesting because for more than one sample thickness range examined by us (from 1 μm to several millimeters), Ferguson[221,222] has obtained indirectly (by the transmission and the excitation spectra) the reflection spectrum ($\mathbf{E}\|\mathbf{b}$) of anthracene crystals of only 47 $(\mathbf{a},\mathbf{b})$ planes. The result of our calculations shows that relative to the near-lorentzian line form of a crystal of $N = 5$ planes, the reflectivity soon ($N = 50$) takes on the form and width of thick crystals, in good agreement with Ferguson's spectra. We see that the low energy oscillations diminish in amplitude as the thickness increases, whereas their frequency increases with the number of planes. While this calculation does not take into account the defects of impurities which can alter the phase lag between the waves reflected off the front and back surfaces, it gives fair agreement with our observations of these oscillations.

We have also profited from the good fit of the theoretical reflectivity given by this $\gamma(\omega)$ function to deduce, via the dispersion formula $k_4(\omega)$, the real and imaginary parts of the bulk permittivity $\tilde{\varepsilon}(\omega)$ and the index $\tilde{n}(\omega)$ (Fig. 56). While these results are not final, the absorption coefficient $n_2(\omega)$ is roughly the same as that deduced from the same spectra by the Kramers—Kronig transformation (see Ref. 33). We are presently develop-

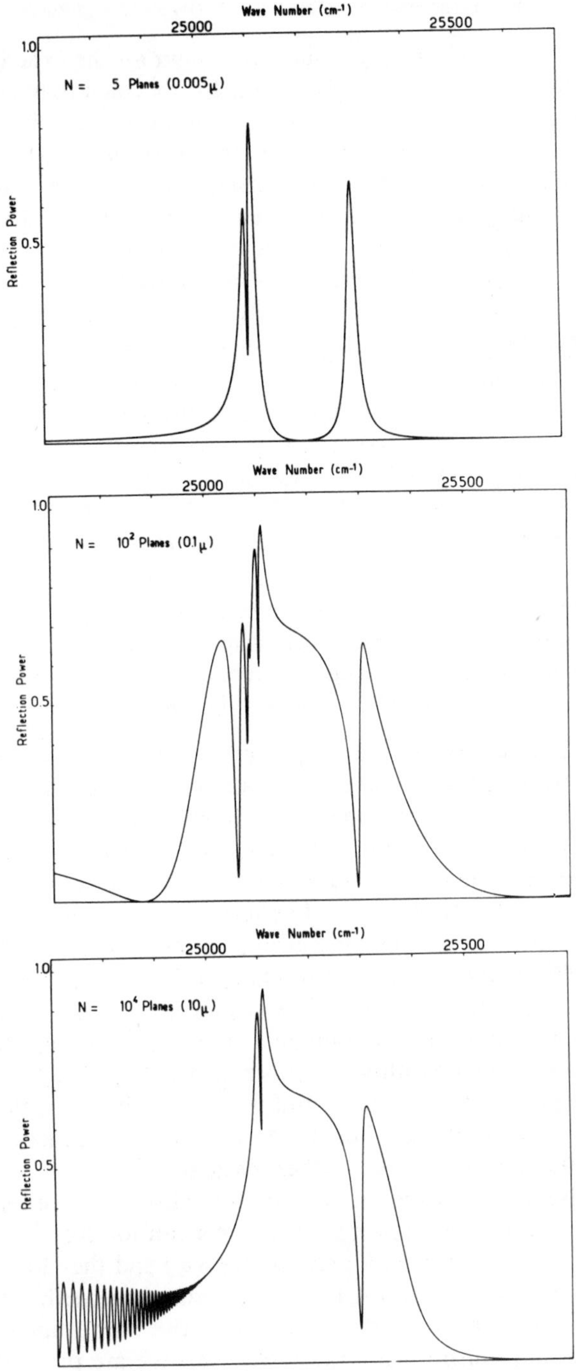

Fig. 55. Dependence of the theoretical absolute reflectivity spectrum on the number of planes, calculated with exponential damping $\gamma(\omega)$.

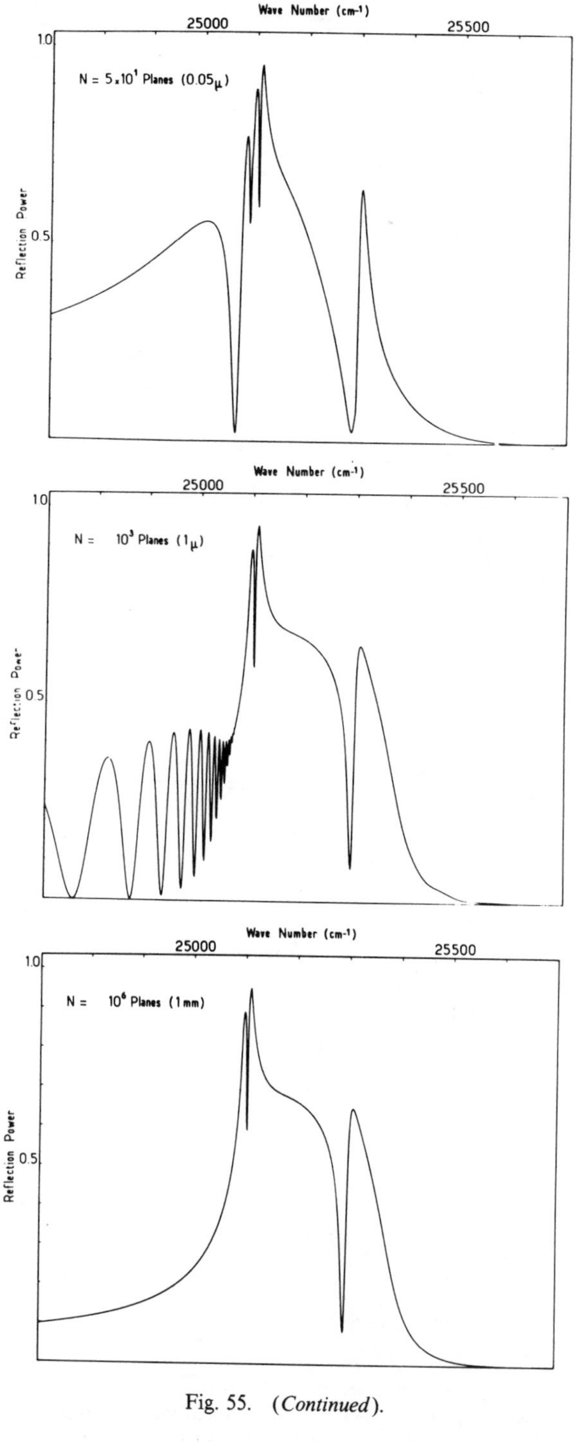

Fig. 55. (*Continued*).

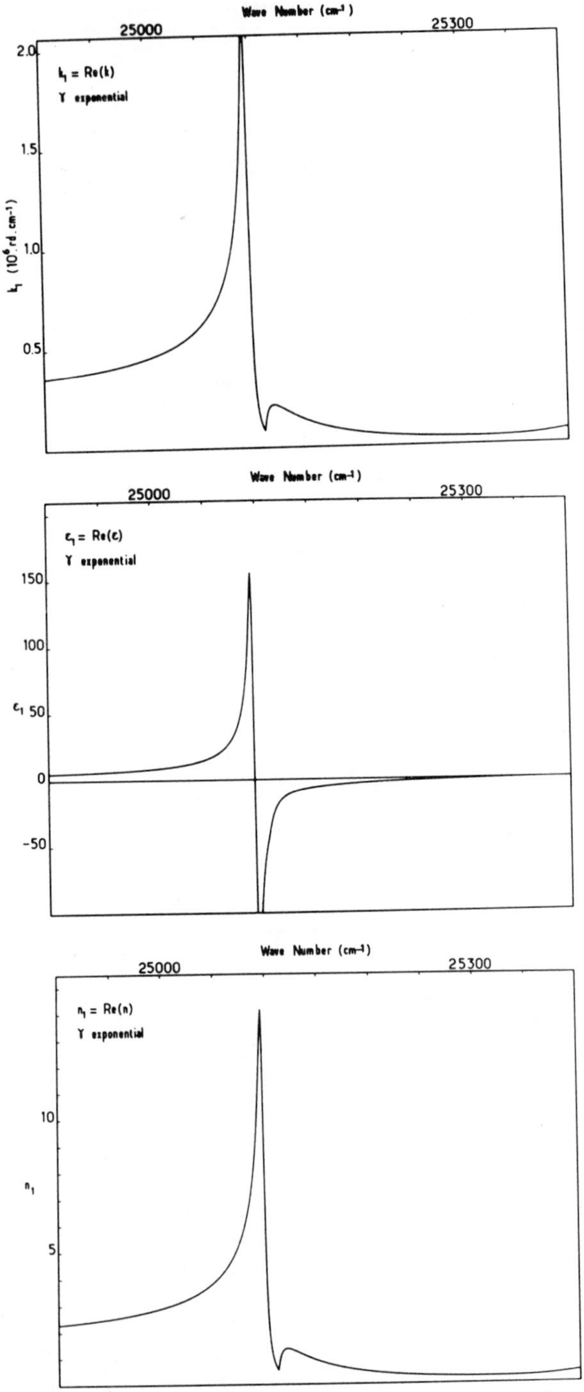

Fig. 56. The real (*left*) and imaginary (*right*) parts of the wave vector $\tilde{k}$, the permittivity $\tilde{\varepsilon}$, and the refractive index $\tilde{n}$ of anthracene, calculated with only one transition ($\omega_m = 25{,}435\ \mathrm{cm}^{-1}$) and an exponential damping $\gamma(\omega)$.

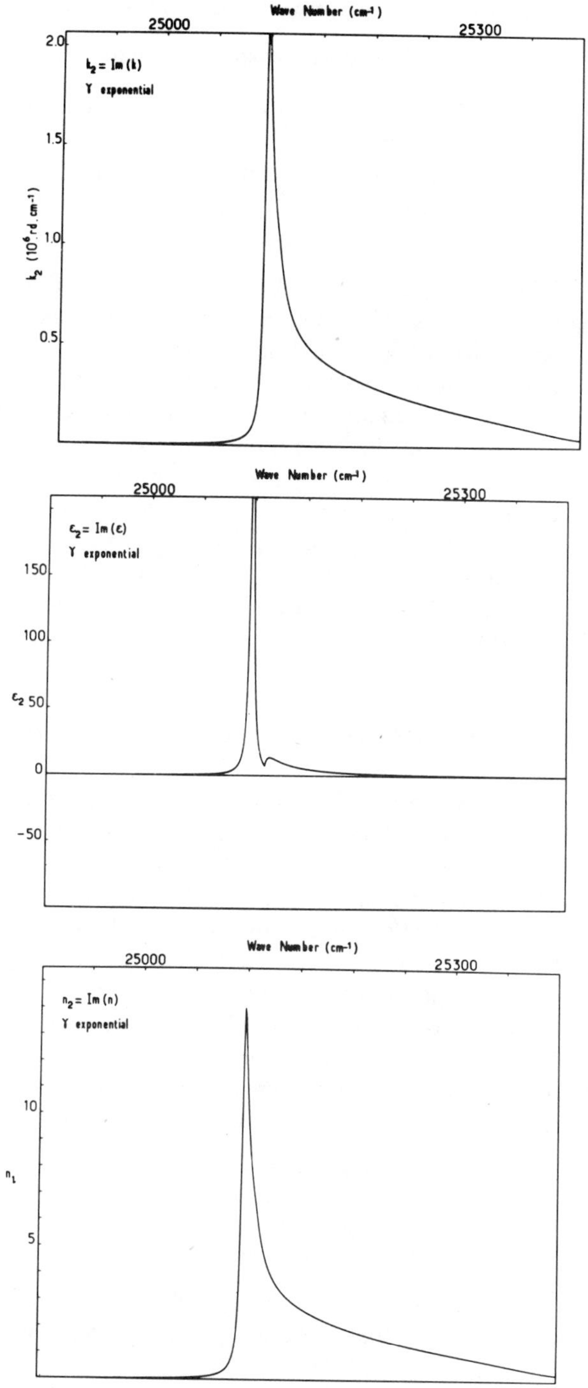

Fig. 56. (*Continued*).

ing these two methods of analysis of the absolute reflectivity spectra:

1. By trying to determine the function $\gamma(\omega)$ producing the best fit between theory and experiment.
2. By trying to obtain an accurate absorption spectrum of anthracene by Kramers–Kronig transformation on a large number of points of the high-resolution reflectivity spectrum.

These two methods and comparisons between them should lead to better understanding of the various relaxation processes in the crystal.

Concluding, we see from Fig. 54 that the model of the finite crystal simulates the main features of the experimental spectrum. It would be futile at present to try to draw exact quantitative information on the optical properties of anthracene because of the large number of experimental parameters involved. However, besides a generally good phenomenological representation of the different modes of propagation in a finite crystal, and of the relaxation processes in it, one of the main points brought out here is that the introduction of excitations localized in the surface and subsurface reproduces the experimental structures I and II. (Note that this model could be easily generalized to a larger number of planes nondegenerate with the bulk.)

VII. INTERPRETATION AND GENERALIZATION

It became clear at the end of Section II that none of the three initially proposed explanations of the "anomalous" structures in the reflectivity and fluorescence spectra of anthracene crystals was entirely satisfactory and that the problem had to be reconsidered as a whole (i.e., both reflectivity and fluorescence simultaneously). We now attempt to relate the various effects observed, in light of the experimental and theoretical results obtained above, and to form an interpretation in which they fit together.

Following the discussion, in which we shall include results reported by other groups during our work or since its publication, we shall give a description of the work undertaken to generalize these observations and of various new directions in which this work could be extended.

A. Discussion and Interpretation of Our Observations on Anthracene

1. *General Conclusion*

Our work on the first singlet transition of anthracene [(001) face] has revealed a number of characteristic features of the reflectivity and fluorescence spectra.

We summarize the conclusions drawn after the description of each experiment (see Section V):

a. b-polarized Spectra (E||b Reflectivity and Fluorescence). Our observations confirm the presence and correlation of structures I (25,301 cm^{-1}), II (25,103 cm^{-1}), and III (25,095 cm^{-1}) in the reflectivity and fluorescence spectra. These structures showed in both types of experiment:

- Almost perfect coincidence in energy [particularly emission I (25,301 cm^{-1}) and the middle of system I in reflectivity (25,294 cm^{-1}, minimum; 25,307 cm^{-1}, maximum)]
- **b** Polarization
- The same temperature dependence
- The same variations with sample purity and surface quality
- The same shifts to low energy when the crystal is covered with a transparent coating (78 cm^{-1} for I and 3 cm^{-1} for II in the case of a nitrogen coating; no shift of III was observed with a resolution of 0.6 cm^{-1})

Conclusion (i). These results show clearly that structures I, II, and III seen in fluorescence are the counterparts of structures I, II, and III in reflectivity and that both effects have a common origin. This is important since it means we must look for a common explanation consistent with emission from or absorption by energy levels above 206 cm^{-1} (I) and 8 cm^{-1} (II), the bulk energy level corresponding to **E**||**b**.

In both reflectivity and fluorescence experiments we found that structures I, II, and III could be ranged in the same order of difficulty of detection, purity of crystals, quality of surface, and low temperature required to exhibit them [I ($T < 100$ K), II ($T < 15$ K), and III ($T < 5$ K)]. We never saw structure II before structure I, nor III before I and II were very pronounced.

Conclusion (ii). We may conclude that structures I, II, and III involve excited states which, apart from their coupling to the radiation field (emission), are more and more (from I and III) efficiently coupled to, and affected by, a nearby manifold with a high density of states. In the case of our experiment, the manifold may be the bulk polariton continuum (three-dimensional), more precisely the **b**-polarized sub-band with an edge excitation energy $h\omega_T = 25{,}095$ cm^{-1}. It has excitation gaps of $\delta_{S_{III}} \sim 0.6$ cm^{-1}, $\delta_{S_{II}} = 8$ cm^{-1}, and $\delta_{S_I} = 206$ cm^{-1}, respectively, below the second subsurface, the first subsurface, and the surface, which are coupled to the bulk by a weak exchange integral. Thus, the experimental pattern provides an additional argument in favor of interpreting I, II, and III structures as resonances of collective excitations localized in the surface and in the two subsurfaces.

Coupling of states II and III to the bulk modes, or to defects, may be enhanced by temperature increase and cause broadening of the corresponding lines. Broadening and vanishing of structures II and III is also very sensitive to the overlap with the bulk thermal broadening on the high energy side.

Our experiments on the effect of surface modifications (crystals grown from solution, surface photochemistry, and condensation of transparent gases on the surface) distinguish structures I, II, and III (by their energy shifts and variations in strength) from the rest of the spectrum, which is unaltered in both reflectivity and fluorescence measurements.

Conclusion (iii). These important experiments establish for the first time the origin of structures I, II, and III in the surface and the first few planes of the crystal. (While extension to structure III is supported by some of our experiments, it is necessary to confirm the gas deposit effect on III by a very high resolution study to check if there is any observable shift on covering the surface.)

Besides establishing the origin of these structures, the shifts to lower energies observed on coating with gas (78 cm^{-1} for I, 3 cm^{-1} for II, and ~ 0 cm^{-1} for III) show that the states responsible for them are progressively less sensitive to modifications of molecular interactions at the surface. This corroborates conclusion (ii) and is a major argument in favor of their interpretation as surface exciton states.

In both fluorescence and reflectivity experiments we saw that surface modifications had no effect on the general envelope and the superposed structures (other than I, II, and III).

Conclusion (iv). We conclude that the form and the structures other than I, II, and III of the spectra are mainly due to bulk transitions unperturbed by the surface and the first few crystal planes. While this may seem obvious, it is an extremely important point to establish.

On introducing surface and subsurface modes and damping $\gamma(\omega)$ in the theoretical model of $\mathbf{E}\|\mathbf{b}$ reflectivity, we were able to simulate the main features of the spectrum, including the oscillations.

Conclusion (v). The theoretical simulation of I and II strengthens the idea of surface and subsurface modes. The general shape of the reflectivity and the oscillations gives a satisfactory qualitative description of the optical properties of finite crystals in terms of the propagation of different polariton modes and their scattering by relaxation processes, principally phonon mechanisms.

b. a-Polarized Spectra (E||a Reflectivity and the Excitation Spectrum of Structure I). The **E||a** reflectivity is weak and generally structureless. However, we have exhibited a weak maximum at 25,523 cm^{-1}, shifted toward higher energy than the bulk **E||a** maximum (25,319 cm^{-1}) by $\delta_a = 204$ cm^{-1}, very nearly equal to the difference ($\delta_b = 206$ cm^{-1}) between structure I and the bulk maximum. It has the same behavior as I regarding temperature, purity, and the like, and it disappears completely when the crystal is covered with a transparent film. In the excitation spectra of emission I, we saw that an **E||a** excitation at the energy of this maximum strongly stimulated emission from I, which is **b** polarized.

Conclusion (vi). This shows that this maximum, already seen by Brodin et al.,[105] is perfectly correlated with structures I and II observed when $E \| b$ and that like these structures its origin is a state related to the first few planes of the crystal.

2. *Interpretation*

In light of the discussion and conclusions above, the main features of the spectra of the (0,0) region of the first singlet transition of anthracene [(001) face] can be interpreted as follows:

a. Bulk Transition. The envelopes of the **E||b** and the **E||a** reflectivity spectra originate mainly in the coupling between the photon and the two subbranches $E_V^b(\mathbf{k})$ and $E_V^a(\mathbf{k})$ of the bulk exciton (excited eigenstate of the crystal bulk in absence of a field—therefore a virtual state). This coupling leads to mixed modes of propagation in the crystal (polaritons) which may (or may not if $\omega \sim \omega_T$) be formed, alone or simultaneously with another quasi particle (e.g. phonon), depending on the energy and the momentum of the incident photon. Propagation in real crystals (finite, nonrigid, impure, etc.) is influenced by various relaxation processes (phonon, surface, etc.) which scatter the polariton, causing absorption and emission (fluorescence) by the crystal.

(i) Davydov splitting of the bulk exciton. We can work backward from our observations and from the theoretical dispersion of the polariton towards the properties of the bulk exciton. Since the dispersion of the two exciton components $E_V^b(\mathbf{k})$ and $E_V^{ac}(\mathbf{k})$ is small (a few cm^{-1}) in the direction (001) in the Brillouin zone, we will write them simply as E_V^b and E_V^{ac}.

The maximum of reflectivity (like the transmission minima) do not necessarily coincide with the absorption maxima [poles of the permittivity $\tilde{\varepsilon}(\omega)$]. In fact, for a Lorentz oscillator with small damping, the reflection maximum occurs at slightly higher energies than ω_T. However the **E||b** reflectivity spectrum (Fig. 29) exhibits a sufficiently steep front and high reflectivity around its maximum for us to say, by polariton theory, that the energy of the **b**-polarized component of the bulk exciton is going to be very close to the position of maximum reflectivity:

$$E_V^b = (25{,}095 \pm 2)\ \text{cm}^{-1}\ (\text{in vacuo})$$

N.B. As we have already seen in the polariton theory, an electronic transition at ω_T does not cause a maximum at this energy, but a steep front [if the dispersion $E(\mathbf{k})$ is small]. The peak in the $\mathbf{E}\|\mathbf{b}$ reflectivity spectrum is due to the minimum of structure III which lies just above. Moreover, the dispersion curve of the polariton shows that emission of the fluorescence origin [(0,0) phononless line] is not observable owing to the large difference at this energy between the polariton wave vector in the crystal and the photon wave vector in vacuum.

This value of E_V^b, which is in good agreement with that (25,097 cm^{-1}) determined from vibronic analysis of the fluorescence spectrum by Glockner and Wolf,[86,87] leads to two-particle[33] interpretations of a number of vibronic reflectivity structures.

We can be less sure that the maximum in the $\mathbf{E}\|\mathbf{a}$ spectrum corresponds to E_V^{ac}, partly because of the generally low reflectivity (~ 30%) and partly because of its rounded form. In absence of a theory of reflection in this case [$\mathbf{k}$ normal to (001) and $\mathbf{E}\|\mathbf{a}$], which is not a pure transverse mode of propagation, and lacking at present an "accurate" Kramers–Kronig transformation of the reflectivity spectrum, we shall nonetheless use the maximum reflectivity to determine the second bulk exciton component at (uncertain accuracy):

$$E_V^{a,c} = 25{,}319\ \text{cm}^{-1}$$

The above values of E_V^b and $E_V^{a,c}$ lead to the following value for the "Reflectivity Davydov splitting:"

$$\Delta_V = E_V^{a,c} - E_V^b = 224\ \text{cm}^{-1}$$

This is in good agreement with more recent estimates and the more "reasonable" values (see Section II) proposed by Wolf[61] (220 cm^{-1}); Brodin and Marisova[48] (230 cm^{-1}), based on the transmission of thin crystals, and Brodin et al.[105] (220 cm^{-1}), based on the reflectivity spectra.

(ii) Influence of phonons. As we have already said, we think the gradual drop in $\mathbf{E}\|\mathbf{b}$ reflectivity on the high side of $E_V^b(\omega_T)$ is due to the simultaneous creation of a polariton and a phonon (two-particle system) conserving the energy and the momentum of the incident photon in what would otherwise be the stopping band.

In particular, the distinct dip in the $\mathbf{E}\|\mathbf{b}$ reflectivity spectrum (Fig. 29) at

$$E_{ph}^b = 25{,}142\ \text{cm}^{-1} = E_V^b + 47\ \text{cm}^{-1}$$

attributed to a bulk transition since it is not shifted by laying gas film on the surface, can be explained by this process involving the a_g mode at 49 cm^{-1} of anthracene.[88]

Other diffuse but reproducible structures may be explained similarly. This interpretation should, however, be reinforced by a careful study of the density of states of the optical and acoustic phonons invoked and of the theory of the mechanism by which they intervene (or do not, cf. absence of the 70 and 132 cm^{-1} modes) in the penetration of light into the crystal.

Since our fluorescence studies dealt only with emissions I and II, we shall not here discuss the means by which a polariton, once created, can get out of the crystal as a photon. (Work on this subject[90, 91] shows that here also phonons play an important role.)

b. Surface and Subsurface Excitons. In view of the conclusions (i) to (vi) above, we shall not prolong the debate, but shall instead propose a global interpretation of structures, I, II, and III (**E**||**b** reflectivity and fluorescence, Fig. 43) and of the maximum at 25,523 cm^{-1} (**E**||**a** reflectivity, Fig. 39) and the excitation spectrum of emission I (Fig. 47).

These structures, seen in reflectivity and fluorescence, are signs of the creation (reflectivity, excitation spectrum) and of the radiative decay (fluorescence) of collective excitations localized on the (**a**, **b**) surface. (structure I: **b** polarized and maximum at 25,523 cm^{-1}; **a** polarized) and the (**a**, **b**) subsurfaces deeper in the crystal (structures II and III). Thus:

Surface exciton S_1

■ **b**-polarized exciton subbranch:

$$E^b_{S_1} = 25{,}301 \text{ cm}^{-1} = E^b_V + 206 \text{ cm}^{-1} \text{ (structure I, Fig. 43)}$$

■ **a**-polarized exciton subbranch:

$$E^a_{S_1} = 25{,}523 \text{ cm}^{-1} = E^a_V + 204 \text{ cm}^{-1} \text{ (arrow, Fig. 39)}$$

■ Davydov splitting:

$$\Delta_{S_1} = E^a_{S_1} - E^b_{S_1} = 222 \text{ cm}^{-1}$$

Subsurface exciton S_2

■ **b**-polarized exciton subbranch:

$$E^b_{S_2} = 25{,}103.7 \text{ cm}^{-1} = E^b_V + 8.7 \text{ cm}^{-1} \text{ (structure II, Figs. 30, 43)}$$

■ a-polarized subbranch: not observed

Subsurface Exciton S_3:

■ b-polarized exciton subbranch:

$$E^b_{S_3} = 25{,}095.7\ \text{cm}^{-1} = E^b_V + 0.7\ \text{cm}^{-1}\ \text{(structure III, Figs. 30, 43)}$$

■ a-polarized subbranch: not observed

N.B. In the **E**||**b** reflectivity spectra, the positions of structures I, II, and III were determined like those of any resonance, by taking the average of the energies of the minimum reflectivity and the nearest maximum on the high energy side. Contrary to the scales on the diagrams, the above values are given in vacuo.

The assignments of the transitions we have observed are shown in Fig. 57 (white represents excitation; black, emission), where B stands for bulk, S_1 for the surface exciton, and S_2 for the first subsurface exciton. Subsurface exciton S_3 has not been shown because it is nearly degenerate with the bulk exciton.

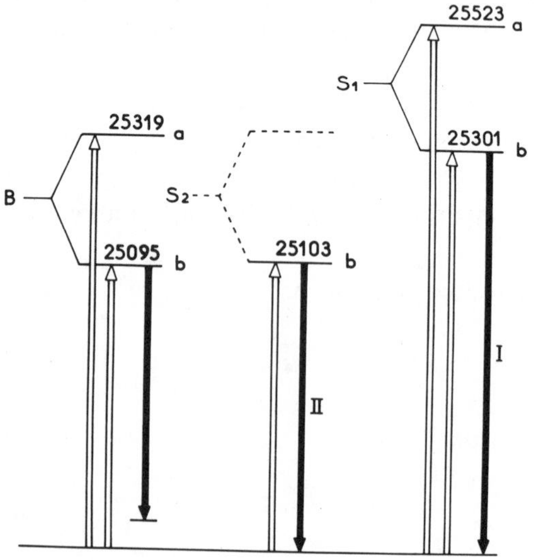

Fig. 57. Representation of the transitions observed, in terms of a bulk exciton (B) and a surface (S_1) and subsurface (S_2) exciton. Energies are in wave numbers in vacuo. Transition III seen at 25,095.7 cm^{-1} is not shown.

3. *Discussion of Our Interpretation*

We must now see how our observations and our interpretation of them fit into the general plan of results published before or during this work.

a. Agreements with Our Observations. We first note that our observations on the presence, characteristics, and evolution of the structures agree perfectly (excepting, however, structure III, which was not seen by these authors) with those made in the same conditions by Marisova et al.[97–99, 103–105] (reflectivity and fluorescence) and of Glockner and Wolf[86,87,100,101] (fluorescence). Our 80 K, **E**||**b** reflectivity spectrum (Fig. 31) is very like that reported by Morris and Sceats[109] at the same temperature, (including the presence of the shoulder). Their observation of the disappearance of the shoulder may be explained, in view of our results, either by the crystal being warmed from 10 to 20 K (*cf.* Ferguson's point of view[222]), or by the presence of a surface coating.

The presence of the structures in reflectivity has been confirmed since publication, by Ferguson's[221,222] work on the transmission and fluorescence of very thin crystals (47 layers), by Syassen and Philpott[228], who studied the reflectivity of six natural faces of anthracene grown by sublimation, and more recently by Tokura, Koda, and Nakada[229] who also checked the presence of structure III.

In attenuated total reflection (A.T.R.) experiments on a PMMA coated anthracene crystal, Sceats et al.[230, 231] have detected an excitation mode at $25{,}310 \pm 6$ cm^{-1} for 41° incidence and at $25{,}335 \pm 6$ cm^{-1} for 55° incidence; they have related this mode to the structure (I) which we observed at 25,301 cm^{-1} under normal incidence on an uncoated crystal surface. Although this equivalence needs to be confirmed by new experiments, the observations of Sceats et al. bring experimental evidences (via the measured dispersion) of the large delocalization of the state involved in structure I and consequently, support its assignment in terms of site shift surface exciton. (More account of the work done by Sceats et al., will be given in the next chapter).

Our gas coating experiments have been confirmed by Ferguson[222] and Tokura et al.[229], who observed shifts of structures I and II, nearly identical to our values. We believe that the shifts Glockner and Wolf saw in their B crystals[100, 101] were due to the same effect of condensed air on the crystal.

At present no other confirmation, by simultaneous reflectivity and fluorescence spectra, exists to show the important fact we established that the emission maximum I falls between the minimum and the maximum of structure I in reflectivity. However, it follows from our discussions with J. Ferguson, that the same result derives from the relative positions of his transmission and excitation spectra.

b. Justification of Our Interpretation. Various explanations of the anomalous structures, proposed before this work by Brodin et al., Glockner and Wolf, and Morris and Sceats, were described at the end of Section II. The conclusions drawn by Ferguson on the one hand and Sceats et al. on the other are sufficiently different from our own to justify critical examination of all these explanations, in light of the experimental and theoretical results now available:

Brodin et al.[103-105] *Surface exciton* (*I*) *and phononless* (0,0) (*II*). While our conclusions concerning the interpretation of structure I and the **E**||**a** maximum at 25,523 cm^{-1} agree well with those of Brodin et al., we differ in our interpretation of structure II. It is hard to justify it as a phononless (0,0) line. The latter hypothesis is incompatible, in reflectivity as well as in fluorescence, with the theory of polariton scattering on phonons, with the persistence of structure II at 15 or 20 K, and with its shift when the crystal surface is coated. We note, moreover, that their work contains no arguments in favor of their assignment of structure II.

Glockner and Wolf.[100,101] *I and II bulk exciton transitions.* The shifts observed for both these structures, in reflectivity and fluorescence, when the crystal surface is covered and the wrong dispersion of the exciton band [**k** normal to face (001)], on which this interpretation was based, render it untenable.

Morris and Sceats.[109] *Shoulder I at* 80 *K induced by crystal dislocations on cooling.* We find this hypothesis hard to accept because its experimental basis (the disappearance of the shoulder by annealing at 80 K) agrees neither with all the observations made since on the presence and shift of structure I, nor with the real chances of disappearance of dislocations in a crystal held at 80 K.

J. Ferguson.[222] Ferguson's work on very thin crystals of anthracene is very interesting, due to its complementarity with the present work, in the study of the processes by which a finite crystal absorbs or emits. Structures I and II are present in Ferguson's reflectivity spectrum deduced indirectly from $R = 1 - T - A$ where T is the transmission and A the absorption (supposed to be the same as the excitation spectrum he measured). He interprets them as follows:

I (~ 395 mm) "anomaly expected near the longitudinal-like frequency as discussed by Hopfield and Thomas[208]".
II (398,35 nm) "Surface induced exciton state".

We first note that Ferguson's interpretation of structure II, despite his prudence, is similar to our own. Regarding structure I, if we understand correctly Hopfield's and Thomas's work[208] on the reflectivity of CdS and ZnTe crystals, the anomalies they found close to ω_L may be correctly simulated (Fig. 8 of their article) by considering surface induced effects. While it is difficult to compare anthracene with such different systems (as regards description of their bulk excitons in which electron and hole do not occupy the same site, and description of their surfaces since these inorganic materials have dead layers corresponding to exciton devoid regions), it would seem that their conclusions, and hence Ferguson's assignment, are not fundamentally different (since they link the anomalies to the surface) from the interpretation which we propose.

Sceats, Tomioka, and Rice.[230] As already mentioned, the observations reported by Sceats and co-workers on the attenuated total reflection of anthracene crystal, support the concept of site-shift surface exciton. However, in an appendix to their paper, these authors have expressed reservations regarding this assignment, on the following three points that we will discuss successively.

The first one concerns the observation that structures I and II shift monotonically as the bulk origin, in the fluorescence of mixed normal and deuterated anthracene,[110] over the complete range of deuteration. As pointed out by these authors, if it is not obvious that the two dimensional surface exciton band shifts upon deuteration identically with the bulk exciton band, there being no accurate account of how the surface can be perturbed by isotopic mixing, it is hard to say that these observations contradict our assignments. Indeed, as noted in,[228] these monotonic shifts are readily explained by the very likely (in view of the shifts of less than 80 cm^{-1}) hypothetical process of amalgamation which cause these surface structures to shift by the same amount as the bulk. More recent theoretical and experimental work by Tokura et al.[229] on these systems, shows that the surface exciton model is perfectly compatible with these observations.

The second objection (no reproducible shift when the surface is "contaminated," based on the observations reported in[109-110], in contradiction with our own observations, has been previously analysed and will not be discussed anymore. Concerning the "uncontaminated" position of the "surface" structure for an anthracene surface coated with PMMA, we can make the following remarks:

1. The surface origin of the structure observed in[230] is not clearly established.
2. One can hardly compare a 41° or 55° incidence ATR spectrum with a normal incidence reflectivity spectrum.
3. The optical quality of the contact at 18 K between the PMMA layer and the anthracene surface is not "transparently" obvious.

Finally, an alternative explanation is proposed in[230] to account for the fluorescence lines I and II (artifacts arising from the reflectivity structure itself). While these authors do not explain neither why the reflectivity spectrum itself is structured, nor the origin of these "bulk" emissions of higher energy than the exciton band origin E_v^b, their hypothesis does explain the correlation of the fluorescence and reflectivity structures and their identical shifts after laying the gas film.

However, even if one assumes that there is such a bulk emission verifying:

1. Continuous over at least 220 cm^{-1} above the fluorescence origin.
2. Completely **b** polarized.
3. At least as strong as emission I and II.
4. Enhanced with crystal purification and crystal quality improvement.
5. With an excitation spectrum complementary of the true bulk emissions,

consideration of Fig. 43 shows that the maximum fluorescence I ($25{,}301 \pm 1$ cm^{-1}) does not coincide with the minimum reflectivity I ($25{,}294 \pm 1$ cm^{-1}) and that the emission line shapes of I and II are not at all these which would occur due to the filtering of a continuous background by variation of reflectivity.

Hypothesis of surface and sub-surface excitons. Contrary to the explanations given above, no experiment to date has contradicted the interpretation in terms of surface and sub-surface excitons which we propose. While this is not proof of its truth, it does show that it is well founded.

This theory gives a consistent explanation of the presence, correlated variations, shifts, and modifications induced by perturbation of the surface concerning structures I, II, and III and the maximum at 25,523 cm^{-1} seen in reflectivity and fluorescence. Besides these confirmed observations which strengthen this interpretation, the following points should be mentioned.

In agreement with Sugakov's theory, according to which the Davydov splitting of surface and volume excitons should be the same, we have indeed found them to be the same, to within experimental error ($\Delta_{S_1} = 222$ cm^{-1}; $\Delta_V = 224$ cm^{-1}). Moreover, the presence of the surface exciton component $E_{S_1}^a$ at 25,523 cm^{-1} (also attributed by Brodin et al.[105]) explains why $\mathbf{E} \| \mathbf{a}$ excitation centred on this energy strongly favours emission I (also related to the surface), polarised along the **b** axis. In our model, the surface is directly excited, and after rapid relaxation between the exciton components, may emit with the energy and polarisation of emission I.

Observations by Brodin et al.[105] and most recently by Syassen and Philpott[228] on the reflectivity of the other natural facets of anthracene show that structures I and II (III was not seen by them on any face) are present in the **E**||**b** reflectivity spectrum of face (001) but not in those of any other face, including (20$\bar{1}$) and ($\bar{1}$01), both of which contain the **b** axis. This important fact clears away any doubts that structures I and II are due to some impurity, and shows they are related to a unique property of the (001) planes and not to the bulk dielectric constant. The absence of surface excitons on the other faces may be explained by the far larger interplane excitation exchange terms when the wave vector is normal to them.

Finally, the hypothesis of surface and subsurface excitons is one of the few explanations of why there are emissions from energies 206, 8.7 and 0.7 cm^{-1} above the bulk exciton band, E_V^b. Semiclassical calculations by Sugakov[215] and Philpott and Sherman,[232] and purely quantum ones by Orrit,[223] concerning the radiative instability of collective excitations in an (**a**, **b**) layer of molecules all lead to a radiative lifetime of the collective state $(\lambda/2\pi a)^2 \simeq 10^{-4}$ smaller than that of the free molecule (i.e., about a picosecond in anthracene). If we suppose the first crystal planes to be like this model, then it is readily understandable why emission competes successfully with different nonradiative relaxation processes involving the quasi continuum of bulk states.

Comparing the reflectivity and fluorescence spectra, one can "estimate" the radiative widths of structures I and II (the only entirely resolved) at 13 and 1 cm^{-1}, respectively, corresponding to lifetimes of 0.4 ps (I) and 5 ps (II), in good agreement with the theoretical order of magnitude. Furthermore, the decreasing linewidth from I to III, which may seem to contradict the idea of states relaxing with increasing ease toward the bulk, can be quite well explained, classically or quantum mechanically,[223] by considering the coupling of surface and subsurface states to ever denser polariton modes as the energy approaches that of the bulk exciton ($\hbar\omega_T = E_v^b$).

Summing up, these emissions are possible because surface to bulk couplings are small along the **c**′ axis perpendicular to the (**a**, **b**) planes and because the surface exciton has a very short lifetime due to the collective coherent emission by all the sites in the plane. In this context, these emissions may well be one of the few examples of coherent exciton emission.

B. Generalization of Our Observations

While the above discussion shows that the concept of surface and subsurface excitons is consistent with all experimental results on anthracene, it is insufficient proof that they exist and that we have detected them. One of the fundamental points remaining to be firmly established is the generality of such observations, by extending the above methods to other systems with suitable crystalline structure for such a facet to harbor surface excitons (interplane excitation exchange terms $I_{(\mathbf{k})}^{(l)}$ must be very small compared to the corresponding $D^{(l)}$). Other aromatic hydrocarbons are interesting in this respect since, as in anthracene, the main part of the exchange term $I(\mathbf{k})$ is intraplanar when the wave vector is normal to the (001) face.[37]

We have therefore studied the reflectivity of the (001) face of the "next neighbor" of anthracene in the aromatic series, tetracene (naphthacene), in the region of its first singlet transition (~ 18,700 cm^{-1}). We give below a brief summary of our results, showing the analogy of these two systems.

It should be noted that the low temperature (2 K) study of tetracene was much harder to carry out because of the tendency of even "freely" mounted crystals to disintegrate despite very slow cooling.

By mounting the samples in "optical contact" with two quartz discs and cooling very slowly (as slowly as 0.25 K/min) we were able to follow the phenomenon already observed by Prikhotko and Skorobogatko[234, 235] and Vaubel and Baessler.[236] The high temperature (~ 186 K) Davydov splitting of 700 cm^{-1} increases abruptly to 930 cm^{-1} at low temperatures (Fig. 58). While no low temperature crystallographic work has been done to confirm this observation, we believe that this sudden variation is most certainly related to a phase transition in the crystal, which is triclinic at room temperature.[237]

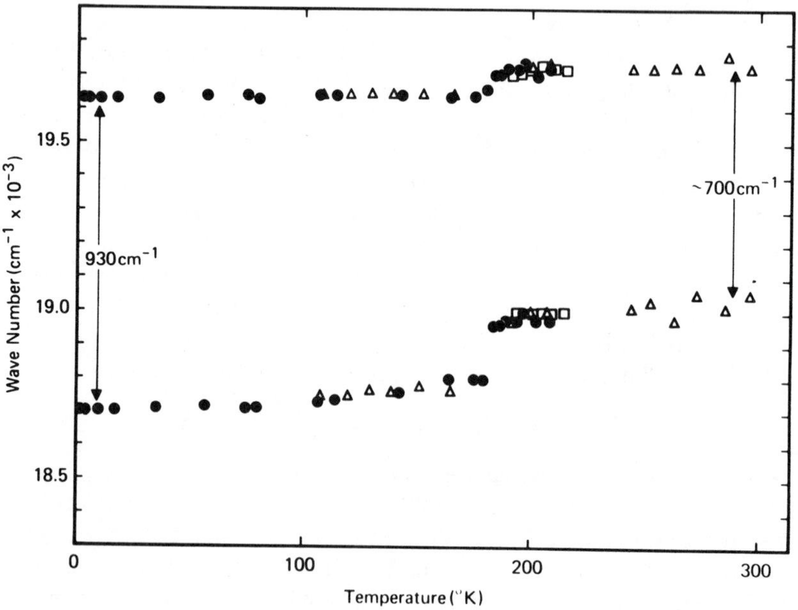

Fig. 58. Temperature dependence (1.7 to 300 K) of the positions of the two Davydov components of the first singlet crystal transition of tetracene (**E**||**b** is the lower curve). The different symbols stand for three different samples. The discontinuity occurs around 186 K.

Figure 59 shows that the **E**||**b** (polarization determined at high temperature) reflectivity spectra of samples, we managed to cool to 2 K, is much like those of anthracene and that they too exhibit "anomalous" structures at 18,694 cm^{-1} and 18,859 to 18,880 cm^{-1} (minimum–maximum), though they are perhaps less distinct owing to probable perturbation of the surface during cooling. Although the fragility of the crystals prevented us from carrying out the condensed gas experiments, the similarities of these results with those of anthracene, as illustrated by the temperature dependence shown in Fig. 60, lead us to consider the presence of surface and subsurface excitons to explain these structures.

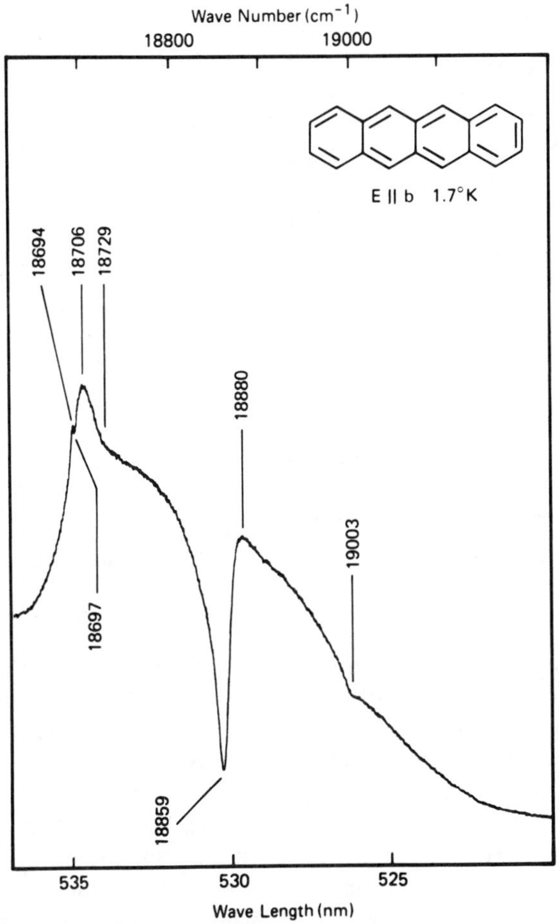

Fig. 59. **E**||**b**-polarized reflectivity of the (001) face of tetracene (sublimation grown), at 1.7 K.

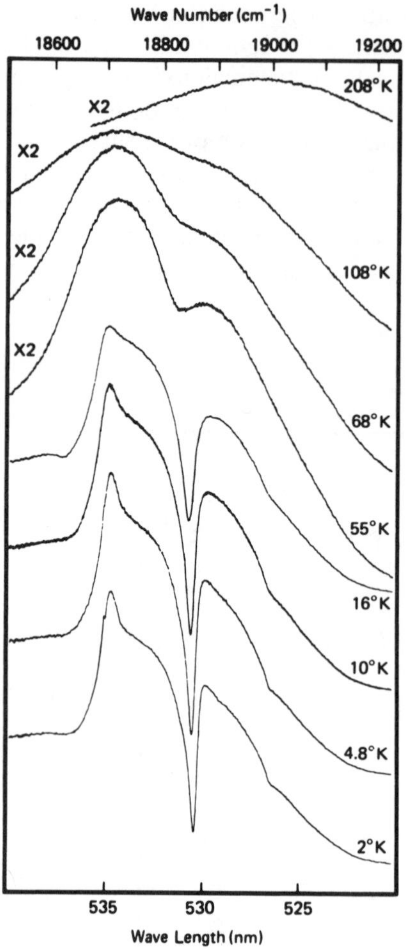

Fig. 60. Temperature dependence of the **E**||**b** reflectivity of the (001) face of tetracene. Note the large shift in the maximum reflectivity before the phase change (top curve).

Besides this study of tetracene, recent work by Ostapenko et al.[238] on the reflectivity of the (001) face of naphthalene at 4.2 K in the region of the first vibronic singlet transition (31,959.3 cm^{-1}) suggests the presence of anomalous structures attributable to collective excitations (vibrons this time) on the surface and the first few planes.

These authors have noticed two structures in the **b**-polarized spectrum. Situated at $\nu_{S_2} = 31{,}960.7\ cm^{-1}$ (narrow minimum) and at $\nu_{S_1} = 32{,}038\ cm^{-1}$ (weak maximum), they are very sensitive to the surface condition of the crystal. [In naphthalene, the smaller values of the Van der Waals shift (molecule–crystal) $D \sim -465\ cm^{-1}$ ($-2300\ cm^{-1}$ in anthracene) and the smaller oscillator strength of the transition (lower and narrower reflectivity) make it possible to observe a surface exciton as a maximum in reflectivity, shifted 78 cm^{-1} from the corresponding bulk transition.]

Finally, structures in the reflectivity of other molecular crystals very different from the aromatic hydrocarbons, have been likewise attributed to surface excitons by Hesse et al.[239] for the dye CTIP and by Saile et al.,[240] who also induced, by surface coating, the reproducible energy shift of some structures, in the reflectivity and transmission spectra of rare gas solids (Ar, Kr, Xe).

VIII. NEW DEVELOPMENTS: EXCITON SURFACE POLARITONS

Besides site shift surface excitons molecular crystals have another class of surface excitations called "surface polaritons" that plays an important role in the energy transfer processes between surface and volume. The term surface polariton, refers to an electromagnetic wave confined to the macroscopic surface of any active medium, via its electric or magnetic polarization and depends on the origin of this polarization. Therefore, there are different sorts of surface polaritons.[127, 130, 249, 250] Although most of the work done in this field concerns metal and semiconductors (surface plasmon polaritons) or inorganic insulators (optical phonon surface polaritons and Wannier-Mott exciton surface polaritons), there is now a growing interest in the properties of exciton surface polaritons in organic materials where microscopic theoretical models of crystal optics may be more easily elaborated. On the other hand, the usual low symmetry of molecular crystals, as well as the spatial and orientational dispersion of their electronic excitations, induce severe although interesting experimental and theoretical complications since macroscopic surface modes may easily couple to bulk polaritons. However, it has been experimentally demonstrated in anthracene at low temperatures,[230, 231] that exciton surface polaritons are observable, and more recently even at room temperatures on highly reflecting organic materials as CTIP, PTS or $TCNQ^{\circ}$[245-248] in spite of the important damping that surface polaritons may endure in this latter case. The work done by Sceats and co-workers on anthracene[230, 231] involves the archetype of organic molecular crystals and, moreover, it connects directly site shift surface excitons to surface polaritons. Therefore, it represents an important breakthrough in the understanding of the coherent energy exchanges between surface and bulk. As a basis for a possible extension to our work, we give in this section, a brief account of the principal observations of these authors with some general principles on exciton surface polaritons.

A. Basic Principles on Exciton Surface Polaritons

At the interface between two semi-infinite media, provided certain conditions on the respective values of the optical constants (ε, μ) of these two media are fulfilled, Maxwell equations and the interface boundary conditions admit as solution, besides the trivial oscillating electromagnetic wave, another electromagnetic mode with the following characteristics:[126, 130]

1. This wave travels parallel to the interface with a propagation bidimensional wave-vector **k** contained in it.
2. It is associated with a surface charge due to the interface.
3. It is *p*-polarized. The electric field is in the sagittal plane containing the surface wave-vector **k** and the normal to the interface (the incidence plane in an ATR experiment).
4. Its field amplitudes decrease exponentially in the direction normal to the interface.

These characteristics effectively define a surface excitation provided the term surface is understood as the *macroscopic surface* extending on one wavelength range into the bulk. This macroscopic surface must not be confused with the *first monomolecular layer* of the crystal (microscopic surface) where site shift surface excitons are localized.

Of course, the conditions of existence of such a surface mode and its dispersion behavior, depend on the polarization origin of the active medium and on the dielectric permittivity $\tilde{\varepsilon}(\mathbf{k}, \omega)$ and magnetic permeability $\tilde{\mu}(\mathbf{k}, \omega)$ tensors of each semi-infinite medium. As our purpose is simply to survey the application of the surface polariton concept to the specific case of an excited molecular crystal bound by a perfect nonabsorbing dielectric, let us consider only the problem of the interface between an "inactive" isotropic medium (with a positive dielectric constant ε_i and a magnetic permeability $\mu_i = 1$) and an "active" medium (with a frequency dependent dielectric constant $\varepsilon_a(\omega)$ and a magnetic permeability $\mu_a = 1$) that we will consider first as isotropic, thicker than one wavelength, and loss-free. (For an extensive review see Reference 130.)

1. *Isotropic and Loss-Free Semi-Infinite Active Medium*

In this simple case, boundary conditions at the interface will impose the following equation

$$\frac{\varepsilon_i \varepsilon_a(\omega)}{\varepsilon_a(\omega) + \varepsilon_i} = \frac{k^2}{\omega^2/c^2}$$

that requires $\varepsilon_a(\omega) + \varepsilon_i < 0$ for the existence of a surface polariton with the characteristics defined above. As ε_i is set positive, this last condition imposes $\varepsilon_a(\omega) < 0$.

As we saw in Section III, a molecular crystal with an excitonic transition at the energy $\hbar\omega_T$, has its dielectric function $\varepsilon_a(\omega)$ becoming negative in the "stopping band" (ω_T, ω_L) where ω_L is the longitudinal exciton frequency. If we adopt for $\varepsilon_a(\omega)$ a dispersion form analogous to the one derived in Section III, we get

$$\varepsilon_a(\omega) = \varepsilon_\infty + \frac{f\omega_T^2}{\omega_T^2 - \omega^2}$$

Reporting this expression in the above equation leads to the dispersion behavior (ω, k) for the exciton surface polariton sketched on Fig. 61, where $\omega_L^2 = \omega_T^2(1 + f/\varepsilon_\infty)$ and ω_{SP}, the asymptotical frequency of the surface

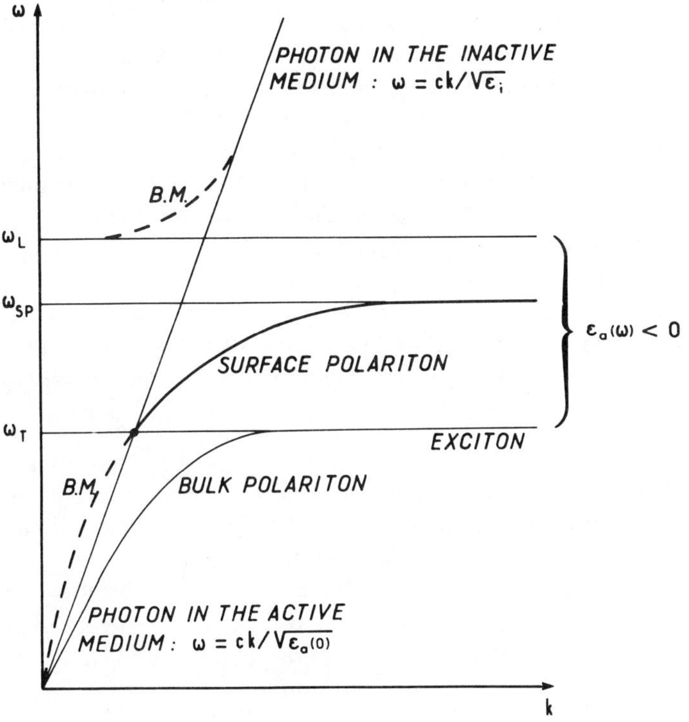

Fig. 61. Dispersion curve of the exciton surface polariton for an isotropic, loss-free active medium. (B.M. stand for Brewster modes).

polariton for $k \to \infty$, is given by

$$\omega_{SP}^2 = \omega_T^2 + \frac{f\omega_T^2}{\varepsilon_\infty + \varepsilon_i}$$

As we assume no spatial and no orientational dispersion, ω_T, ω_L, and ω_{SP} are independent from the amplitude and direction of the wavevector $\mathbf{k}$; there is no possibility for the surface exciton to couple to a bulk polariton mode. Consequently, this surface polariton whose dispersion is bound in frequency to ω_{SP} but is spatially ($\mathbf{k}$) unlimited, is referred to as a *REAL* surface polariton.

As bulk polaritons are mixed exciton (frequency ω_T)-photon states (Section III), from this dispersion curve we may interpret surface polaritons as resulting from the coupling of an incident photon coming from the inactive medium, with the surface electric dipole excitation (frequency ω_{SP}) due to the charge density induced at the interface by the discontinuity of the normal component of the polarization. However, at the difference of bulk polaritons, this surface excitation ω_{SP} depends not only on the active medium (ω_T), but also on the inactive medium (ε_i). Moreover, we may note that, except in the two extreme points of the dispersion curve, the fields associated with the surface polariton wave are transverse and exist in both active and inactive media with opposite elliptical polarizations and an amplitude decaying exponentially along the normal to the interface. From the integration of the time-averaged Poynting vector, one can see that the energy flows in opposite directions in the interface media, but shows a resulting nonzero flow in the direction of propagation.

While the "knee region" of this dispersion curve exhibits the reciprocal continuous variations of the photon and dipolar excitation components of the surface polariton, the two extreme points of this curve correspond to the uncoupled cases:

1. When $k \to k_{\text{lim}}$ and $\omega \to \omega_{SP}$ $[\varepsilon_a(\omega) = -\varepsilon_i]$, the photon component of the surface polariton goes to zero (null group velocity) and the fields are essentially electrostatic in character (zero-retardation limit) analogous to the excitonic limit for bulk polaritons.
2. When $k = \omega_T\sqrt{\varepsilon_i}/c$, $\omega = \omega_T$, and $\varepsilon_a(\omega) \to -\infty$, the matter electric dipole excitation component of the surface polariton goes to zero and the latter becomes essentially a pure photon propagating along the interface only in the inactive medium.

The penetration depth of the surface polariton is null for these two extreme cases and maximum in the knee region, which means that the surface polariton is strongly localized at the interface when $k = \omega_T\sqrt{\varepsilon_i}/c$ and $k \to \infty$, but much less ($\sim k^{-1}$) between these limits.

We may note on this dispersion curve that the surface polariton is entirely situated in the nonradiative region (at the right of the photon curve $\omega = ck/\sqrt{\varepsilon_i}$), which implies that it will be impossible to optically excite directly surface polariton with light coming from the inactive medium, due to the wave-vector mismatching at a given energy.

Fano and Brewster Modes. At this stage it is important to point out the difference that exists between the nonradiative surface polaritons that we defined above (the so-called Fano modes in a loss-free active medium), with the radiative Brewster modes that arise when *p*-polarized electromagnetic waves imping an interface at a particular incidence θ_B (Brewster angle), with subsequent zero reflection in the incident medium. As the Brewster incidence θ_B is given by the relation $tg\theta_B = \sqrt{\varepsilon_a/\varepsilon_i}$, with $k_B = (\omega/c)\sqrt{\varepsilon_i}\sin\theta_B$ one can easily check that Brewster modes verify, in the radiative region, the same dispersion relation than surface polariton Fano modes. In the case of the two isotropic and no-damped interface media we consider, there is a clear distinction between Brewster and Fano modes, the transition occuring at the point $\omega = \omega_T$.

Although Brewster modes may also be considered as interface modes (there is only one electromagnetic wave in each medium). However, they are not bound to the macroscopic surface since their field amplitudes do not decrease along the normal at the interface. For Brewster modes, the surface and normal components of the electric field are in phase in each medium ($\pi/2$ out of phase for Fano modes) with the same sign for the two normal components (opposite signs for Fano modes).

2. *Damped Active Medium*

With the same dielectric isotropic inactive medium (ε_i positive and constant), we consider now possible relaxation channels in the isotropic active medium, either nonradiative (as exciton-phonon coupling) or radiative (as site-shift surface exciton emission), leading to a complex permittivity $\varepsilon_a(\omega)$. As we saw in Section III, one may account for this relaxation by introducing a phenomenlogical damping parameter $\gamma(\omega)$ in such a way that $\varepsilon_a(\omega)$ becomes:

$$\tilde{\varepsilon}_a(\omega) = \varepsilon_\infty + \frac{\omega_T^2 f}{\omega_T^2 - \omega^2 - i\omega\gamma}$$

This expression for $\tilde{\varepsilon}_a(\omega)$ makes the dispersion relation complex. As fully discussed in References (130 and 249), we may have situations that range, depending on the experimental configuration, from pure temporal damping ($\text{Im}(k) = 0$, $\text{Im}(\omega) < 0$) to pure spatial damping ($\text{Im}(k) > 0$, $\text{Im}(\omega) = 0$). As a consequence, there is no longer a clear distinction between Brewster modes and surface polaritons. In an angle scan experiment (reflectivity vs. θ at ω constant) the dispersion curve has a backbending which links the radiative and nonradiative regions in the frequency range of high damping with a large imaginary part of the wave vector (shaded area on Fig. 62). Consequently, in this case it will be possible to find ATR maxima in both radiative and nonradiative regions.

Depending on the relative values of the real and imaginary parts of $\tilde{\varepsilon}_a(\omega)$, we may have, for instance: 1) when $\text{Re}(\tilde{\varepsilon}_a(\omega)) < 0$ with $\text{Re}(\tilde{\varepsilon}_a(\omega)) \gg \text{Im}(\tilde{\varepsilon}_a(\omega))$, a surface mode analogous in character to the Fano surface polaritons described above (lossy Fano modes) while, 2) for $\text{Re}(\tilde{\varepsilon}_a(\omega)) \ll \text{Im}(\tilde{\varepsilon}_a(\omega))$, we may have particular surface modes (Zenneck modes). Although we may find these different types of modes in the same solid, organic crystals exhibit more often the latter type of modes. The occurrence of a particular type of surface polaritons in damped active media may be evaluated from its location in the complex plane ($\text{Im}(\tilde{\varepsilon}_a(\omega))$ vs. $\text{Re}(\tilde{\varepsilon}_a(\omega))$); it depends on the respective values of the oscillator strength f, of the energy ω_T of the excitonic transition and of the magnitude of the damping parameter γ (see References 130, 246, 247, 249).

3. *Anisotropic Active Medium*

A general orientation of the interface relative to the principal directions of the dielectric tensor $\tilde{\varepsilon}_a(\omega)$ characterizing the anisotropic active medium, will now involve two electromagnetic waves in this medium, leading to surface polaritons with two different attenuation constants. Even in the case where $\tilde{\varepsilon}_a(\omega)$ is real (no damping), these attenuation constants may now be complex, which means that propagation and attenuation directions will be coupled with, as a physical consequence, a leak of the surface polariton energy into the bulk. These difficulties, due to birefringence, will be combined to those caused by the following:

Spatial dispersion (variation of $\tilde{\varepsilon}_a$ with $\mathbf{k}$),
Axial dispersion (variations of the principal axis directions with ω),
Damping (the dielectric tensor $\tilde{\varepsilon}_a(\omega)$ is complex with in general no simultaneous diagonalization of the real and imaginary parts),

and will make a general treatment of surface polaritons in biaxial crystals very cumbersome. To our knowledge, this treatment has not been undertaken as yet.

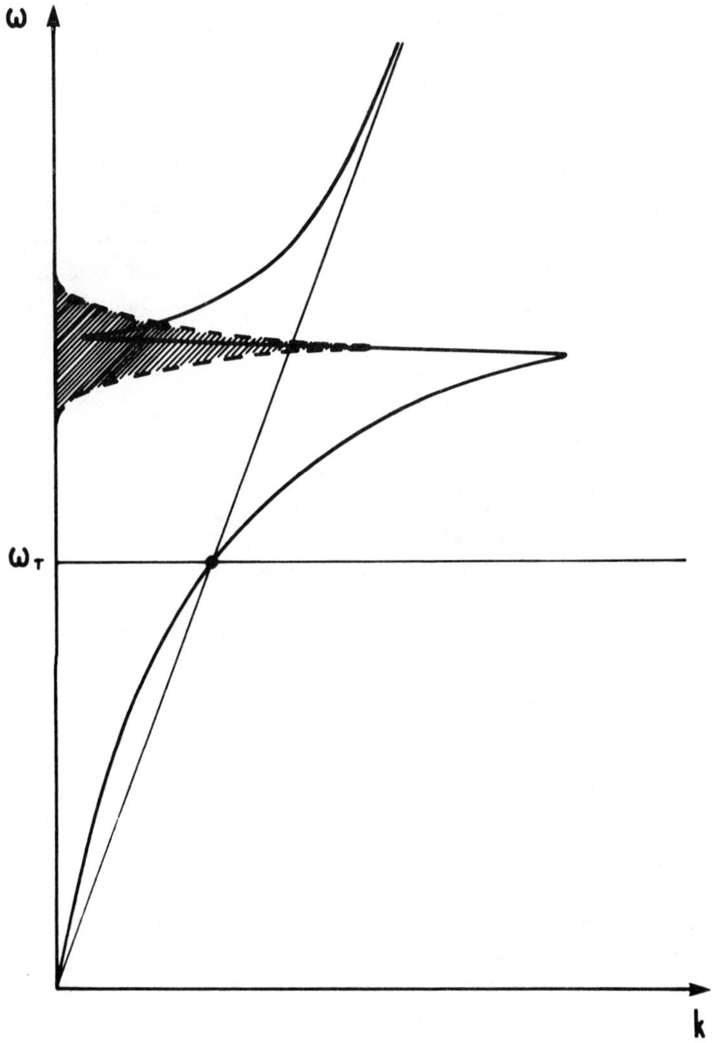

Fig. 62. Dispersion curve of the exciton surface polariton for an isotropic damped active medium; full line: Re(k), broken line: Im(k).

However, when the normal at the interface or the direction of propagation (**k**) are oriented along a principal dielectric direction, the attenuation remains uncoupled to the propagation and we may have a surface polariton as defined previously, although some of these characteristics may be altered; then we speak of *REGULAR* surface polaritons.

For the particular case where the principal axes coincide with the cartesian axes ($\hat{x}$, $\hat{y}$, $\hat{z}$) setting the direction of propagation along the $\hat{x}$ axis, the normal to the interface along the $\hat{z}$ axis and assuming no damping and a positive real dielectric constant ε_i for the inactive medium, one may easily demonstrate that the dispersion relation for the surface polariton is:

$$k^2 = \left(\frac{\omega^2}{c^2}\right)\varepsilon_i\varepsilon_{az}\frac{\varepsilon_{ax}-\varepsilon_i}{\varepsilon_{ax}\varepsilon_{az}-\varepsilon_i^2}$$

This relation shows that, in this case, the existence of surface polaritons (exponential decay of the field amplitudes along $\hat{z}$) depends on the relative values of ε_i, $\varepsilon_{ax}(\omega)$, and $\varepsilon_{az}(\omega)$:

When $\varepsilon_{ax}(\omega)$ *and* $\varepsilon_{az}(\omega)$ *are both negative* we get a surface polariton with a dispersion curve analogous to the one previously found for an isotropic medium. The surface polariton is frequency bound but it exists in the nonradiative region for all **k** spanning the Brillouin zone. We speak in this case of a *REAL* surface polariton.

When $\varepsilon_{ax}(\omega)$ *negative with* $\varepsilon_{az}(\omega)$ *positive* we get also a surface polariton provided the additional relation $\varepsilon_i < k^2c^2/\omega^2 < \varepsilon_{az}$ is fulfilled. This last relation limits in general the existence of the surface polariton to a small **k** domain in the nonradiative region, near the light line ($\omega = ck/\sqrt{\varepsilon_i}$). We speak then usually of *VIRTUAL* surface polariton although, as pointed out by Kliever and Fuchs[130], this term is inappropriate since these modes have not a particularly short lifetime, as the term virtual usually implies: the term **k**-limited surface polariton could have been more justified.

The stop point Q, where the dispersion curve of the virtual surface polariton crosses the dispersion curve of the bulk $\hat{z}$ polariton, may be interpreted as the decay of the surface polariton into a transverse bulk polariton and a longitudinal bulk exciton mode. Consequently, Q is the limit in energy of the surface polariton quenching by the bulk.

The above results may be made explicit in the case of an uniaxial crystal for different orientations of the optical axes relative to the interface. Noting $\varepsilon_{a\|}(\omega)$ the dielectric component along the optical axis and $\varepsilon_{a\perp}(\omega)$ the two components normal to it, we separate the following cases:

1. The optical axis is along $\hat{y}$ (i.e., within the interface but normal to the propagation $\hat{x}$). We have $\varepsilon_{ax} = \varepsilon_{az} = \varepsilon_{a\perp}$ and the dispersion relation reduces

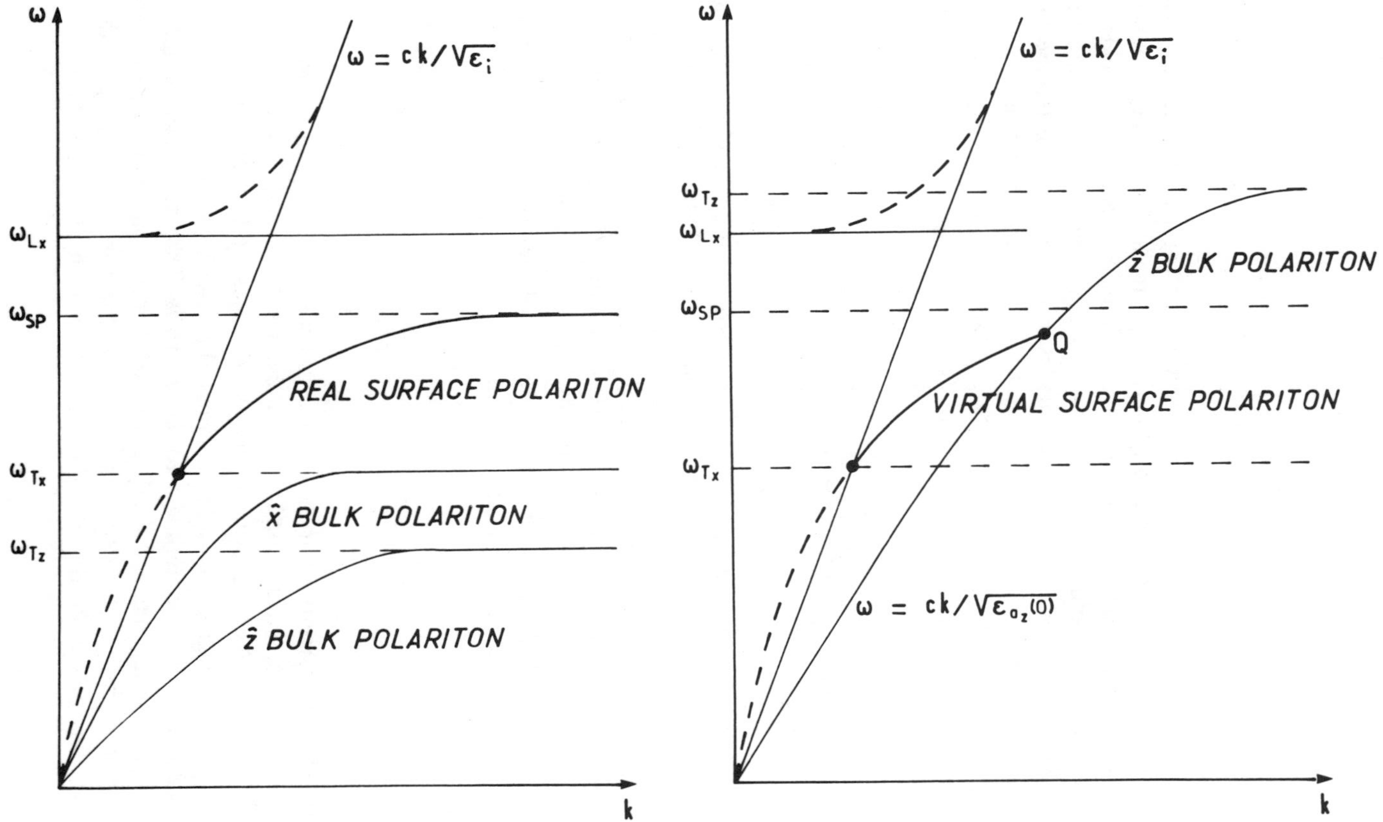

Fig. 63. Dispersion curve of real and virtual surface polaritons for an anisotropic active medium.

to:

$$k_x^2 = \left(\frac{\omega^2}{c^2}\right)\frac{\varepsilon_i \varepsilon_{a\perp}}{\varepsilon_i + \varepsilon_{a\perp}}$$

This relation corresponds to the propagation of a regular real surface polariton analogous to those encountered in isotropic media, since it depends only on one dielectric function for the active medium. In this case, the fields are purely transverse and the energy propagates along the wave vector direction $\mathbf{k}_x$. By analogy with bulk polariton modes in uniaxial crystals (see Section III), this mode is referred to as an *ORDINARY* surface polariton.

2. The optical axis is along $\hat{x}$ (i.e., within the interface and along the propagation $\hat{x}$). We have $\varepsilon_{ax} = \varepsilon_{a\parallel}$ and $\varepsilon_{az} = \varepsilon_{a\perp}$, with the dispersion relation:

$$k_x^2 = \left(\frac{\omega^2}{c^2}\right)\varepsilon_i \varepsilon_{a\perp}\frac{(\varepsilon_{a\parallel} - \varepsilon_i)}{(\varepsilon_{a\parallel}\varepsilon_{a\perp} - \varepsilon_i^2)}$$

Its solutions correspond to real surface polaritons if $\varepsilon_{a\perp}$ and $\varepsilon_{a\parallel}$ are both negative, and to virtual surface polaritons if $\varepsilon_{a\perp} < 0$ and $\varepsilon_{a\parallel} > 0$ with $\varepsilon_i < k^2c^2/\omega^2 < \varepsilon_{a\perp}$. As they depend on the two dielectric functions of the active medium, these modes have a longitudinal component and consequently, do not propagate the energy along the wave-vector direction $\mathbf{k}_x$. They are called *EXTRAORDINARY* surface polaritons.

3. The optical axis is along $\hat{z}$ (i.e., normal to the interface). Then $\varepsilon_{ax} = \varepsilon_{a\perp}$ and $\varepsilon_{az} = \varepsilon_{a\parallel}$ and the dispersion relation is:

$$k_x^2 + k_y^2 = \left(\frac{\omega^2}{c^2}\right)\varepsilon_i \varepsilon_{a\parallel}\frac{(\varepsilon_{a\perp} - \varepsilon_i)}{(\varepsilon_{a\parallel}\varepsilon_{a\perp} - \varepsilon_i)}$$

Its solutions correspond to real surface polaritons if $\varepsilon_{a\perp}$ and $\varepsilon_{a\parallel}$ are both negative, and to virtual surface polaritons if $\varepsilon_{a\perp} < 0$ and $\varepsilon_{a\parallel} > 0$ with $\varepsilon_i < k^2c^2/\omega^2 < \varepsilon_{a\parallel}$. As the precedent case these modes are extraordinary.

4. The optical axis is within the sagittal plane $(\hat{x}, \hat{z})$ and makes an angle $0 < \theta < \pi/2$ with the normal $\hat{z}$ to the interface. Then the dielectric tensor $\tilde{\varepsilon}_a(\omega)$ is nondiagonal, and the directions of propagation and attenuation are coupled (complex attenuation constant). At the opposite of the precedent cases, the surface polaritons, if they exist, are no longer regular. Although it

is not possible to express the dispersion of these modes within a simple relation, this case can, however, be treated and the conditions of existence for the surface polaritons may be extracted. (See References 126, 248, 251.)

Organic molecular crystals present generally a low symmetry (monoclinic or triclinic) that may lead one to believe that exciton surface polaritons are unobservable in these materials or, if they are, that their quantitative analysis is too difficult. However, as these organic candidates must be "quasi-metallic" reflectors in order to have a chance to exhibit surface polariton effect, the high strength of their electronic transitions induce large variations in the magnitude and the dispersion of their dielectric components. Consequently,for particular crystal faces, organic molecular crystals may be more or less approximated in the excitonic region, by an orthorombic or even an uniaxial crystal model. For example, assuming for ε_{az} a constant value and for $\varepsilon_{ax}(\omega)$ a dispersion relation analogous to the one derived in Sections III and VI, quantitative analysis of the exciton surface polariton may be carried out, even for a damped active medium. Besides the work done on anthracene[230] that will be described later, such an experimental and theoretical study has been done, in the excitonic region, for three organic crystals as different as CTIP (a cationic dye), PTS (a polymer) and TCNQ° (a neutral crystal) at room temperature.[246, 248] Due to the difference in the damping strength and, consequently, to the relative values of the real and imaginary parts of their dielectric function, these last three systems exhibit different types of exciton surface polaritons impinging largely the nonradiative region in CTIP (Fano modes) or quasi-limited to the radiative region in PTS (Zenneck and Brewster modes). With different degrees of applicability, the uniaxial model accounts for the dispersion of these modes and allows to calculate the electromagnetic field intensities in the crystal and, consequently, to understand how the surface polariton energy relaxes in the bulk.[248]

4. *Thin Active Medium (Slab): Guided Wave Modes*

As we saw in a previous section, surface polaritons have their field magnitudes decreasing exponentially along the normal at the interface with a penetration depth ($\sim k^{-1}$) maximum in the knee region. Consequently, as long as the thickness of the active medium is larger than k^{-1}, this medium may be considered as semi-infinite and the surface polariton modes remain those previously described.

At the opposite, for a very thin crystal with a thickness smaller or of the order of magnitude of the wavelength, the two interfaces limiting this crystal couple via the electromagnetic field and give rise to surface polaritons "dimeric" modes exhibiting dispersion curves ($\omega_+(k)$, $\omega_-(k)$) which differ from the semi-infinite case.[130, 249, 250] As the crystal works like a plane dielectric wave-guide for the surface polaritons, these modes are often referred to as guided wave modes. Although such modes have been observed in the anthracene crystal by Rice and co-workers,[231] their study is far beyond the scope of this survey and will not be developed here.

B. Experimental Excitation and Detection of Surface Polaritons

As we saw in the previous section, for a perfect plane interface between an isotropic inactive medium (ε_i) and an undamped active medium [$\varepsilon_a(\omega)$], the surface polariton, when it exists, is a pure nonradiative mode with a dispersion curve entirely located on the right side of the light line ($\omega = kc/\sqrt{\varepsilon_i}$) in the inactive medium. Consequently, for any incidence angle θ on the interface, a photon coming from the inactive medium will never be able to excite directly a surface polariton because of the mismatch (at a given energy) of their respective wave-vectors. As direct optical excitation is not possible, different methods have been developed to create and detect surface polaritons:

1. Nonplanar Interface Method

It involves changing the geometry of the interface in order to introduce new boundary conditions that will create a nonzero radiative component in the surface modes. As these modes now couple to the radiation field, they will induce some structures in a simple reflectivity experiment. Different theoretical models equivalent to the planar case have been proposed for spherical and cylindrical interfaces: they may be useful to account for the optical properties of molecular crystals with particular growing shapes as, for example, thin needles.

2. Grating Coupling and Surface Roughness Experiments

Another method for surface polariton excitation is based on the use of a rough interface, either periodic (grating coupling) or random (surface roughness experiments). A perfect periodic grating interface, with a period c on the order of magnitude of the wavelength and a groove depth small compared to the wavelength, will diffract an incident electromagnetic wave in either a bulk or a surface polariton mode, depending on the incidence and diffraction angles θ_{inc} and θ_{diff}. The surface components of the wave-vectors $\mathbf{k}_{\text{inc}}$ and $\mathbf{k}_{\text{diff}}$ of the incident and diffracted electromagnetic waves, satisfy with the lattice grating wave-vector $\mathbf{k}_g$ ($k_g = \pm n2\pi/c$ with n being the grating order), a relation of the Bragg type:

$$\mathbf{k}_{\text{inc}} \sin\theta_{\text{inc}} \pm \mathbf{k}_g = \mathbf{k}_{\text{diff}} \cdot \sin\theta_{\text{diff}} = \mathbf{k}_s$$

such as, for high values of the grating order n, $\mathbf{k}_s$ may be sufficiently large to fit the surface polariton wave-vector in the nonradiative region. Inverse processes of surface to bulk mode conversion, as well as intrasurface mode conversion may be explained on the same diffraction basis.[250] One of the striking consequences of such effects is the "Wood anomalies" in the efficiency curve of an optical grating, observed a long time ago but explained only recently on the basis of surface plasmon resonances.[117]

As it is possible to rule a periodic grating on a metallic surface or to deposit a metallic layer on a ruled substrat, this grating coupling method may be used to study surface plasmon dispersion, but with more difficulties exciton surface polaritons in organic solids; this is due to the important surface damages that ruling or etching may induce, or, to the non-monocrystalline state of a deposited layer. However, as a rough surface may be considered as a Fourier superposition of periodic waves, the above method may be extrapolated to more random interfaces, and it is possible to couple exciton surface polariton with light, by a "gentle" roughening of the crystal surface. Some experimental evidences of this possibility have been reported by Morris and Sceats[244] for the anthracene crystal. Although surface roughness experiments are not the best way to do reproducible dispersion studies on exciton surface polariton experiments, these authors pointed out the possibility of excitation and radiative deactivation of surface and bulk nonradiative modes (via surface defects), demonstrating the need of a high surface quality, even in a pure bulk experiment, to overcome coupling to such nonradiative modes.

3. *Prism Coupling*

The most commonly used method for excitation and detection of nonradiative surface modes is now the attenuated total reflection (ATR.) method, first introduced by Otto.[249] Using above the inactive medium (ε_i), a more refringent isotropic medium (generally a prism) with a dielectric constant $\varepsilon_p > \varepsilon_i$, by total reflection at the plane interface between these two media, this method creates an evanescent wave propagating along this interface with field amplitudes decaying exponentially normal to it. Provided the depth of the inactive medium (ε_i) is less than the penetration depth ($\sim \lambda$), this wave is able to excite the surface mode at the inactive–active medium interface (ε_i, $\varepsilon_a(\omega)$). At a given energy $\hbar\omega_0$, scanning the incidence angle θ of the photon at the (ε_p, ε_i) interface, above the critical angle θ_c (given by $\sin\theta_c = \sqrt{\varepsilon_i}/\sqrt{\varepsilon_p}$), one observes in the total reflection spectrum (R, θ), an attenuation band corresponding to the surface mode excitation and from which one can deduce one point of the surface polariton dispersion curve (ω, k). Changing the energy value of the incident light, and with different inactive

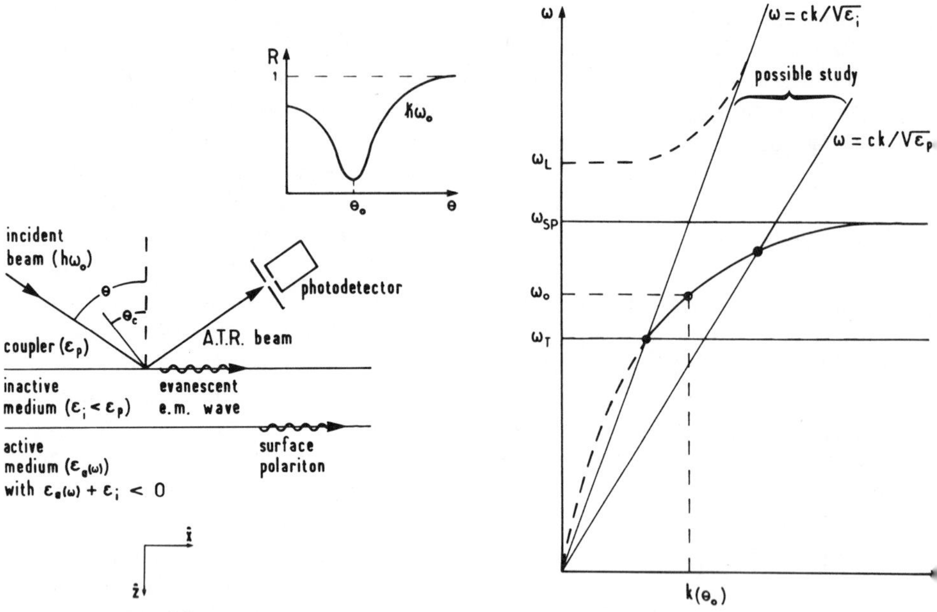

Fig. 64. Principle of surface polariton excitation and dispersion study by A.T.R.

medium depths, one can get point by point, the nonradiative domain of this dispersion curve between the two light lines $\omega = ck/\sqrt{\varepsilon_i}$ and $\omega = ck/\sqrt{\varepsilon_p}$.

Consequently, in order to get the largest accessible domain of the dispersion curve, one must use an inactive medium with ε_i as low as possible and to have the best excitation efficiency, one must work with an inactive medium as thin as possible, although for a too small depth the surface polariton energy may leak out through the prism in a radiative form (radiation coupling).[249]

Beside the experimental arrangement formerly introduced by Otto who used a well collimated beam (~ mm wide) with variable incidence and a precise control of the inactive medium depth,[249] Kretchman proposed to sharply focus the incident beam on a small point (~ μm wide) of the interface and to scan the detection angle into the totally reflected cone of light. The principal advantage of this last technique is the possibility to select a small region of the interface with suitable inactive medium gap without any precise control requirement.

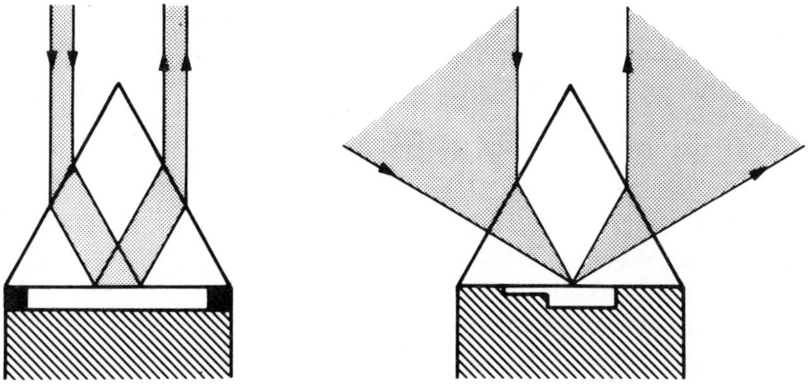

Fig. 65. Otto (*left*) and Kretchman (*right*) optical configurations for A.T.R. study.

4. *Particle Beams*

Finally, we mention that surface polariton modes may be excited with particles of large wave-vectors (as compared to the photon wave-vector) using, for example, electron energy loss techniques. Measuring the energy loss and the scattering angle of high or low energy electrons will allow information collection on the elementary surface excitations. Due to the probable damages that particle beams may induce on organic surfaces, this method seems apparently not well-suited in our case. However some interesting results have been reported for anthracene by Kunstreich and Otto with this method.[163]

C. Study of Exciton Surface Polaritons on Anthracene Crystal

Although anthracene is probably not the best choice to get well-defined and easily detectable exciton surface polariton effects in organic molecular crystals, the ATR. study done by Tomioka, Sceats, and Rice[231] on anthracene crystal is of fundamental interest since it links all the numerous information already obtained on the surface and on the bulk excitations of this material. Since that work is (to our knowledge) the first experimental study in the field of organic solids and to date the only one carried out at low temperatures (18 K), we report and discuss here some of its principal results in order to stimulate further insights and investigations on this subject.

1. *Experimental Conditions [See Fig. 1 in (230)]*

As anthracene is not a high quasi-metallic reflector at room temperature, it is susceptible to exhibit exciton surface polariton effect only at low temperatures, when exciton-phonon coupling becomes negligible relative to the exciton-photon interaction. Consequently, the ATR. experiment, already not very trivial at room temperature, must be realized under low temperature conditions where it becomes very difficult to control the interface quality as well as the inactive layer depth. The solution adopted by the authors was to use an IRG2 glass prism coupler ($\sqrt{\varepsilon_p}=1.95$ at 4000 Å) with its base coated with a PMMA layer ($\sqrt{\varepsilon_i}=1.51$ at 3900 Å) and anthracene sublimation (001) flakes in optical contact with this layer, the **b** crystal axis being oriented along the propagation direction $\hat{x}$. The optical arrangement chosen was of the Otto-type (see above), using two collimated beams (~ 1 mm in diameter) striking respectively the crystal covered and uncovered regions of the prism base. The lowest temperature reached in a cold finger type cryostat was about 18 K. The different thicknesses of the inactive PMMA layer (measured by interference fringes method) were 3,560 Å, 3,470 ± 30 Å, 1,000 Å, 970 ± 60 Å, and 0 Å (no PMMA film).

2. *Results Concerning Exciton Surface Polaritons. [See Figs. 6 and 8 in (231)]*

Varying the incidence angle θ on the IRG2 prism-PMMA interface below and above the critical value $\theta_c=50.75°$ the ATR. anthracene spectra present at low enough temperatures ($T<100$ K) structures of variable intensities and widths. They correspond to modes that generally span the radiative and non-radiative regions of the dispersion curve and which Sceats and co-workers interpret as the following.

1. The principal structure (well-defined below 100 K for all θ values) is assigned to the vibrationless surface polariton propagating along the **b** axis that corresponds to the pure electronic **b** polarized exciton mode. Consistently with this assignment, the dispersion curve found for this mode crosses the light line ($\omega=ck/\sqrt{\varepsilon_i}$) at 25097 ± 6 cm^{-1} (that is the **b** bulk exciton frequency found in the present work) and when extrapolated to the $\mathbf{k}\rightarrow\infty$ limit, gives a dipolar surface excitation energy limit $\hbar\omega_{SP}$ in agreement with the value reported by Morris and Sceats[244] from surface roughness experiments (25420 ± 10 cm^{-1}). The question as to whether this mode is *REAL* (that is to say, not k limited) is however left unanswered. The dephasing time and the group velocity of this mode, derived respectively from the linewidth and from the dispersion curve, may be combined to give a scattering length varying from 2 to 0.3 μm at 16 K. This relatively large value indicates the weak efficiency of phonon to scatter surface polaritons into bulk modes and, possibly, the weakness of the matter (excitonic) component of the modes.

2. On the high energy side of the precedent structure, the ATR. spectrum exhibits for $41°<\theta<70°$, a band less intense but still well-defined,

followed on its low energy side by a diffuse shoulder, and these are interpreted respectively as a single and a two particle surface polaritons involving the 391 cm^{-1} intramolecular vibration mode, the first one exhibiting, at the opposite of the second, a relatively large dispersion of about 60 cm^{-1}.

3. Finally, it has been observed, on the low energy side of the principal structure, that there are three other bands with very different dispersion behavior. Since these bands may be observed only for thin samples, they have been interpreted as "wave guided" surface polaritons modes analogous to those discussed in A.4 of this Section (VIII).

3. Dispersion of Site Shift Surface Excitons and Their Coupling to Surface Polaritons [See Fig. 2 and 3 in (230)]

Besides the surface polariton structures associated to bulk dipolar excitations and discussed above, the ATR. spectrum at 45 K (with a 3,650 Å PMMA layer) presents a diffuse band of weak intensity at an energy position increasing from $25{,}310 \pm 6$ cm^{-1} at 41° angle of incidence, to $25{,}335 \pm 6$ cm^{-1} at 55° angle of incidence. From sample to sample this structure is not always observed and may disappear with time. It is also present in an ATR. spectrum taken at 18 K (prism without PMMA layer) at the same energy positions (although the crystal interface was different) and without any increase in intensity and definition (although the temperature was lower).

Because of its energy position close to structure I observed in the normal reflectivity spectrum (see previous Section V), and to its unreproducible character (although to our sense it is not a good criterion to characterize surface excitation), this band is related to the **b** component for the site shift surface exciton resonance on the first monomolecular layer. However, the relatively large spatial dispersion (25 cm^{-1} for $41° < \theta < 55°$), observed from this band, is in apparent contradiction with this assignment since negligible dispersion is theoretically expected for the **b** polarized surface exciton. To lift this ambiguity, Sceats and co-workers emit either the idea of a possible change in the nature of the interference effect giving rise to the observed resonance, or that of a possible mixing between site shift surface exciton and surface polariton modes, which seems to us more realistic. Indeed, although there is a discrepancy in the incidence angle value for such a mixing, the theoretical model of energy dispersion for $\varepsilon_a(\omega)$, analogous to the one presented in Section V of this work, predicts a spatial dispersion in the domain where site shift surface excitons and surface polaritons are degenerate.

4. Discussion

Despite difficult low temperature experimental conditions, the successfull observations of exciton surface polaritons on anthracene crystal done by Tomioka, Sceats, and Rice, represent an important step towards the understanding of the organic molecular crystal photodynamics. The surface polaritons appear as a fundamental link for the energy exchange between the surface and the bulk since, on one hand, they probe within one wavelength depth the macroscopic surface region of the crystal and, on the other hand, they may couple to the site shift surface excitons (localized on the first monomolecular layers) and to the bulk excitons. Moreover, even in apparently pure radiative experiments these nonradiative modes may be excited via scattering processes (phonons, surface irregularities, etc.) and will contribute to the energy relaxation or to the re-emission of light. It is an important reason for continuing their study in easily modelized molecular crystals such as anthracene, even if in these materials their effects are more difficult to observe and to interpret. The information derived from the observations of Tomioka and co-workers may be certainly improved by new ATR. experiments:

1. *At lowest temperatures.* This would reduce the damping of surface polaritons, increase their spectral definition, and, consequently, facilitate their interpretation.
2. *Without a* PMMA *layer in the Kretschmann configuration.* As the frequency limit ω_{SP} of the real surface polaritons depends on the dielectric constant ε_i of the inactive medium, lowering its value near unity (vacuum gap) would increase ω_{SP} and consequently extend the possible domain of study of the dispersion. It is particularly important to know whether the detected surface polaritons span the entire Brillouin zone and to get information on the exciton band structure. Providing there are some gaps between the prism and the crystal in optical contact with it, a focused beam would allow to select good quality interface regions and variable gap depths. The adaptation of the Kretschmann configuration at low temperatures is however not trivial due to the aperture limitation of the cryostat windows.

Concerning the dispersion of site shift surface excitons, the band observed by Sceats and co-workers remains unresolved and of very low intensity. This low spectral definition is probably due to the relatively high temperature ($T = 45$ K) and also to the poor surface quality (since the band for $T = 18$ K is not better resolved) certainly induced by the different thermal contractions of the interface materials. New experiments, done at much lower temperatures, would certainly permit to bring more conclusive informations on the relation of this band to site shift surface exciton resonances and, if this is verified, on their effective dispersion. Indeed, calculations reported in (127) predict a large intensity effect as well as an energy variation for the site shift surface exciton resonances in the ATR anthracene spectrum.

The question, raised by Sceats and co-workers, of a possible mixing of site shift surface excitons with surface polaritons, is of prime interest for the understanding of these two types of surface excitations. An experimental improvement of the spectral definition of the band as well as a better account for the anthracene anisotropy in the theoretical model of dispersion, would certainly bring more elements to solve this problem.

Finally, at the light of the observations done by Sceats, Tomioka, and Rice on surface polaritons and of those described in the previous sections on site shift surface excitons, we conclude by emphasizing the following:

1. The determinant role played by surface excitations in specific surface experiments (photo-conductivity, charge generation, photochemistry, etc.) as well as in "bulk" experiments (fluorescence, Raman scattering, nonlinear spectroscopy, etc.) where their importance should not be though less obvious.
2. The ability of these nondestructive optical experiments to detect sensitively surface effects with an application range going from the first crystal layers to the bulk, via the macroscopic surface region.

CONCLUSION AND PROSPECTS

In this work we have contributed to a better theoretical and experimental description of the photodynamic properties of organic molecular crystals close to their resonances by examining conjointly the influence of matter–radiation coupling and the finite bounds of the crystal. Compared to previous spectroscopic studies in this area, which sought above all to describe the energy of the crystal, supposedly isolated and perfectly periodic, our approach, based on the determination of the modes of electromagnetic propagation in a real crystal, is relatively novel. It is justified by the many difficulties which appeared in interpreting the data of most systems, due principally to oversimplification of the model crystal. Significantly, an archetype of the organic molecule crystal (such as anthracene), which has polarized most work in this field, is still "mysterious." By studying the region of its first singlet ($S_0 \rightarrow S_1^*$) transition, we wanted to solve some of the mystery and, above all, to lay new foundations for the study of the optical response of this type of crystal.

We have treated these dual problems (nonnegligible crystal–field coupling and finite crystal) by an analysis of the optical reflectivity since this is stronger for stronger transitions and, in addition, contains the complex optical constants of the surface and the bulk. In fact, we have taken advantage of a strong bulk effect to exhibit a weak surface effect. In our theoretical approach to the problem as well as in the experimental study, we found two principal components (volume and surface) which we have linked in a model of the reflectivity of a finite crystal, so enabling us to understand better the structure of the real crystal. We would like finally to underline two important, complementary aspects of our work:

The first aspect concerns the consideration of the complete crystal–field system.

Coupling between "virtually" excited eigenmodes of the isolated crystal (excitons) and the incident photon creates a mixed (polariton) state whose character varies continuously with energy between a pure photon state (at high and low energies) and a pure exciton state (at resonance). This polariton concept combines the complementary aspects of collective excitation of the crystal and electromagnetic propagation, characteristic of the interaction between an ordered system and light. Thus Davydov's original representation of the crystal–field system as exciton OR photon should be replaced by the exciton AND photon representation, which because of its continuity gives a much better (zero-order) description of the optical properties of a crystal.

Clearly, the difference between these two representations is merely conceptual if we are talking about weak transitions with small oscillator strengths. But the mixed state representation must be used in the case of medium to strong transitions like those we have examined above. One may thus explain the quasi-metallic reflection associated with these transitions. Obviously, however, the optical properties of the crystal (propagation, degradation, reemission, and scattering of the incident photon) cannot be described by the simple process photon ↔ polariton, but by polariton scattering on other elementary excitations (phonons, surface modes, other polaritons) created simultaneously.

In our discussion of the theory of the polariton in infinite crystals (Section III) we tried to give as much meaning as possible to these concepts, which are only just being incorporated in the study of molecular crystals. After a general introduction of this concept, illustrated at the end by the simple example of a set of harmonic oscillators coupled to the field, we gave semiclassical and quantum treatments of matter–field coupling and the dispersion of anthracene. Taking advantage of these complementary representations of the mixed state, we were able to check their equivalence through the identical forms they predict for the dispersion $\mathbf{k}(\omega)$. Finally, we applied the dispersion relation to find the real and imaginary parts of the "optical constants" of anthracene in several models with one or more transitions.

The second aspect concerns the experimental evidence of collective states of the surface and first few planes of the crystal.

Starting from the rather confusing situation existing prior to our work, we have tried to exhibit surface excitons by the fullest possible experimental study and to provide the experimental information which had always been lacking in order to take the theory seriously.

This led us to develop careful experimental techniques for measuring almost simultaneously the low temperature (1.7 K), high resolution (0.3 cm^{-1}), and reflectivity and fluorescence spectra and to invent new experimental techniques (like condensing a gas on the crystal surface) for the preparation and study of our samples.

We thus obtained accurate information on the reflectivity and fluorescence spectra of the (001) face of anthracene and were able to confirm, in both types of experiments, the presence of "anomalous" structures already observed. By irreversible (photochemical) and reversible (gas films) perturbations of the surface we were able to differentiate these structures from parts of the spectra due to the bulk and to link them with surface and subsurface excited states. Our interpretation of these structures in terms of surface and subsurface excitons appears at present to be the only one consistent with all the experimental data on them.

Thus, for the right incidence of the wave vector on a finite crystal, the destabalization (difference in the Van der Waals term D) of surface molecules may be large at the same time as the probability of these molecules exchanging their excitation with those of the bulk (interplane $I_{(k)}^{(l>0)}$ exchange terms) is very small. One may then find a surface mode like that which we have described in anthracene, at 200 cm^{-1} or so above the bulk state, which consequently absorbs or emits independently from the bulk. Study of these modes is worthwhile because of the information it can give on the electronic and structural properties of the as yet little known surface regions of organic compounds and on their environmental perturbation.

Finally we have combined these two concepts, of the polariton and of the surface exciton, in a model of the reflectivity of a finite crystal. Introducing relaxation processes for these modes, we have been able to simulate satisfactorily our experimental results. While this description of a real crystal is still relatively naive, we believe that through constant comparison with experiment it will improve itself and thus lead to a greater understanding of the structure and optical properties of these systems.

In conclusion, while we have attained our main goals in shedding some light on the role that collective surface excitations and polaritons play in the optical properties of real molecular crystals, it is obvious that the subject is far from exhausted. We should like to terminate with a brief description of some possible extensions of this work in order, on the one hand, to verify and complete some hypothesis we have made and, on the other, to take advantage of our results. There are two complementary surface and bulk research axes, sketched on Fig. 66, which are linked together by the funda-

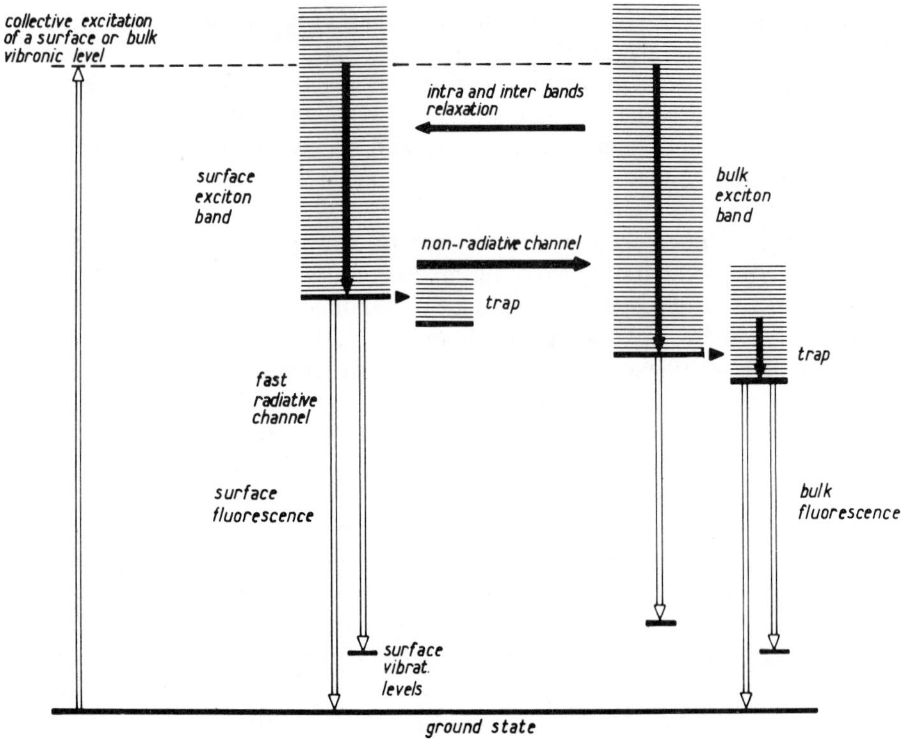

Fig. 66. Schema of the transitions involved in the radiative and nonradiative decay of surface excitons.

mental and quasi unexplored subject on surface-volume coherent and incoherent energy transfer:

Study of collective surface excitations and of their radiative and nonradiative decay

Our results offer new possibilities for the exploration of the superficial (monolayer) region of organic molecular crystals through the investigation of excitation and emission surfaces modes. There being no means of carrying a nondestructive microscopic analysis of organic surfaces, we know very little at present about their molecular arrangement. Therefore, it would be important to test the concept of surface and sub-surface states, since information on the excitation energies and on the coupling of the first few layers, would add to our knowledge on the energy dynamics on the superficial region of the crystal. To do so, it will be necessary to get a better definition of the collective surface excitations and in particular more information on:

Vibronic and two particle surface states (through U.V. reflectivity, fluorescence excitation spectra, etc.).
Inter and intra molecular vibrational surface modes (study of reflectivity, fluorescence, resonance Raman scattering of coated, and uncoated crystal surface).
Dispersion of the surface exciton band and coupling to exciton surface polaritons (by low temperature attenuated total reflection experiment).
Collective nature of the surface states (study of mixed crystals).

Moreover, as shown in Fig. 66, once excited, these surface states are effectively coupled to a one-dimensional photon continuum and present a strong radiative instability ($\tau \sim 10^{-12}$ second). This surface fluorescence is one of the rare observed cases of collective emission in molecular crystals. Therefore, it is of prime interest to extend the study of the emissive properties of the surface (life-time, directivity, excitation spectra, etc.) and to observe how these emissions depend on surface coherence (temperature, trap and impurity inclusion, mixed crystals, etc.).

Beside the radiative channel, surface excitons may decay into the bulk coherently in the polariton continuum or incoherently through phonon scattering into localized states (traps, impurities, etc.). Our understanding of the optical behavior of molecular crystals near strong transitions is of course highly dependent on a study of these fundamental surface relaxation mechanisms. One way to explore these nonradiative channels may be to monitor the transmission and the bulk fluorescence in pure and mixed crystals at different concentrations and for different host–guest energy gaps.

Study of polariton modes in finite crystals and of their coupling to other elementary excitations.

In order to increase our understanding of the optical properties of real molecular crystals near their resonances, one needs to develop, parallel to the study of their surface, an analysis of the polariton modes which can exist in finite crystal and their scattering. While we have been able to glimpse the mixed nature (exciton–photon) of the eigenmodes of a crystal–radiation system, our simple description does not give any real understanding of absorption and emission processes in the crystal. It is necessary, therefore, to use this zero-order basis (polaritons) to consider possible couplings with other elementary modes of the crystal. As we saw in our interpretation, polariton–phonon coupling plays a determining role in the absorption and emission mechanism in the crystal. We envisage parallel theoretical study of this coupling with direct (absorption deduced from reflectivity and transmission) and indirect (numerical analysis of reflectivity spectra) experimental approaches in order to better understand these relaxation channels. A quantitative study (lifetime, temperature dependence) of the "intrinsic" bulk fluorescence certainly will also bring new information on polariton scattering.

On the other hand, low temperature studies of very thin and "perfect" crystals might reveal the existence of polaritons in nonabsorbing coherent quantum states.[209,220,221]

Finally, the effects of polaritons in the photodynamics of molecular crystals may be also evaluated through two-photon absorption experiments and other nonlinear spectroscopy techniques. Thus, it is possible to create polariton modes directly into the bulk and to have access to parity forbidden states.[243] There is no doubt that this new field will soon bring drastic complementary information on this subject.

APPENDIX

TABLE A.1
Data Used for Anthracene[a]

Transitions	Polarization	ω_m (cm^{-1})	F_m^2	d_m (A)
$S_0 \rightarrow S_1^*(0,0)$	M	25,614	0.324	0.61
(0, 1)	M	27,014	0.316	0.61
(0,2)	M	28,414	0.218	0.61
(0,3)	M	29,814	0.092	0.61
(0,4)	M	31,214	0.050	0.61
$S_0 \rightarrow S_2^*$	L	39,000	1	1.87

[a]Constants:

F_m^2 = Franck–Condon factor, d_m; transition moment

$f_m^2 = 2m\omega_m F_m^2 d_m^2/\hbar$; oscillator strength of transition m (m in kg, d_m in m, ω_m in rad·s^{-1}

e = electronic charge	$= 1.602 \times 10^{-19}$ C
m = electronic mass	$= 9.110 \times 10^{-31}$ kg
c = speed of light	$= 2.998 \times 10^{8}$ m·s^{-1}
ε_0 = permittivity of empty space	$= 8.854 \times 10^{-12}$ Å^2·s^4·kg^{-1} m^{-3}
h = Planck's constant	$= 6.626 \times 10^{-34}$ J·s
$\hbar = h/2\pi$	$= 1.055 \times 10^{-34}$ J·s
$e^2/(\varepsilon_0 \cdot m)$	$= 3.182 \times 10^{3}$ s^{-2}·m^{-3}
V_c = volume of the unit cell	$= 474.3 \times 10^{-30}$ m^3 (300 K)
of anthracene	$= 457.9 \times 10^{-30}$ m^3 (95 K)
ω_p = plasma frequency $\lvert e^2/\varepsilon_0 m V_c \rvert^{1/2}$	$= 2.590 \times 10^{15}$ rad·s^{-1}
	$= 1.375 \times 10^{4}$ cm^{-1}

TABLE A.2

Dipole Sums (in $cm^{-1}/\text{Å}^2$) Calculated for Anthracene by Ewald's Method:[154]

$|\mathbf{k}| = 0$; $\hat{k} \perp$ face (001)

	Equivalent molecules $\alpha = 1, \beta = 1$			Nonequivalent molecules $\alpha = 1, \beta = 2$		
Polarization of transitions r and s	$r = M$ $s = M$	$r = M$ $s = L$	$r = L$ $s = L$	$r = M$ $s = M$	$r = M$ $s = L$	$r = L$ $s = L$
$t_{\alpha r \beta s}$	−2076	321	−168	664	984	394
$(\hat{d}_{\alpha r} \cdot \hat{k})(\hat{k} \cdot \hat{d}_{\beta s})$	296	826	2278	296	826	2278
$\hat{d}_{\alpha r} \cdot \hat{d}_{\beta s}$	3075	0	3075	−1844	672	2983

[a] The retarded dipole sums [defined in Eqs. (54), (58), and (65)] appearing in the dispersion equation of the polariton (67), are calculated from the table above by the relations:

$$\underset{\text{(table)}}{\phi^{\pm}_{rs}} = (t_{1r1s} \pm t_{1r2s}) + \frac{\hat{k}^2\left[(\hat{d}_{1r}\cdot\hat{k})(\hat{k}\cdot\hat{d}_{1s}) \pm (\hat{d}_{1r}\cdot\hat{k})(\hat{k}\cdot\hat{d}_{2s})\right] - (\omega/c)^2\left[(\hat{d}_{1r}\cdot\hat{d}_{1s}) - (\hat{d}_{1r}\cdot\hat{d}_{2s})\right]}{k^2 - (\omega/c)^2}$$

$$\phi^{\pm}_{rs} = \frac{V_c}{4\pi}\cdot\frac{4\pi\varepsilon_0}{e^2}\cdot\frac{1.6\times10^{-19}\times10^{20}}{8067}\cdot\underset{(\text{in cm}^{-1}/\text{Å}^2)}{\phi^{\pm}_{rs(\text{table})}} = 3144.10^{-4}\cdot\underset{\text{(table)}}{\phi^{\pm}_{rs}}$$

References

1. J. Becquerel, *J. Phys. Radium* **4**, 328 (1907).
2. J. Frenkel, *Phys. Rev.* **37**, 17 (1931).
3. R. Peierls, *Ann. Physik.* **13**, 905 (1932).
4. G. H. Wannier, *Phys. Rev.* **52**, 191 (1937).
5. P. Prinsheim and A. Kronenberger, *Z. Physik* **40**, 75 (1926).
6. I. V. Obreimov, *Zh. Russ. Fiz. Khim. Obshch.* **59**, 548 (1927).
7. I. V. Obreimov and A. F. Prikhot'ko, *Physik. Z. Sowjetunion* **1**, 203 (1932).
8. I. V. Obreimov and A. F. Prikhot'ko, *Physik. Z. Sowjetunion* **9**, 34 (1936).
9. I. V. Obreimov and A. F. Prikhot'ko, *Physik. Z. Sowjetunion* **9**, 48 (1936).
10. A. F. Prikhot'ko, *J. Phys. USSR*, **8**, 257, (1944).
11. A. S. Davydov, *Zh. E.T.F.* **18**, 210 (1948).
12. A. S. Davydov, *Theory of Molecular Excitons*, McGraw-Hill, New York, 1962.
13. R. S. Knox, *Theory of Excitons*, Academic Press, New York, 1963.
14. R. M. Hochstrasser, *Molecular Aspect of Symmetry*, Benjamin, New York, 1966.

15. D. P. Craig and S. M. Walmsley, *Excitons in Molecular Crystals*, Benjamin, New York, 1968.
16. A. S. Davydov, *Theory of Molecular Excitons*, Plenum Press, New York, 1971.
17. D. S. McClure, *Solid State Physics* **8**, 1 (1958).
18. H. C. Wolf, *Solid State Physics* **9**, 1 (1959).
19. J. Tanaka, *Suppl. Progr. Theo. Phys.* **12**, 183 (1959).
20. R. M. Hochstrasser, *Rev. Mod. Phys.* **34**, 531 (1962).
21. A. S. Davydov, *Soviet Phys. Uspekhi*, **82**, 145 (1964).
22. T. N. Misra, *Rev. Pure Appl. Chem.* **15**, 39 (1965).
23. D. P. Craig and S. H. Walmsley, in *Physics and Chemistry of the Organic Solid State*, Vol. I, 1963, 585.
24. S. A. Rice and J. Jortner, in *Physics and Chemistry of the Organic Solid State*, Vol. III, 1967, p. 199.
25. W. H. Wright, *Chem. Rev.* **67**, 581 (1967).
26. G. W. Robinson, *Ann. Rev. of Phys. Chem.* **21**, 429 (1970).
27. M. R. Philpott, *Adv. Chem. Phys.* **23**, 226 (1973).
28. V. L. Broude and E. I. Rashba, *Rev. Pure Appl. Chem.* **37**, 21 (1974).
29. C. Aslangul, J. P. Lemaistre, and Ph. Kottis, in *Localization and Delocalization in Quantum Chemistry*, Vol. II, Reidel, 1976, p. 209.
30. J. P. Lemaistre, C. Aslangul, and Ph. Kottis, in *Localization and Delocalization in Quantum Chemistry*, Vol. II, Reidel, 1976, p. 239.
31. C. Aslangul and Ph. Kottis, *Phys. Rev.* **10**, 4364 (1974).
32. J. P. Lemaistre and Ph. Kottis, *J. Chem. Phys.* **68**, 2730 (1978).
33. M. R. Philpott and J. M. Turlet, *J. Chem. Phys.* **64**, 3852 (1976).
34. I. Tamm, *Phys. Z. Sowjetunion* **1**, 733 (1932).
35. G. Borstel, H. I. Falge, and A. Otto, *Springer Tracts Mod. Phys.* **74**, 107 (1974).
36. F. W. de Wette and G. E. Schacher, *Phys. Rev.* **137**, A78 (1965).
37. M. R. Philpott, *J. Chem. Phys.* **58**, 588 (1973).
38. A. S. Davydov and E. F. Sheka, *Phys. Stat. Sol.* **11**, 877 (1965).
39. M. R. Philpott, *J. Chem. Phys.* **54**, 111 (1971).
40. A. Honma, *J. Phys. Soc. Japan* **42**, 1129 (1977).
41. V. I. Sugakov, *Ukr. Fiz. Zh.* **14**, 1425 (1969).
42. F. Abeles, *Optical Properties of Solids*, North-Holland Publs., Amsterdam, 1972.
43. B. O. Seraphin, *Optical Properties of Solids: New Developments*, North-Holland Publs., Amsterdam, 1976.
44. V. I. Sugakov, *Ukr. Fiz. Zh.* **15**, 2060 (1970).
45. M. R. Philpott, *J. Chem. Phys.* **60**, 1410 (1974).
46. J. P. Byrne and I. G. Ross, *Can. J. Chem.* **43**, 3253 (1965).
47. R. M. Macnab and K. Sauer, *J. Chem. Phys.* **53**, 2805 (1970).
48. M. S. Brodin and S. V. Marisova, *Opt. Spectrosc.* **10**, 242 (1960).
49. H. L. Jetter and H. C. Wolf, *Phys. Stat. Sol.* **22**, K39 (1967).
50. V. L. Broude, D. S. Pakhomova, and A. F. Prikhot'ko, *Opt. Spectrosc.* **2**, 323 (1954).
51. D. P. Craig and P. C. Hobbins, *J. Chem. Soc.* p. 2309 (1955).

52. J. Sidman, *Phys. Rev.* **102**, 96 (1956).
53. A. Bree and L. E. Lyons, *J. Chem. Soc.* p. 2662 (1956).
54. J. Ferguson and W. G. Schneider, *J. Chem. Phys.* **28**, 761, (1958).
55. A. R. Lacey and L. E. Lyons, *J. Chem. Soc.* p. 5393 (1964).
56. M. R. Philpott and J. W. Lee, *J. Chem. Phys.* **58**, 595 (1973).
57. D. P. Craig, *J. Chem. Soc.* p. 2302 (1955).
58. D. P. Craig and J. R. Walsh, *J. Chem. Soc.* p. 1613 (1958).
59. A. R. Lacey and L. E. Lyons, *Proc. Chem. Soc.* p. 414 (1960).
60. R. Silbey, J. Jortner and S. A. Rice, *J. Chem. Phys.* **42**, 1515 (1965).
61. H. C. Wolf, *Z. Naturforsch.* **13a**, 414 (1958).
62. T. A. Claxton, D. P. Craig, and T. Thirunamachandran, *J. Chem. Phys.* **35**, 1525, (1961).
63. G. D. Mahan, *J. Chem. Phys.* **41**, 2930 (1964).
64. M. R. Philpott, *J. Chem. Phys.* **50**, 5117 (1969).
65. J. Schroeder and R. Silbey, *J. Chem. Phys.* **55**, 5418 (1971).
66. M. Sceats and S. A. Rice, *Chem. Phys. Lett.* **44**, 425 (1976).
67. M. D. Borisov, *Izd. Akad. Nauk. Ukr. SSR.* **4**, 102 (1953).
68. B. G. Anex and W. T. Simpson, *Rev. Mod. Phys.* **32**, 466 (1960).
69. L. B. Clark, *J. Chem. Phys.* **51**, 5719 (1969).
70. L. B. Clark, *J. Chem. Phys.* **53**, 4092 (1970).
71. L. B. Clark, and M. R. Philpott, *J. Chem. Phys.* **53**, 3790 (1970).
72. S. V. Marisova, *Opt. Spectrosc.* **22**, 310 (1966).
73. S. V. Marisova, *Ukr. Fiz. Zh.* **12**, 521 (1967).
74. M. S. Brodin, S. V. Marisova and S. A. Shturkhetskaya, *Ukr. Fiz. Zd.* **13**, 249 (1968).
75. G. C. Morris, S. A. Rice and A. E. Martin, *J. Chem. Phys.* **52**, 5149 (1970).
76. G. C. Morris and M. G. Sceats, *Chem. Phys.* **3**, 164 (1974).
77. G. C. Morris and M. G. Sceats, *Chem. Phys.* **1**, 259 (1973).
78. M. R. Philpott, *J. Chem. Phys.* **54**, 2120 (1971).
79. M. R. Philpott, *Chem. Phys. Letters* **24**, 418 (1974).
80. P. W. Alexander, A. R. Lacey and L. E. Lyons, *J. Chem. Phys.* **34**, 2200 (1961).
81. M. T. Shpak and N. I. Sheremet, *Opt. Spectrosc.* **17**, 374 (1964).
82. W. Helfrich and F. R. Lipsett, *J. Chem. Phys.* **43**, 4368 (1965).
83. L. E. Lyons and L. J. Warren, *Aust. J. Chem.* **25**, 1411 (1972).
84. N. J. Bridge and D. Vincent, *J. Chem. Soc.* **68**, 1522 (1972).
85. J. O. Williams, B. P. Clarke, J. M. Thomas and M. J. Shaw, *Chem. Phys. Lett.* **38**, 41 (1976).
86. E. Glockner, Diplomarbeit, Univ. of Stuttgart (1969).
87. E. Glockner and H. C. Wolf, *Z. Naturforsch.* **24a**, 943 (1969).
88. M. Susuki, T. Yokoyama and M. Ito, *Spectrochim. Acta* **24A**, 1091 (1968).
89. J. B. Birks, *Mol. Cryst. Liq. Cryst.* **28**, 117 (1974).
90. M. S. Brodin, M. A. Dudinskii, S. V. Marisova and E. N. Myasnikov, *Phys. Stat. Sol.* **b74**, 453 (1976).
91. M. D. Galanin and Sh. D. Khan-Magometova, *Proc. 1978 Int. Conf. on Luminescence, Paris*, published in *J. Luminescence*.

92. R. J. Bateman, R. R. Chance and J. F. Hornig, *Chem. Phys.* **4**, 402 (1974).
93. M. D. Galanin, Sh. D. Khan-Magometova, Z. A. Chizhikova, M. I. Demchuk and A. F. Chernyavskii, *J. Luminescence* **9**, 459 (1975).
94. A. G. Bale, N. J. Bridge and D. B. Smith, *Chem. Phys. Lett.* **42**, 166 (1976).
95. S. I. Pekar, *Zh. Eksp. Theor. Fiz* **33**, 1022 (1957).
96. A. S. Selivanenko, *J. Exptl. Teor. Phys. (USSR)* **32**, 75 (1957).
97. S. V. Marisova, *Opt. Spectrosc.* **22**, 310 (1966).
98. S. V. Marisova, *Ukr. Fiz. Zh.* **12**, 521 (1967).
99. M. S. Brodin, S. V. Marisova and S. A. Shturkhetskaya, *Ukr. Fiz. Zh.* **13**, 353 (1968).
100. E. Glockner, Thesis, Univ. of Stuttgart (1974).
101. E. Glockner and H. C. Wolf, *Chem. Phys. Lett.* **27**, 161 (1974).
102. V. I. Sugakov and V. N. Tovstenko, *Ukr. Fiz. Zh.* **18**, 1495 (1973).
103. M. S. Brodin, M. A. Dudinskii, and S. V. Marisova, *Opt. Spectrosc.* **31**, 401 (1971).
104. M. S. Brodin, M. A. Dudinskii, and S. V. Marisova, *Izv. Akad. Nauk. SSRR, Ser. Fiz.* **36**, 1047 (1972).
105. M. S. Brodin, M. A. Dudinskii, and S. V. Marisova, *Opt. Spectrosc.* **34**, 651 (1973).
106. S. A. Rice, G. C. Morris, and W. L. Greer, *J. Chem. Phys.* **52**, 4279 (1970).
107. G. C. Morris, S. A. Rice and A. E. Martin, *J. Chem. Phys.* **52**, 5149 (1970).
108. M. G. Sceats, Thesis, Univ. of St Lucia, Queensland, Australia (1973).
109. G. C. Morris and M. G. Sceats, *Chem. Phys.* **1**, 376 (1973).
110. G. C. Morris and M. G. Sceats, *Chem. Phys.* **3**, 164 (1974).
111. D. Pines, *Elementary Excitations in Solids*, Benjamin, New York 1963.
112. E. Burstein, in *Polaritons*, Pergamon Press, Elmsford, N.Y., 1974, p. 1.
113. J. J. Hopfield, *Phys. Rev.* **112**, 1555 (1958).
114. C. H. Henry and J. J. Hopfield, *Phys. Rev. Lett.* **15**, 964 (1965).
115. K. Huang, *Proc. Roy. Soc.* (London) **A208**, 308 (1951).
116. M. Born and K. Huang, *Dynamical Theory of Crystal Lattices*, Oxford Univ. Press, New York, 1956, p. 89.
117. U. Fano, *Phys. Rev.* **103**, 1202 (1956).
118. S. I. Pekar, *Zh. Eksp. Teor. Fiz.* **33**, 1022 (1957); *Soviet Phys.—JETP* **6**, 785 (1958).
119. V. M. Agranovich, *Zh. Eksp. Teor. Fiz* **37**, 340 (1959); *Soviet Physic—JETP* **10**, 307 (1960).
120. R. Claus, L. Merten, and J. Brandmüller, *Springer Tracts Mod. Phys.* **75**, (1975).
121. E. Burstein, *Atomic Structure and Properties of Solids*, F. I. S. E.Fermi, No. 52, Academic Press, New York, 1972.
122. R. S. Knox, *Theory of Excitons*, Academic Press, New York, 1963, p. 103.
123. J. J. Hopfield, in *Quantum Optics*, F. I. S. E.Fermi, No. 42, Academic Press, New York, 1969, p. 340.
124. M. R. Philpott, *Adv. Chem. Phys.* **23**, Academic Press, New York, 1973, p. 227.
125. V. M. Agranovich in *Optical Properties of Solids*, North-Holland Publs., Amsterdam, 1972, p. 315.
126. G. Borstel, H. J. Falge, and A. Otto, *Springer Tracts Mod. Phys.* **74**, 107 (1974).
127. M. R. Philpott, in *Topics in Surface Chemistry*, Plenum Press, New York, 1978, p. 329.

128. H. Raether, *Phys. Thin Films* **9**, 145 (1977).
129. G. D. Mahan, in *Elementary Excitations in Solids, Molecules and Atoms*, Pt. B, Plenum Press, New York, 1973, p. 93.
130. K. L. Kliever and R. Fuchs, *Adv. Chem. Phys.* **27**, 355 (1974).
131. D. L. Mills and E. Burstein, *Rep. Progr. Phys.* **37**, 817 (1974).
132. E. Burstein, in *Dynamical Process in Solid State Optics*, Benjamin, New York, 1967.
133. E. Burstein and F. de Martini, *Polaritons*, Pergamon Press, Elmsford, N.Y., 1974.
134. J. Frenkel, *Phys. Rev.* **37**, 1276 (1931).
135. W. Heitler, *Quantum Theory of Radiation*, Oxford Univ. Press, New York, 1954.
136. W. M. Agranovich and V. L. Ginzburg, *Spatial Dispersion in Crystal Optics and the Theory of Excitions*, Wiley (Interscience), New York, 1966, p. 242.
137. C. Kittel, *Introduction à la physique de l'état solide*, Dunod, Paris, 1972, p. 454.
138. S. A. Rice and J. Jortner, in *Physics and Chemistry of the Organic Solid State*, Vol. III, New York, (Interscience), 1967, p. 199.
139. A. S. Davydov and E. F. Sheka, *Phys. Stat. Sol.* **11**, 877 (1965).
140. M. R. Philpott, *J. Chem. Phys.* **54**, 111 (1971).
141. R. H. Lyddane, R. G. Sachs, and E. Teller, *Phys. Rev.* **59**, 673 (1941).
142. L. Merton, *Phys. Stat. Solid.* **30**, 449 (1968).
143. R. Claus, in *Lattice Dynamics and Intermolecular Forces*, F. I. S. E.Fermi, No. 55, Academic Press, New York, 1975.
144. E. E. Koch, A. Otto, and K. L. Kliever, *Chem. Phys.* **3**, 362 (1974).
145. E. E. Koch and A. Otto, *Chem. Phys.* **3**, 370 (1974).
146. L. B. Clark and M. R. Philpott, *J. Chem. Phys.* **53**, 3790 (1970).
147. G. C. Morris and M. G. Sceats, *Chem. Phys.* **1**, 259 (1973).
148. A. T. Amos, *Mol. Phys.* **6**, 393 (1963).
149. M. A. Ball and A. D. McLachlan, *Proc. R. Soc.* (London) **A282**, 433 (1964).
150. G. D. Mahan, *J. Chem. Phys.* **41**, 2930 (1964).
151. G. D. Mahan, *J. Chem. Phys.* **43**, 1569 (1965).
152. R. Silbey, J. Jortner, and S. A. Rice, *J. Chem. Phys.* **42**, 1515 (1965).
153. M. R. Philpott, *J. Chem. Phys.* **50**, 5117 (1969).
154. M. R. Philpott, *J. Chem. Phys.* **50**, 3925 (1969).
155. M. R. Philpott, *J. Chem. Phys.* **52**, 5842 (1970).
156. M. R. Philpott, *J. Chem. Phys.* **54**, 2120 (1971).
157. M. R. Philpott and J. W. Lee, *J. Chem. Phys.* **58**, 595 (1973).
158. M. R. Philpott, *J. Chem. Phys.* **58**, 588 (1973).
159. P. P. Ewald, *Ann. Physik* (Leipzig) **49**, 117 (1916).
160. M. Born and E. Wolf, *Principles of Optics*, Pergamon Press, Elmsford, N.Y., 1959, p. 79.
161. D. Fox and Sh. Yatsiv, *Phys. Rev.* **108**, 938 (1957).
162. J. C. Slater, *Quantum Theory of Molecules and Solids*, Vol. 3, McGraw-Hill, New York, 1967, p. 102.
163. S. Kuntreich and A. Otto, *Chem. Phys.* **3**, 384 (1974).
164. A. S. Davydov, *Theory of Molecular Excitons*, Plenum Press, New York, 1971.

165. S. V. Tyablikov, *Methods in the Quantum Theory of Magnetism*, Plenum Press, New York, 1967.

166. J. Gernet, Thesis, Univ. of Bordeaux (1977).

167. V. C. Sinclair, J. M. Robertson and A. McL. Mathieson, *Acta Cryst.* **3**, 251 (1950).

168. J. M. Robertson, *Rev. Mod. Phys.* **30**, 155 (1958), and references therein.

169. R. Mason, *Acta Cryst.* **17**, 547 (1964).

170. R. M. Hochstrasser, *Molecular Aspects of Symmetry*, Benjamin, New York, 1966, p. 301.

171. D. P. Craig and S. H. Walmsley, *Excitons in Molecular Crystals* Benjamin, New York, 1968, p. 38.

172. P. M. Robinson and H. G. Scott, *J. Cryst. Growth* **1**, 187 (1967).

173. P. M. Robinson and H. G. Scott, *Phys. Stat. Solid.* **20**, 461 (1967).

174. P. M. Robinson and H. G. Scott, *Mol. Cryst. Liq. Cryst.* **11**, 13 (1970).

175. J. O. Williams and J. M. Thomas, *Trans. Farad. Soc.* **63**, 1720 (1967).

176. J. O. Williams and J. M. Thomas, *Mol. Cryst. Liq. Cryst.* **16**, 223 (1972).

177. G. J. Sloan, J. M. Thomas and J. O. Williams, *Mol. Cryst. Liq. Cryst.* **30**, 167 (1975).

178. W. G. Pfann, *Zone Melting*, Wiley, New York, (1958).

179. E. F. G. Herington, *Zone Melting of Organic Compounds*, Wiley, New York, 1963.

180. G. J. Sloan, *Physics and Chemistry of the Organic Solid State*, Vol. I, Wiley (Interscience), New York, 1963, p. 179.

181. G. J. Sloan, *Mol. Cryst.* **1**, 161 (1966).

182. Y. Lupien and D. F. Williams, *Mol. Cryst.* **5**, 1 (1968).

183. J. N. Sherwood, *Purification of Inorganic and Organic Materials*, Dekker, New York, 1969, p. 157.

184. J. M. Thomas and J. O. Williams, *Prog. Solid. State Chem.* **6**, 119 (1971).

185. J. M. Thomas, E. L. Evans and J. O. Williams, *Proc. Roy. Soc.* (London) **A331**, 417 (1972).

186. M. O'Donnel, *Nature* **218**, 460 (1968).

187. D. Donati, G. G. T. Guarini, and P. Sarti-Fantoni, *Mol. Cryst. Liq. Cryst.* **21**, 289 (1973).

188. S. E. Morsi and J. O. Williams, *Mol. Cryst. Liq. Cryst.* **39**, 13 (1977).

189. J. O. Williams and J. M. Thomas, *Mol. Cryst. Liq. Cryst.* **16**, 371 (1972).

190. D. M. Burland and U. Konzelmann, *J. Chem. Phys.* **67**, 1, 1977.

191. F. R. Lipsett, *Can. J. Phys.* **35**, 284 (1957).

192. N. N. Spendiarov and B. S. Aleksandrov, *Growth of Crystals*, Vol. 2, Consultants Bureau, New York, (1959); English transl. of *Rost Kristallov* **2**, 78 (1975)).

193. H. Kallmann and M. Pope, *Rev. Sci. Inst.* **29**, 993 (1958).

194. G. F. Reynolds, *Physics and Chemistry of the Organic Solid State*, Vol. 1, Wiley (Interscience), New York, 1963, p. 223.

195. K. H. Jones, *Mol. Cryst.* **3**, 393 (1968).

196. Y. Lupien, J. O. Williams and D. F. Williams, *Mol. Cryst. Liq. Cryst.* **18**, 129 (1972).

197. E. Glockner, Diplomarbeit, Univ. of Stuttgart (1969).

198. J. Bernard, Thesis, Univ. of Bordeaux (1978).

199. A. Eichiss, *Zh. Eksp. Teor. Fiz.* **20**, 471 (1950).

200. A. N. Winchell, *The Optical Properties of Organic Compounds*, Academic Press, New York, 1954.

201. I. Nakada, *J. Phys. Soc. Jap.* **17**, 113 (1962).
202. D. P. Craig and P. C. Hobbins, *J. Chem. Soc.* p. 2309 (1955).
203. *Bull. Tech. Zeiss* **46**, p. 118.
204. *Rev. Iena* **2**, 135 (1965).
205. E. E. Wahlstrom, *Optical Crystallography*, Wiley, New York, 1979.
206. A. Bree and L. E. Lyons, *J. Chem. Soc.* p. 2658 (1956).
207. G. Castro, Thésis p. 113.
208. J. J. Hopfield and D. G. Thomas, *Phys. Rev.* **132**, 563 (1963).
209. A. A. Demidenko and S. I. Pekar, *Soviet Phys. Solid State* **6**, 2204 (1965).
210. C. W. Deutsche and C. A. Mead, *Phys. Rev.* **138**, A63 (1965).
211. Y. Osaka, Y. Imal, and Y. Takeuti, *J. Phys. Soc. Japan* **24**, 236 (1968).
212. G. D. Mahan and G. Obermair, *Phys. Rev.* **183**, 834 (1969).
213. V. I. Sugakov, *Ukr. Fiz. Zh.* **11**, 59 (1966).
214. V. I. Sugakov, *Ukr. Fiz. Zh.* **15**, 2060 (1970).
215. V. I. Sugakov, *Soviet Phys. Solid State* **14**, 1711 (1973).
216. M. R. Philpott, *Chem. Phys. Lett.* **24**, 418 (1974).
217. M. R. Philpott, *J. Chem. Phys.* **60**, 1410 (1974).
218. M. R. Philpott, *J. Chem. Phys.* **65**, 3599 (1976).
219. P. E. Rimbey, *J. Chem. Phys.* **67**, 698 (1978).
220. M. R. Philpott, *Chem. Phys. Lett.* **30**, 387 (1975).
221. J. Ferguson, *Chem. Phys. Lett.* **36**, 316 (1975).
222. J. Ferguson, *Zeit. Phys. Chemie* **101**, S45 (1976).
223. M. Orrit, Thesis, Univ. of Bordeaux (1978), and M. Orrit, C. Aslangul and Ph. Kottis *Phys. Rev.* **25**, 7263 (1982).
224. J. Hoshen and R. Kopelman, *J. Chem. Phys.* **61**, 330 (1974).
225. H. Ueba and S. Ichimura, *J. Phys. Soc. Japan* **41**, 1974 (1976).
226. H. Ueba and S. Ichimura, *J. Phys. Soc. Japan* **42**, 355 (1977).
227. H. Ueba, *J. Phys. Soc. Japan*, **43**, 353 (1977).
228. K. Syassen and M. R. Philpott, *J. Chem. Phys.* **68**, 4870 (1978).
229. Y. Tokura, T. Koda, and I. Nakada, *J. Phys. Soc. Jap.* **47**, 1936 (1979).
230. M. G. Sceats, K. Tomioka, and S. A. Rice, *J. Chem. Phys.* **66**, 4486 (1977).
231. K. Tomioka, M. G. Sceats, and S. A. Rice, *J. Chem. Phys.* **66**, 2984 (1977).
232. M. R. Philpott and P. G. Sherman, *Phys. Rev.* **B12**, 5381 (1976).
233. J. M. Turlet and M. R. Philpott, *J. Chem. Phys.* **62**, 4260 (1975).
234. A. F. Prikhotko and A. F. Skorobogatko, *Opt. Spectrosc.* **20**, 33 (1966).
235. A. F. Prikhotko and A. F. Skorobogatko, *Fiz. Tverd. Tela* **7**, 1259 (1965).
236. G. Vaubel and H. Baessler, *Mol. Cryst. Liq. Cryst.* **12**, 39 (1970).
237. R. B. Campell, J. M. Robertson, and J. Trotter, *Acta Cryst.* **15**, 289 (1962).
238. N. I. Ostapenko, M. P. Chernomorets, and M. T. Shpak, *Phys. Stat. Sol.* **B72**, K 117 (1975).
239. H. J. Hesse, W. Fuchs, G. Weiser, and L. von Szentpaly, *Phys. Stat. Sol.* **B76**, 817 (1976).
240. V. Saile, P. Gürtler, E. E. Koch, et al. *Phys. Rev. Letters.* **37**, 305 (1976).

241. E. F. Sheka, *Soviet Physics USPEKHI* **14**, 484 (1972).
242. J. Bernard, M. Orrit, J. M. Turlet, and Ph. Kottis, *J. Chem. Phys.* (1983) and *Chem. Phys. Lett.* (1983).
243. R. M. Hochstrasser and G. R. Meredith, *J. Chem. Phys.* **67**, 1273 (1977).
244. G. C. Morris and M. G. Sceats, *Mol. Cryst. Liq. Cryst.* **25**, 339 (1974).
245. I. Pockrand, A. Brillante, M. R. Philpott, and J. D. Swalen, *Opt. Comm.* **27**, 91 (1978).
246. M. R. Philpott and J. D. Swalen, *J. Chem. Phys.* **69**, 2912 (1978).
247. M. R. Philpott A. Brillante, I. R. Pockrand, and J. D. Swalen, *Mol. Cryst. Liq. Cryst.* **50**, 139 (1979).
248. M. R. Philpott, I. Pockrand, A. Brillante, and J. D. Swalen, *J. Chem. Phys.* **72**, 2774 (1980).
249. A. Otto, in *Optical properties of solids, new developments*, B. O. Seraphin, ed., North-Holland, 1976, p. 677.
250. E. Burstein et al. in *Polaritons*, E. Burstein and F. de Martini, eds., Pergamon, 1974, p. 89.
251. A. Hartstein et al. in *Polaritons*, E. Burstein and F. de Martini, eds., Pergamon, 1974, p. 111.
252. E. Kretschmann, *Opt. Comm.* **26**, 41 (1978).

AUTHOR INDEX

Numbers in parentheses are reference numbers and indicate that the author's work is referred to although his name is not mentioned in the text. Numbers in *italics* show the pages on which the complete references are listed.

SUBJECT INDEX